VIRUSES AND HUMAN DISEASE

Preface

This book grew out of a course at Caltech that we have taught for the past 30 years. During this time, the interests of the students have changed. In the 1970s, students considered viruses to be interesting objects for learning about the molecular biology of living systems, and the course focused on molecular biology. Within the past decade, however, student interest has shifted to viruses as infectious disease agents rather than as tools with which to study molecular biology. There has been a rising awareness in the general population of the continuing importance of infectious diseases, and the interests of the current students reflect that. The course changed to reflect the interests of the students, which necessitated that more of the teaching be in the form of lecture notes, eventually leading to this book. Here we cover viruses as agents of human disease. But to understand their biology, epidemiology, and pathology, knowledge of their molecular biology of replication is essential. Thus, we attempt to integrate what is known about the molecular biology of viruses with what is known about the diseases they cause. Because we also believe that a key to the future is an understanding of the past, we also cover the history of viral diseases to the extent feasible.

Current knowledge of viruses is vast, and two people cannot hope to keep current in all fields relating to viruses. We are extremely grateful to many of our colleagues who have read individual chapters, commented on the accuracy of presentation, and offered suggestions for improving the presentation. In alphabetical order, the following people are gratefully acknowledged: Tom Benjamin, Pamela Bjorkman, Tara Chapman, Bruce Chesebro, Marie Csete, Diane Griffin, Jack Johnson, Bill Joklik, Dennis O'Callaghan, James Ou, Ellen Rothenberg, Gail Wertz, Eckard Wimmer, and William Wunner. We are also grateful to the students in our course during the past few years for feedback on the text in its various incarnations.

James H. Strauss
Ellen G. Strauss

1

Overview of Viruses
and Virus Infection

INTRODUCTION

The Science of Virology

The science of virology is relatively young. We can recognize specific viruses as the causative agents of epidemics that occurred hundreds or thousands of years ago from written descriptions of disease or from study of mummies with characteristic abnormalities. Furthermore, immunization against smallpox has been practiced for more than a millennium. However, it was only approximately 100 years ago that viruses were shown to be filterable and therefore distinct from bacteria that cause infectious disease. It was only about 60 years ago that the composition of viruses was described, and even more recently before they could be visualized as particles in the electron microscope. Within the last 20 years, however, the revolution of modern biotechnology has led to an explosive increase in our knowledge of viruses and their interactions with their hosts. Virology, the study of viruses, includes many aspects: the molecular biology of virus replication, the structure of viruses, the interactions of viruses and hosts, the evolution and history of viruses, virus epidemiology, and the diseases caused by viruses. The field is vast and any treatment of viruses must perforce be selective.

Viruses are known to infect most organisms, including bacteria, blue-green algae, fungi, plants, insects, and vertebrates, but we attempt here to provide an overview of virology that emphasizes their potential as human disease agents. Because of the scope of virology, and because human viruses that cause disease, especially epidemic disease, are not uniformly distributed across virus families, the treatment is not intended to be comprehensive. Nevertheless, we feel that it is important that the human viruses be presented in the perspective of viruses as a whole so that some overall understanding of this fascinating group of agents can emerge. We also consider many nonhuman viruses that are important for our understanding of the evolution and biology of viruses.

Viruses Cause Disease But Are Also Useful as Tools

Viruses are of intense interest because many cause serious illness in humans or domestic animals, and others damage crop plants. During the last century, progress in the control of infectious diseases through improved sanitation, safer water supplies, the development of antibiotics and vaccines, and better medical care have dramatically reduced the threat to human health from these agents, especially in developed countries. This is illustrated in Fig. 1.1, in which the death rate from infectious disease in the United States during the last century is shown. At the beginning of the 20th century, 0.8% of the population died each year from infectious diseases. Today the rate is less than one-tenth as great. The use of vaccines has led to effective control of the most dangerous of the viruses. Smallpox virus has been eradicated worldwide by means of an ambitious and concerted effort, sponsored by the World Health Organization, to vaccinate all people at risk for the disease. Poliovirus has been eliminated from the Americas, and measles virus eliminated from North America, by intensive vaccination programs. There is hope that these two viruses can also be eradicated worldwide in the near future. Vaccines exist for the control of many other viral diseases, such as mumps, rabies, rubella, yellow fever, and Japanese encephalitis.

The dramatic decline in the death rate from infectious disease has led to a certain amount of complacency. There is a small but vocal movement in the United States to eliminate immunization against viruses, for example. However,

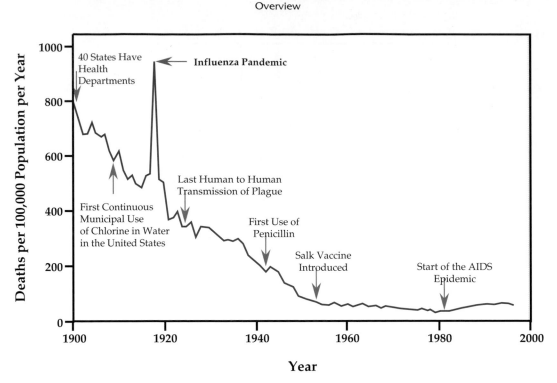

FIGURE 1.1   Death rate from infectious disease in the United States, 1900–1996. The death rate has dropped over this century from around 800 deaths per 100,000 population per year to about 50. Significant milestones in public health are shown. After dropping steadily for 80 years, interrupted only by the influenza pandemic of 1918–1919, the death rate began to rise in 1980 with the advent of the AIDS epidemic. From *Morbidity and Mortality Weekly Report (MMWR)*, Vol. 48, No. 29, p. 621 (1999).

viral diseases continue to plague humans, as do infectious diseases caused by bacteria, protozoa, fungi, and multicellular parasites. Deaths worldwide due to infectious disease are shown in Fig. 1.2, divided into six categories. In 1998 more than 3 million deaths occurred as a result of acute respiratory disease, many of which are caused by viruses. More than 2 million deaths were attributed to diarrheal diseases, about half of which are due to viruses. AIDS killed 2 million people worldwide in 1998, and measles is still a significant killer in developing countries. Recognition is growing that infectious diseases, of which viruses form a major component, have not been conquered by the introduction of vaccines and drugs. The overuse of antibiotics has resulted in an upsurge in antibiotic-resistant bacteria, and viral diseases continue to resist elimination.

The persistence of viruses is in part due to their ability to change rapidly and adapt to new situations. Human immunodeficiency virus (HIV) is the most striking example of the appearance of a virus that has recently entered the human population and caused a plague of worldwide importance. The arrival of this virus in the United States caused a significant rise in the number of deaths from infectious disease, as seen in Fig. 1.1. Other, previously undescribed viruses also continue to emerge as serious pathogens. Sin Nombre virus caused a 1994 outbreak in the United States of hantavirus pulmonary syndrome with a

50% case fatality rate, and it is now recognized as being widespread in North America. Junin virus, which causes Argentine hemorrhagic fever, and related viruses have become a more serious problem in South America with the spread of farming. Ebola virus, responsible for several small African epidemics with a case fatality rate of 70%, was first described in the 1970s. Nipah virus, previously unknown, appeared in 1998 and caused 258 cases of encephalitis, with a 40% fatality rate, in Malaysia and Singapore. As faster and more extensive travel becomes ever more routine, the potential for rapid spread of all viruses increases. The possibility exists that any of these viruses could become more widespread, as has HIV since its appearance in Africa perhaps half a century ago, and as has West Nile virus, which spread to the Americas in 1999.

Newly emerging viruses are not the only ones to plague humans, however. Many viruses that have been known for a long time, and for which vaccines may exist, continue to cause widespread problems. Respiratory syncytial virus, as an example, is a major cause of pneumonia in infants. Despite much effort, it has not yet been possible to develop an effective vaccine. Even when vaccines exist, problems may continue. For example, influenza virus changes rapidly and the vaccine for it must be reformulated yearly. Because the major reservoir for influenza is birds, it is not possible to eradicate the virus. Thus, to control influenza would

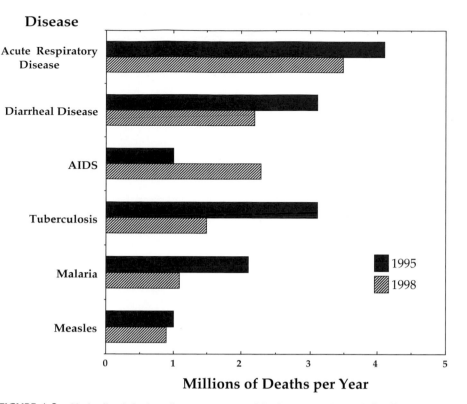

Disease

FIGURE 1.2   Six leading infectious diseases as causes of death. Data are the totals for all ages worldwide in 1995 and in 1998. The data came from the World Health Organization web site: http://www:who.int/infectious-disease-report/pages/graph5.html.

require that the entire population be immunized yearly. This is a formidable problem and the virus continues to cause annual epidemics with a significant death rate (Chapter 4). Although primarily a killer of the elderly, the potential of influenza to kill the young and healthy was shown by the worldwide epidemic of influenza in 1918 in which 20–100 million people died worldwide. In the United States, perhaps 1% of the population died during the epidemic (Fig. 1.1). Continuing study of virus replication and virus interactions with their hosts, surveillance of viruses in the field, and efforts to develop new vaccines as well as other methods of control are still important.

The other side of the coin is that viruses have been useful to us as tools for the study of molecular and cellular biology. Further, the development of viruses as vectors for the expression of foreign genes has given them a new and expanded role in science and medicine, including their potential use in gene therapy (Chapter 9). As testimony to the importance of viruses in the study of biology, numerous Nobel Prizes have been awarded in recognition of important advances in biological science that resulted from studies that involved viruses (Table 1.1). To cite a few examples, Max Delbrück received the prize for pioneering studies in what is now called molecular biology, using bacteriophage T4. Cellular oncogenes were first discovered from their

presence in retroviruses that could transform cells in culture, a discovery that resulted in a prize for Francis Peyton Rous for his discovery of transforming retroviruses, and for Michael Bishop and Harold Varmus, who were the first to show that a transforming retroviral gene had a cellular counterpart. As a third example, the development of the modern methods of gene cloning have relied heavily on the use of restriction enzymes and recombinant DNA technology, first developed by Daniel Nathans and Paul Berg working with SV40 virus, and on the use of reverse transcriptase, discovered by David Baltimore and Howard Temin in retroviruses. As another example, the study of the interactions of viruses with the immune system has told us much about how this essential means of defense against disease functions, and this resulted in a prize for Rolf Zinkernagel and Peter Doherty. The study of viruses and their use as tools has told us as much about human biology as it has told us about the viruses themselves.

In addition to the interest in viruses that arises from their medical and scientific importance, viruses form a fascinating evolutionary system. There is debate as to how ancient are viruses. Some argue that RNA viruses contain remnants of the RNA world that existed before the invention of DNA. All would accept the idea that viruses have been present for hundreds of millions of years and have helped to shape the

TABLE 1.1   Nobel Prizes Involving Virology[a]

Year	Names	Nobel citation; *virus group or family*
1946 [chemistry]	Wendell Stanley	Isolation, purification, and crystallization of tobacco mosaic virus; *Tobamovirus*
1951	Max Theiler	Development of yellow fever vaccine; *Flaviviridae*
1954	John F. Enders Thomas Weller Frederick C. Robbins	Growth and cultivation of poliovirus; *Picornaviridae*
1958	Joshua Lederberg	Transforming bacteriophages
1965	Francois Jacob André Lwoff Jacques Monod	Operons; bacteriophages
1966	Francis Peyton Rous	Discovery of tumor-producing viruses; *Retroviridae*
1969	Max Delbrück Alfred D. Hershey Salvador E. Luria	Mechanism of virus infection in living cells; bacteriophages
1975	David Baltimore Howard M. Temin Renato Dulbecco	Discoveries concerning the interaction between tumor viruses and the genetic material of the cell; *Retroviridae*
1976	D. Carleton Gajdusek Baruch S. Blumberg	New mechanisms for the origin and dissemination of infectious diseases; Blumberg with *Hepadnaviridae*, Gajdusek with prions
1978[b]	Daniel Nathans	Application of restriction endonucleases to the study of the genetics of SV40; *Polyomaviridae*
1980 [chemistry]	Paul Berg	Studies of the biochemistry of nucleic acids, with particular regard to recombinant DNA (SV40); *Polyomaviridae*
1982 [chemistry]	Aaron Klug	Development of crystallographic electron microscopy and structural elucidation of biologically important nucleic acid–protein complexes; *Tobamovirus* and *Tymovirus*
1988[b]	George Hitchings Gertrude Elion	Important principles of drug treatment using nucleotide analogs (acyclovir)
1989	J. Michael Bishop Harold E. Varmus	Discovery of the cellular origin of retroviral oncogenes; *Retroviridae*
1993	Phillip A. Sharp Richard J. Roberts	Discoveries of split (spliced) genes; *Adenoviridae*
1996	Rolf Zinkernagel Peter Doherty	Presentation of viral epitopes by major histocompatibility complex molecules
1997	Stanley Prusiner	Prions

[a]All prizes listed are in physiology or medicine except those three marked [chemistry].
[b]In these two instances, the prize was shared with unlisted recipients whose work did not involve viruses.

evolution of their hosts. Viruses are capable of very rapid change, both from drift due to nucleotide substitutions that may occur at a rate 10^6-fold greater than that of the plants and animals that they infect, and from recombination that leads to the development of entirely new families of viruses. This makes it difficult to trace the evolution of viruses back more than a few millennia or perhaps a few million years. The development of increasingly refined methods of sequence analysis, and the determination of more structures of virally encoded proteins, which change far more slowly than do the amino acid sequences that form the structure, have helped identify relationships among viruses that were not at first obvious. The coevolution of viruses and their hosts remains a study that is intrinsically interesting and has much to tell us about human biology.

The Nature of Viruses

Viruses are subcellular, infectious agents that are obligate intracellular parasites. They infect and take over a host cell in order to replicate. The mature, extracellular virus particle is called a virion. The virion contains a genome that may be DNA or RNA wrapped in a protein coat called a capsid or nucleocapsid. Some viruses have a lipid envelope surrounding the nucleocapsid (they are "enveloped"). In such viruses, glycoproteins encoded by the virus are embedded in the lipid envelope. The function of the capsid or envelope is to protect the viral genome while it is extracellular and to promote the entry of the genome into a new, susceptible cell. The structure of viruses is covered in detail in Chapter 2.

The nucleic acid genome of a virus contains the information needed by the virus to replicate and produce new virions after its introduction into a susceptible cell. Virions bind to receptors on the surface of the cell, and by processes described below the genome is released into the cytoplasm of the cell, sometimes still in association with protein ("uncoating"). The genome then redirects the cell to the replication of itself and to the production of progeny virions. The cellular machinery that is in place for the production of energy (synthesis of ATP) and for macromolecular synthesis, such as translation of mRNA to produce proteins, is essential.

It is useful to think of the proteins encoded in viral genomes as belonging to three major classes. First, most viruses encode enzymes required for replication of the genome and the production of mRNA from it. RNA viruses must encode an RNA polymerase or replicase, since cells do not normally replicate RNA. Most DNA viruses have access to the cellular DNA replication machinery in the nucleus, but even so, many encode new DNA polymerases for the replication of their genomes. Even if they use cellular DNA polymerases, many DNA viruses encode at least an initiation protein for genome replication. An overview of the replication strategies used by different viruses is presented below, and details of the replication machinery used by each virus are given in the chapters that describe individual viruses. Second, viruses must encode proteins that are used in the assembly of progeny viruses. For simpler viruses, these may consist of only one or a few structural proteins that assemble with the genome to form the progeny virion. More complicated viruses may encode scaffolding proteins that are required for assembly but are not present in the virion. In some cases, viral proteins required for assembly may have proteolytic activity. Assembly of viruses is described in Chapter 2. Third, the larger viruses encode proteins that interfere with defense mechanisms of the host. These defenses include, for example, the immune response and the interferon response of vertebrates, which are highly evolved and effective methods of controlling and eliminating virus infection; and the DNA restriction system in bacteria, so useful in molecular biology and genetic engineering, that prevents the introduction of foreign DNA. Vertebrate defenses against viruses, and the ways in which viruses counter these defenses, are described in Chapter 8.

It is obvious that viruses that have larger genomes and encode larger numbers of proteins, such as the herpesviruses (family Herpesviridae), have more complex life cycles and assemble more complex virions than viruses with small genomes, such as poliovirus (family Picornaviridae). The smallest known nondefective viruses have genomes of about 3 kb (1 kb = 1000 nucleotides in the case of single-stranded genomes or 1000 base pairs in the case of double-stranded genomes). These small viruses may encode as few as three proteins (for example, the bacteriophage MS2). At the other extreme, the largest known RNA viruses, the coronaviruses (family Coronaviridae), have genomes somewhat larger than 30 kb, whereas the largest DNA viruses, poxviruses belonging to the genera Entomopoxvirus A and C (family Poxviridae), have genomes of up to 380 kb. These large DNA viruses encode hundreds of proteins. It is the larger viruses that can afford the luxury of encoding proteins that interfere effectively with host defenses such as the immune system. It is worthwhile remembering that even the largest viral genomes are small compared to the size of the bacterial genome (2000 kb) and miniscule compared to the size of the human genome (2×10^6 kb).

There are other subcellular infectious agents that are even "smaller" than viruses. These include satellite viruses, which are dependent for their replication on other viruses; viroids, small (~300 nucleotide) RNAs that are not translated and have no capsid; and prions, infectious agents whose identity remains controversial, but which may consist of only protein. These agents are covered in Chapter 7.

CLASSIFICATION OF VIRUSES

The Many Kinds of Viruses

Three broad classes of viruses can be recognized, which may have independent evolutionary origins. One class, which includes the poxviruses and herpesviruses among many others, contains DNA as the genome, whether single stranded or double stranded, and the DNA genome is replicated by direct DNA → DNA copying. During infection, the viral DNA is transcribed by cellular and/or viral RNA polymerases, depending on the virus, to produce mRNAs for translation into viral proteins. The DNA genome is replicated by DNA polymerases that can be of viral or cellular origin. Replication of the genomes of most eukaryotic DNA viruses and assembly of progeny viruses occur in the nucleus, but the poxviruses replicate in the cytoplasm.

A second class of viruses contains RNA as their genome and the RNA is replicated by direct RNA → RNA copying. Some RNA viruses, such as yellow fever virus (family Flaviviridae) and poliovirus, have a genome that is a messenger RNA, defined as plus-strand RNA. Other RNA viruses, such as measles virus (family Paramyxoviridae) and rabies virus (family Rhabdoviridae), have a genome that is anti-messenger sense, defined as minus strand. The arenaviruses (family Arenaviridae) and some of the genera belonging to the family Bunyaviridae have a genome that has regions of both messenger and anti-messenger sense and are called ambisense. The replication of these viruses follows a minus-sense strategy, however, and they are classified with the minus-sense viruses. Finally, some RNA viruses, for example, rotaviruses (family Reoviridae), have double-strand RNA genomes. In the case of all RNA viruses, virus-encoded proteins are required to form a replicase to replicate the viral RNA, since cells do not possess (efficient) RNA → RNA copying enzymes. In the case of the minus-strand RNA viruses and double-strand RNA viruses, these RNA synthesizing enzymes also synthesize mRNA and are packaged in the virion, because their genomes cannot function as messengers. Replication of the genome proceeds through RNA intermediates that are complementary to the genome in a process that follows the same rules as DNA replication.

The third class of viruses encodes the enzyme reverse transcriptase (RT), and these viruses have an RNA → DNA step in their life cycle. The genetic information encoded by these viruses thus alternates between being present in RNA and being present in DNA. Retroviruses (e.g., HIV, family Retroviridae) contain the RNA phase in the virion; they have a single-stranded RNA genome that is present in the virus particle in two copies. Thus, the replication of their genome occurs through a DNA intermediate (RNA → DNA → RNA). The hepadnaviruses (e.g., hepatitis B virus, family Hepadnaviridae) contain the DNA phase as their genome, which is circular and largely double stranded. Thus their genome replicates through an RNA intermediate (DNA → RNA → DNA). Just as the minus-strand RNA viruses and double-strand RNA viruses package their replicase proteins, the retroviruses package active RT, which is required to begin the replication of the genome in the virions. Although in many treatments the retroviruses are considered with the RNA viruses and the hepadnaviruses with the DNA viruses, we consider these viruses to form a distinct class, the RT-containing class, and in this book references to RNA viruses, or to DNA viruses are not meant to apply to the retroviruses.

All viruses, with one exception, are haploid; that is, they contain only one copy of the genomic nucleic acid. The exception is the retroviruses, which are diploid and contain two identical copies of the single-stranded genomic RNA. The nucleic acid genome may consist of a single piece of DNA or RNA or may consist of two or more nonidentical fragments. The latter can be considered analogous to chromosomes and can reassort during replication. In the case of animal viruses, when a virus has more than one genome segment, all of the different segments are present within a single virus particle. In the case of plant viruses with multiple genome segments, it is quite common for the different genome segments to be separately encapsidated into different particles. In this case, the infectious unit is multipartite: Infection to produce a complete replication cycle requires simultaneous infection by particles containing all of the different genome segments. Although this does not seem to pose a problem for the transmission of plant viruses, it must pose a problem for the transmission of animal viruses since such animal viruses have not been found. This difference probably arises because of different modes of transmission, the fact that many plant viruses grow to exceptionally high titers, and the fact that many plants grow to very high density.

The ICTV Classification of Viruses

The International Committee on Taxonomy of Viruses (ICTV), a committee organized by the Virology Division of the International Union of Microbiological Societies, is attempting to devise a uniform system for the classification and nomenclature of all viruses. Viruses are classified into species on the basis of a close, but not necessarily identical, relationship. The decision as to what constitutes a species is arbitrary because a species usually contains many different strains that may differ significantly (10% or more) in nucleotide sequence. Whether two isolates should be considered as being the same species rather than representing two different species can be controversial. Virus species that exhibit close relationships are then grouped into a genus. Species within a genus usually share significant nucleotide sequence identity demonstrated by antigenic cross reaction or by direct sequencing of the genome. Genera are grouped into families, which can be considered the fundamental unit of virus taxonomy. Classification into families is based on the type and size of the nucleic acid genome, the structure of the virion, and the strategy of replication used by the virus, which is determined in part by the organization of the genome. Groupings into families are not always straightforward because little or no sequence identity is present between members of different genera. However, uniting viruses into families attempts to recognize evolutionary relationships and is valuable for organizing information about viruses.

Higher taxonomic classifications have not been recognized for the most part. To date only three orders have been established that group together a few families. Taxonomic classification at higher levels is difficult because viruses evolve rapidly and it can be difficult to prove that any two given families are descended from a common ancestor, although it is almost certain that higher groupings based on common evolution do exist and will be elucidated with time. Viral evolution involves not only sequence divergence, however, but also the widespread occurrence of recombination during the rise of the modern families, a feature that blurs the genetic relationships between viruses. Two families may share, for example, a related polymerase gene but have structural protein genes that appear unrelated; how should such viruses be classified?

The ICTV has recognized almost 4000 viruses as species (more than 30,000 strains of viruses exist in collections around the world), and classified these 3954 species into 203 genera belonging to 56 families plus 30 "floating" genera that have not yet been assigned to a family. An overview of these families, in which viruses that cause human disease are emphasized, is shown in Table 1.2.

Included in the table is the type of nucleic acid that serves as the genome, the genome size, the names of many families, and the major groups of hosts infected by viruses within each grouping. For many families the names and detailed characteristics are not shown here, but a complete listing of families can be found in the reports of the ICTV on virus taxonomy or in *The Encyclopedia of Virology* (2nd

TABLE 1.2 Major Virus Families

Nucleic acid	Genome size	Segments	Family	Major hosts (number of members infecting that host)[a]	
DS DNA	130–375 kbp	1	Poxviridae	Vertebrates (38), insects (25), plus 15 U[b]	
	170–190 kbp	1	Asfariviridae	Vertebrates (1)	
	170–400 kbp	1	Iridoviridae	Vertebrates (11), insects (6)	
	120–220 kbp	1	Herpesviridae	Vertebrates (56 + 65T)[c]	
	90–230 kbp	1	Baculoviridae	Insects (17 + 7T)	
	36–48 kbp	1	Adenoviridae	Vertebrates (26 + 35T)	
	5 kbp	1	Polyomaviridae	Vertebrates (12)	
	6.8–8.4 kbp	1	Papillomaviridae	Vertebrates (7 + 88T)	
	Various	1	Several families	Bacteria (42 + 368T)	
SS DNA	6–8 kb	1	Parvoviridae	Vertebrates (31), insects (7), plus 15U	
	Various	1	Several families	Bacteria (43 + 38T), plants (98 + 11T)	
DS RNA	20–30 kbp	10–12	Reoviridae	Vertebrates (174 + 22T), insects (66), plants (10)	
	5.9 kbp	2	Birnaviridae	Chickens (1), fish (2), drosophila (1)	
	4.6–7.0 kbp	1 or 2	Three families	Fungi (7 + 7T), plants (30 + 15T), protozoans (14)	
SS(+) RNA	28–33 kb	1	Coronaviridae	Vertebrates (16 + 1T)	
	13–16 kb	1	Arteriviridae	Vertebrates (4)	
	10–13 kb	1	Togaviridae	Vertebrates (insect vectors) (23)	
	10–12 kb	1	Flaviviridae	Vertebrates (some insect vectors) (57 + 6T)	
	7–8.5 kb	1	Picornaviridae	Vertebrates (16 + 137T)	
	7–8 kb	1	Astroviridae	Vertebrates (6)	
	8 kb	1	Caliciviridae	Vertebrates (6 plus 8T)	
	Various	1	Many families	Plants (331 + 219T)	
SS (–) RNA	15–16 kb	1	Paramyxoviridae	Vertebrates (31 + 2T)	
	13 kb	1	Filoviridae	Vertebrates (5)	
	13–16	1	Rhabdoviridae	Vertebrates (23 + 81T), plants (14 + 61T)	
	9 kb	1	Bornaviridae	Vertebrates (1)	
	13 kb	8	Orthomyxoviridae	Vertebrates (5)	
	12–23 kb	3	Bunyaviridae	Vertebrates and insect vectors (91 + 66T), plants (2)	
	11 kb	2	Arenaviridae	Vertebrates (19 + 2T)	
SS RNA RT	7–10 kb	dimer	Retroviridae	Vertebrates (59 + 2T)	DNA intermediate
DS DNA RT	3 kbp	1	Hepadnaviridae	Vertebrates (5 + 2T)	RNA intermediate
	8 kbp	1	Caulimoviridae	Plants (26 + 8T)	RNA intermediate

Source: This is according to Granoff and Webster (1999).
[a]Where underlined, humans are among the vertebrates infected.
[b]U = assigned to the family, but not to any particular genus within the family.
[c]T = tentatively assigned to a particular genus.

ed.). Tables that describe the members of families that infect humans are presented in the chapters that follow in which the various virus families are considered in some detail.

AN OVERVIEW OF THE REPLICATION CYCLE OF VIRUSES

Receptors for Virus Entry

The infection cycle of an animal virus begins with its attachment to a receptor expressed on the surface of a susceptible cell, followed by penetration of the genome, either naked or complexed with protein, into the cytoplasm. Binding often occurs in several steps. For many viruses, the virion first binds to an accessory receptor that is present in high concentrations on the surface of the cell. These accessory receptors are usually bound with low affinity, and binding often has a large electrostatic component. Use of accessory receptors seems to be fairly common among viruses adapted to grow in cell culture, but less common in primary isolates of viruses from animals. This first stage binding to an accessory receptor is not required for virus entry even where used, but such binding does accelerate the rate of binding and uptake of the virus.

Following this initial binding, the virus is transferred to a high-affinity receptor. Binding to the high-affinity receptor is required for virus entry, and cells that fail to express the appropriate receptor cannot be infected by the virus. These receptors are specifically bound by one or more of the external proteins of a virus. Each virus uses a specific receptor (or perhaps a specific set of receptors) expressed on the cell surface, and both protein receptors and carbohydrate receptors are known. In some cases, unrelated viruses make use of identical receptors. A protein called CAR (*C*oxsackie-*a*denovirus *r*eceptor), a member of the immunoglobulin (Ig) superfamily, is used by the RNA virus Coxsackie B virus (Picornaviridae) and by many adenoviruses (Adenoviridae), which are DNA viruses. Sialic acid, a carbohydrate attached to most glycoproteins, is used by influenza virus (family Orthomyxoviridae), human coronavirus OC3 (family Coronaviridae), reovirus (Reoviridae), bovine parvovirus (Parvoviridae), and many other viruses. Conversely, members of the same viral family may use widely disparate receptors. Fig. 1.3 illustrates a number of receptors used by different retroviruses (family Retroviridae). These receptors differ widely in their structures and in their cellular functions. Where known, the region of the cellular receptor that is bound by the virus is indicated. Table 1.3 lists receptors used by different herpesviruses (Herpesviridae) and different coronaviruses.

In addition to the requirement for a high-affinity or primary receptor, many viruses also require a coreceptor in order to penetrate into the cell. In the current model for virus entry, a virus first binds to the primary receptor and then binds to the coreceptor. Only on binding to the coreceptor can the virus enter the cell.

The nature of the receptors utilized by a virus determines in part its host range, tissue tropism, and the pathology of the disease caused by it. Thus, the identification of

TABLE 1.3   Viruses within a Family That Use Unrelated Receptors

	Virus	High-affinity receptor	Accessory receptor
Herpesviridae			
Alpha	Herpes simplex	HIgR (CD155 family)	Heparan sulfate
		HVEM (TNF receptor family)	
	Pseudorabies	140-kDa heparan sulfate proteoglycan	
		85-kDa integral membrane protein	
		CD155 and related proteins	
Beta	Cytomegalovirus	Protein?? unidentified	Heparan sulfate
Gamma	Bovine herpesvirus	56-kDa protein	Heparan sulfate
	Epstein–Barr	CD21 (CR2 receptor)	
Coronaviridae			
Human	OC43	HLA class I	
	229e	Human aminopeptidase N	
Murine	Mouse hepatitis	Carcinoembryonic antigens (Ig superfamily)	
Porcine	TGEV[a]	Porcine aminopeptidase N	
Bovine	Bovine coronavirus	Sialic acid residues on glycoproteins and glycolipids	

[a] *Virus abbreviation:* TGEV, transmissible gastroenteritis virus (of swine).

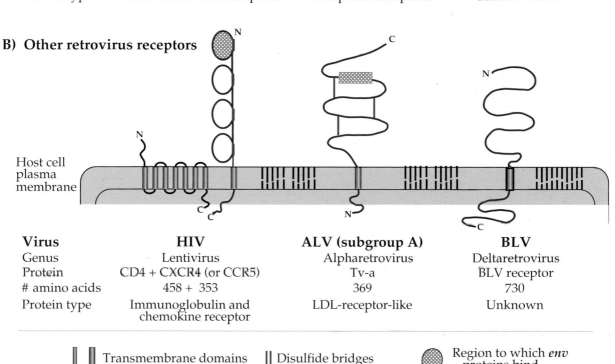

A) Gammaretrovirus receptors

Host cell plasma membrane

Virus	MoMLV	GALV/FeLV	MLV
Type	Ecotropic		Amphotropic
Protein	mCAT-1	hPiT-1	rPiT-2
# amino acids	622	679	652
Protein type	Basic amino acid transporter	Phosphate transporter	Similar to hPiT-1

B) Other retrovirus receptors

Host cell plasma membrane

Virus	HIV	ALV (subgroup A)	BLV
Genus	Lentivirus	Alpharetrovirus	Deltaretrovirus
Protein	CD4 + CXCR4 (or CCR5)	Tv-a	BLV receptor
# amino acids	458 + 353	369	730
Protein type	Immunoglobulin and chemokine receptor	LDL-receptor-like	Unknown

Transmembrane domains   Disulfide bridges   Region to which *env* proteins bind

FIGURE 1.3 Cellular receptors for retroviruses. The structures of various retrovirus receptors are shown schematically to illustrate their orientation in the cell plasma membrane. The receptors for the gammaretroviruses contain multiple transmembrane domains and have known cellular functions. The HIV receptor consists of a molecule of CD4 plus a chemokine receptor such as CXCR4. The receptor for alpharetroviruses is a type II membrane protein similar to the LDL receptor, with the N terminus in the cytoplasm. Little is known about the cellular function of the BLV receptor, other than its orientation as a Type I membrane protein. MLV, murine leukemia virus; GALV, gibbon/ape leukemia virus; FeLV, feline leukemia virus; HIV, human immunodeficiency virus; ALV, avian leukosis virus; BLV, bovine leukemia virus; LDL, low-density lipoprotein. [Adapted from Fields *et al.* (1996, p. 1788) and Coffin *et al.* (1997, pp. 76–82).]

virus receptors is important, but identification of receptors is not always straightforward.

Primary (High-Affinity) Receptors

Many members of the Ig superfamily are used by viruses as high-affinity receptors, as illustrated in Fig. 1.4. The Ig superfamily contains thousands of members, which play important roles in vertebrate biology. The best known members are found in the immune system (Chapter 8), from which the family gets its name. Members of this superfamily contain one or more Ig domains of about 100 amino acids that arose by duplication of a prototypical gene. During evolution of the superfamily, thousands of dif-

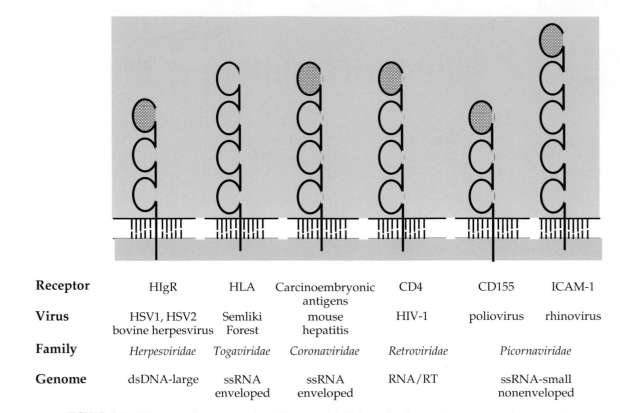

Receptor	HIgR	HLA	Carcinoembryonic antigens	CD4	CD155	ICAM-1
Virus	HSV1, HSV2 bovine herpesvirus	Semliki Forest	mouse hepatitis	HIV-1	poliovirus	rhinovirus
Family	*Herpesviridae*	*Togaviridae*	*Coronaviridae*	*Retroviridae*	*Picornaviridae*	
Genome	dsDNA-large	ssRNA enveloped	ssRNA enveloped	RNA/RT	ssRNA-small nonenveloped	

FIGURE 1.4 Diagrammatic representation of immunoglobulin superfamily membrane proteins that are used as receptors by viruses. The domains indicated by cross-hatching have been shown to be required for receptor activity. ssRNA, single-strand RNA; dsRNA, double-strand RNA; RNA/RT, RNA reverse transcribed into DNA.

ferent proteins arose by a combination of continuing gene duplication, sequence divergence, and recombination. Many proteins belonging to this superfamily are expressed on the surface of cells, where they serve many functions, and many have been usurped by animal viruses for use as receptors.

Other surface proteins used as receptors include the vibronectin receptor $\alpha_v\beta_3$, used by several members of the Picornaviridae; aminopeptidase N, used by some coronaviruses; CD55, used by Coxsackie A21 virus; the different proteins illustrated in Fig. 1.3; and other proteins too numerous to describe here. The receptors used by four viruses are described in more detail as examples of the approaches used to identify receptors and their importance for virus pathology.

One well-characterized receptor is that for poliovirus, which attaches to a cell surface molecule that is a member of the Ig superfamily (Fig. 1.4). The normal cellular function of this protein is unknown. It was first called simply the poliovirus receptor or PVR, but has now been renamed CD155, following a scheme for the designation of cell surface proteins. Poliovirus will bind only to the version of this molecule that is expressed in primates, but not to the

version expressed in rodents, for example. Thus, in nature, poliovirus infection is restricted to primates. Although chicken cells or most mammalian cells that lack CD155 are resistant to poliovirus infection, they can be transfected with the viral RNA by a process that bypasses the receptors. When infected in this way, they produce a full yield of virus, showing that the block to replication is at the level of entry.

Cells lacking CD155 have been transfected with expression clones so that they express CD155, and these modified cells are sensitive to infection with poliovirus. This was, in fact, the way the receptor was identified. Such a system also allows the testing of chimeric receptors, in which various domains of CD155 come from the human protein and other parts come from the homologous mouse protein, or even from entirely different proteins like CD4. In this way it was shown that only the distal Ig domain from human CD155 (shaded in Fig. 1.4) is required for a chimeric protein to function as a receptor for poliovirus.

In humans, CD155 is expressed on many cells, including cells of the gut, nasopharynx, and the central nervous system (CNS). Infection begins in the tonsils, lymph nodes of the neck, Peyer's patches, and the small intestine. In more than 98% of cases, the infection progresses no further and no illness, or only minor illness, results. In some cases, however, virus spreads to the CNS, probably both by passing through the blood–brain barrier and through retrograde axonal transport. Once in the CNS, the virus expresses an astounding preference for motor neurons, whose destruction leads to paralysis or even death via a disease called poliomyelitis. This preference for motor neurons, and the failure of the virus to grow in other tissues, is not understood. Athough CD155 is required for virus entry, other factors within the cell are also important for efficient virus replication.

Making use of the CD155 gene, transgenic mice have been generated in which the syndrome of poliomyelitis can be faithfully reproduced. Although these transgenic mice can be infected only by injection of virus and not by ingestion, the normal route of poliovirus infection in humans, a small animal model for poliomyelitis is valuable for the study of virus pathology or for vaccine development. To date our information on the pathology of poliovirus in the CNS was obtained only from experimental infection of nonhuman primates, which are very expensive to maintain, or from humans naturally infected with the virus.

As a second example of virus–receptor interactions, HIV utilizes as its receptor a cell surface molecule known as CD4, wich is also a member of the Ig superfamily (Figs 1.3 and 1.4). As described below, a coreceptor is also required. CD4 is primarily expressed on the surface of certain lymphocytes (described in more detail in Chapter 8). Furthermore, the virus has a narrow host range and will bind with high efficiency only to the human version of CD4 (Fig. 1.3). Thus, humans are the primary host of HIV. Immune function is impaired over time as helper $CD4^+$ T cells, which are required for an immune response directed against infectious agents, are killed by virus infection, leading to the observed syndrome of AIDS (*acquired immunodeficiency syndrome*). The virus can also infect cells of the monocyte-macrophage lineage, and possibly other cells in the CNS, leading to neurological manifestations.

As a third example of virus–receptor interaction, among the receptors used by Sindbis virus (family Togaviridae) is the high-affinity laminin receptor. Sindbis virus is an arbovirus, that is, it is arthropod-borne. In nature it alternates between replication in mosquitoes, which acquire the virus when they take a blood meal from an infected vertebrate, and higher vertebrates, which acquire the virus when bitten by an infected mosquito. The high-affinity laminin receptor is a cell adhesion molecule that binds to laminin present in basement membranes. It has been very highly conserved during evolution, and Sindbis virus will bind to both the mosquito version and the mammalian version of this protein. Viruses with broad host ranges, such as arboviruses, must use receptors that are highly conserved, or must have evolved the ability to use different receptors in different hosts.

Finally, as a fourth example, the receptor for influenza virus (family Orthomyxoviridae) is sialic acid covalently linked to glycoproteins or glycolipids present at the cell surface. Because sialic acid is expressed on many different cells and in many different organisms, the virus has the potential to have a very wide host range. The virus infects many birds and mammals, and its maintenance in nature depends on its ability to infect such a broad spectrum of animals. The epidemiology of influenza virus will be considered in Chapter 4.

Accessory Receptors and Coreceptors

The process by which a virus binds to a cell and penetrates into the cytoplasm may be complicated by the participation of more than one cellular protein in the process. Some viruses may be able to use more than one primary receptor, which thus serve as alternative receptors. Second, many viruses appear to first bind to a low-affinity receptor or accessory receptor before transfer to a high-affinity receptor by which the virus enters the cell. Third, many viruses absolutely require a coreceptor, in addition to the primary receptor, for entry.

Many viruses, belonging to different families, have been shown to bind to heparan sulfate (Table 1.4), which is expressed on the surface of many cells. In at least some cases, however, best studied for human herpes simplex virus (HSV) (family Herpesviridae), heparan sulfate is not absolutely required for the entry of the virus. Cells that do not express heparan sulfate or from which it has been removed can still be infected by HSV. Heparan sulfate does dramatically increase the efficiency of infection, however. The current model is that HSV first binds to heparan sulfate with low affinity and is then transferred to the primary receptor for entry. In this model, heparan sulfate serves an accessory function, which can be dispensed with.

TABLE 1.4   Viruses That Bind to Heparin-like Glycosaminoglycans

Virus	Family	High-affinity receptor
RNA viruses		
Sindbis	Togaviridae	High-affinity laminin receptor
Dengue	Flaviviridae	???
Foot-and-mouth disease	Picornaviridae	$\alpha_v\beta_3$ Integrin
HIV-1	Retroviridae	CD4 (Ig superfamily)
DNA viruses		
Vaccinia	Poxviridae	EGF receptor???
Herpes simplex	Herpesviridae	HIgR (CD155 family)
Adeno-associated type 2	Parvoviridae	FGFRI

Abbreviations: EGF receptor, epidermal growth factor receptor; HIgR, herpes immunoglobulin-like receptor; CD155, the poliovirus receptor; FGFR1, human fibroblast growth factor receptor 1.

The primary receptor for HSV has now been identified as a protein belonging to the Ig superfamily (Fig. 1.4). This protein is closely related to CD155, and, in fact, CD155 will serve as a receptor for some herpesviruses, but not for HSV. The story is further complicated by the fact that more than one protein can serve as a receptor for HSV. Two of these proteins, one called HIgR (for *h*erpesvirus *Ig*-like *r*eceptor) and the other called either PRR-1 (for *p*oliovirus *r*eceptor *r*elated) or HveA (for *h*erpes*v*irus *e*ntry mediator *A*), appear to be splice variants that have the same ectodomain.

Heparan sulfate may serve as an accessory receptor for the other viruses shown in Table 1.4, or it may serve as a primary receptor for some or all. It is thought that it may be a primary receptor for dengue virus (family Flaviviridae). In the case of Sindbis virus, the situation is complex and interesting. Primary isolates of the virus do not bind to heparan sulfate. Passage of the virus in cultured cells selects for viruses that do bind to heparan sulfate, which infect cultured cells more efficiently. It is thought that heparan sulfate serves as an accessory receptor in this case.

Many viruses absolutely require a coreceptor for entry, in addition to the primary receptor to which the virus first binds. The best studied example is HIV, which requires one of a number of chemokine receptors as a coreceptor. Thus a mouse cell that is genetically engineered to express human CD4 will bind HIV, but binding does not lead to entry of the virus into the cell. Only if the cell is engineered to express both human CD4 and a human chemokine receptor can the virus both bind to and enter into the cell.

The requirement for a coreceptor has important implications for the pathology of HIV. Chemokines are small proteins, secreted by certain cells of the immune system, that serve as chemoattractants for lymphocytes. They are important regulators of the immune system and are described in Chapter 8. Different classes of lymphocytes express receptors for different chemokines at their surface. To simplify the story, macrophage-tropic (M-tropic) strains of HIV, which is the virus most commonly transmitted sexually to previously uninfected individuals, require a coreceptor called CCR5 (a receptor for β chemokines). Human genetics has shown that two mutations can block the expression of CCR5. One is a 32-nucleotide deletion in the gene, the second is a mutation that results in a stop codon in the CCR5 open reading frame. The deletion mutation is fairly common, present in about 20% of Caucasians of European descent, whereas the stop codon mutation has been reported in only one individual. Individuals who lack functional CCR5 because they are homozygous for the deleted form, or in the case of one individual, heterozygous for the deletion but whose second copy of CCR5 has the stop codon, are resistant to infection by HIV. Heterozygous individuals who have only one functional copy of the CCR5 gene appear to be partially resistant. Although they can be infected with HIV, the probability of transmission has been reported to be lower, and once infected, pro-gression to AIDS is slower. During the course of infection by HIV, T-cell-tropic strains (T-tropic) of HIV arise that require a different coreceptor, called CXCR4 (a receptor for α chemokines). After the appearance of T-tropic virus, both M-tropic and T-tropic strains cocirculate. The requirement for a new coreceptor is associated with mutations in the surface glycoprotein of HIV. The presence of T-tropic viruses is associated with more rapid progression to severe clinical disease.

Entry of Plant Viruses

Many plant viruses are important pathogens of food crops and have been intensively studied. No specific receptors have been identified to date, and it has been suggested that virus penetration of plant cells requires mechanical damage to the cell in order to allow the virus entry. Such mechanical damage can be caused by farm implements or by damage to the plant caused by insects such as aphids or leafhoppers that feed on the plants. Many plant viruses are transmitted by insect or fungal pests, in fact, with which the virus has a specific association. There remains the possibility that specific receptors will be identified in the future, however, for at least some plant viruses.

Penetration

After the virus binds to a receptor, the next step toward successful infection is the introduction of the viral genome into the cytoplasm of the cell. In some cases, a subviral particle containing the viral nucleic acid is introduced into the cell. This particle may be the nucleocapsid of the virus or it may be an activated core particle. For other viruses, only the nucleic acid is introduced. The protein(s) that promotes entry may be the same as the protein(s) that binds to the receptor, or it may be a different protein in the virion.

For enveloped viruses, penetration into the cytoplasm involves the fusion of the envelope of the virus with a cellular membrane, which may be either the plasma membrane or an endosomal membrane. Fusion is promoted by a fusion domain that resides in one of the viral surface proteins. Activation of the fusion process is thought to require a change in the structure of the viral fusion protein that exposes the fusion domain. For viruses that fuse at the plasma membrane, interaction with the receptor appears to be sufficient to activate the fusion protein. In the case of viruses that fuse with endosomal membranes, the virus is first endocytosed into clathrin-coated vesicles and proceeds through the endosomal pathway. During transit, the clathrin coat is lost and the endosomes become progressively acidified. On exposure to a defined acidic pH (often ~5–6), activation of the fusion protein occurs and results in fusion of the viral envelope with that of the endosome.

A dramatic conformational rearrangement of the HA protein of influenza virus, a virus that fuses with internal membranes, has been observed by X-ray crystallography of HA following its exposure to low pH. HA, which is cleaved into two disulfide-bonded fragments HA_1 and HA_2, forms trimers that are present in a spike on the surface of the virion. The atomic structure of an HA monomer is illustrated in Fig. 1.5. HA_1 (shown in blue) contains the domain (indicated with a star in the figure) that binds to sialic acid receptors, whereas HA_2 (shown in red) has the fusion domain (yellow). The fusion domain, present at the N terminus of HA_2, is hidden in a hydrophobic pocket within the spike near the lipid bilayer of the virus envelope. Exposure to low pH results in a dramatic rearrangement of HA that exposes the hydrophobic peptide and transports it more than 100 Å upward, where it is thought to insert into the cellular membrane and promote fusion. It is assumed that similar events occur for all enveloped viruses, whether fusion is at the cell surface or with an internal membrane.

After fusion of the viral envelope with a cellular membrane, the virus nucleocapsid is present in the cytoplasm of the cell. Virus entry by fusion can be very efficient. In some well-studied cases using cells in culture, almost all particles succeed in initiating infection.

For nonenveloped viruses, the mechanism by which the virus breaches the cell membrane is less clear. After binding

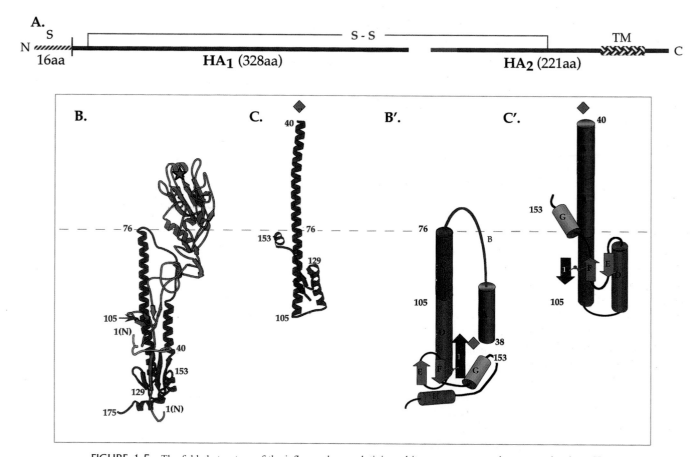

FIGURE 1.5 The folded structure of the influenza hemagglutinin and its rearrangement when exposed to low pH. (A) A schematic of the cleaved HA molecule. S is the signal peptide, TM is the membrane-spanning domain. HA_1 is in blue, HA_2 is in red, and the fusion peptide is shown in yellow. The same color scheme is used in (B) and (C). (B) X-ray crystallographic structure of the HA monomer. TM had been removed by proteolytic digestion prior to crystallization. The receptor binding pocket in HA_1 is shown with a green star. In the virion HA occurs as a trimeric spike. (C) The HA_2 monomer in the fusion active form. The fragment shown is produced by digesting with thermolysin, which removes most of HA_1 and the fusion peptide of HA_2. Certain residues are numbered to facilitate comparison of the two forms. The approximate location of the fusion peptide before thermolysin digestion is indicated with a yellow diamond. (B′) Diagrammatic representation of the HA_2 shown in (B), with α helices shown as cylinders and β sheets as arrows. The disulfide link between HA_1 and HA_2 is shown in ochre. The domains of HA_2 are color coded from N terminus to C terminus with a rainbow. (C′) Diagrammatic representation of the fusion-active form shown in (C). [Redrawn from Fields *et al.* (1996, p. 1361).]

to a receptor, somehow the virus or some subviral component ends up on the cytoplasmic side of a cellular membrane, the plasma membrane for some viruses or the membrane of an endosomal vesicle for others. It is believed that the interaction of the virus with a receptor, perhaps potentiated by the low pH in endosomes for those viruses that enter via the endosomal pathway, causes conformational rearrangements in the proteins of the virus capsid that lead to the penetration of the membrane. In the case of poliovirus, it is known that interactions with receptors *in vitro* will lead to conformational rearrangements of the virion that result in the release of one of the virion proteins, called VP4. The N terminus of VP4 is myristylated and thus hydrophobic [myristic acid = $CH_3(CH_2)_{12}COOH$]. It is proposed that the conformational changes induced by receptor binding result in the insertion of the myristic acid on VP4 into the cell membrane and the formation of a channel through which the RNA can enter the cell. It is presumed that other viruses also have hydrophobic domains that allow them to enter. A number of other viruses also have a structural protein with a myristilated N terminus that might promote entry. In some viruses, there is thought to be a hydrophobic fusion domain in a structural protein that provides this function. The entry process may be very efficient. The specific infectivity of reoviruses assayed in cultured cells can be almost one (all particles can initiate infection). In other cases, entry may be less efficient. The specific infectivity of poliovirus in cultured cells is usually less than 1%. Whether the specific infectivity of the virus is low when infecting humans is not clear, however.

Following initial penetration into the cytoplasm, further uncoating steps must often occur. It has been suggested that, at least in some cases, translation of the genomic RNA of plus-strand RNA viruses may promote its release from the nucleocapsid. In other words, the ribosomes may pull the RNA into the cytoplasm. In other cases, specific factors in the host cell, or the translation products of early viral transcripts, have been proposed to play a role in further uncoating.

It is interesting to note that bacteriophage face the problem of penetrating a rigid bacterial cell wall, rather than one of simply penetrating a plasma membrane or endosomal membrane. Many bacteriophage have evolved a tail by which they attach to the cell surface, drill a hole into the cell, and deliver the DNA into the bacterium. In some phage, the tail is contractile, leading to the analogy that the DNA is injected into the bacterium. Tailless phage are also known that introduce their DNA into the bacterium by other mechamisms.

Replication and Expression of the Virus Genome

The replication strategy of a virus, that is, how the genome is organized and how it is expressed so as to lead to the formation of progeny virions, is an essential component in the classification of a virus and in understanding the basic details of how it replicates. Moreover, it is necessary to understand the replication strategy in order to decipher the pathogenic mechanisms of a virus and, therefore, to design strategies to interfere with viral disease.

DNA Viruses

A simple schematic representation of the replication of a DNA virus is shown in Fig. 1.6. After binding to a receptor and penetration of the genome into the cell, the first event in the replication of a DNA virus is the production of mRNAs from the viral DNA. For all animal DNA viruses except poxviruses, the infecting genome is transported to

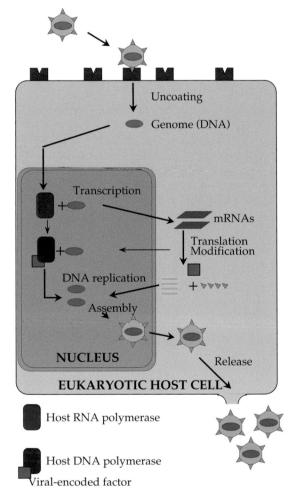

FIGURE 1.6 General replication scheme for a DNA virus. After a DNA virus attaches to a cellular membrane receptor, the virus DNA enters the cell and is transported to the cell nucleus. There it is transcribed into mRNA by host RNA polymerase. Viral mRNAs are translated by host ribosomes in the cytoplasm, and newly synthesized viral proteins, both structural and nonstructural, are transported back to the nucleus. After the DNA genome is replicated in the nucleus, either by the host DNA polymerase or by a new viral-encoded polymerase, progeny virus particles are assembled and ultimately released from the cell. [Adapted from Mims *et al.* (1993, p. 2.3).]

the nucleus where it is transcribed by cellular RNA polymerase. The poxviruses replicate in the cytoplasm and do not have access to host cell polymerases. Thus, in poxviruses, early mRNA is transcribed from the incoming genome by a virus-encoded RNA polymerase that is present in the virus core. For all animal viruses, translation of early mRNA is required for DNA replication to proceed. Early gene products may include DNA polymerases, proteins that bind to the origin of replication and lead to initiation of DNA replication, proteins that stimulate the cell to enter S phase and thus increase the supply of materials required for DNA synthesis, or products required for further disassembly of subviral particles.

The initiation of the replication of a viral genome is a specific event that requires an origin of replication, a specific sequence element that is bound by cellular and (usually) viral factors. Once initiated, DNA replication proceeds, catalyzed by either a cellular or a viral DNA polymerase. The mechanisms by which replication is initiated and continued are different for different viruses.

DNA polymerases, in general, are unable to initiate a polynucleotide chain. They can only extend an existing chain, following instructions from a DNA template. Replication of cellular DNA, including that of bacteria, requires the initiation of polynucleotide chains by a specific RNA polymerase called DNA polymerase α-primase, or primase for short. The resulting RNA primers are then extended by DNA polymerase. The ribonucleotides in the primer are removed after extension of the polynucleotide chain as DNA. Removal requires the excision of the ribonucleotides by a $5' \rightarrow 3'$ exonuclease, fill-in by DNA polymerase, and sealing of the nick by ligase. Because DNA polymerases can synthesize polynucleotide chains only in a $5' \rightarrow 3'$ direction, and cannot initiate a DNA chain, removal of the RNA primer creates a problem at the end of a linear chromosome. How is the 5' end of a DNA chain to be generated? The chromosomes of eukaryotic cells have special sequences at the ends, called telomeres, that function in replication to regenerate ends. The telomeres become shortened with continued replication, and eukaryotic cells that lack telomerase to repair the telomeres can undergo only a limited number of replication events before they lose the ability to divide.

Viruses and bacteria have developed other mechanisms to solve this problem. The chromosomes of bacteria are circular, so there is no 5' end to deal with. Many DNA viruses have adopted a similar solution. Many have circular genomes (e.g., poxviruses, polyomaviruses, papillomaviruses). Others have linear genomes that cyclize before or during replication (e.g., herpesviruses). Some DNA viruses manage to replicate linear genomes, however. Adenoviruses use a virus-encoded protein as a primer, which remains covalently linked to the 5' end of the linear genome. The single-stranded parvovirus DNA genome replicates via

a foldback mechanism in which the ends of the DNA fold back and are then extended by DNA polymerase. Unit sized genomes are cut from the multilength genomes that result from this replication scheme and packaged into virions.

Once initiated, the progression of the replication fork is different in different viruses, as illustrated in Fig. 1.7. In SV40 (family Polyomaviridae), for example, the genome is circular. An RNA primer is synthesized by primase to initiate replication, and the replication fork then proceeds in both directions. The product is two double-strand circles. In the herpesviruses, the genome is circular while it is replicating but the replication fork proceeds in only one direction. A linear double-strand DNA is produced by what has been called a rolling circle. For this, one strand is nicked by an endonuclease and used as a primer. The strand displaced by the synthesis of the new strand is made double stranded by the same mechanism used by the host cell for lagging strand synthesis. In adenoviruses, in contrast, the genome is linear and the replication fork proceeds in only one direction. A single-strand DNA is displaced during the progression of the fork and coated with viral proteins. It can be made double stranded by an independent synthetic event. These different mechanisms will be described in more detail in the discussions of the different DNA viruses in Chapter 6.

As infection proceeds, most DNA viruses undergo a regular developmental cycle, in which transcription of early genes is followed by the transcription of late genes. Activation of the late genes may result from production of a new RNA polymerase or the production of factors that change the activity of existing polymerases. The developmental cycle is, in general, more elaborate in the larger viruses than in the smaller viruses.

Plus-strand RNA Viruses

A simple schematic of the replication of a plus-strand RNA virus is shown in Fig. 1.8. The virus example shown is enveloped and gives rise to subgenomic RNAs (see below). Although the details of RNA replication and virus release are different for other viruses, this scheme is representative of the steps required for gene expression and RNA replication.

Following entry of the genome into the cell, the first event in replication is the translation of the incoming genomic RNA, which is a messenger, to produce proteins required for synthesis of antigenomic copies, also called minus strands, of the genomic RNA. Because the replication cycle begins by translating the RNA genome to produce the enzymes for RNA synthesis, the naked RNA is infectious, that is, introduction of the genomic RNA into a susceptible cell will result in a complete infection cycle. The antigenomic copy of the genome serves as a template for the production of more plus-strand genomes. For some

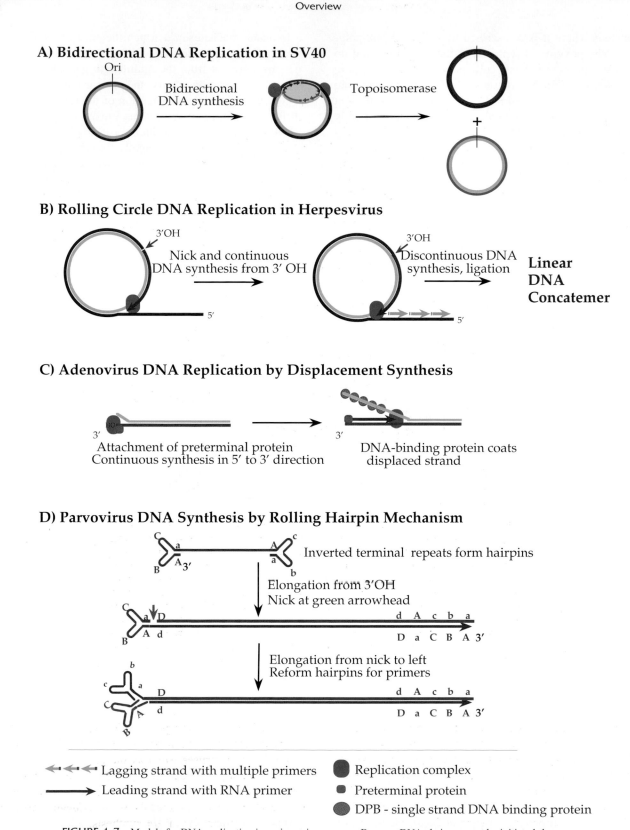

A) Bidirectional DNA Replication in SV40

Ori

Bidirectional
DNA synthesis

Topoisomerase

+

B) Rolling Circle DNA Replication in Herpesvirus

3'OH

Nick and continuous
DNA synthesis from 3' OH

5'

3'OH

Discontinuous DNA
synthesis, ligation

5'

**Linear
DNA
Concatemer**

C) Adenovirus DNA Replication by Displacement Synthesis

3'

Attachment of preterminal protein
Continuous synthesis in 5' to 3' direction

3'

DNA-binding protein coats
displaced strand

D) Parvovirus DNA Synthesis by Rolling Hairpin Mechanism

C   c
    a   A
 A3'   a
B   b

Inverted terminal repeats form hairpins

Elongation from 3'OH
Nick at green arrowhead

C D   d A c b a
 A d   D a C B A 3'
B

Elongation from nick to left
Reform hairpins for primers

   b
c   a
      D   d A c b a
C   d   D a C B A 3'
 A
B

Lagging strand with multiple primers   ⬤ Replication complex

Leading strand with RNA primer   ⬤ Preterminal protein

                                             ⬤ DPB - single strand DNA binding protein

FIGURE 1.7   Models for DNA replication in various virus groups. Because DNA chains cannot be initiated *de novo*, viruses have used a variety of ways to prime new synthesis, such as (A) using RNA primers generated by a primase, (B) elongation from a 3'OH formed at a nick in a circular molecule, (C) priming by an attached protein, and (D) priming by hairpins formed of inverted terminal repeats. [Adapted from Flint *et al.* (2000, Figs. 9.8, 9.16, 9.10, and 9.9, respectively).]

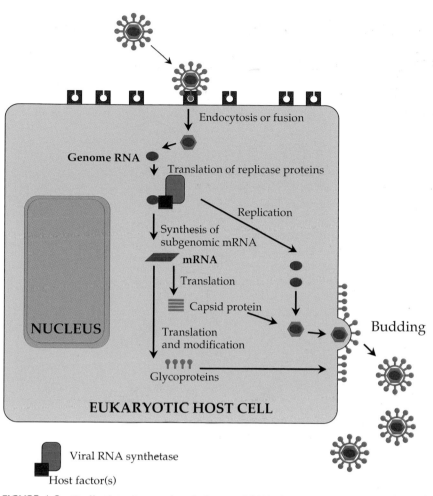

Endocytosis or fusion

Genome RNA

Translation of replicase proteins

Replication

Synthesis of
subgenomic mRNA

mRNA

Translation

Capsid protein

NUCLEUS

Translation
and modification

Budding

Glycoproteins

EUKARYOTIC HOST CELL

Viral RNA synthetase

Host factor(s)

FIGURE 1.8   Replication of an enveloped, plus-strand RNA virus. After the virus attaches to a
cellular receptor, fusion of the virus envelope with the cell plasma membrane or with an endocytic
vesicle releases the nucleocapsid into the cytoplasm. The genome RNA is an mRNA, and is translated on cytoplasmic ribosomes into the proteins required for RNA synthesis. The synthetase
complex can both replicate the RNA to produce new genomes and synthesize viral subgenomic
mRNAs from a minus-strand copy of the genome. The viral structural proteins are then translated
from these subgenomic mRNAs. In the example shown, the capsid protein assembles with the
genome RNA to form a capsid, while the membrane glycoproteins are transported to the cell
plasma membrane. In the final maturation step the nucleocapsid buds out through areas of
modified membrane to release the enveloped particle. [Adapted from Mims *et al.* (1993, p. 2.3)
and Strauss and Strauss (1997, Fig. 2.2).]

plus-strand viruses, the genomic RNA is the only mRNA
produced, as illustrated schematically in Fig. 1.9A. It is
translated into a polyprotein, a long, multifunctional protein
that is cleaved by viral proteases, and sometimes also by
cellular proteases, to produce the final viral proteins. For
other plus-strand RNA viruses, one or more subgenomic
mRNAs are also produced from the antigenomic template
(Fig. 1.9B). For these viruses, the genomic RNA is translated into a polyprotein required for RNA replication (i.e.,
the synthesis of the antigenomic template and synthesis of
more genomic RNA) and for the synthesis of the subge-

nomic mRNAs. The subgenomic RNAs are translated into
the structural proteins required for assembly of progeny
virions. Some viruses, such as the coronaviruses (family
Coronaviridae), which produce multiple subgenomic
RNAs, also use subgenomic RNAs to produce nonstructural
proteins that are required for the virus replication cycle, but
that are not required for RNA synthesis.

The replication of the genome and synthesis of subgenomic RNAs require recognition of promoters in the viral
RNAs by the viral RNA synthetase. This synthetase contains several proteins encoded by the virus, one of which is

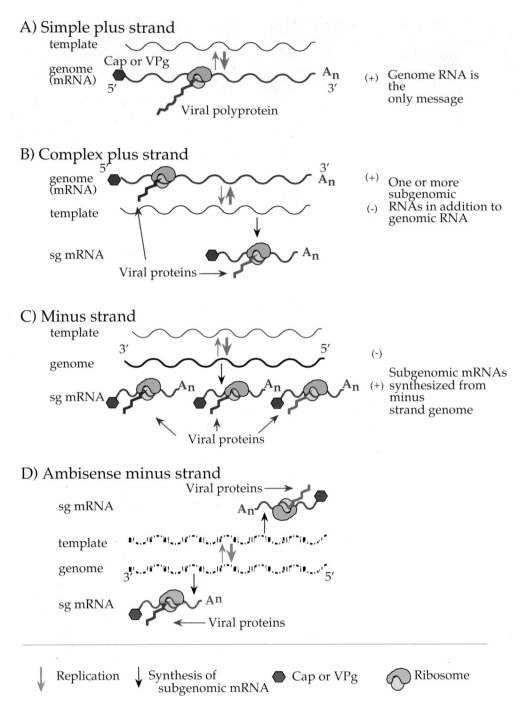

FIGURE 1.9   Schematic of mRNA transcription and translation for the four major types of RNA viruses.

an RNA polymerase. Cellular proteins are also components of the synthetase.

All eukaryotic plus-strand RNA viruses replicate in the cytoplasm. There is no known nuclear involvement in their replication. In fact, where examined, plus-strand viruses will even replicate in enucleated cells. However, it is known that for many viruses, virus-encoded proteins are transported to the nucleus. These proteins may inhibit nuclear functions. For example, a poliovirus protein cleaves transcription factors in the nucleus.

Minus-Sense and Ambisense RNA Viruses

The ambisense RNA viruses and the minus-sense viruses are closely related. One family, the Bunyaviridae, even contains both types of viruses as members. The ambisense strategy is, in fact, a simple modification of the minus-sense strategy, and these viruses are generally lumped together as "negative-strand" or "minus-strand" RNA viruses (Table 1.2).

A simple schematic of the replication of a minus-sense or ambisense RNA virus is shown in Fig. 1.10. All of these viruses are enveloped. After fusion of the virus envelope with a host cell membrane (some enter at the plasma membrane, some via the endosomal pathway), the virus nucleocapsid enters the cytoplasm. The nucleocapsid is helical (Chapter 2). It remains intact and the viral RNA is never released from it. Because the viral genome cannot be translated, the first event after entry of the nucleocapsid must be the synthesis of mRNAs. Thus, the minus-sense or ambisense strategy requires that the viral RNA synthetase be incorporated into the virion if it is to be infectious, and the naked RNA is not infectious if delivered into a cell.

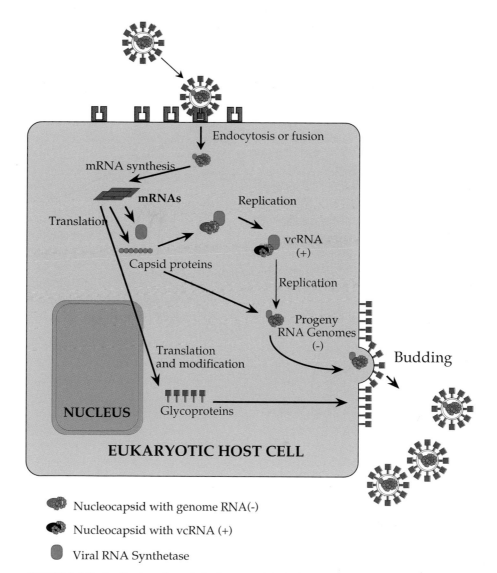

FIGURE 1.10 Replication of a typical minus-strand RNA virus. After the virus attaches to a cellular receptor, the nucleocapsid, containing the viral RNA synthetase, is released into the cytoplasm. The viral synthetase first synthesizes mRNAs, which are translated into the viral proteins required for synthesis of full-length complementary RNAs (vcRNAs). These vcRNAs are the templates for minus-strand genome RNA synthesis. Throughout replication minus-strand genomes and plus-strand vcRNAs are present in nucleocapsids. Viral mRNAs are also translated into membrane glycoproteins that are transported to the cell plasma membrane (or in some cases specialized internal membranes). In the final maturation step, the nucleocapsid buds out through areas of modified membrane to release the enveloped particle. [Adapted from Strauss and Strauss (1997, Fig. 2.3).]

Multiple mRNAs are synthesized by the enzymes present in the nucleocapsid. Each mRNA is usually monocistronic in the sense that it is translated into a single protein, not into a polyprotein (illustrated schematically in Fig. 1.9C). mRNAs are released from the nucleocapsid into the cytoplasm, where they are translated. The newly synthesized proteins are required for the replication of the genome.

Replication of the RNA requires the production of a complementary copy of the genome, as is the case for all RNA viruses, but the antigenomic or vcRNA (for virion complementary) is distinct from mRNA (Fig. 1.9C). Although technically plus sense, it is not translated and is always present in nucleocapsids with the associated RNA synthetic machinery. Replication requires ongoing protein synthesis to supply protein for encapsidation of the nascent antigenomic RNA during its synthesis. In the absence of such protein, the system defaults to the synthesis of mRNAs. The antigenomic RNA in nucleocapsids can be used as a template to synthesize genomic RNA, if proteins for the encapsidation of the nascent genomic RNA are available.

In the ambisense viruses, the antigenomic RNA can also be used as a template for mRNA (Fig. 1.9D). Thus, ambisense viruses modify the minus-sense strategy by synthesizing mRNA from both the genome and the antigenome. Neither the genome nor the antigenome serves as mRNA. The effect is to delay the synthesis of mRNAs that are made from the antigenomic RNA, and thus to introduce a timing mechanism into the virus life cycle.

The mRNAs synthesized by minus-sense or ambisense viruses differ in several key features from their templates. First, the mRNAs lack the promoters required for encapsidation or replication of the genome or antigenome. Thus, they are not encapsidated and do not serve as templates for the synthesis of minus strand. Second, as befits their function as messengers, the mRNAs of most of these viruses are capped and polyadenylated, whereas genomic and antigenomic RNAs are not. Third, the mRNAs of the viruses in the families Orthomyxoviridae, Arenaviridae, and Bunyaviridae have 5′ extensions that are not present in the genome or antigenome, which are obtained from cellular mRNAs. Fourth, although most minus-strand and ambisense RNA viruses replicate in the cytoplasm, influenza virus RNA replication occurs in the nucleus. Thus, influenza RNAs have access to the splicing enzymes of the host, and two of the mRNAs of influenza viruses are exported in both an unspliced and a singly spliced version.

Double-Stranded RNA Viruses

The Reoviridae, the best studied of the double-strand RNA viruses, comprise a very large family of viruses that infect vertebrates, insects, and plants (Table 1.2). The genome consists of 10–12 pieces of double-strand RNA. The incoming virus particle is only partially uncoated. This partial uncoating activates an enzymatic activity within the resulting subviral particle or core that synthesizes an mRNA from each genome fragment. These mRNAs are extruded from the subviral particle and translated by the usual cellular machinery. Thus, the reoviruses share with the minus-strand RNA viruses the attribute that the incoming virus genome remains associated with virus proteins in a core that has the virus enzymatic machinery required to synthesize RNA, and the first step in replication, following entry into a cell, is the synthesis of mRNAs.

The mRNAs also serve as intermediates in the replication of the viral genome and the formation of progeny virions. After translation, the mRNAs become associated with virus proteins. At some point, complexes are formed that contain double-stranded forms of the mRNAs; in these complexes, the 10–12 segments are found in equimolar amounts. These complexes can mature into progeny virions. In other words, mRNAs eventually form the plus strands of the double-strand genome segments.

Retroviruses

An overview of the replication cycle of a retrovirus is shown in Fig. 1.11. The retroviruses are enveloped and enter the cell by fusion, some at the plasma membrane, some at an endosomal membrane. After entry, the first event is the production of a double-strand DNA copy of the RNA genome. This requires the activities of the enzymes RT and RNase H, which are present in the virion. RT synthesizes DNA from either a DNA or an RNA template. RNase H degrades the RNA strand of a DNA–RNA hybrid and is essential for reverse transcription of the genome. The mechanism by which the genome is reverse transcribed is complicated and is described in detail in Chapter 5.

The double-strand DNA copy of the genome is transported to the nucleus, where it integrates into host DNA. Integration is essentially random within the host genome and requires a recombinational event that is catalyzed by another protein present in the virus, called integrase. The integrated DNA copy, called a provirus, is transcribed by cellular RNA polymerases to produce an RNA that is identical to the viral genome. This RNA is exported to the cytoplasm either unspliced or as one or more spliced mRNAs.

The genomic RNA is a messenger for the translation of a series of polyproteins. These polyproteins contain the translation products of genes called *gag*, *pro*, and *pol*. Gag (group-specific *a*ntigen) proteins form the capsid of the virus. Pro is a protease that processes the polyprotein precursors. Pol contains RT, RNase H, and integrase. The three genes are immediately adjacent in the genomic RNA, sepa-

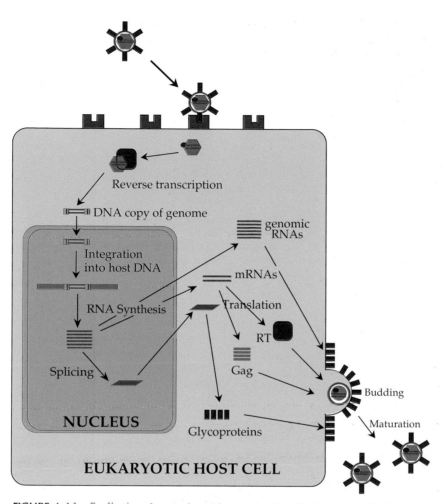

FIGURE 1.11 Replication of a retrovirus. After entering the cell, the retrovirus RNA genome is reverse transcribed into double-stranded DNA by RT present in the virion. The DNA copy migrates to the cell nucleus and integrates into the host genome as the "provirus." Viral mRNAs are transcribed from proviral DNA by host cell enzymes in the nucleus. Both spliced and unspliced mRNAs are translated into viral proteins in the cytoplasm. The capsid precursor protein, "Gag," and RT are translated from full-length RNA. The glycoproteins are translated from spliced mRNA and transported to the cell plasma membrane. Immature virions containing Gag, RT, and the genome RNA assemble near the modified cell membrane. The final maturation step involves proteolytic cleavage of Gag by the viral protease and budding to produce enveloped particles. [Adapted from Fields *et al.* (1996, p. 1786) and Coffin *et al.* (1997, p. 8).]

rated by translation stop codons whose arrangement and number depend on the virus. The arrangement of stop codons is described in detail in Chapter 5, and the mechanisms by which readthrough of stop codons occurs to produce longer polyproteins are described below. A simple diagram of one retrovirus arrangement is shown in Fig. 1.12 as an example. In this example, the polyproteins translated from the genomic RNA are Gag and Gag–Pro–Pol. These two polyproteins assemble with two copies of the virus genome to form the capsid of the virus, usually at regions of the plasma membrane where virus glycoproteins are present. During and immediately after virus assem-

bly by budding, the viral protease cleaves Gag into several components and also separates the enzymes. Cleavage is essential for the assembled virion to be infectious. Thus, current inhibitors of HIV target the protease of the virus as well as the RT, both of which are required for replication, but neither of which is present in the uninfected cell.

The simple retroviruses also produce one spliced mRNA, which is translated into a precursor for the envelope glycoproteins. In some retroviruses, notably the lentiviruses, of which HIV is a member, differential splicing can also lead to the production of mRNAs for a number of regulatory proteins.

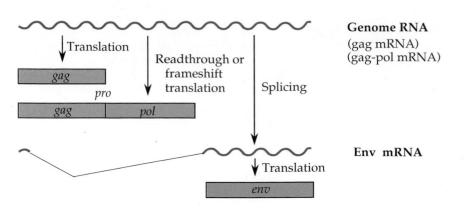

FIGURE 1.12 Transcription and translation of the retroviral genome. Three major polyproteins (shown as colored blocks) are produced. **Gag** is processed to form the nucleocapsid proteins, **Pol** contains RT, RNase H, and integrase, and **Env** is the precursor to the membrane glycoproteins. The protease, *pro*, lies between *gag* and *pol* and may be in either the *gag* reading frame or the *pol* reading frame.

Hepadnaviruses

A schematic of the replication of an hepadnavirus is shown in Fig. 1.13. Hepadnaviruses, which are enveloped, have a life cycle that also involves alternation of the information in the genome between DNA and RNA. The incoming genome is circular, partially double-stranded DNA. The genome is transported to the nucleus where it is converted to a covalently closed, circular, double-strand DNA (cccDNA). Unlike the retroviruses, the DNA does not integrate into the host genome but persists in the nucleus as a nonreplicating episome. It is transcribed by cellular RNA polymerases, using several different promoters in the cccDNA, to produce a series of RNAs. These RNAs are exported to the cytoplasm where they serve as mRNAs. One of these RNAs, called the pregenomic (pg) RNA, is slightly longer than unit length and serves as a template for reverse transcription into DNA. Reverse transcription is performed by RT and RNase H that are translated from the viral mRNAs. It occurs in a core particle assembled from viral capsid proteins, the viral enzymes, and pgRNA. Reverse transcription, described in detail in Chapter 5, resembles that which occurs in the retroviruses, but differs in details.

A complete minus-sense DNA is first synthesized by reverse transcription. Second strand plus-sense DNA is then initiated but only partially completed, so that the core contains partially double-stranded, circular DNA (i.e., the genome). The core with its DNA can proceed through one of two pathways. Early in infection, newly assembled cores may serve to amplify the cccDNA present in the nucleus. These cores contain genomic DNA and are essentially indistinguishable from cores that enter the cytoplasm upon infection by a virion. Thus, their genomic DNA can be transported to the nucleus and be converted into cccDNA.

Note that amplification of cccDNA occurs only through an RNA intermediate, using the pathway just described; there is no direct replication of the DNA in the nucleus. Later in infection, the cores mature into virions by budding through the endoplasmic reticulum. The switch to budding appears to be driven by the presence of viral envelope proteins.

Cellular Functions Required for Replication and Expression of the Viral Genome

The relationship between a virus and its host is an intimate one, shaped by a long evolutionary history. Viruses have small genomes and cannot encode all the functions required for successful replication, and have borrowed many cellular proteins as components of their replication machinery. The nature of the interactions between virus proteins and cellular proteins is an important determinant of the host range and pathology of a virus.

All animal DNA viruses, with the exception of the poxviruses, replicate in the nucleus. They make use of the cellular machinery that exists there for the replication of their DNA and the transcription of their mRNAs. Some viruses use this machinery almost exclusively, whereas others, particularly the larger ones, encode their own DNA or RNA polymerases. However, almost all DNA viruses encode at least a protein required for the recognition of the origin of replication in their DNA. The interplay between the viral proteins and the cellular proteins can affect the host range of the virus. The monkey virus SV40 (family Polyomaviridae) will replicate in monkey cells but not in mouse cells, whereas the closely related mouse polyomavirus (also family Polyomaviridae) will replicate in mouse cells but not in monkey cells. The basis for the host restriction is an incompatibility between the DNA polymerase α-primase of the nonpermissive host and the T antigen of the restricted virus.

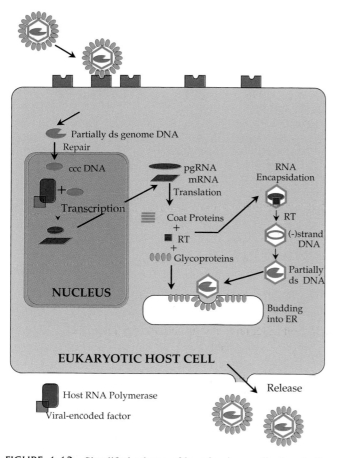

FIGURE 1.13 Simplified scheme of hepadnavirus replication. In the virion, the partially double-stranded DNA genome consists of one full-length minus-strand DNA, and a plus-strand DNA of variable length. After the virus enters the cell, the genome is repaired to a closed circular coiled form (cccDNA) in the nucleus. RNAs of several sizes, including pregenomic (pg) RNA of greater than unit length, are transcribed by host RNA polymerases. These mRNAs are exported and translated into viral proteins in the cytoplasm. The pgRNA is encapsidated, then reverse transcribed into minus-strand DNA. The last step before budding into the endoplasmic reticulum is the partial synthesis of plus-strand DNA. [Scheme derived from Fields *et al.* (1996, p. 2709).]

T antigens are large multifunctional proteins, one of whose functions is to bind to the origin of replication. The T antigens of the viruses form a preinitiation complex on the viral origin of replication, which then recruits the primase into the complex. Because the preinitiation complex containing SV40 T antigen cannot recruit the mouse primase to form an initiation complex, SV40 DNA replication does not occur in mouse cells. However, replication will occur in mouse cells if they are transfected with the gene for monkey primase. Similarly, monkey primase is not recruited into the complex containing mouse polyoma virus T antigen and mouse polyoma virus does not replicate in monkey cells.

In the case of RNA viruses, there is no preexisting cellular machinery to replicate their RNA, and all RNA viruses must encode at least an RNA-dependent RNA polymerase. This RNA polymerase associates with other viral and host proteins to form an RNA replicase complex, which has the ability to recognize promoters in the viral RNA as starting points for RNA synthesis. Early studies on the RNA replicase of RNA phage Qβ showed that three cellular proteins were associated with the viral RNA polymerase and were required for the replication of Qβ RNA. These three proteins, ribosomal protein S1 and two translation elongation factors EF-Ts and EF-Tu, all function in protein synthesis in the cell. The virus appropriates these three proteins in order to assemble an active replicase complex. Recent studies on animal and plant RNA viruses have shown that a variety of cellular proteins also appear to be required for their transcription and replication. One interesting finding is that the animal equivalents of EF-Ts and EF-Tu are required for the activity of the replicase of vesicular stomatitis virus. This suggests that the association of these two translation factors with viral RNA replicases is ancient. Several other cellular proteins have also been found to be associated with viral RNA polymerases or with viral RNA during replication, but evidence for their functional role is incomplete.

Although our knowledge of the nature of host factors involved in the replication of viral genomes and the interplay between virus-encoded and host cell proteins is incomplete, it is clear that such factors can potentially limit the host range of a virus. The restriction of SV40 in cells that do not make a compatible primase was cited above. As a second example, the replication of poliovirus in a restricted set of cells in the gastrointestinal tract and its profound tropism for neurons if it reaches the CNS was also described earlier. These tropisms exhibited by poliovirus do not correlate with the distribution of receptors for the virus, and are a result of restrictions on growth after entry of the virus. Thus in addition to a requirement for a specific receptor for the virus to enter a cell, there may be a need for specific host factors to permit replication once a virus enters a cell. The permissivity of a cell for virus replication after its entry, as well as the distribution of receptors for a virus, are major determinants of viral pathogenesis.

Translation and Processing of Viral Proteins

Viral mRNAs are translated by the cellular translation machinery. Most mRNAs of animal viruses are capped and polyadenylated. Thus, the translation pathways are the same as those that operate with cellular mRNAs, although many viruses interfere with the translation of host mRNAs to give the viral mRNAs free access to the translation machinery. However, there are mechanisms of translation and processing used by some viruses that have no known cellular counterpart. These appear to have evolved because of the special problems faced by viruses, and are described below.

Cap-Independent Translation of Viral mRNAs

A 5′-terminal cap on an mRNA is normally required for its translation. A cellular cap-binding protein binds the cap as part of the translation initiation pathway. The mRNA is then scanned by the initiation complex, starting at the cap, and translation begins at a downstream AUG start codon that is present in a favorable context. However, some viruses, such as poliovirus and hepatitis C virus (genus *Hepacivirus*, family Flaviviridae), have uncapped mRNAs and use another mechanism for the intiation of translation. The 5′ nontranslated region of poliovirus RNA, illustrated schematically in Fig. 1.14, is long—more than 700 nucleotides. Within this 5′ region is a sequence of about

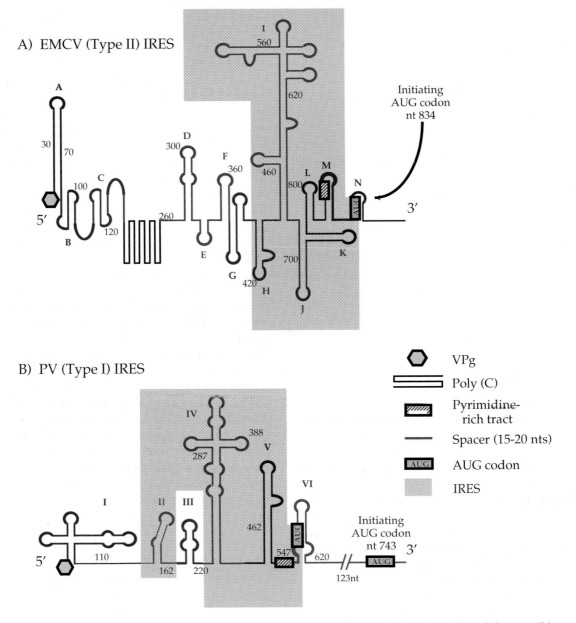

FIGURE 1.14 Schematic diagram of the secondary structures in the 5′-nontranslated regions of encephalomyocarditis virus (EMCV) and poliovirus (PV). The internal ribosome entry sites, or IRES elements, are shaded. There is an element at the 3′ border of the IRES consisting of a pyrimidine-rich tract Yn, a spacer Xm (usually 15–20 nucleotides), and an AUG codon. In the Type II IRES this AUG is the initiating codon, but in a Type I IRES a second codon, in this case 123 nucleotides further down, is used for initiation. Predicted hydrogen-bonded stem-loop structures within the IRES element are labeled A through N for EMCV and I through VI for PV. [Adapted from Wimmer *et al.* (1993, p. 374).]

400 nucleotides called an IRES (Internal Ribosome Entry Site). Ribosomes bind to the IRES and initiate translation in a cap-independent fashion. It is known that the secondary structure of the IRES is critical for its function and that the position of the initiating AUG with respect to the IRES is important.

Interestingly, IRES elements may have entirely different sequences in different viruses and, hence, different apparent higher order structures. The IRES elements of poliovirus and encephalomyocarditis virus (EMCV), both members of the family Picornaviridae, are compared in Fig. 1.14, illustrating the difference in apparent structure. Yet they promote internal initiation in a similar fashion, and the IRES elements of poliovirus and EMCV can be exchanged to yield a viable virus.

The IRES elements of viruses are always found within the 5′ nontranslated region. However, if an IRES is placed in the middle of an mRNA, it will function to initiate translation. True polycistronic mRNAs have been constructed by placing multiple IRES elements into mRNAs, or by combining cap-dependent translation of a 5′ gene with an IRES-dependent translation of a 3′ gene. Since it is possible to construct polycistronic mRNAs using IRES elements, it is somewhat puzzling that animal viruses have never used them to do so.

Using an IRES allows a virus to preferentially translate viral mRNAs. Thus, for example, poliovirus blocks cap-dependent translation after infection by cleaving the cellular cap-binding protein. By blocking cap-dependent translation, host mRNAs cannot be translated, whereas viral mRNAs are not affected. This reserves the translation machinery for translation of viral mRNAs and also blocks many host defense mechanisms (Chapter 8).

Ribosomal Frameshifting

Retroviruses, most plus-strand RNA viruses, and certain other viruses expand the range of polyproteins produced by having long open reading frames (ORFs) interrupted by stop codons. Termination of the polyprotein at the stop codon leads to production of a truncated polyprotein having certain functions, whereas readthrough leads to the production of a longer polyprotein with additional functions. This was illustrated in Fig. 1.12 for a retrovirus. Two mechanisms are used by different viruses to ignore the stop codon. The first is simply to read the stop codon as sense. In this case the downstream sequences are in the same reading frame as the upstream sequences. The mechanism of readthrough is thought to involve wobble in the third codon position that allows a tRNA to bind to and insert an amino acid at the stop codon position. Both amber (UAG) and opal (UGA) codons have been found in different readthrough positions. Readthrough efficiency is variable, although usually between 5 and 20%.

The second mechansim for ignoring the stop codon is ribosomal frameshifting, in which the reading frame is shifted into the plus 1 or minus 1 frame upstream of the stop codon. In this case, the downstream sequences are in a different reading frame from the upstream sequences. Fig. 1.15 shows the -1 frameshifting that occurs between the *gag* and *pol* genes of avian leukosis virus (ALV). Frameshifting occurs at a precise sequence known as a "slippery" sequence, shown in green. Slippery sequences have short strings of the same nucleotide and are often rich in A and U. For ALV, termination at the UAG indicated with the asterisk produces a polyprotein that contains the sequences for Gag and Pro. Frameshifting results in movement into the -1 frame at this point so that the codon following UUA (leucine) becomes AUA (isoleucine) rather than UAG (stop), and translation of the downstream gene (*pol*) follows to produce the polyprotein Gag–Pro–Pol.

Frameshifting usually requires a structural feature, such as a hairpin, downstream of the slippery sequence in order to slow the ribosome at this point and allow more time for the frameshifting to take place. The structure that is required for frameshifting in a yeast RNA virus called L-A is shown in Fig. 1.16. The sizable hairpin just downstream of the slippery sequence is further stabilized by a pseudoknot. In this case the slippery sequence, in which frameshifting takes place, occurs about 110 nucleotides upstream of the stop codon for the upstream reading frame (Gag protein reading frame). Thus, these 110 nucleotides are translated in two different reading frames: the Gag reading frame if frameshifting does not occur, or the Pol reading frame if frameshifting does occur. The sequence illustrated in the figure is the minimum sequence required for frameshifting. It can be placed in the middle of an unrelated mRNA, and -1 frameshifting will occur.

The efficiency of frameshifting is variable and depends on the frameshifting sequences. Thus, the efficiency of frameshifting can be modulated by mutations in the virus. For example, the frameshifting sequence in Fig. 1.16 results in frameshifting about 2% of the time. Changing the slippery sequence from GGGUUU to UUUUUU results in frameshifting 12% of the time. Frameshifting efficiencies of from 2% to more than 20% have been observed at different frameshift sites in different viruses.

Processing of Viral Polyproteins

Many viruses produce polyproteins, as described above. These polyproteins are cleaved into individual proteins by viral enzymes, with the exception of precursors to viral envelope proteins, which have access to cellular enzymes present in subcellular compartments. Thus, the use of viral enzymes to process polyproteins may be due in part to a lack of appropriate proteases in the cytoplasm of eukaryotes. However, the use of a viral protease also allows the

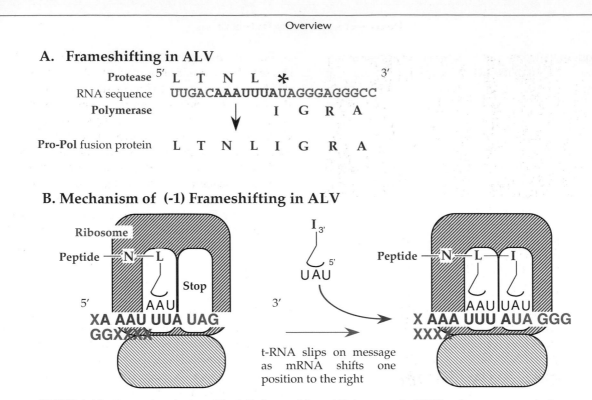

A. Frameshifting in ALV

Protease 5′ L T N L **✳** 3′

RNA sequence UUGACAAAUUUAUAGGGAGGGCC

Polymerase I G R A

Pro-Pol fusion protein L T N L I G R A

B. Mechanism of (-1) Frameshifting in ALV

FIGURE 1.15 Proposed mechanism of the (−1) ribosomal frameshift that occurs in ALV. The slippery sequence is shown in green. The asterisk identifies the UAG codon that terminates the upstream (Gag–Pro) ORF. Frameshifting is thought to require a pseudoknot downstream of the "slippery sequence" (illustrated in Fig. 1.16). [Adapted from Fields *et al.* (1996, p. 577) and Goff (1997, p. 156).]

virus to fine-tune the processing events, and this control is often used to regulate the viral replication cycle.

The viral proteases are components of the translated polyproteins. Some cleavages catalyzed by these proteases

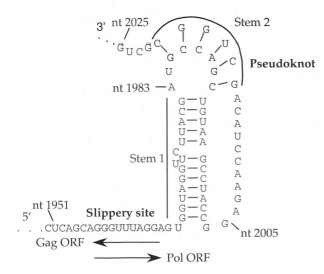

FIGURE 1.16 The site in L-A RNA that promotes −1 ribosomal frameshifting to form the Gag–Pol fusion protein. The pseudoknot makes the ribosome pause over the slippery site (nt 1958–1964 shown in green). The *gag* and *pol* ORFs overlap for 130 nucleotides, and in the absence of frameshifting the Gag protein terminates at nt 2072. [Adapted from Fields *et al.* (1996, p. 566).]

can occur in *cis*, in a monomolecular reaction in which the polyprotein cleaves itself. Other cleavages occur in *trans*, whereby a polyprotein, or a cleaved product containing the protease, cleaves another polyprotein in a bimolecular reaction. In many cases the series of cleavages effected by a protease proceeds in a defined manner and serves to regulate the virus life cycle.

Four types of virus proteases have been found in different viruses. Proteases related to the serine proteases of animals are common in animal RNA viruses. The animal serine proteases, of which chymotrypsin is a well-studied example, have a catalytic triad composed of histidine, aspartic acid, and serine, which form the active site of the enzyme. Where structures have been determined, the viral proteases possess a fold related to that of chymotrypsin and possess an active site with geometry identical to that of chymotrypsin. The structure of protease 3C[pro] of a human rhinovirus (family Picornaviridae) is shown in Fig. 1.17 as an example (another example is shown in Chapter 2, Fig. 2.15). Although most viral serine-type proteases have a catalytic triad composed of histidine, aspartic acid, and serine, the rhinovirus protease has an active site composed of histidine, glutamic acid, which replaces aspartic acid, and cysteine which replaces serine. The replacement of the active site serine by cysteine causes problems with nomenclature. Here we use serine protease (or serine-type protease) to refer to a protease whose structure and active site

Effects of Virus Infection on the Host Cell

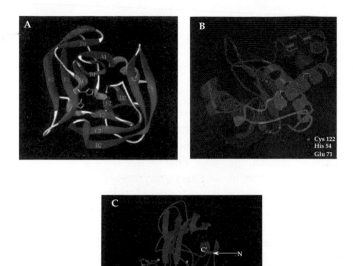

FIGURE 1.17 Divergent structures of viral proteases. (A) Ribbon diagram of human rhinovirus 3C^pro. The beta strands are shown in blue and the helical secondary structure is shown in yellow. The side chains that make up the catalytic triad are shown: Cys-146 (with the sulfur atom shown in yellow), His-40, and Glu-71 (with the atoms of the charged carboxyl group shown in red). The N and C termini are indicated "N" and "C," respectively in red. [From Matthews *et al.* (1994) with permission.] (B) Ribbon diagram of the papain-like protease of adenovirus complexed with its 11-amino-acid cofactor (red arrow at bottom). The protein trace is colored from N terminus to C terminus as the visible spectrum from red to violet. The locations of the amino acids making up the catalytic triad (Cys-122, His-54, and Glu-71) are indicated with dots. [From Ding *et al.* (1996), with permission.] (C) The HIV protease, a dimer of identical subunits, is shown complexed with a non-amino-acid inhibitor in the active site. One monomer is colored from red to green (N terminus to C terminus) while the second one is colored from green to dark blue. [Adapted from Rutenber *et al.* (1993).]

geometry are related to those of the animal serine proteases, rather than as a description of the catalytic amino acid. These similarities in structure make it likely that the viruses acquired the protease from a host and modeled it to fit their own needs.

A second group of proteases is related to the papain-like enzymes of plants and animals. These proteases have a catalytic dyad consisting of histidine and cysteine. A third residue is sometimes considered a part of the active site (as described below) and the active site is then referred to as a catalytic triad. Model folding studies have suggested that these viral papain-like proteases are related to the cellular counterparts. Such proteases are found in many plus-strand RNA viruses and in adenoviruses. The structure of the adenovirus papain-like protease is shown in Fig. 1.17B. Notice that the structure of this enzyme is very different from the

serine-type protease of rhinoviruses in Fig. 1.17A, even though the active site is composed of the same three residues, Cys, His, and Glu. In addition, the sequence of the active site residues in the linear amino acid sequence of serine-type proteases is His, Glu, Cys, whereas the order is Cys, His, Glu in the adenovirus protease and in papain (where the active site contains Asn rather than Glu).

A third type of protease, known to be present only in hepatitis C virus, is a metalloprotease. A divalent cation is a component of the active site of this enzyme.

The retroviruses encode a protease to process the Gag and Gag-Pol polyproteins during virus maturation. The protease is related to the aspartate proteases of animals, which include pepsin, renin, and cathepsin D. The structure of the HIV protease is shown in Fig. 1.17C. The active site of aspartate proteases consists of two asparate residues. In the animal enzymes, the two asparate residues are present in a single polypeptide chain that folds to bring the aspartate residues together to form the active site. In HIV, the active site is formed by dimerization of two Pro protein monomers of about 100 residues, each of which contributes one of the asparate residues to the active site.

Assembly of Progeny Virions

The last stage in the virus life cycle is the assembly of progeny virions and their release from the infected cell. The assembly of viruses will be discussed in Chapter 2, after the structure of viruses is described.

EFFECTS OF VIRUS INFECTION ON THE HOST CELL

Cells can be described as permissive, semipermissive, or nonpermissive for virus replication. Semipermissive or nonpermissive cells lack factors required for a complete replication cycle. The term *nonpermissive* usually refers to a cell in which no progeny virus are produced. A cell may be nonpermissive because it lacks receptors for the virus or because it lacks factors required by the virus after entry. In the latter case, an abortive infection may occur in which virus replication begins but does not result in the production of progeny virus. The term *semipermissive* usually refers to a cell in which a small yield of progeny virus may be produced.

Several types of viral infection cycle can be distinguished. The infection may be lytic, latent, persistent, or chronic. In some cases, virus infection results in the transformation of a cell.

Lytic Infection or Latent Infection

In a lytic infection, the virus replicates to high titer, host cell macromolecular synthesis is shut down, and the host cell dies. Bacterial cells are usually actively lysed by the elaboration of a specific lysis product during bacteriophage infection. Animal viruses, in contrast, usually cause cell death by inducing apoptosis or programmed cell death. Apoptosis is a suicide pathway in which the mitochrondria cease to function, the cell destroys its DNA, and the cell fragments into small vesicles (Chapter 8). Cell death may also be due to necrosis, a generalized loss of cell integrity caused by virus interference with activities necessary for the upkeep of the cell. Membrane integrity is lost during necrotic cell death and cytoplasmic contents leak out of the cell. Apoptosis is a normal event in animals and does not result in an inflammatory response. In contrast, necrosis does result in an inflammatory response.

During lytic infection, profound changes in the condition of the cell occur well before it dies and fragments. These changes may result in alterations that are observable in the light microscope, such as changes in the morphology of the cell, the formation of vacuoles within the cell, or the fusion of cells to form syncytia. Such changes are given the name cytopathic effect, or CPE. CPE is often an early sign that the cell is infected.

In latent infections, no virus replication occurs. The best understood case of latent infection is that of temperate bacteriophage, which express genes that repress the replication of the virus. Once the lysogenic state is established, in which viral replication is repressed, it can persist indefinitely. Among vertebrate viruses, many of the herpesviruses are capable of latently infecting specialized cells that are nonpermissive or semipermissive for virus replication. As one example, herpes simplex virus type 1 establishes a lifelong, latent infection of neurons of the trigeminal ganglia. In this case it is thought that latent state arises because the neuron lacks cellular factors required for the transcription and replication of the viral DNA, rather than because of the production of a herpes protein that suppresses replication. Reactivation of the virus at times leads to active replication of the virus in epithelial cells innervated by the infected neuron, resulting in fever blisters usually at the lip margin.

Persistent versus Chronic Infection

Persistent infection and chronic infection are often used interchangeably, but these two terms will be distinguished here in order to describe two types of infections that persist by different mechanisms. In what is here referred to as a persistent infection, an infected cell lives and produces progeny virus indefinitely. The retroviruses represent the best studied case of persistent infection. During infection by retroviruses, the DNA copy of the genome is integrated into the host cell genome. Continual transcription of the genome and assembly of progeny virus, which bud from the cell surface, occur without apparent ill effects on the host cell. The infected cell retains its normal functions and can divide. However, although infection by most retroviruses does not lead to cell death, active replication of HIV can result in cell death.

Chronic infection is a property of a group of cells or of an organism in which lytic infection is established in many cells, but many potentially susceptible cells escape the infection at any particular time, for whatever reason. The infection is not cleared and the continual appearance of susceptible cells in the population leads to the continued presence of replicating virus. One well-known example of a chronic infection in humans is HIV, in which the infection cannot be cleared by the immune system and virus continues to replicate. AIDS results when the immune system is finally overwhelmed by the virus. Hepatitis B and hepatitis C viruses are also well known for their ability to establish chronic liver infections that can persist for life.

Transformation of Cells

The normal outcome of the infection of a cell by a virus is the death of the cell and the release of progeny virus. The major exceptions are the persistent infection of cells by retroviruses and the latent infection of cells by viruses such as herpesviruses, in which the cell survives with its properties little altered except for the new ability to produce virus. However, another possible outcome is the transformation of the cell, which involves not only the survival of the cell but an alteration in its growth properties caused by deregulation of the cell cycle. Transformed cells may be able to induce the formation of a tumor if they are produced within an animal or are injected into an animal after formation *ex vivo*. Transformation of a cell needs to be distinguished from tumorigenicity, the ability of the transformed cell to cause a tumor. Transformed cells may fail to cause a tumor because they are rejected by the host's immune system or because the transformed cells lack some properties required for the growth of a tumor in an animal, in which case additional mutations may eventually allow tumors to form.

The avian and mammalian sarcoma viruses, specialized retroviruses that arise when a cellular oncogene is incorporated into the retroviral genome, can transform cells in culture and cause tumors in animals. It was this feature that led to their discovery in the first place and resulted in intensive study of the retroviruses. Cellular oncogenes encode proteins that regulate the cell cycle. They induce the cell to enter S phase, in which DNA replication occurs, on receipt of appropriate signals. Following infection by a sarcoma virus and intergration of the provirus into the host genome, the overexpression of the incorporated oncogene, or expression of a mutated oncogene that continuously induces the cell to multiply, results in transformation of the infected cell. The incorporation of cellular oncogenes into a retrovirus is an accident that results from recombination of the viral genome with a cellular mRNA encoding an oncogene.

The oncogene replaces viral genes in the genome and almost all sarcoma viruses are defective, unable to undergo a complete replication cycle without a helper. In nature sarcoma viruses are able to cause a tumor in the animal in which they arise but are not passed on, and thus die out. The subject of sarcoma viruses will be covered in more detail in Chapter 5, after a discussion of the genome organization of retroviruses and the details of their replication.

Many DNA viruses also encode proteins that are capable of transforming cells. These viral oncogenes induce cycling in infected cells, providing an environment suitable for the replication of the viral DNA. Whereas the cellular oncogenes present in sarcoma viruses serve no function in viral replication, the viral oncogenes in DNA viruses are essential for viral replication. If the cell is not induced to enter S phase, the virus replicates poorly or not at all. Infection of a cell by a DNA virus normally leads to the death of the cell caused by replication of the virus. Thus, although transformed, the infected cell does not survive. However, if the virus is unable to undergo a complete lytic cycle, either because it is defective or because the cell infected is nonpermissive or semipermissive, the cell may survive as a transformed cell if the early (transforming) genes continue to be expressed. We return to the subject of viral oncogenes in Chapter 6, because they are an ingredient of the replication cycle of DNA viruses.

Transformation of cells is accompanied by a number of phenotypic changes. Cells maintained in culture are normally derived from solid tissues and require anchorage to a solid substrate as well as a supply of nutrients, including growth factors, for growth. Transformed cells have a decreased requirement for a solid substrate and may have lost such a requirement entirely, and thus may grow in soft agar or in liquid culture. They have decreased requirements for growth factors and will continue to divide in medium with lowered amounts of such factors. They grow to higher densities in culture than do normal cells, and tend to pile up in multilayers on the surface of the culture dish (loss of contact inhibition or density-dependent growth). Transformed cells may be immortal and able to divide indefinitely in culture, whereas normal cells stop dividing after a limited number of divisions. Many of these changes in growth properties are reflected in changes in their cytoskeleton, their metabolism, and in their interactions with the extracellular matrix.

Lymphocytes, which do not require anchorage for growth, can also be transformed by the appropriate viruses. They may become immortal and able to divide indefinitely in culture. In the animal this may lead to leukemias or lymphomas.

EPIDEMIOLOGY: THE SPREAD OF VIRUSES FROM PERSON TO PERSON

Viruses must be able to pass from one infected organism to another. The spread of specific viruses will be considered together with their other attributes in the chapters that follow, but it is useful to consider virus epidemiology in overview at this point. The tissues infected by a virus and the seriousness of the disease caused by it are attributes that determine in part the mechanism of spread of a virus. Thus, knowledge of the epidemiology of a virus is important for understanding the biology of its replication and pathology.

We can discriminate several general ways in which animal viruses are spread: oral–fecal, airborne, blood-borne (including viruses that are spread by blood-sucking arthropods), sexual, and congenital. We can also distinguish human viruses that have humans as their major or only host, and viruses that are also associated with other animals.

Viruses spread by an oral-fecal route are disseminated by ingestion of contaminated food or water. Infection begins in the gut, and it may or may not spread to other organs. Many of these viruses cause gastroenteritis. Virus is excreted in feces or urine to continue the cycle. Such viruses are usually fairly stable outside the organism because they may have to persist in an infectious form for long periods of time before being ingested by the next victim.

Airborne or respiratory viruses are spread when virus present in the respiratory tract is expelled as aerosols or in mucus. Infection begins when contaminated air is inhaled or when virus present in mucosal secretions, for example on doorknobs or on a companion's hands, is contacted and the virus is transferred to mucosal surfaces in the nose, mouth, or eyes. These viruses are often unstable outside the body; spread thus requires close person-to-person contact. Infection begins on mucosal surfaces in the upper respiratory tract or the eye. Many of these viruses are restricted to growth in the upper respiratory tract, but some are able to spread to other organs. Respiratory tract disease is caused by many of these viruses.

Blood-borne viruses establish a viremia in which infectious virus circulates in the blood. Some are transmitted by blood-sucking arthropods, which act as vectors, whereas others are transmitted by exposure to contaminated blood or other bodily fluids. Arboviruses (e.g., yellow fever virus, genus *Flavivirus*, family Flaviviridae) can replicate in both arthropods, such as ticks or mosquitoes, and in vertebrates. The arthropod may become infected when it takes a blood meal from a viremic vertebrate. After replication of the virus in the arthropod, it can be transmitted to a vertebrate when the arthropod takes another blood meal. Although arboviruses tend to have broad host ranges, a virus is usually maintained in only one or a few vertebrate hosts and vectored by a limited set of arthropods.

Therapeutic blood transfusion, use of hypodermic injections, and intravenous drug use are methods of spread of many blood-borne viruses. HIV and hepatitis B (family Hepadnaviridae) and C (genus *Hepacivirus*, family Flaviviridae) viruses, for example, are commonly spread among drug users through sharing of contaminated needles. Transfusion with contaminated blood is still possible despite diagnostic tests to identify infected blood products.

In developed countries, the blood supply is screened for HIV and hepatitis B and C viruses, as well as other viral agents for which tests exist, but in developing countries contaminated blood is often still a major problem. Blood-borne viruses that are not arboviruses are often spread sexually as well as by the methods above, but in some cases it is not clear how the viruses were spread before the introduction of blood transfusion and hypodermic needles.

Because of the need to establish a significant viremia, which requires extensive viral production in organs that can shed virus into the bloodstream, blood-borne viruses often cause serious disease. Furthermore, because spread is direct, these viruses need not be stable outside the body and usually have a short half-life outside an organism.

Many viruses are transmitted by sexual contact. Virus may be present in warts in the genital area (e.g., herpes simplex virus type 2 and human papillomaviruses) or in semen or vaginal secretions (e.g., HIV, hepatitis B virus). Infection begins in the genital mucosa but may spread to other organs. Because the opportunity for spread by sexual contact is much more restricted than for spread by other routes, viruses spread by sexual transmission almost invariably set up long-term persistent infections that cause only mild disease, at least early in infection. This allows the virus to be disseminated over long periods of time.

Many viruses can be spread vertically. Congenital infection of the fetus *in utero* or during passage of the infant through the birth canal occurs with viruses such as HIV, cytomegalovirus (family Herpesviridae), and rubella virus (genus *Rubivirus*, family Togaviridae). Vertical transmission can also occur shortly after birth, by breast feeding for example. HTLV I (family Retroviridae), which causes leukemia in humans, is such a virus.

Many of the viruses that cause human disease infect only humans in nature, or are maintained only in humans (e.g., all of the human herpesviruses, HIV, hepatitis B, poliovirus). Thus, spread is from one person to another. Many others are associated with wildlife and spread is often from animal to man (e.g., most of the arboviruses, rabies virus, the hantaviruses, and the arenaviruses). The hantaviruses (family Bunyaviridae) and arenaviruses (family Arenaviridae) are associated with small rodents, in which they cause little disease. Humans, in which these viruses cause serious illness, can become infected by inhaling aerosols containing excreta from such rodents (e.g., Lassa fever virus; Junin virus, the causative agent of Argentine hemorrhagic fever; and Sin Nombre virus, the causative agent of hantavirus pulmonary syndrome). Rabies virus (family Rhabdoviridae) is associated with unvaccinated domestic dogs and with many species of wildlife, including foxes, coyotes, skunks, racoons, and bats. It is spread by the bites of infected animals. The virus is present in salivary fluid of the infected animal, and the disease induces an infected animal to become aggressive and bite potential hosts. Interestingly, although many humans die worldwide of rabies each year contracted from the bites of rabid animals, human-to-human transmission does not occur. Some recent human infections have been contracted from wildlife only indirectly. Nipah virus (family Paramyxoviridae) is associated with flying foxes, large fruit-eating bats. The virus recently caused an epidemic of disease in pigs in Malaysia and Singapore, and pig farmers contracted the disease from the pigs. Before the epidemic died out, 258 humans developed encephalitis, of whom 40% died.

FURTHER READING

Effects of Infectious Disease on Human History

Crosby, A. W. (1989). "America's Forgotten Pandemic: The Influenza of 1918." Cambridge University Press, Cambridge, England.

Gould, T. (1995). "A Summer Plague: Polio and its Survivors." Yale University Press, New Haven, CT.

Hopkins, D. R. (1983). "Princes and Peasants: Smallpox in History." University of Chicago Press, Chicago.

Kolata, G. (1999). "Flu: The Story of the Great Influenza Pandemic of 1918 and the Search for the Virus that Caused It." Farrar, Straus, & Giroux, New York.

Koprowski, H., and Oldstone, M. B. A., eds. (1996). "Microbe Hunters—Then and Now." Medi-Ed Press, Bloomington, IL.

Oldstone, M. B. A. (1998). "Viruses, Plagues, and History." Oxford University Press, New York.

Virus Taxonomy

van Regenmortel, M. H. V., Fauquet, C. M., Bishop, D, H, L., *et al.*, eds. (2000). "Virus Taxonomy, Seventh Report of the International Committee on Taxonomy of Viruses." Academic Press, New York.

Virus Receptors and Entry

Bella, J., and Rossmann, M. G. (1999). Rhinoviruses and their ICAM receptors. *J. Struct. Biol.* **128**; 69–74.

Bullough, P. A., Hughson, F. M., Skehel, J. J. *et al.* (1994). Structure of influenza hemagglutinin at the pH of membrane-fusion. *Nature* **371**; 37–43.

Cocchi, F., Menotti, L., Mirandola, P. *et al.* (1998). The ectodomain of a novel member of the immunoglobulin subfamily related to the poliovirus receptor has the attributes of a bona fide receptor for herpes simplex virus types 1 and 2 in human cells. *J. Virol.* **72**; 9992–10002.

Rostand, K. S., and Esko, J. D. (1997). Microbial adherence to and invasion through proteoglycans. *Infect. Immun.* **65**; 1–8.

Saragovi, H. U., Rebai, N., Roux, E. *et al.* (1998). Signal transduction and antiproliferative function of the mammalian receptor for type 3 reovirus. *Curr. Top. Microbiol. Immunol.* **233**; 155–166.

Sommerfelt, M. (1999). Retrovirus receptors. *J. Gen. Virol.* **80**; 3–049–3064.

Weissenhorn, W., Dessen, A., Calder, L. J. *et al.* (1999). Structural basis for membrane fusion by enveloped viruses. *Mol. Membr. Biol.* **16**; 3–9.

Wilson, I., Skehel, J., and Wiley, D. (1981). Structure of the hemagglutinin membrane glycoprotein of influenza virus at 3 Å resolution. *Nature (London)* **289**; 366–373.

Wimmer, E., ed. (1994). "Cellular Receptors for Animal Viruses." Cold Spring Harbor Press, Cold Spring Harbor, NY.

Viral Proteases

de Francesco, R., and Steinkuhler, C. (2000). Structure and funtion of the hepatitis C virus NS3–NS4A proteinase. *Curr. Top Microbiol.* **242**; 149–169.

Ding, J., McGrath, W. J., Sweet, R. M. *et al.* (1996). Crystal structure of the human adenovirus proteinase with its 11 amino acid cofactor. *EMBO J.* **15**; 1778–1783.

Matthews, D. A., Smith, W. W., Ferre, R. A. *et al.* (1994). Structure of the human rhinovirus 3C protease reveals a trypsin-like polypeptide fold, RNA-binding site, and means for cleaving precursor polyprotein. *Cell* (Cambridge, Mass.) **77**; 761–771.

Rutenber, E., Fauman, E. B., Keenan, R. J. *et al.* (1993). Structure of a non-peptide inhibitor complexed with HIV-1 protease. Developing a cycle of structure-based drug design. *J. Biol. Chem.* **268**; 15343–15346.

Strauss, J. H., ed. (1990). Viral proteinases. *Semin. Virol.* **1**; 307–384.

Frameshifting, Pseudoknots, and IRESes

Deiman, B. A. L. M., and Pleij, C. W. A. (1997). Pseudoknots: A vital feature in viral RNA. *Semin. Virol.* **8**; 166–175.

Dinman, J. D., Icho, T., and Wickner, R. B. (1991). A-1 ribosomal frameshift in a double-stranded RNA virus of yeast forms a gag-pol fusion protein. *Proc. Natl. Acad. Sci. U.S.A.* **88**; 174–178.

Kim, Y.-G., Su, L., Maas, S. *et al.* (1999). Specific mutations in a viral RNA pseudoknot drastically change ribosomal frameshifting efficiency. *Proc. Natl. Acad. Sci. U.S.A.* **96**; 14234–14239.

Lemon, S. M., and Honda, M. (1997). Internal ribosome entry sites within the RNA genomes of hepatitis C virus and other flaviviruses. *Semin. Virol.* **8**; 274–288.

Stewart, S. R. and Semler, B. L. (1997). RNA determinants of picornavirus cap-independent translation initiation. *Semin. Virol.* **8**; 242–255.

2

The Structure of Viruses

INTRODUCTION

Virus particles, called virions, contain the viral genome encapsidated in a protein coat. The function of the coat is to protect the genome of the virus in the extracellular environment as well as to bind to a new host cell and introduce the genome into it. Viral genomes are small and limited in their coding capacity, which requires that three-dimensional virions be formed using a limited number of different proteins. For the smallest viruses, only one protein may be used to construct the virion, whereas the largest viruses may use 30 or more proteins. To form a three-dimensional structure using only a few proteins requires that the structure must be regular, with each protein subunit occupying a position at least approximately equivalent to that occupied by all other proteins of its class in the final structure (the principle of quasi-equivalence). Such a regular three-dimensional structure can be formed from repeating subunits using either helical symmetry or icosahedral symmetry principles. In the case of the smallest viruses, the final structure is simple and quite regular. Larger viruses with more proteins at their disposal can build more elaborate structures. Enveloped viruses may be quite regular in construction or may have irregular features, because the use of lipid envelopes allows some irregularities in construction.

Selected families of vertebrate viruses are listed in Table 2.1 grouped by the morphologies of the virions. Also shown for each family is the presence or absence of an envelope in the virion, the triangulation number (defined below) if the virus is icosahedral, the morphology of the nucleocapsid or core, and figure numbers where the structures of members of a family are illustrated. Electron micrographs of five DNA viruses belonging to different families and of five RNA viruses belonging to different families are shown in Fig. 2.1. The viruses chosen represent viruses that are among the largest known and the smallest known, and are all shown to the same scale for comparison. For each virus, the top micrograph is of a virus that has been negatively stained, the middle micrograph is of a section of infected cells, and the bottom panel shows a schematic representation of the virus. The structures of these and other viruses are described below.

HELICAL SYMMETRY

Helical viruses appear rod shaped in the electron microscope. The rod can be flexible or stiff. The best studied example of a simple helical virus is tobacco mosaic virus (TMV). The TMV virion is a rigid rod 18 nm in diameter and 300 nm long (Fig. 2.2B). It contains 2130 copies of a single capsid protein of 17.5 kDa. In the right-hand helix, each protein subunit has six nearest neighbors and each subunit occupies a position equivalent to every other capsid protein subunit in the resulting network (Fig. 2.2A), except for those subunits at the very ends of the helix. Each capsid molecule binds three nucleotides of RNA within a groove in the protein. The helix has a pitch of 23 Å and there are $16^{1}/_{3}$ subunits per turn of the helix. The length of the TMV virion (300 nm) is determined by the size of the RNA (6.4 kb).

Many viruses are constructed with helical symmetry and often contain only one protein or a very few proteins. The popularity of the helix may be due in part to the fact that the length of the particle is not fixed and RNAs or DNAs of different sizes can be readily accommodated. Thus the genome size is not fixed, unlike that of icosahedral viruses.

TABLE 2.1   Selected Vertebrate Virus Families by Morphology

Morphology of particle	Enveloped	Triangulation number	Morphology of nucleocapsid	Figure numbers
Icosahedral				
Adenoviridae	No	T = 25	Not applicable	Figs. 2.1, 2.5, 2.12
Reoviridae	No	T = 131[a]	Icosahedral	Figs. 2.1, 2.5, 2.11
Papillomaviridae	No	P = 7d[a,b]	Not applicable	Fig. 2.5
Polyomaviridae	No	P = 7d[a,b]	Not applicable	Figs. 2.5, 2.9
Parvoviridae	No	T = 1	Not applicable	Figs. 2.1, 2.5
Picornaviridae	No	P = 3[b]	Not applicable	Figs. 2.1, 2.5, 2.7, 2.8
Astroviridae	No	T = 3	Not applicable	
Caliciviridae	No	T = 3	Not applicable	
Herpesviridae	Yes	T = 16	Icosahedral	Figs. 2.1, 2.5, 2.16
Togaviridae	Yes	T = 4	Icosahedral	Figs. 2.5, 2.14, 2.15
Flaviviridae	Yes	T = 3	Icosahedral	Fig. 2.5
Irregular				
Poxviridae (ovoid)	Yes		Dumbbell	Figs. 2.1, 2.20
Rhabdoviridae (bacilliform)	Yes		Coiled helix	Figs. 2.1, 2.19
Spherical				
Retroviridae	Yes	?	Icosahedral?	Figs. 2.1, 2.17
Round[c]				
Coronaviridae	Yes		Helical or tubular	
Paramyxoviridae	Yes		Helical	Fig. 2.18
Orthomyxoviridae	Yes		Helical	Figs. 2.1, 2.18
Bunyaviridae	Yes		Helical	
Arenaviridae	Yes		Helical	
Hepadnaviridae	Yes	T = 3 & 4	Icosahedral	
Filamentous[c]				
Filoviridae	Yes		Helical	Fig. 2.19

[a]Two mirror image structures can be formed using T = 13 or T = 7, a symmetry referred to as "d" or "I".

[b]Because the subunits are not exactly equivalent, *Papillomaviridae*, *Polyomaviridae*, and as well as poliovirus have "pseudo-triangulation numbers" so are referred to as P = 7, P = 7, and P = 3 symmetries respectively.

[c]Virions are often pleiomorphic

ICOSAHEDRAL SYMMETRY

Virions can be approximately spherical in shape, based on icosahedral symmetry. Since the time of Euclid, there have been known to exist only five regular solids in which each face of the solid is a regular polygon: the tetrahedron, the cube, the octahedron, the dodecahedron, and the icosahedron. The icosahedron has 20 faces, each of which is a regular triangle, and each face thus has threefold rotational symmetry (Fig. 2.3A). There are 12 vertices where 5 faces meet, and thus each vertex has fivefold rotational symmetry. There are 30 edges in which 2 faces meet, and each edge possesses twofold rotational symmetry. Thus the icosahedron is characterized by twofold, threefold, and fivefold symmetry axes. The dodecahedron, the next

simpler regular solid, has the same symmetry axes as the icosahedron and is therefore isomorphous with it in symmetry: the dodecahedron has 12 faces which are regular pentagons, 20 vertices where three faces meet, and 30 edges with twofold symmetry. The three remaining regular solids have different symmetry axes. The vast majority of regular viruses that appear spherical have icosahedral symmetry.

In an icosahedron, the smallest number of subunits permissible to form the three-dimensional structure is 60 (5 subunits at each of the 12 vertices, or viewed slightly differently, 3 units on each of the 20 triangular faces). Some viruses do in fact use 60 subunits, but most use more subunits in order to provide a larger shell capable of holding more nucleic acid. The number of subunits in an icosahedral

Contents

7. Subviral Agents

9. Gene Therapy

Appendix A

8. Host Defenses against Viral Infection and Viral Counterdefenses

Appendix B

VIRUSES AND HUMAN DISEASE

JAMES H. STRAUSS
ELLEN G. STRAUSS
Division of Biology
California Institute of Technology
Pasadena, California

ACADEMIC PRESS

A Division of Harcourt, Inc.

San Diego   San Francisco   New York   Boston   London   Sydney   Tokyo

Cover illustration: The binding of a neutralizing antibody to the surface of a human rhinovirus. Cryoelectron microscopy was used to determine the structure of human rhinovirus 14 with bound monoclonal antibody to about 25 Å resolution [Smith, T. J., Olson, N. H., Chen, R. H., Chase, E. S., and Baker, T. S. (1993). Structure of a human rhinovirus–bivalent antibody complex: Implications for virus neutralization and antibody flexibility. *Proc. Natl. Acad. Sci. U.S.A.* **90**; 7015–7018]. The upper left shows the surface of a virion, in blue, with a stylized model of the antibody, in gold (heavy chain) and silver (light chain) bound to it, based upon this 25-Å structure. The lower right presents another rendition of the virus with bound antibody that is based on the structure of the complex determined to 4 Å resolution by X-ray crystallography [Smith, T. J., Chase, E. S., Schmidt, T. J., Olson, N. H., and Baker, T. S. (1996). Neutralizing antibody to human rhinovirus 14 penetrates the receptor-binding canyon. *Nature* **383**; 350–354]. The polypeptide chains that make up the antibody (gold and silver) and the virion (blue, red, green) are contained within a ghost image of the molecular surface. The antibody binds in a groove in the surface of the virion called the canyon, which contains the receptor binding site of the virus. Rhinoviruses are described in Chapter 3 and antibodies are described in Chapter 8. This montage is courtesy of Thomas J. Smith, Donald Danforth Plant Sciences Center, St. Louis, MO.

209 08741

Academic Press
A division of Harcourt, Inc.
525 B Street, Suite 1900, San Diego, California 92101-4495, USA
http://www.academicpress.com

Academic Press
Harcourt Place, 32 Jamestown Road, London NW1 7BY, UK
http://www.academicpress.com

Library of Congress Catalog Card Number: 2001088200

International Standard Book Number: 0-12-673050-4

PRINTED IN CANADA
01 02 03 04 05 06 FR 9 8 7 6 5 4 3 2 1

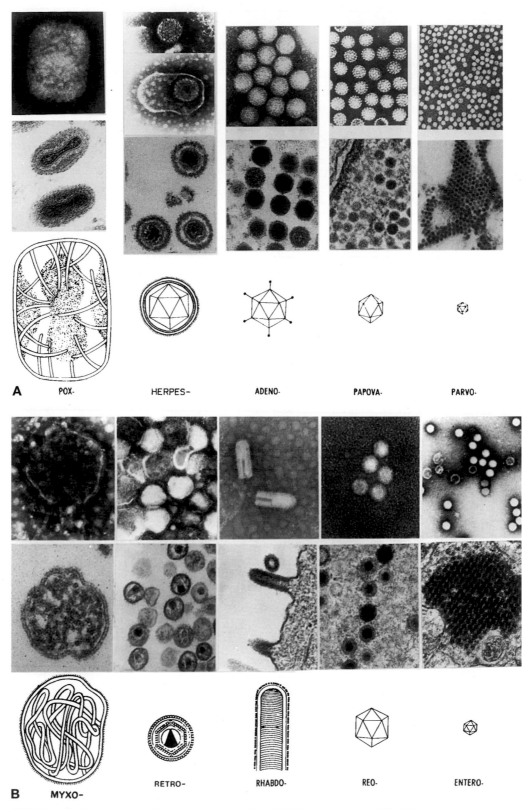

FIGURE 2.1 Relative size and shape of representative (A) DNA-containing and (B) RNA-containing viruses. In each panel the top row shows negatively stained virus preparations, the second row shows thin sections of virus-infected cells, and the bottom row illustrates schematic diagrams of the viruses. Magnification of the electron micrographs is 50,000. [From Granoff and Webster (1999, Vol. 1, p. 401).]

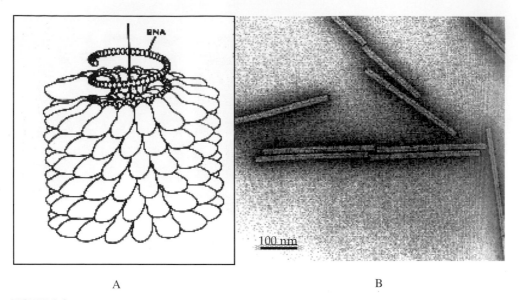

| A | B |

FIGURE 2.2   Structure of TMV, a helical plant virus. (A) Schematic diagram of a TMV particle showing about 5% of the total length. (B) Electron micrograph of negatively stained TMV rods. [From Murphy *et al.* (1995, p. 434).]

structure is 60T, where the permissible values of T are given by $T = H^2 + HK + K^2$, where H and K are integers, and T is called the triangulation number. Permissible triangulation numbers are 1, 3, 4, 7, 9, 12, 13, 16, and so forth. Note that a subunit defined in this way is not necessarily formed by one protein molecule, although in most cases this is how a structural subunit is in fact formed. Some viruses that form regular structures that are constructed using icosahedral symmetry principles do not possess true icosahedral sym-

metry. In such cases they are said to have pseudo-triangulation numbers. Examples are described below.

Structural studies of viruses have shown that the capsid proteins that form the virions of many plant and animal icosahedral viruses have a common fold. This fold, an eight-stranded antiparallel β sandwich, is illustrated in Fig. 2.3B. The presence of a common fold suggests that these capsid proteins have a common origin even if no sequence identity is detectable. The divergence in sequence

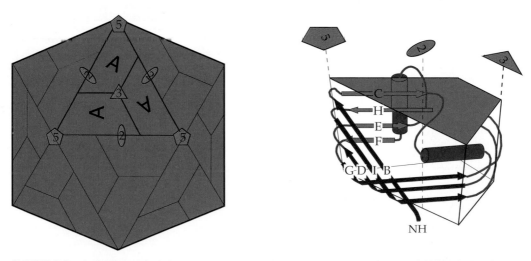

FIGURE 2.3   A simple icosahedral virus. (A) Diagram of an icosahedral capsid made up of 60 identical copies of a protein subunit, shown as blue trapezoids labeled "A." The twofold, threefold, and fivefold axes of symmetry are shown in yellow. This is the largest assembly in which every subunit is in an identical environment. (B) Schematic representation of the subunit building block found in many RNA viruses, known as the eight-stranded antiparallel β sandwich. The β sheets, labeled B though I from the N terminus of the protein, are shown as yellow and red arrows; two possible α helices joining these sheets are shown in green. Some proteins have insertions in the C–D, E–F, and G–H loops, but insertions are uncommon at the narrow end of the wedge (at the fivefold axis). [From Plate 31 in color, Granoff and Webster (1999, Vol. 3), contributed by Johnson (1996).]

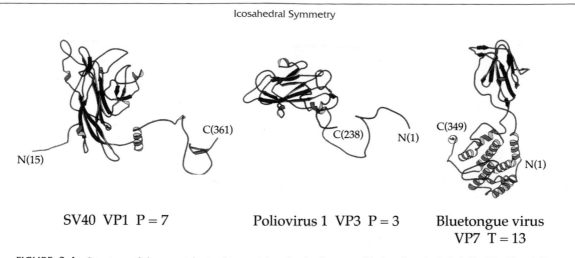

SV40 VP1 P = 7   Poliovirus 1 VP3 P = 3   Bluetongue virus
                                                                VP7 T = 13

FIGURE 2.4 Structure of three vertebrate virus protein subunits that assemble into icosahedral shells. The N and C termini are labeled with the residue number in parentheses. The β barrels are shown as red arrows, α helices are gray coils, and the subunit regions involved in quasi-symmetric interactions that are critical for assembly are colored green. SV40 and poliovirus have triangulation numbers of "pseudo-T = 7" or P = 7 and "pseudo-T = 3" or P = 3, respectively. [Adapted from Plate 32, Granoff and Webster (1999, Vol. 3).]

while maintaining this basic fold is illustrated in Fig. 2.4, where capsid proteins of three viruses are shown. SV40 (family Polyomaviridae), poliovirus (family Picornaviridae), and bluetongue virus (family Reoviridae) are a DNA virus, a single-strand RNA virus, and a double-strand RNA virus, respectively. Their capsid proteins have insertions into the basic eight-stranded antiparallel β-sandwich structure and serve important functions in virus assembly. However, they all possess a region exhibiting the common β-sandwich fold, and may have originated from a common ancestral protein. Thus, once a suitable capsid protein arose that could be used to construct simple icosahedral particles, it may ultimately have been acquired by many viruses. The viruses that possess capsid proteins with this fold may be related by descent from common ancestral viruses, or recombination may have resulted in the incorporation of this successful ancestral capsid protein into many lines of viruses.

Because the size of the icosahedral shell is fixed by geometric constraints, it is difficult for a change in the size of a viral genome to occur. A change in size will require a change in the triangulation number or changes in the capsid proteins sufficient to produce a larger or smaller internal volume. In either case, the changes in the capsid proteins required are relatively slow to occur on an evolutionary timescale and the size of an icosahedral virus is "frozen" for long periods of evolutionary time. For this reason, as well as for other reasons, most viruses have optimized the information content in their genomes, as will be clear when individual viruses are discussed in the following chapters.

Comparison of Icosahedral Viruses

Cryoelectron microscopy has been used to determine the structure of numerous icosahedral viruses to about 20- to 25-Å resolution. For this, a virus-containing solution on an electron microscope grid is frozen very rapidly so that the sample is embedded in amorphous frozen water. The sample must be maintained at liquid nitrogen temperatures so that ice crystals do not form and interfere with imaging. Unstained, slightly out-of-focus images of the virus are captured on film using a low dose of electrons. These images are digitized and the density measured. Mathematical algorithms that take advantage of the symmetry of the particle are used to reconstruct the structure of the particle.

A gallery of structures of viruses determined by cryoelectron microscopy is shown in Fig. 2.5. All of the images are to scale so that the relative sizes of the virions are apparent. The largest particle is the nucleocapsid of herpes simplex virus, which is 1250 Å in diameter and has T=16 symmetry (the virion is enveloped but only the nucleocapsid is regular). The rotavirus and reovirus virions are smaller and have T=13. Human papillomavirus and mouse polyoma virus are pseudo-T=7. Ross River virus (family Togaviridae) is enveloped but has regular symmetry, with T=4. Several examples of viruses with T=3 or pseudo-T=3 are shown (dengue 2, flock house, rhino-, polio-, and cowpea mosaic viruses, of which dengue 2 is enveloped but regular and the rest are not enveloped). B19 parvovirus has T=1. The general correlation is that larger particles are constructed using higher triangulation numbers, which allows the use of larger numbers of protein subunits. Larger particles accommodate larger genomes.

Atomic Structure of T=3 Viruses

Because the simple viruses are regular structures, they will often crystallize, and such crystals may be suitable for X-ray diffraction. Several viruses formed using icosahedral symmetry principles have now been solved to atomic resolution. Among T=3 viruses, the structures of

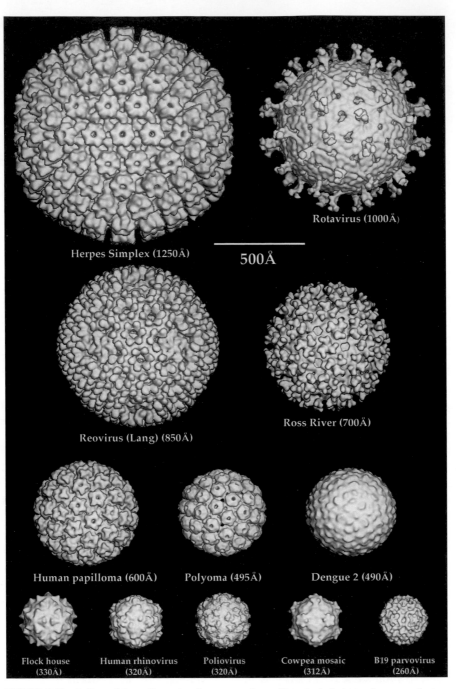

FIGURE 2.5 Gallery of three-dimensional reconstructions of icosahedral viruses from cryoelectron micrographs. All virus structures are surface shaded and are viewed along a twofold axis of symmetry. All of the images are of intact virus particles except for the herpes simplex structure, which is of the nucleocapsid of the virus. [Most of the images are taken from Baker *et al.* (1999), except the images of Ross River virus and of dengue virus, which were kindly provided by Drs. R. J. Kuhn and T. S. Baker.]

several plant viruses, including tomato bushy stunt virus (TBSV), turnip crinkle virus (TCV), and Southern bean mosaic virus (SBMV), which represent more than one family, have been solved. All three of these viruses have capsid proteins possessing the eight-stranded antiparallel β sandwich. $T=3$ means that 180 identical molecules of

capsid protein are utilized to construct the shell. The structures of two insect viruses have also been solved. The $T=3$ capsid of the insect nodavirus, flock house virus, is illustrated in Fig. 2.6.

The 180 subunits in these $T=3$ structures interact with one another in one of two different ways, such that the

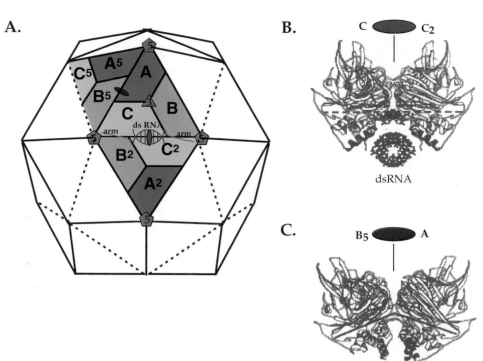

FIGURE 2.6 Diagrammatic representation of a $T = 3$ virus, flock house virus. The positions of the three identical proteins that make up a triangular face are only quasi-equivalent. The angle between the A and B5 units (shown with an red oval and in diagram C) is more acute than that along the C–C2 edge, shown with a blue oval, and diagram B. This difference in the angles is due to the presence of an RNA molecule located under the C–C2 edge. [From Johnson (1996, Fig. 4).]

protein shell can be thought of as being composed of an assembly of 60 AB dimers and 30 CC dimers (Fig. 2.6A). The bond angle between the two subunits of the dimer is more acute in the AB dimers than in the CC dimers (Figs. 2.6B and C). For the plant viruses, there are N-terminal and C-terminal extensions from the capsid proteins that are involved in interactions between the subunits and with the RNA. The N-terminal extensions have a positively charged, disordered domain for interacting with and neutralizing the charge on the RNA and a connecting arm that interacts with other subunits. In the case of the CC dimers, the connecting arms interdigitate with two others around the icosahedral threefold axis to form an interconnected internal framework. In the case of the AB conformational dimer, the arms are disordered, allowing sharper curvature. For flock house virus, the RNA plays a role in controlling the curvature of the CC dimers, as illustrated in Fig. 2.6.

Atomic Structure of Viruses Having Pseudo-T=3 Symmetry

The structures of several picornaviruses and of a plant comovirus (cowpea mosaic virus) have also been solved to atomic resolution. The structures of these viruses are similar to those of the plant T=3 viruses, but the 180 subunits that form the virion are not all identical. A comparison of the structure of a T=3 virus with those of poliovirus and of cowpea mosaic virus is shown in Fig. 2.7. Poliovirus has 60 copies of each of three different proteins, whereas the comovirus has 60 copies of an L protein (each of which fills the niche of two units) and 60 copies of an S protein. All three poliovirus capsid proteins have the eight-stranded antiparallel β-sandwich fold. In the comoviruses, the L protein has two β-sandwich structures fused to form one large protein, and the S protein is formed from one β sandwich. The structures of the picornavirus and comovirus virions are called pseudo-T=3 or P=3, since they are not true T=3 structures.

The picornavirus virion is 300 Å in diameter. The 60 molecules of each of the three different proteins have different roles in the final structure, as illustrated in Fig. 2.8, in which the structure of a rhinovirus is shown. Notice that five copies of VP1 are found at each fivefold axis (compare Fig. 2.7 with Fig. 2.8). VP1, VP2, and VP3 are structurally related to one another, as stated, all possessing the common β-sandwich fold. There exists a depression around each fivefold axis of rhinoviruses that has been termed a "canyon." This depression is believed to be the site at which the virus interacts with the cellular receptor during entry, as illustrated in Fig. 2.9. This interaction is thought to lead to conformational changes that open a channel at the fivefold axis, through which VP4 is extruded, followed by the viral RNA.

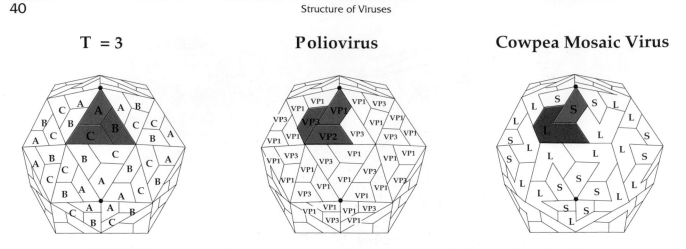

FIGURE 2.7 Arrangement of the coat protein subunits of comoviruses compared with those of simple $T = 3$ viruses and picornaviruses. In simple viruses, the asymmetric unit contains three copies of a single protein β sandwich, labeled A, B, and C in order to distinguish them. In picornaviruses such as poliovirus the asymmetric unit is made up of three similar but not identical proteins, all of which have the β-sandwich structure. In comoviruses such as cowpea mosaic virus, two of the β-sandwich subunits are fused to give the L protein. [Adapted from Granoff and Webster (1999, p. 287).]

Atomic Structure of Polyomaviruses

The structures of both mouse polyoma virus and of SV40 virus, two members of the family Polyomaviridae, have now been solved to atomic resolution. Both viruses possess pseudo-T=7 icosahedral symmetry. Although T=7 symmetry would require 420 subunits, these viruses contain only 360 copies of a major structural protein known as VP1. These 360 copies are assembled as 72 pentamers. Twelve of

the 72 pentamers lie on the fivefold axes and the remaining 60 fill the intervening surface in a closely packed array (Fig. 2.10). These latter pentamers are thus sixfold coordinated and the proteins in the shell are not all in quasi-equivalent positions, a surprising finding for our understanding of the principles by which viruses can be constructed. The pentamers are stabilized by interactions of the β sheets between adjacent monomers in a pentamer (Fig. 2.10C). The pentamers are then tied together by C-terminal arms of VP1 that invade monomers in an adjacent pentamer (Figs. 2.10B and C). Because each pentamer that is sixfold coordinated has five C-terminal arms to interact with six neighboring pentamers, the interactions between monomers in different pentamers are not all identical (Fig. 2.10B). Flexibility in the C-terminal arm allows it to form contacts in different ways.

Atomic Structure of Bluetongue Virus

Members of the reovirus family are regular T=13 icosahedral particles. They are composed of two or three concentric protein shells. Cryoelectron microscopy has been used to solve the structure of one or more members of three genera within the Reoviridae, namely, *Reovirus, Rotavirus,* and *Orbivirus*, to about 25-Å resolution. Structures of a reovirus and of a rotavirus are shown in Fig. 2.5. The complete structure of virions has not been determined because of their large size, but in a remarkable feat the atomic structure of the core of bluetongue virus (genus *Orbivirus*) has now been solved. This is the largest structure determined to atomic resolution to date. Solution of the structure was possible because the virus particle had been solved to 25 Å by cryoelectron microscopy, and the structure of a number of virion proteins had been solved to atomic resolution. Fitting

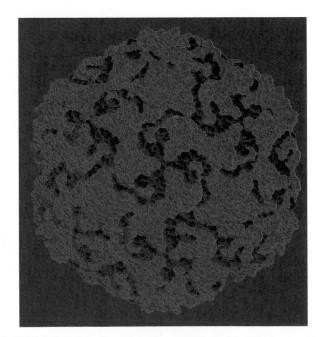

FIGURE 2.8 Three-dimensional space-filling model of the human rhinovirus 14 virion, based on X-ray crystallographic data. VP1 is shown in blue, VP2 in yellow-green, and VP3 in red. VP4 is interior and not visible in this view. (This figure was kindly provided by Dr. Michael Rossmann.)

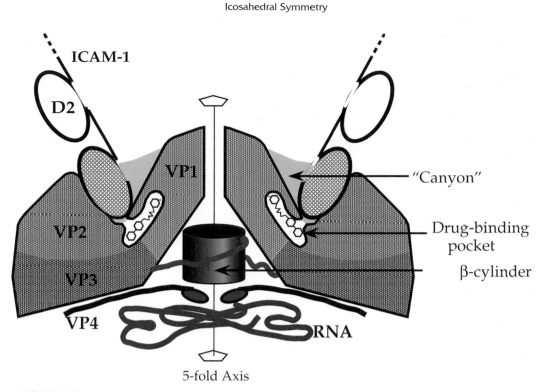

FIGURE 2.9 Binding of the rhinovirus receptor, ICAM-1, to the "canyon" at a fivefold vertex of a virion of a major group human rhinovirus. The colors of the three virion proteins are the same as those shown in the surface view in Fig. 2.8. The distal two domains of ICAM-1 are represented schematically as they were in Fig. 1.4. The amino-terminal domains of the five VP3 molecules around the fivefold axis form a five-stranded β cylinder on the virion's interior and are thought to stabilize the pentamer. Below the canyon is the hydrophobic pocket where certain antiviral drugs (indicated schematically in black) are known to bind. [Adapted from Kolatkar *et al.* (1999).]

the atomic structure of the proteins into the 25-Å structure gave a preliminary reconstruction at high resolution, which allowed the interpretation of the X-ray data to atomic resolution.

Core particles are formed following infection, when the outer layer is proteolytically cleaved (described in more detail in Chapter 3). The structure of the inner surface of the bluetongue virus core is shown in Fig. 2.11A and of the outer surface in Fig. 2.11B. The outer surface is formed by 780 copies of a single protein, called VP7, in a regular T=13 icosahedral lattice. The inner surface is surprising, however. It is formed by 120 copies of a single protein, called VP3. These 120 copies have been described as forming a T=2 lattice. Because T=2 is not a permitted triangulation number, these 120 copies, strictly speaking, form a T=1 lattice in which each unit of the lattice is composed of two copies of VP3. However, the interactions are not symmetrical, leading to the suggested terminology of T=2.

It has been suggested that the inner core furnishes a template for the assembly of the T=13 outer surface. The reasoning is that a T=13 structure may have difficulty in forming, whereas the T=2 (or T=1) structure could form readily. In this model, the threefold symmetry axis of the

inner surface could serve to nucleate VP7 trimers and organize the T=13 structure.

Structure of Adenoviruses

Cryoelectron microscopy has also been applied to adenoviruses, which have a triangulation number of 25 or pseudo-25. Various interpretations of the structure of adenoviruses, both schematic and as determined by microscopy or crystallography, are shown in Fig. 2.12. Three copies of a protein called the hexon protein, whose structure has been solved to atomic resolution, associate to form a structure called a hexon (Fig. 2.12C). The hexon is the basic building block of adenoviruses. Five hexons, called peripentonal hexons, surround each of the 12 vertices of the icosahedron (which, as has been stated, have fivefold rotational symmetry). Between the groups of peripentonal hexons are found groups of 9 hexons, which are sixfold coordinated. Each group of 9 hexons forms the surface of one of the triangular faces. Thus, there are 60 peripentonal hexons and 180 hexons in groups of nine.

The structure of the hexon trimer has been solved to atomic resolution by X-ray crystallography (Fig. 2.12C).

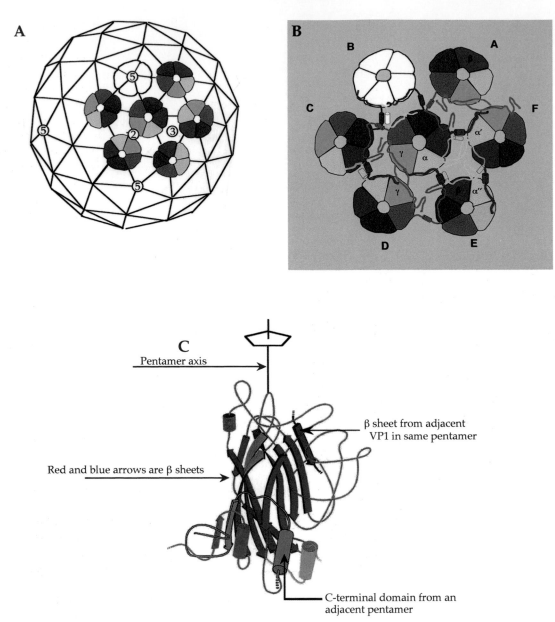

FIGURE 2.10 Organization of the capsid of the polyomavirus SV40. (A) Arrangement of the strict pentamers (white) and quasi-pentamers (colored) on the $T = 7d$ icosahedral lattice. (B) Schematic showing the pattern of interchange of arms in the virion. The central pentamer shares "arms" with six neighboring pentamers. (C) A single VP1 subunit, viewed normal to the pentamer axis. The N-terminal domain is green, the C-terminal domain is yellow, and the β sandwich is shown as arrows of blue and red. The central yellow C-terminal domain (outlined in black) comes from an adjacent pentamer and the β sheet outlined in black comes from the neighboring VP1 within the same pentamer. [From Stehle *et al.* (1996, Fig. 1) and Fields *et al.* (1996, Colorplate 4).]

The hexon protein has two eight-strand β sandwiches to give the trimer an approximately sixfold symmetry. There are long loops that intertwine to form a triangular top. These structures can be fitted uniquely into the envelope of density determined by cryoelectron microscopy, which produces a structure refined to atomic resolution for most of the capsid. The minor proteins can be fitted into this structure.

From the 12 vertices of the icosahedron project long fibers. Each fiber terminates in a spherical extension that forms an organ of attachment to a host cell (Fig. 2.12A). The length of the fiber differs in the different adenoviruses.

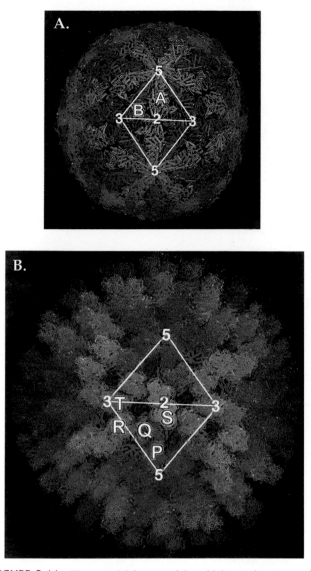

FIGURE 2.11 The essential features of the orbivirus native core particle. The asymmetric unit is indicated by the white lines forming a triangle and the fivefold, threefold, and twofold axes are marked. (A) The inner capsid layer of the bluetongue virus (BTV) core is composed of 120 molecules of VP3, arranged in what has been called $T = 2$ symmetry. Note the green subunit A and the red subunit B, which fill the asymmetric unit. (B) The core surface layer is composed of 780 copies of VP7 arranged as 260 trimers, with $T = 13$ symmetry. The asymmetric unit contains 13 copies of VP7, arranged as five trimers, labeled P, Q, R, S, and T, with each trimer a different color. Trimer T in blue sits on the icosahedral threefold axis and thus contributes only a monomer to the asymmetric unit. [From Plate 17, Granoff and Webster (1999).]

NONENVELOPED VIRUSES WITH MORE COMPLICATED STRUCTURAL FEATURES

In addition to the nonenveloped viruses that possess relatively straightforward icosahedral symmetry or helical symmetry, many viruses possess more complicated symme-

tries made possible by the utilization of a large number of structural proteins to form the virion. The tailed bacteriophages are prominent examples of this (Fig. 2.13). Some of the tailed bacteriophages possess a head that is a regular icosahedron (or, in at least one case, an octahedron) connected to a tail that possesses helical symmetry. Other appendages, such as baseplates, collars, and tail fibers, may be connected to the tail. Other tailed bacteriophages have heads that are assembled using more complicated patterns. For example, the T-even bacteriophages have a large head, which can be thought of as being formed of two hemi-icosahedrons possessing regular icosahedral symmetry, which are elongated in the form of a prolate ellipsoid by subunits arranged in a regular net connecting the two icosahedral ends of the head of the virus.

ENVELOPED VIRUSES

Many animal viruses and some plant viruses are enveloped; that is, they have a lipid-containing envelope surrounding a nucleocapsid. The lipids are derived from the host cell. Although there is some selectivity and reorganization of lipids during virus formation, the lipid composition in general mirrors the composition of the cellular membrane from which the envelope was derived. However, the proteins in the nucleocapsid, which may possess either helical or icosahedral symmetry, and the proteins in the envelope are encoded in the virus. The protein–protein interactions that are responsible for assembly of the mature enveloped virions differ among the different families and the structures of the resulting virions differ. The virions of alphaviruses, and possibly of flaviviruses, are uniform structures that possess icosahedral symmetry. Poxviruses, rhabdoviruses, and retroviruses also appear to have a regular structure, but there is flexibility in the composition of the particle and the mature virions do not possess icosahedral symmetry. The herpesvirus nucleocapsid is a regular icosahedral structure (Fig. 2.5), but the enveloped herpesvirions possess irregularities. Other enveloped viruses appear irregular, often pleiomorphic, and are heterogeneous in composition to a greater or lesser extent. The structures of different enveloped viruses that illustrate these various points are described below.

The Nucleocapsid

The nucleocapsids of enveloped RNA viruses are fairly simple structures that contain only one major structural protein, often referred to as the nucleocapsid protein. This protein is usually quite basic or has a basic domain. It binds to the viral RNA and encapsidates it to form the nucleocapsid. For most RNA viruses, nucleocapsids can be recognized as distinct structures within the infected

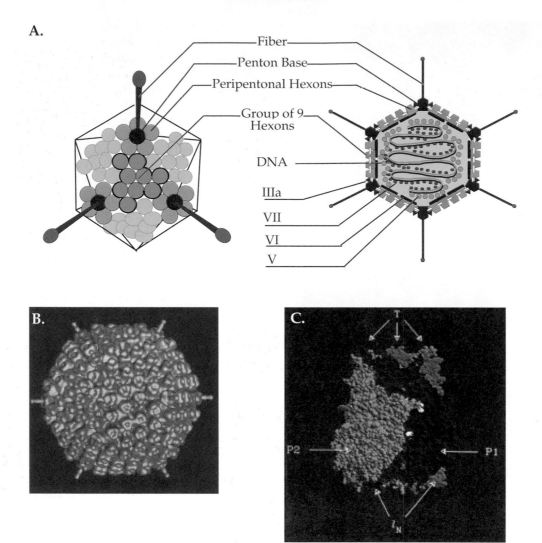

FIGURE 2.12 Structure of adenovirus particles. (A) Schematic drawing of the outer shell of an adenovirus (left), and a schematic cross section through an adenovirus particle, showing the locations of minor polypeptide components (right). The virus is composed of 60 peripentonal hexons at the bases of the fibers at the 5-fold vertices, and groups of nine hexons, one on each triangular face of the icosahedron. (B) Cryoeletron microscopic reconstruction of an adenovirus virion, viewed down the threefold axis. (C) Space-filling model of the hexon trimer, with each subunit in a different color. The atomic structure of the hexon has been solved and fitted into the cryoelectron microscopic reconstruction. The locations of the minor constitutents, indicated schematically in (A), were deduced by subtraction. [(A) is from Fields *et al.* (1996, p. 80), (B) is from Stewart *et al.* (1991), and (C) is from Athappilly *et al.* (1994).]

cell, and can be isolated from virions by treatment with detergents that dissolve the envelope. The nucleocapsids of alphaviruses, and probably flaviviruses and arteriviruses as well, are regular icosahedral structures, and there are no other proteins within the nucleocapsid other than the nucleocapsid protein. In contrast, the nucleocapsids of all minus-strand viruses are helical and contain, in addition to the major nucleocapsid protein, two or more minor proteins that possess enzymatic activity. As described, the nucleocapsids of minus-strand RNA viruses remain intact within the cell during the entire infection cycle and serve as machines that make viral RNA. The coronaviruses also have helical nucleocapsids, but being plus-strand RNA viruses they do not need to carry enzymes in the virion to initiate infection. The helical nucleocapsids of RNA viruses appear disordered within the envelope of all viruses except the rhabdoviruses, in which they are coiled in a regular fashion (see below).

The nucleocapsids of retroviruses also appear to be fairly simple structures. They are formed from one major precursor protein, the Gag polyprotein, that is cleaved

A. Enterobacteria phage T4

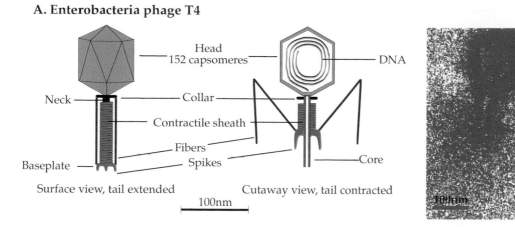

Head
152 capsomeres — DNA

Neck — Collar —

Contractile sheath

Fibers

Baseplate — Spikes

Core

Surface view, tail extended   Cutaway view, tail contracted

100nm

100nm

Electron micrograph

B. Enterobacteria phage T7

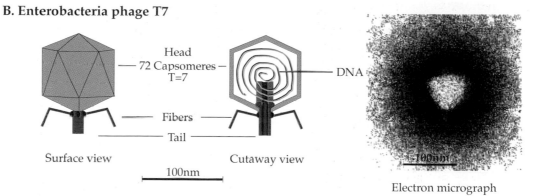

Head
72 Capsomeres —
T=7

DNA

Fibers

Tail

Surface view   Cutaway view

100nm

100nm

Electron micrograph

C. Lambda-like Phage

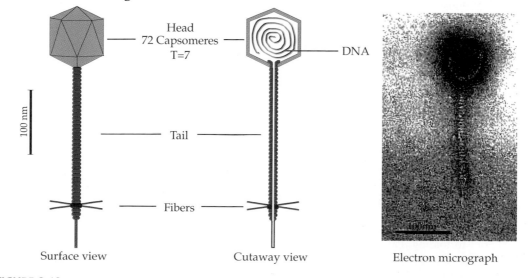

Head
72 Capsomeres —
T=7

DNA

100 nm

Tail

Fibers

Surface view   Cutaway view

100nm

Electron micrograph

FIGURE 2.13 Morphology of some bacteriophages (members of the *Caudovirales*). (A) Enterobacteria phage T4, in the family Myoviridae. The head is an elongated pentagonal structure. (B) Enterobacteria phage T7, a member of the Podoviridae. (C) Enterobacteria phage λ, a member of the Siphoviridae. All electron micrographs are stained with uranyl acetate, and all bars shown are 100 nm. [Adapted from Murphy *et al*. (1995, pp. 51, 60, 55).]

during maturation into four or five components. The precursor nucleocapsid is spherically symmetric but lacks icosahedral symmetry, as does the mature nucleocapsid produced by cleavage of Gag. The nucleocapsid also contains minor proteins, produced by cleavage of Gag-pro-pol, as described in Chapter 1. These minor proteins include the protease, RT, RNase H, and integrase that are required to cleave the polyprotein precursors, to make a cDNA copy of the viral RNA, and to integrate this cDNA copy into the host chromosome.

The two families of enveloped DNA viruses that we consider here, the poxviruses and the herpesviruses, contain large genomes and complicated virus structures. The nucleocapsids of herpesviruses are regular icosahedrons but those of poxviruses are complicated structures containing a core and associated lateral bodies.

Envelope Glycoproteins

The external proteins of enveloped virions are virus-encoded proteins that are anchored in the lipid bilayer of the virus or whose precursors are anchored in the lipid bilayer. In the vast majority of cases these proteins are glycoproteins, although examples are known that do not contain bound carbohydrate. These proteins are translated from viral mRNAs and transported by the usual cellular processes to reach the membrane at which budding will occur. When budding is at the cell plasma membrane, the glycoproteins are transported via the Golgi apparatus to the cell surface. Some enveloped viruses mature at intracellular membranes, and in these cases the glycoproteins are directed to the appropriate place in the cell. Both Type I integral membrane proteins, in which the N terminus of the protein is outside the lipid bilayer and the C terminus is inside the bilayer, and Type II integral membrane proteins, which have the inverse orientation with the C terminus outside, are known for different viruses.

Following synthesis of viral glycoproteins, during which they are transported into the lomen of the endoplasmic reticulum in an unfolded state, they must fold to assume their proper conformation, and assume their proper oxidation state by formation of the correct disulfide bonds. This process often occurs very quickly, but for some viral glycoproteins it can take hours. Folding is often assisted by chaperonins present in the endoplasmic reticulum. It is believed that at least one function of the carbohydrate chains attached to the protein is to increase the solubility of the unfolded glycoproteins in the lumen of the ER, so that they do not aggregate prior to folding. During folding, the solubility of the proteins is increased by hiding hydrophobic domains within the interior of the protein and leaving hydrophilic domains at the surface.

The glycoproteins possess a number of important functions in addition to their structural functions. They carry the attachment domains by which the virus binds to a susceptible cell. This activity is thought to be related to the ability of many viruses, nonenveloped as well as enveloped, to bind to and agglutinate red blood cells, a process called hemagglutination. The protein possessing hemagglutinating activity is often called the hemagglutinin or HA. The viral glycoproteins also possess a fusion activity that promotes the fusion of the membrane of the virus with a membrane of the cell. The protein possessing this activity is sometimes called the fusion protein, or F. The glycoproteins, being external on the virus, are also primary targets of the humoral immune system, in which circulating antibodies are directed against viruses; many of these are neutralizing antibodies that inactivate the virus.

The glycoproteins of some enveloped viruses also contain enzymatic activities. Many orthomyxoviruses and paramyxoviruses possess a neuraminidase that will remove sialic acid from glycoproteins. The primary receptor for these viruses is sialic acid. The neuraminidase may allow the virus to penetrate through mucus to reach a susceptible cell. It also removes sialic acid from the viral glycoproteins, so that these glycoproteins or the mature virions do not aggregate, and from the surface of an infected cell, preventing virions from binding to it. The viral protein possessing neuraminidase activity may be called NA, or in the case of a protein that is both a neuraminidase and hemagglutinin, HN.

The structure of most enveloped viruses is not as rigorously constrained as that of icosahedral virus particles. The glycoproteins are not required to form an impenetrable shell, which is instead a function of the lipid bilayer. They appear to tolerate mutations more readily than do proteins that must form a tight icosahedral shell and appear to evolve rapidly in response to immune pressure. However, the integrity of the lipid bilayer is essential for virus infectivity, and enveloped viruses are very sensitive to detergents.

Other Structural Proteins in Enveloped Viruses

In some enveloped viruses, there is a structural protein that underlies the lipid envelope but which does not form part of the nucleocapsid. Several families of minus-strand RNA viruses possess such a protein, called the matrix protein. This protein may serve as an adapter between the nucleocapsid and the envelope. It may also have regulatory functions in viral RNA replication. The herpesviruses also have a protein underlying the envelope that is called the tegument protein. This protein forms a thick layer, the tegument, between the nucleocapsid and the envelope. The thickness of the tegument is not uniform within a virion, giving rise to some irregularity in its structure.

Structure of Alphaviruses

The alphaviruses, a genus in the family Togaviridae, are exceptional among enveloped RNA viruses in the regularity of their virions, which are uniform icosahedral particles. Virions of two alphaviruses have been crystallized, and the crystals diffract to 30–40 Å. Higher resolution has been obtained from cryoelectron microscopy, which has been used to determine the structures of several alphaviruses to 9–25 Å (Fig. 2.5).

A more detailed reconstruction of Ross River virus, derived from a combination of cryoelectron microscopy of the intact virion and X-ray crystallography of the virus capsid protein, is shown in Fig. 2.14. The nucleocapsid has a diameter of 400 Å, and is a regular icosahedron with $T=4$ symmetry. It is formed from 240 copies of a single species of capsid protein of size 30 kDa. The structure of the the capsid protein itself has been solved to atomic resolution by conventional X-ray crystallography (Fig. 2.15). It has a structure that is very different from the eightfold β sandwich described above (compare Fig. 2.15 with Figs 2.3B and 2.4). Instead, its fold resembles that of chymotrypsin, and it has an active site that consists of a catalytic triad whose geometry is identical to that of chymotrypsin. The capsid protein is an active protease that cleaves itself from a polyprotein precursor. The interactions between the capsid protein subunits that lead to formation of the $T=4$ icosahedral lattice have been deduced by fitting the electron density of the capsid protein at 2.5-Å resolution into the electron density of the nucleocapsid found by cryoelectron microscopy (Fig. 2.14B). The combined approaches of X-ray crystallography and cryoelectron microscopy thus define the structure of the shell of the nucleocapsid to atomic resolution.

The envelopes of alphaviruses contain 240 copies of each of two virus-encoded glycoproteins. These two glycoproteins form a heterodimer and both span the lipid bilayer as Type I integral membrane proteins (having a membrane-spanning anchor at or near the C terminus). The C-terminal cytoplasmic extension of one of the glycoproteins interacts in a specific fashion with a nucleocapsid protein, and the 240 glycoprotein heterodimers form a $T=4$ icosahedral lattice on the surface of the particle. Three glycoprotein heterodimers associate to form a trimeric structure called a spike, easily seen in Figs. 2.5 and 2.14. The apex of the spike contains the domains that attach to receptors on a susceptible cell. The structure of the virion is so ordered that it can be thought of as being composed of a protein lattice in which a lipid bilayer is embedded.

Structure of Other Enveloped Viruses with Icosahedral Nucleocapsids

The flaviviruses, like the alphaviruses, appear to be regular icosahedral stuctures (Fig. 2.5). The arteriviruses

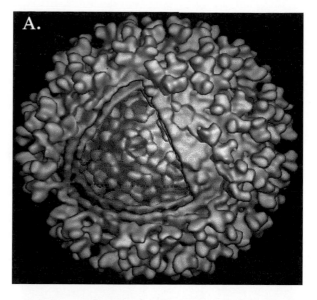

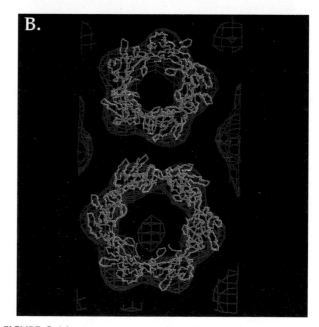

FIGURE 2.14 Structure of Ross River virus reconstructed from cryoelectron microscopy. (A) Cutaway view of the cryoelectron microscopic reconstruction illustrating the multilayered structure of the virion. Envelope glycoproteins are shown in blue, the lipid bilayer in green, the ordered part of the nucleocapsid (Fig. 2.15) in yellow, and the remainder of the nucleocapsid in orange. (B) The fit of the Sindbis capsid protein $C\alpha$ trace (yellow), determined by X-ray diffraction (Fig. 2.15), into the electron density of Ross River virus (blue) determined by cryoelectron microscopy. [From Strauss et al. (1995, Fig. 4).]

possess icosahedral nucleocapsids, but the mature virion does not appear to be regular in structure.

The herpesviruses are large DNA viruses that have a $T=16$ icosahedral nucleocapsid (Fig. 2.5). A schematic diagram of an intact herpesvirion is shown in Fig. 2.16A. Underneath the envelope is a protein layer called the tegu-

---

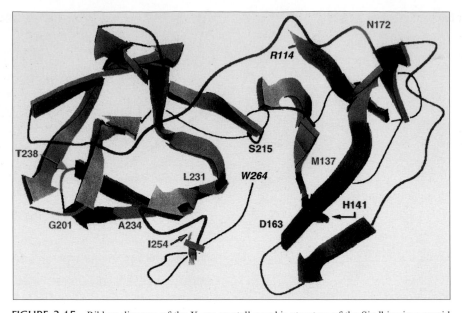

FIGURE 2.15   Ribbon diagram of the X-ray crystallographic structure of the Sindbis virus capsid protein, with β sheets represented by large arrows. Only the carboxy-terminal domain, which starts at Arg-114, is ordered in crystals. The active site residues of the autoprotease, Ser-215, His-141, and Asp-163, are shown in red. The carboxy-terminal Trp-264, which is the P1 residue of the cleavage site, lies within the active site of the enzyme. The seven residues shown in yellow-green may interact with the cytoplasmic domain of glycoprotein E2 during budding of the nucleocapsid. [From Strauss *et al.* (1995, Fig. 3).]

ment. The tegument does not have a uniform thickness, and thus the virion is not uniform. An electron micrograph of a negatively stained nucleocapsid is shown in Fig. 2.16B, which can be compared with the cryoelectron microscopic reconstruction in Fig. 2.5.

The retroviruses have a nucleocapsid that forms initially using spherical symmetry principles. Cleavage of Gag during virus maturation results in a nucleocapsid that is not icosahedral and that is often eccentrically located in the virion. Figure 2.17A presents a schematic of a retrovirus particle that illustrates the current model for the location of the various proteins after cleavage of Gag and Gag–Pol. Figures 2.17B and C show electron micrographs of budding virus particles and of mature extracellular

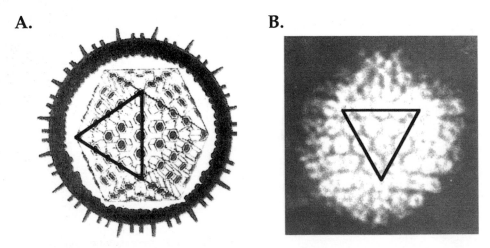

A.   **B.**

FIGURE 2.16   Two views of herpes simplex virus. (A) Cutaway schematic representation showing the outer envelope with projecting spikes, the irregular inner margin of the envelope due to the tegument, and the icosahedral core containing 162 capsomeres in a $T = 16$ arrangement. (B) Negatively stained electron micrograph of an intracellular particle without the outer envelope. One of the triangular faces of the icosahedron is outlined in red in each case. [Adapted from Murphy *et al.* (1995, p. 114) and Dalton and Haguenau (1973, p. 92), respectively.]

virions for two genera of retroviruses. Betaretrovirus particles usually mature by the formation of a nucleocapsid within the cytoplasm that then buds through the plasma membrane. This process is shown in Fig. 2.17B for mouse mammary tumor virus. In the top micrograph in Fig. 2.17B, preassembled capsids are seen in the cytoplasm. In the middle micrograph, budding of the capsid through the plasma membrane is illustrated. In the bottom micrograph, a mature virion with an eccentrically located capsid is shown.

In gammaretroviruses, the capsid forms during budding, and the nucleocapsid is round and centrally located in the mature virion. This process is illustrated in Fig. 2.17C for murine leukemia virus. The top micrograph shows a budding particle with a partially assembled capsid. The bottom micrograph shows a mature virion.

In the lentiviruses, of which HIV is a member, the capsid also forms as a distinct structure only during budding. After cleavage of Gag, the capsid usually appears cone shaped or bar shaped.

Mason–Pfizer monkey virus is a betaretrovirus whose capsid is cone shaped and centrally located in the mature virion. A single amino acid change in the matrix protein MA determines whether the capsid preassembles and then buds, or whether the capsids assemble during budding. Thus, the point at which capsids assemble does not reflect a fundamental difference in retroviruses. Preassembly of capsids or assembly during budding appears to depend on the stability of the capsid in the cell. Stable capsids can preassemble. Unstable capsids require interactions with other viral components to form as a recognizable structure.

Enveloped Viruses with Helical Nucleocapsids

The coronaviruses and the minus-strand RNA viruses have nucleocapsids with helical symmetry. The structures of the mature virions are irregular, with the exception of the rhabdoviruses, and the glycoprotein composition is not invariant. Because of the lack of regularity in these viruses, as well as the lack of symmetry, detailed structural studies

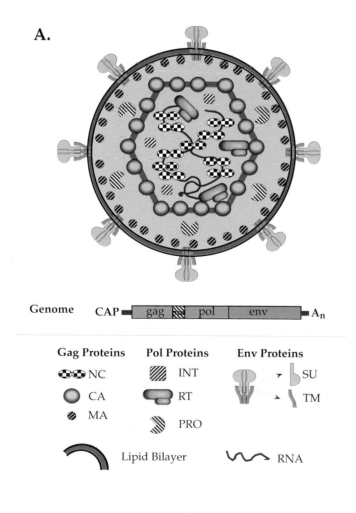

A.

Genome   CAP ■—□ gag ▨ pol │ env □ ■—A_n

Gag Proteins   **Pol Proteins**   **Env Proteins**

NC   INT   SU

CA   RT   TM

MA   PRO

Lipid Bilayer   RNA

B.   C.

FIGURE 2.17 Structure of retrovirus particles. (A) Schematic cross section through a retrovirus particle. The lipid bilayer surrounds the particle and has imbedded in it trimeric spikes composed of surface (SU) and transmembrane (TM) envelope proteins. The internal nonglycosylated proteins are encoded by the *gag* gene and include NC, the nucleocapsid protein complexed with the genomic RNA; CA, the major capsid protein; and MA, the matrix protein that lines the inner surface of the membrane. Other components include RT, the reverse transcriptase; IN, the integrase; and PR, the protease. [Adapted from Coffin *et al.* (1997, p. 2).] (B) Electron micrographs of mouse mammary tumor virus particles (betaretrovirus). Top: intracytoplasmic particles; middle: budding particles; bottom: mature extracytoplasmic particles. (C) Electron micrographs of murine leukemia virus particles (gammaretrovirus). Top: budding particles; bottom: mature extracytoplasmic particles. [Adapted from Coffin *et al.* (1997, p. 30).]

of virions have not been possible. The lack of regularity arises in part because in these viruses there is no direct interaction between the nucleocapsid and the glycoproteins. The lack of such interactions permits these viruses to form pseudotypes, in which glycoproteins from other viruses substitute for those of the virus in question. Pseudotypes are also formed by retroviruses.

The structure of orthomyxoviruses and paramyxoviruses is illustrated schematically in Fig. 2.18. The helical nucleocapsids contain a major nucleocapsid protein called N or NP, and the minor proteins P (NS1) and L (PB1, PB2, PA), as shown. There is a matrix protein M (M1) lining the inside of the lipid bilayer and also two glycoproteins anchored in the

bilayer that form external spikes. Electron micrographs of virions are shown in Figs. 2.18C and D. The particles in the preparations shown are round and reasonably uniform, but in other preparations the virions are pleomorphic bag-like structures that are not uniform in appearance. In fact, clinical specimens of some orthomyxoviruses and paramyxoviruses are often filamentous rather than round, illustrating the flexible nature of the structure of the virion. Notice that the paramyxovirus particle in Fig. 2.18C has been penetrated by the strain, revealing the lack of higher order structure in the internal helical nucleocapsid.

The structures of rhabdoviruses and filoviruses are illustrated in Fig. 2.19. The rhabdoviruses assemble into bullet-

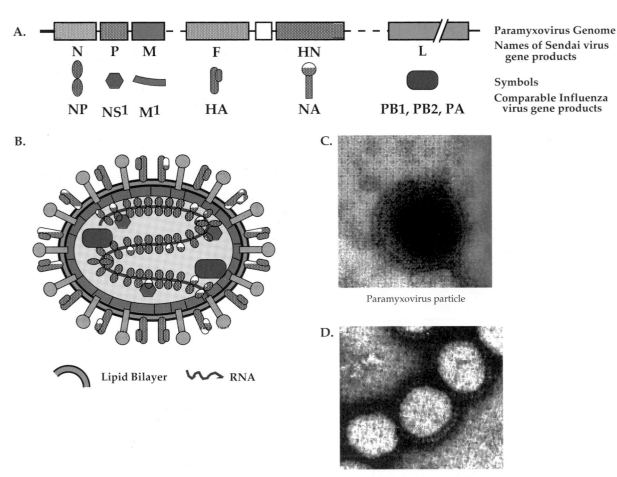

FIGURE 2.18 Morphology of orthomyxoviruses and paramyxoviruses. (A) Schematic of the genome organization of a paramyxovirus, Sendai virus. The names of the gene products and symbols to be used in the diagram below are indicated. Also shown are the comparable gene products of influenza virus, an orthomyxovirus. (B) Schematic cutaway view of an orthomyxovirus or paramyxovirus particle. The nucleocapsid consists of a helical structure made up of the RNA complexed with many copies of the nucleocapsid protein. This internal structure also contains a few molecules of the RNA polymerase L (or PB1, PB2, PA in influenza virus) and P (or NS1). The nucleocapsid is enveloped in a lipid bilayer derived from the host cell in which are embedded two different glycoproteins, F and HN (or HA and NA in influenza virus), and which is lined on the inner surface with the matrix protein M. (C) Electron micrograph of a negatively stained Sendai virion, which was sufficiently permeabilized to reveal the helical nucleocapsid. (D) Electron micrograph of negatively stained intact influenza virions. [Electron micrographs are from Dalton and Haguenau (1973, p. 217) and Granoff and Webster (1999, p. 1136).]

shaped or bacilliform particles in which the helical nucleo-capsid is wound in a regular elongated spiral conformation (Figs. 2.19B and C). The virus encodes only five proteins (Fig. 2.19A), all of which occur in the virion (Fig. 2.19B). The nucleocapsid contains the major nucleocapsid protein N and the two minor proteins L and NS. The matrix protein M lines the inner surface of the envelope, and G is an external glycoprotein that is anchored in the lipid bilayer of the envelope. Budding is from the plasma membrane (Fig. 2.19D).

The filoviruses are so named because the virion is filamentous. A schematic diagram of a filovirus is shown in

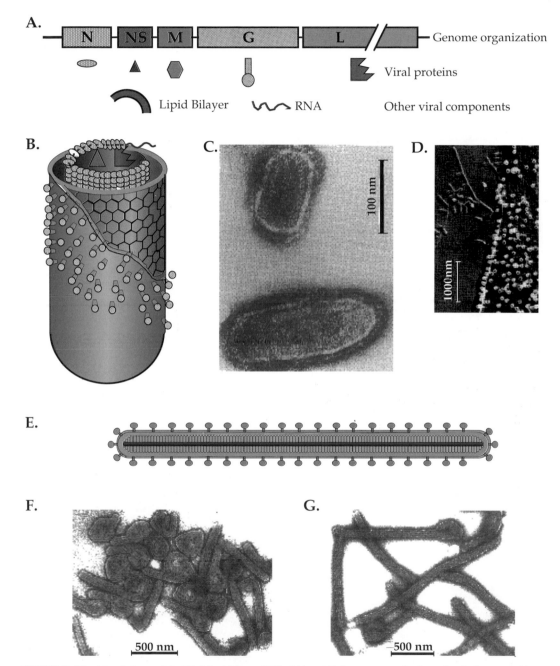

FIGURE 2.19 Morphology of the Rhabdoviridae and Filoviridae. (A) Genome organization of vesicular stomatitis virus (VSV), a vesiculovirus, with the symbols for the various viral components shown below. (B) Cutaway diagram of a VSV particle. (C) A negatively stained electron micrograph of VSV virions. (D) Surface replica of a chicken embryo fibroblast infected with VSV. Note that the magnification is approximately 1/10 of that shown in (C). (E) Diagram of a filovirus, using the same color code for the components. (F) Negatively stained preparation of Marburg virus. (G) Filamentous forms of Ebola (Reston) virus. [(B) and (C) are adapted from Murphy *et al.* (1995, p. 275); (F) and (G) From Murphy *et al.* (1995, p. 289); and (D) is from Birdwell and Strauss (1974).]

Fig. 2.19E, and electron micrographs of two filoviruses, Marburg virus and Ebola virus, are shown in Figs. 2.19F and G. Notice that in the electron microscope, filovirus virions often take the shape of the number 6.

Vaccinia Virus

The poxviruses, large DNA-containing viruses, also have lipid envelopes. In fact, they may have two lipid-containing envelopes. The structures of poxviruses belonging to two different genera, *Orthopox* and *Parapox,* are illustrated in Fig. 2.20. Electron micrographs of the orthopox virus vaccinia virus and of a parapox virus are also shown.

Vaccinia virus has been described as brick shaped. The interior of the virion consists of a nucleoprotein core and two proteinaceous lateral bodies. Surrounding these is a lipid-containing surface membrane, outside of which are several virus-encoded proteins present in structures referred to as tubules.

This particle is called an intracellular infectious virion. As its name implies, it is present inside an infected cell, and if freed from the cell it is infectious. A second form of the virion is found outside the cell and is called an extracellular enveloped virion. This second form has a second, external lipid-containing envelope with which are associated five additional vaccinia proteins. This form of the virion is also infectious.

Parapox virions are similiar to orthopox virions. However, their morphology is detectably different, as illustrated in Fig. 2.20.

ASSEMBLY OF VIRIONS

Self Assembly

Virions self-assemble within the infected cell. In most cases, assembly appears to begin with the interaction of one

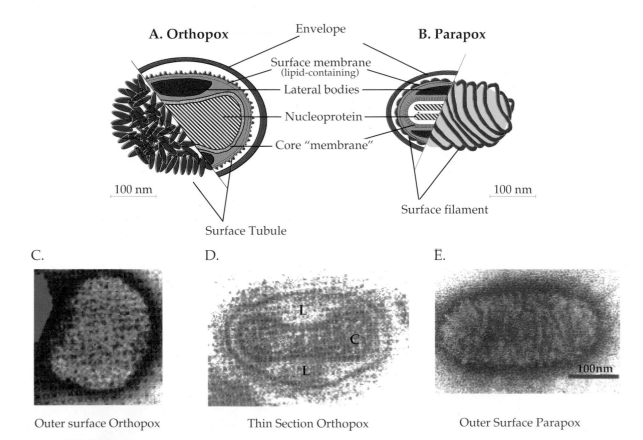

FIGURE 2.20 Morphology of orthopox and parapox virions. (A, B) Diagrams of orthopox and parapox virions. At the far left and right are shown the surfaces of the particles as they are isolated from infected cells, with the outer tubules or the outer filament shown in turquoise. The inner parts of each diagram show the enveloped particle in cross section illustrating the core membrane, the lateral bodies, and the nucleoprotein. [Adapted from Fenner and Nakano (1988).] (C) Purified vaccinia virus negatively stained with phosphotungstate. Magnification is 120,000. [From Dalton and Haguenau (1973, p. 111).] (D) Thin section of a particle in an agglutinated clump of vaccinia. L, lateral bodies; C, core. The membrane is completely coated with antibody. [From Dalton and Haguenau (1973, p. 116).] (E) Outer surface of nonenveloped parapox virus with a single long filament wound around the particle. [From Murphy *et al.* (1995, p. 79).]

or more of the structural proteins with an encapsidation signal in the viral genome, which ensures that viral genomes are preferentially packaged. After initiation, encapsidation continues by recruitment of additional structural protein molecules until the complete helix or icosahedral structure has been assembled. Thus, packaging of the viral genome is coincident with assembly of the virion, or of the nucleocapsid in the case of enveloped viruses. The requirement for a packaging signal may not be absolute. In many viruses that contain an encapsidation signal, RNAs or DNAs lacking such a signal may be encapsidated, but with lower efficiency. For some viruses, there is no evidence for an encapsidation signal.

Assembly of the TMV rod (Fig. 2.2) has been well studied. Several coat protein molecules, perhaps in the form of a disk, bind to a specific nucleation site within TMV RNA to initiate encapsidation. Once the nucleation event occurs, additional protein subunits are recruited into the structure and assembly proceeds in both directions until the RNA is completely encapsidated. The length of the virion is thus determined by the size of the RNA.

The assembly of the icosahedral turnip crinkle virion has also been well studied. Assembly of this $T=3$ structure is initiated by formation of a stable complex that consists of six capsid protein molecules bound to a specific encapsidation signal in the viral RNA. Additional capsid protein dimers are then recruited into the complex until the structure is complete.

It is probable that most other viruses assemble in a manner similar to these two well-studied examples. At least some viruses deviate from this model, however, and assemble an empty particle into which the viral genome is later recruited. It is also known that many viruses will assemble empty particles if the structural proteins are expressed in large amounts in the absence of viral genomes, even if assembly is normally coincident with encapsidation of the viral genome in infected cells.

Enveloped Viruses

The nucleocapsids of most enveloped viruses form within the cell by pathways assumed to be similar to those described above. They can often be isolated from infected cells, and for many viruses the assembly of nucleocapsids does not require viral budding or even the expression of viral surface glycoproteins. After assembly, the nucleocapsids bud through a cellular membrane, which contains viral glycoproteins, to acquire their envelope. Budding retroviruses were illustrated in Fig. 2.17 and budding rhabdoviruses in Fig. 2.19. A gallery of budding viruses belonging to other families is shown in Fig. 2.21. The membrane chosen for budding depends on the virus and depends, in part if not entirely, on the membrane to which the viral glycoproteins are directed by signals within those glycoproteins. Many viruses bud through the cell plasma membrane (Figs. 2.21B–F); in polarized cells, only one side of the cell may be used. Other viruses, such as the coronaviruses and the bunyaviruses, use the endoplasmic reticulum or other internal membranes. The herpesviruses replicate in the nucleus and the nucleocapsid assembles in the nucleus; in this case, budding is through the nuclear membrane (Fig. 2.21A).

Although the nucleocapsid of most enveloped viruses assembles independently within the cell and then buds to acquire an envelope, exceptions are known. The example of retroviruses, some of which assemble a nucleocapsid during virus budding, was discussed above. In these viruses, morphogenesis is a coordinated event.

The forces that result in virus budding are not well understood for most enveloped viruses. In the case of the alphaviruses, there is evidence for specific interactions between the cytoplasmic domains of the glycoproteins and binding sites on the nucleocapsid proteins. The model for budding of these viruses is that the nucleocapsid first binds to one or a few glycoprotein heterodimers at the plasma membrane. By a process of lateral diffusion, additional glycoprotein heterodimers move in and are bound until a full complement is achieved and the virus is now outside the cell. Additional free energy for budding is furnished by lateral interactions between the glycoproteins, which form a contiguous layer on the surface (Fig. 2.14). This model accounts for the regularity of the virion, the one-to-one ratio of the structural proteins in the virion, and the requirement of the virus for its own glycoproteins in order to bud.

In other enveloped viruses, however, there is little evidence for nucleocapsid–glycoprotein interactions. The protein composition of the virion is usually not fixed, but can vary within limits. In fact, glycoproteins from unrelated viruses can often be substituted. In the extreme case of retroviruses, noninfectious virus particles will form that are completely devoid of glycoprotein. The matrix proteins appear to play a key role in the budding process, as do other protein–protein interactions that are yet to be determined.

Maturation Cleavages in Viral Structural Proteins

For most animal viruses, there are one or more cleavages in structural protein precursors during assembly of virions that are required to activate the infectivity of the virion. Interestingly, these cleavages may either stabilize or destabilize the virion in the extracellular environment, depending on the virus. Many of these cleavages are effected by viral proteases, whereas others are performed by cellular proteases present in subcellular organelles. Virions are formed by the spontaneous assembly of components in the infected cell, sometimes with the aid of assembly factors ("scaffolds") that do not form components of the mature virion. For most nonenveloped viruses, complete

A) Herpes

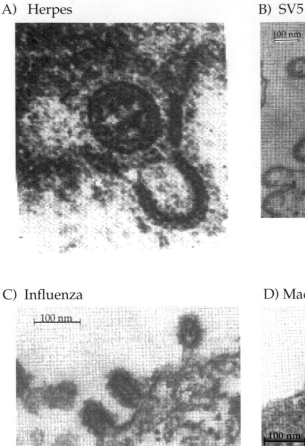

B) SV5

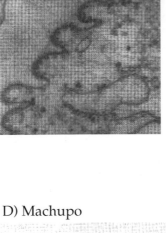

C) Influenza

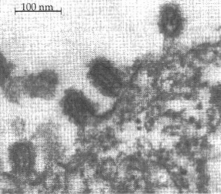

D) Machupo

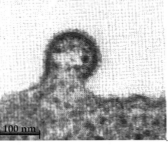

E) Sindbis

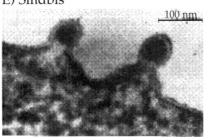

F) Rubella

FIGURE 2.21 Gallery of budding figures of viruses representing several different families. (A) Thin section of a herpes simplex virion (Herpesviridae) in an infected Hep-2 cell. The particle is apparently coated with an inner envelope and is in the process of acquiring its outer envelope from the nuclear membrane. [From Roizman (1969).] (B) Row of SV5 virions (Paramyxoviridae) budding from the surface of a monkey kidney cell. Cross sections of the nucelocapsid can be seen within several of the particles. [From Compans *et al.* (1966).] (C) Influenza virions (Orthomyxoviridae) budding at the surface of a chicken embryo fibroblast. There are distinct projections on the surface of the budding virions, but not on the adjacent membrane. [From Compans and Dimmock (1969).] (D) Machupo virus (Arenaviridae) budding from a Raji cell. [From Murphy *et al.* (1969).] (E) Sindbis virus (Togaviridae) budding from the plasma membrane of an infected chicken cell. [From Strauss *et al.* (1995).] (F) Rubella virions (Togaviridae) budding from the surface of a BHK cell. [From Higashi *et al.* (1976).]

assembly occurs within the cell cytoplasm or nucleoplasm. For enveloped viruses, final assembly of the virus occurs by budding through a cellular membrane. In either case, the virion must subsequently disassemble spontaneously on infection of a new cell. The cleavages that occur during assembly of the virus potentiate penetration of a susceptible cell after binding of the virus to it, and the subsequent disassembly of the virion on entry into the cell. A few examples will be described below that illustrate the range of cleavage events that occurs in different virus families.

In the picornaviruses, a provirion is first formed that is composed of the viral RNA complexed with three viral proteins, called VP0, VP1, and VP3. During maturation to form the virion, VP0 is cleaved to VP2 and VP4. No protease has been found that performs this cleavage, and it has been postulated that the virion RNA may catalyze it. Cleavage to produce VP4, which is found within the interior of the capsid shell, as illustrated schematically in Fig. 2.9, is required for the virus to be infectious. As described in Chapter 1, VP4 appears to be required for entry of the virus into the cell. This maturation cleavage has another important consequence. Whereas the provirion is quite unstable, the mature virion is very stable. The poliovirus virion will survive treatment with proteolytic enzymes and detergents, and survives exposure to the acidic pH of less than 2 that is present in the stomach. Only on binding to its receptor (Figs. 1.4 and 2.9) is poliovirus destabilized such that VP4 can be released for entry into the host cell.

Similarly, the insect nodaviruses first assemble as a procapsid containing the RNA and 180 copies of a single protein species called α(44 kDa). Over a period of many hours, spontaneous cleavage of α occurs to form β (40 kDa) and γ(4 kDa). This cleavage is required for the particle to be infectious. These events in nodaviruses have been well studied because it has been possible to assemble particles *in vitro,* and the structures of both cleaved and uncleaved particles have been solved to atomic resolution.

Rotaviruses, which form a genus in the family Reoviridae, must be activated by cleavage with trypsin after release from an infected cell in order to be infectious. Trypsin is present in the gut, where the viruses replicate, and activation occurs normally during the infection cycle of the virus in animals. When the viruses are grown in cultured cells, however, trypsin must be supplied exogenously.

A different type of cleavage event occurs during assembly of retroviruses and adenoviruses, as well as of a number of other viruses. During assembly of retroviruses, the Gag and Gag–Pol precursor polyproteins are incorporated, together with the viral RNA, into a precursor nucleocapsid. These polyproteins must be cleaved into several pieces by a protease present in the polyprotein if the virus is to be infectious. An analogous situation occurs in adenoviruses, where a viral protease processes a protein precursor in the core of the immature virion.

In most enveloped viruses, one of the envelope proteins is produced as a precursor whose cleavage is required to activate the infectivity of the virus. This cleavage may occur prior to budding, catalyzed by a host enzyme called furin, or may occur after release of the virus, catalyzed by other host enzymes. The example of the hemagglutinin of influenza virus was described in Chapter 1. This protein is produced as a precursor called HA_0, which is cleaved to HA_1 and HA_2 (Fig. 1.5). Cleavage is required to potentiate the fusion activity present at the N terminus of HA_2. As a second example, alphaviruses produce two envelope glycoproteins that form a heterodimer. One of the glycoproteins is produced as a precursor. The heterodimer containing the uncleaved precursor is quite stable, so that a particle containing uncleaved heterodimer is not infectious. The cleaved heterodimer, which is required for virus entry is much less stable and dissociates readily during infection. Thus, in contrast to the poliovirus maturation cleavage, the alphavirus cleavage makes the virion less stable rather than more stable. Maturation cleavages also occur in the envelope glycoproteins of retroviruses, paramyxoviruses, flaviviruses, and coronaviruses.

Neutralization of Charge on the Virion Genome

DNA or RNA has a high net negative charge, and there is a need for counterions to neutralize this charge in order to form a virion. In many viruses, positively charged polymers are incorporated that neutralize half or so of the nucleic acid charge. The DNA in the virions of the polyomaviruses is complexed with cellular histones. The viral genomes in these viruses have been referred to as minichromosomes. In contrast, the adenoviruses encode their own basic proteins that complex with the genome in the core of the virion. Another strategy is used by the herpesviruses, which incorporate polyamines into the virion. Herpes simplex virus has been estimated to incorporate 70,000 molecules of spermidine and 40,000 molecules of spermine, which would be sufficient to neutralize about 40% of the DNA charge. Among RNA viruses, the nucleocapsid proteins are often quite basic and neutralize part of the charge on the RNA. As one example, the N-terminal 110 amino acids of the capsid protein of Sindbis virus have a net positive charge of 29. The positive charges within this domain of the 240 capsid proteins in a nucleocapsid would be sufficient to neutralize about 60% of the charge on the RNA genome. This charged domain is thought to penetrate into the interior of the nucleocapsid and complex with the viral RNA.

STABILITY OF VIRIONS

Virions differ greatly in stability, and these differences are often correlated with the means by which viruses infect new hosts. Viruses that must persist in the extracellular environment for considerable periods, for example, must be more stable than viruses that pass quickly from one host to the next. As an example of such requirements, consider the closely related polioviruses and rhinoviruses, members of two different genera of the family Picornaviridae. These viruses shared a common ancestor in the not too distant past and have structures that are very similar. The polioviruses are spread by an oral–fecal route and have the ability to persist in a hostile extracellular environment for some time where they may contaminate drinking water or food. Furthermore, they must pass through the stomach, where the pH is less than 2, to reach the intestinal tract where they begin the infection cycle. It is not surprising, therefore, that the poliovirion is stable to storage and to treatments such as exposure to mild detergents or to pH < 2. In contrast, rhinoviruses are spread by aerosols or contaminated mucus, and spread normally requires close contact. The rhinovirion is less stable than the poliovirion. It survives for only a limited period of time in the external environment and is sensitive to treatment with detergents or exposure to pH 3.

FURTHER READING

General

Chiu, W., Burnett, R. M., and Garcea, R. L, (1997). "Structural Biology of Viruses." Oxford University Press, Oxford.

Dalton, A. J., and Haguenau, F., eds. (1973). "Ultrastructure of Animal Viruses and Bacteriophages: An Atlas," Ultrastructure in Biological Systems. Academic Press, New York.

Harrison, S. C., Skehel, J. J., and Wiley, D. C. (1996). Virus structure. *In* "Virology" (N. B, Fields, D. M. Knipe, and P. K. Howley, eds.), pp. 59–99, Lippincott-Raven, New York.

Johnson, J. E. (1996). Functional implications of protein-protein interactions in icosahedral viruses. *Proc. Natl. Acad. Sci. U.S.A.* **93**; 27–33.

Johnson, J. E., and Speir, J. A.,(1999). Principles of virus structure. *In* "Encyclopedia of Virology." (A. Granoff and R. G. Webster, eds.) Vol. 3, pp. 1946–1956. Academic Press, San Diego.

Cryoelectron Microscopy

Baker, T. S., Olson, N. H., and Fuller, S. D. (1999). Adding the third dimension to virus life cycles: Three-dimensional reconstruction of icosahedral viruses from cryo-electron micrographs. *Microbiol. Mol. Biol. Rev.* **63**; 862–922.

Cheng, R. H., Kuhn, R. J., Olson N. H., *et al.* (1995). Nucleocapsid and glycoprotein organization in an enveloped virus. *Cell* (Cambridge, Mass.) **80**; 621–630.

Strauss, J. H., Strauss, E. G., and Kuhn, R. J. (1995). Budding of alphaviruses. *Trends Microbiol.* **3**; 346–350.

X-ray Crystallography

Athappilly, F. K., Murali, R., Rux, J. J., *et al.* (1994). The refined crystal structure of hexon, the major coat protein of adenovirus type 2, at 2.9Å resolution. *J. Mol. Biol.* **242**; 430–455.

Hogle, J. M., Chow, M., and Filman, D. J., (1985). Three-dimensional sructure of poliovirus at 2.9Å resolution. *Science* **229**; 1358–1365.

Prasad, B. V., Hardy, M. E., Dokland, T., *et al.* (1999). X-ray crystallographic structure of the Norwalk virus capsid. *Science* **286**; 287–290.

Rossmann, M. G., Arnold, E., Erickson, J. W., *et al.* (1985). Structure of a human common cold virus and functional relationship to other picornaviruses. *Nature (London)* **317**; 145–153.

Stehle T., Gamblin, S. J., Yan, Y., *et al.* (1996). The structure of simian virus 40 refined at 3.1Å resolution. *Structure* **4**; 165–182.

Canyon Hypothesis

Rossmann, M. G. (1989). The canyon hypothesis. *Viral Immunol.* **2**; 143–161.

3

Plus-Strand RNA and Double-Strand RNA Viruses

INTRODUCTION

The plus-strand RNA [(+)RNA] viruses comprise a very large group of viruses belonging to many families. Among these are viruses that cause epidemic disease in humans, including encephalitis, hepatitis, polyarthritis, yellow fever, dengue fever, poliomyelitis, and the common cold. The number of cases of human disease caused by these viruses each year is enormous. As examples, dengue fever afflicts more than 100 million people each year; most humans suffer at least one rhinovirus-induced cold each year, with the cases therefore numbering in the billions; and most humans during their lifetime will suffer several episodes of gastroenteritis caused by astroviruses or caliciviruses. In terms of frequency and severity of illness, the (+)RNA viruses contain many serious human pathogens, and we will begin our description of viruses with this group.

The human (+)RNA viruses belong to six families (Table 3.1). These six families also contain numerous non-human viruses, of which many are important pathogens of domestic animals. Large numbers of (+)RNA viruses that infect plants are also known; in fact, most plant viruses contain (+)RNA genomes. The plant viruses, however, belong to different families and are currently classified by the International Committee on Taxonomy of Viruses (ICTV) into six families plus many unassigned genera. Because of their importance as disease agents of domestic crops, much is known about these viruses. Other families of (+)RNA viruses include a family of bacterial viruses and two families of insect viruses (the nodaviruses, in particular, have been intensively studied). Thus, the (+)RNA viruses have evolved into many distinctly different families and must have arisen long ago. In this chapter, the six families of viruses that include human viruses as members are

TABLE 3.1 Families of Plus-Strand RNA Viruses That Contain Human Pathogens

Family	Size of genome (nucleotides)	Other vertebrate hosts	Representative human pathogens[a]
Picornaviridae	~7500	Cattle, monkeys, mice	Poliovirus, human rhinovirus, hepatitis A virus
Caliciviridae	~7500	Rabbits, swine, cats	Norwalk virus[b]
Astroviridae	6800–7900	Cattle, ducks, sheep, swine	Human astrovirus
Togaviridae	~11,600	Mammals, birds, horses	Ross River virus,
			WEE, VEE, EEE, Mayaro, rubella
Flaviviridae	9500–12,500	Swine, cattle, primates, birds	Dengue, yellow fever, JE, MVE, TBE, hepatitis C
Coronaviridae	20,000–30,000	Mice, birds, swine, cattle	Human coronavirus

[a]*Abbreviations:* WEE, VEE, EEE, Western, Venezuelan, Eastern equine encephalitis viruses, respectively; JE, Japanese encephalitis virus; MVE, Murray Valley encephalitis virus; TBE, tick-borne encephalitis virus.

[b]Hepatitis E virus was formerly classified as a member of Caliciviridae, on the basis of genome size and morphology of particle, but is currently unassigned as to family, and belongs to the floating genus "hepatitis E-like" viruses.

considered, followed by a brief discussion of the relationships of these viruses to the plant viruses and what this means in terms of virus evolution.

Viruses that contain double-stranded (ds) RNA as their genome will also be considered in this chapter. Although several families of dsRNA viruses are known, only the Reoviridae contain members that infect humans and cause human disease. Of the diseases caused by viruses in this family, gastroenteritis induced by rotaviral infection is the most widespread and significant. Infection by rotaviruses results in more than 1 billion cases of human gastroenteritis yearly with 1 million or more deaths, particularly of children in developing countries.

FAMILY PICORNAVIRIDAE

The picornaviruses are so named because they are small (*pico* = small), *RNA*-containing viruses. Nine genera of picornaviruses, five of which contain human pathogens, are currently recognized (Table 3.2), and more will probably be recognized as new viruses are isolated and described. A dendrogram that illustrates the relationship of seven genera to one another, as well as the relationships of a number of viruses within the various genera, is shown in Fig 3.1. This dendrogram makes clear that all picornaviruses are closely related. They share significant nucleotide and amino acid sequence identity and form a well-defined taxon.

TABLE 3.2   Picornaviridae

Genus/members	Virus name abbreviation[a]	Usual host(s)	Transmission	Disease	World distribution
Enterovirus					
Poliovirus (3)	PV	Humans	Oral/fecal, contact	Paralysis, aseptic meningitis	Originally worldwide, extirpated in Americas
Echovirus (29)		Humans	Oral/fecal, contact	Aseptic meningitis, paralysis, encephalitis	Worldwide
Coxsackie (23A and 6B)		Humans	Oral/fecal, contact	Common cold, myocarditis	Worldwide
Enterovirus (4 human, 30 other)		Humans, cattle, monkeys, swine	Oral/fecal, contact	Aseptic meningitis, conjunctivitis (type 70)	Worldwide
Parechovirus					
Human parechovirus (formerly echovirus 22)	HPeV	Humans	Oral/fecal	Gastroenteritis	Worldwide
Rhinovirus					
Human rhinoviruses (>100 serotypes)	HRV-A, HRV-B	Humans	Aerosols, contact	Common cold	Worldwide
Aphthovirus					
Foot-and-mouth disease	FMDV	Cattle, swine	Oral/fecal, contact	Lesions on mouth and feet	Worldwide (except U.S.)
Cardiovirus					
Encephalomyocarditis	EMCV	Mice	Oral/fecal, contact	Encephalitis, myocarditis	Worldwide
Kobuvirus					
Aichi	AiV	Humans	Oral/fecal	Gastroenteritis	Isolated in Japan (oysters)
Hepatovirus					
Hepatitis A	HAV	Humans	Oral/fecal	Hepatitis	Endemic worldwide
Teschovirus					
Porcine teschovirus (formerly porcine enterovirus 1 or PEV-1)	PTV	Swine	Oral/fecal	Paralysis, porcine encephalomyelitis	Britain, Central and Eastern Europe
Erbovirus					
Equine rhinotracheitis B	ERBV	Horses	?	?	?

[a]Standard abbreviations are given for either the virus listed (such as poliovirus) or for the type member of the genus.

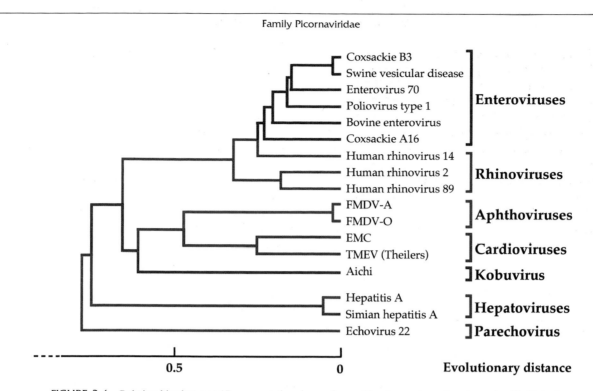

FIGURE 3.1   Relationships between 17 representative picornaviruses. The viruses shown have been classified into six recognized genera, and one provisional genus, Kobuvirus. Comparable data are not available for members of the two other provisional genera, Teschovirus and Erbovirus; therefore, these genera are not included in the tree. The tree was generated from the amino acid sequences of the 3D^{pol} proteins. [Adapted from Yamashita *et al.* (1998).]

As described in Chapter 2, the structures of several picornaviruses have been solved to atomic resolution by X-ray crystallography. The picornavirus virion is composed of 60 copies of each of four different proteins (called VP1–4) that form an icosahedral shell having $T=3$ symmetry (or pseudo-$T=3$) and a diameter of approximately 30 nm (see Figs. 2.1, 2.5, 2.7 and 2.8).

Organization and Expression of the Genome

The structure of the genome of poliovirus and comparison of it with the genomes of the other genera are shown in Fig. 3.2. The picornaviral genome is a single RNA molecule of about 8 kb. It contains one open reading frame (ORF) and is translated into one long polyprotein. This polyprotein is cleaved by two virus-encoded proteinases to form more than 25 different polypeptides, including processing intermediates (not all of which are shown in the figure) as well as final cleavage products. The ORF in the genome contains three regions, called 1 (the 5′ region), 2 (the middle region), and 3 (the 3′ region). Region 1 encodes the structural proteins and regions 2 and 3 encode proteins required for RNA replication. The genome organization of all picornaviruses is similar, but each genus differs in important details. For example, the aphthoviruses and the

cardioviruses have a poly(C) tract near the 5′ end of the RNA that is important for virus replication. These two genera also have a leader polypeptide that precedes the structural protein region. The aphthovirus leader peptide is a papain-like protease that cleaves itself off the polyprotein and has a role in the shutoff of cellular protein synthesis. The function of the cardiovirus leader is not known. Hepatitis A virus, Aichi virus, and echovirus 22, representatives of three other genera, also have leaders.

The picornaviral genome has a small protein, VPg, covalently bound to the 5′ end, which is the primer for initiation of RNA synthesis. VPg is normally removed from RNA that serves as mRNA by a cellular enzyme, but its removal is not required for its translation. The 3′ end of the RNA is polyadenylated. As described in Chapter 1, the 5′ nontranslated region of a picornaviral RNA possesses an IRES (internal ribosome entry site) and the RNA is translated by a cap-independent mechanism. The translation of picornaviral RNA is greatly favored in the infected cell because picornaviruses interfere with host cell macromolecular synthesis and, in particular, interfere with host protein synthesis. Infection with entero-, rhino-, and aphthoviruses leads to proteolytic cleavage of cap-binding protein by 2A^{pro} (entero- and rhinoviruses) or the leader protease (aphthoviruses), which results in inhibition of the translation of

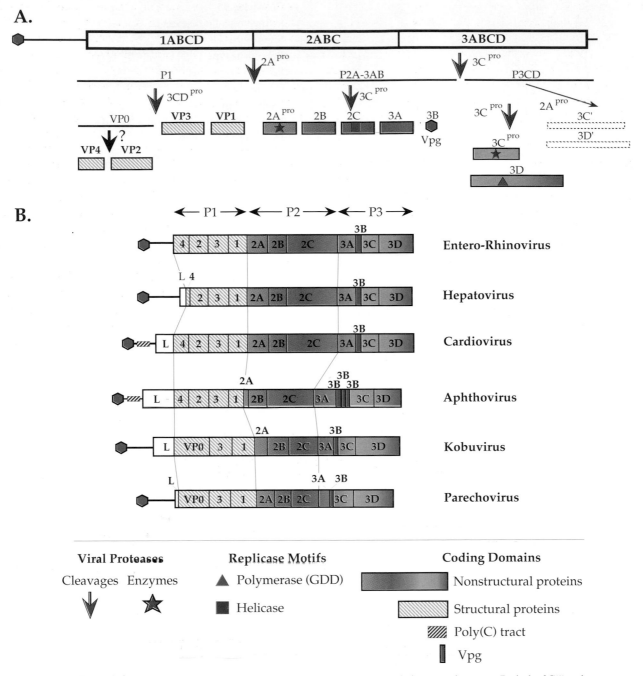

FIGURE 3.2 (A) Genome organization of poliovirus showing the proteolytic processing steps. Both the 3C^{pro} and 2A^{pro} proteases are "serine-like proteases" with cysteine in the catalytic site. (B) Comparative genome organizations of representatives of seven of the nine genera of Picornaviridae. [Adapted from Murphy *et al.* (1995, p. 300) and Yamashita *et al.* (1998).]

RNAs that require the cap-binding protein complex, that is, capped cellular mRNAs. The cardioviruses, which are also cap-independent, interfere with translation of host mRNAs in a different way, by interfering with phosphorylation of cap-binding protein. Poliovirus may also interfere with other initiation factors in addition to the cap-binding

protein in order to achieve the profound inhibition of host protein synthesis that is observed.

The viral 3C^{pro} and its precursor 3CD^{pro} make multiple cleavages in the polyprotein translated from the genome, as illustrated for poliovirus in Fig. 3.2. Some cleavages occur in *cis* and some in *trans*. The crystal structures of 3C^{pro} of poliovirus and of a rhinovirus have been solved to atomic resolution, and their core structure resembles that of chymotrypsin (Fig. 1.17). The catalytic center has the same geometry as that of chymotrypsin, but in 3C^{pro} the catalytic serine has been replaced by cysteine. Moreover, in many, but not all, picornaviruses the aspartic acid in the catalytic triad has been replaced by glutamic acid. Thus 3C^{pro} is related to cellular serine proteases and may have originated by the capture of a cellular serine protease during evolution.

Poliovirus possesses a second protease, 2A^{pro}. 2A^{pro}, like 3C^{pro}, is thought to be a serine protease in which the active site serine has been replaced by cysteine. 2A^{pro} catalyzes one essential cleavage in the polyprotein, that between P1 and P2. This cleavage occurs in *cis*. 2A^{pro} also cleaves the cellular translation initiation factor eIF-4G, which results in the shutoff of cellular protein synthesis. The proteolytic activity of 2A^{pro} is also required for other functions during poliovirus replication, the nature of which have not been established. An interesting experiment is illustrated in Fig. 3.3 because it illustrates the power of molecular genetics and the tricks that modern virologists can play with

viruses. This experiment will serve as a prelude to the discussion of the uses of viruses as vectors in Chapter 9. A poliovirus was constructed in which a stop codon was placed after the structural protein domain (region 1), so that 2A^{pro} was not needed to remove P1 from the polyprotein precursor. The stop codon was followed by an IRES and a new AUG start codon, so that P2A and the rest of the genome could be translated from the polycistronic RNA. This virus was viable. However, when the 2A proteinase was inactivated by changing the active site cysteine, the resulting virus was dead, showing that the proteolytic activity of P2A is required not only to separate regions 1 and 2 of the polyprotein but also for other function(s).

Protein 2A of rhinoviruses is also a protease. The crystal structure of protein 2A of human rhinovirus type 2 reveals that this protease is unrelated to 2A^{pro} of polioviruses, however. Thus, these two proteases in closely related viruses have different origins, and the viruses have solved the problem of how to separate regions 1 and 2 in the polyprotein in different ways. Still another solution to this problem has been adopted by the cardio- and aphthoviruses. Protein 2A is not a protease in these viruses. Indeed, 2A is only 18

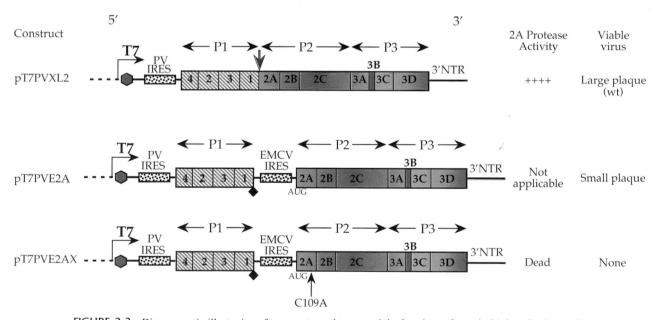

FIGURE 3.3 Diagrammatic illustration of constructs used to unravel the functions of protein 2A in poliovirus replication. cDNA copies of the virus RNA can be manipulated by genetic engineering to insert IRES elements or make specific mutations. RNA can be transcribed from the clones *in vitro* and used to infect cells, which is possible for plus-strand RNA viruses because the first event after infection is translation of the genomic RNA. The wild-type construct pT7PVXL2 is shown in the top line. The 2A proteolytic activity normally cleaves the bond between domains P1 and P2 of the translated polyprotein. If this function is rendered nonessential, as in construct pT7PVE2A, by the insertion of a stop codon at the C terminus of P1 (solid diamond), followed by a second IRES and an initiation AUG at the beginning of 2A, virus is still produced, but forms small plaques. Thus, separation of the structural region and the nonstructural region in this way results in viable virus. However, if the proteolytic activity of 2A is inactivated by mutation of the catalytic cysteine to alanine as in pT7PVE2AX, no virus is produced, demonstrating that the proteolytic activity of 2A^{pro} is necessary for other functions in addition to the P1/P2 cleavage. The pink hexagon is the VPg encoded in 3B and linked to the 5′ end of the RNA. [Adapted from Lu *et al.* (1995) and Molla *et al.* (1993).]

residues long, and cleavage between P1 and P2 is catalyzed by $3C^{pro}$. Another interesting feature of these viruses is that the cleavage 2A/2B occurs spontaneously, catalyzed by the specific amino acid sequence at the scissile bond.

In addition to these cleavages catalyzed by $2A^{pro}$ and $3C^{pro}$, VP0 is cleaved to VP2 and VP4 during virion maturation. Available evidence suggests that this cleavage is not catalyzed by a protease.

Functions of the Picornavirus Proteins

The cleavage product P1 consists of a polyprotein precursor for the four structural proteins of the virus, VP1–4. P1 is first cleaved in *trans* to VP0, VP1, and VP3 by $3CD^{pro}$ (Fig. 3.2A). VP0 is later cleaved to VP2 and VP4 during virus assembly. The cleavage products of P2 and P3 are required for RNA replication. $2A^{pro}$ has been described. $2C^{ATPase}$ has been shown to be an ATPase, not a GTPase, and contains sequence motifs characteristic of helicases. Many, but not all, RNA viruses encode helicases to unwind duplex RNA during replication, and it is assumed that $2C^{ATPase}$ performs such a function. The precursor to $2C^{ATPase}$, a protein called $2BC^{ATPase}$, has a different role in RNA replication. It is required for proliferation of membranous structures in poliovirus-infected cells that serve as sites for RNA replication.

Region 3 encodes VPg, $3CD^{pro}/3C^{pro}$, and the viral RNA polymerase $3D^{pol}$. Cleavages effected by $3C^{pro}$ are illustrated in Fig. 3.2. $3C^{pro}$ may also have a regulatory role in the virus life cycle, because the cleavage intermediate $3CD^{pro}$, which is fairly long lived, has properties that differ from $3C^{pro}$. One function of $3CD^{pro}$ is to bind the viral RNA in conjunction with 3AB, the precursor for VPg, or with a cellular protein, poly(C)-binding protein. Formation of a complex with the viral RNA is essential for its replication, and differential cleavage of the 3C–3D bond during the infection cycle may regulate replication. A strategy in which precursor proteins perform different functions than those performed by the final cleavage products, such as those illustrated by $2BC^{ATPase}$ and $3CD^{pro}$, allows the virus to optimize the coding capacity of its small genome, because a given sequence is used for more than one function.

Replication of Picornaviruses

The replication of poliovirus has been particularly well studied and the virus has served as a model for the replication of RNA viruses. All nonstructural poliovirus proteins, including cleavage intermediates, have been purified and studied for their possible function as enzymes or RNA-binding proteins. These studies have been complemented by studies of replication complexes isolated from infected cells, studies using replicons in which the luciferase gene replaces the P1 coding region, and studies of processes that occur in infected cells.

During replication of poliovirus RNA, a full-length complementary copy of the genomic RNA is produced that serves as a template for the synthesis of genomic RNA (illustrated schematically in Fig. 1.9A). This complementary RNA template has been variously called minus-strand RNA [abbreviated (−)RNA], antigenomic RNA, or virion-complementary (vc) RNA. Much more (+)RNA than (−)RNA is produced, since (+)RNA is needed for translation and encapsidation into progeny as well as for replication, whereas (−)RNA is needed only as a template for making (+)RNA. The mechanisms by which disproportionate amounts of (+) and (−) strands are made are not yet understood.

The RNA-dependent RNA polymerase $3D^{pol}$ is strictly primer dependent. In the presence of template, $3D^{pol}$ can uridylate VPg on a specific tyrosine residue. This nucleotidyl peptide then functions as a primer for the initiation of RNA synthesis. It is of interest that several viruses belonging to other families, such as hepatitis B virus (a virus that uses reverse transcription during the replication of its genome) and adenovirus (a DNA virus), have also adopted the strategy of using a protein primer for initiation of nucleic acid synthesis.

The nature and function of the promoters in the poliovirus genome that are involved in the initiation of RNA replication are still poorly understood. A 5′-terminal sequence that forms a cloverleaf, which binds protein complexes containing $3CD^{pro}$, is essential for RNA replication. A second essential *cis*-acting sequence element has been discovered recently in rhinoviruses and poliovirus; intriguingly, this element is internal, located within the ORF of the polyprotein.

It has been possible to achieve a complete replication cycle of poliovirus in an extract of uninfected HeLa cells. RNA from poliovirus virions added to such an extract will direct the synthesis of all the poliovirus proteins, and these in turn will replicate the input RNA and encapsidate the progeny genomes. This cell-free, *de novo* synthesizing system for poliovirus, is as yet unique in virology.

In cell culture, most picornaviruses complete their replication cycle in about 6 hr. The infection is cytolytic, and large quantities of virus are produced. An exception is hepatitis A virus, which establishes chronic infections in cell culture and grows to very low titers.

Genus Enterovirus

Enteroviruses replicate primarily in the enteric tract where they usually cause only mild disease. More serious enteroviral disease may develop after spread to other organs, for example, the central nervous system or the heart. Enteroviruses are normally contracted though ingestion of the virus, either in contaminated food or water or by exposure to the virus through contacts with individuals that are excreting the virus. The epidemiology of poliovirus has been the most intensively studied among the enteroviruses. Poliovirus is present in oropharyngeal secretions early after infection and is excreted in feces over a period of weeks following infection. The virus spreads readily and rapidly through households, which demonstrates the importance of close contacts in virus spread. The virus also has the ability to persist in the external environment for weeks under favorable conditions, and this may represent another source of infection during epidemics. Sewage surveys, for example, have been used to follow poliovirus epidemics, and poliovirus has been found in lakes and swimming pools.

In general, enteroviruses have a fairly narrow host range. Most of the well-studied viruses are human viruses, because humans take a particular interest in the viruses that cause them the most trouble, but enteroviruses of nonhuman primates, pigs, cattle, and insects are known. The more than 70 known human enteroviruses, many of which are important pathogens, normally infect only humans, but poliovirus will infect Old World monkeys. It has been suggested that the virus may be a natural pathogen of these monkeys but it is unlikely that nonhuman primates constitute a reservoir for it, which is important in relation to efforts spearheaded by the World Health Organization to eradicate poliovirus globally.

Polioviruses

The best known of the enteroviruses are the three serotypes of poliovirus. These viruses are the causative agents of poliomyelitis, a disease characterized by the death of motor neurons in the spinal cord. Most poliovirus infections of susceptible humans are inapparent or result in a mild febrile illness in which cells of the pharynx and the gut are infected and recovery is uncomplicated. However, a transient viremia is established following infection (viremia = virus present in the blood), and in a small percentage (<2%) of infections the virus invades the central nervous system (CNS), where it infects motor neurons in the spinal cord and, in severe cases, other regions of the CNS. Such infection of the CNS can result in paralysis, which can be severe enough to be fatal because of respiratory muscle paralysis. The name *poliomyelitis* comes from the Greek words *polio* = gray and *myelo* = spinal cord, from the pathology caused by damage to the motor neurons in the spinal cord, which are located in the gray matter.

Polioviruses have been important pathogens of humans for a long time. The depiction of a lame priest on an Egyptian stele that dates from 3500 years ago suggests that poliovirus was present in ancient Egypt, and references to clubfoot in ancient Greek and Roman writings probably signifies that polio was present at these early times. However, although it very likely that poliovirus has been widespread in humans for thousands of years, there is no firm evidence for poliomyelitis in human populations until about 200 years ago, when the virus appears to have been (or to have become) widespread. Serosurveys in the United States in the 1930s and 1940s, before the introduction of the Salk and Sabin vaccines, indicated that 80–100% of adults had been infected by poliovirus at some time in their lives. Studies in other areas of the world, including studies of lameness in populations, also suggest that, at least in the 1900s, the majority of the world's population had been infected with poliovirus.

Paradoxically, even though poliovirus was surely widespread earlier, poliomyelitis epidemics of large proportions evolved only during the 20th century and they were concentrated at first in countries practicing the highest standards of hygiene. This startling phenomenon has been explained as resulting from changes in human behavior. Originally, the highly infectious virus was contracted by infants shortly after birth when they were still protected by maternal antibodies. This natural infection served to immunize the infant, protecting it from poliomyelitis for life. However, when the chain of immunization was interrupted upon removal of the virus from the environment by the development of hygienic conditions, unprotected children grew up, giving rise to susceptible populations. If the virus invades such populations, epidemics rapidly evolve.

Notice that this scenario requires that infants be infected very early, while still protected by maternal antibodies. After these antibodies wane, the infant is susceptible to poliomyelitis, although it has been thought that infection of susceptible but very young children is less likely to cause poliomyelitis. Statistics of the fraction of young children who contract poliomyelitis in societies in which the virus is endemic, rather than epidemic, are not well defined, in part because of the high death rate of children in such societies due to many infectious diseases. However, surveys conducted in the 20th century of lameness in populations, most of which is probably due to paralytic polio, found similar extents of lameness whether the virus was endemic or epidemic.

In any event, it is clear that changes in human behavior can bring about serious complications relating to infectious disease, and such scenarios have recurred many times during the last century. However, it is important to note that although higher standards of hygiene led eventually to epidemics of poliomyelitis, these standards also led to a reduction in diseases caused by numerous other infectious agents, both viral and bacterial (see Fig. 1.1).

Before it was controlled with vaccines, epidemic poliomyelitis was greatly feared, and it is hard now for people to realize the extent of fear that the disease induced. A great puzzle was the fact that the epidemics struck during the summer only, during the summer breaks of schools or universities. Many human pathogenic viruses are known to prefer a season for attack on humans: influenza during the winter, measles in early spring, enteroviruses during the summer. It is thought that this phenomenon relates to air temperature and humidity. For example, poliovirus infections are correlated with humidity in the Americas and in Europe.

In the United States, there were huge poliovirus epidemics every summer in the 1950s in which more than 50,000 people, mostly children or adolescents, became ill. Of these cases, about 20,000 were paralytic and 2000–3000 people died (Fig. 3.4). Death was often the result of the paralysis of the muscles required for breathing, and iron lungs were introduced for mechanical ventilation of poliomyelitis patients until their muscles recovered sufficiently that they could breathe on their own. Wards containing dozens of patients in iron lungs became a common sight in the large epidemics of the 1950s (Fig. 3.5). Of the survivors of poliomyelitis, many were permanently paralyzed and confined to wheelchairs or required the use of crutches for walking. One of the best known poliomyelitis cases is that of Franklin D. Roosevelt, who contracted poliovirus in 1921 at the age of 39 and was in a wheelchair for the rest of his life, although he continued to lead an active political life.

Introduction of the Salk and Sabin vaccines in the 1950s and 1960s led to the elimination of poliovirus in the United States over a period of about two decades (Fig. 3.4) and more recently has led to the elimination of poliovirus throughout the Americas. The Salk vaccine, which was the first to be developed, is an inactivated virus vaccine that is given as a series of injections. Introduction of this vaccine resulted in a rapid decrease in the number of poliovirus cases. However, because the vaccine induces circulating antibodies but little in the way of mucosal immunity (see Chapter 8), it prevents poliomyelitis, the disease, by preventing spread of the virus from the gastrointestinal (GI) tract to the CNS, but not infection of the GI tract by the virus. The virus thus remained in circulation. The Sabin vaccine, introduced shortly thereafter, is a live attenuated vaccine that is given orally. Its introduction led to a further rapid decline in paralytic poliomyelitis. This vaccine has the drawback that it induces a very small number of cases of paralytic disease, about 10 per year in the United States since 1980 (Fig. 3.4), but its efficacy is very high because it induces mucosal immunity as well as other forms of immunity. It prevents subsequent infection by the wild-type virus, thus allowing eradication of the wild-type virus if coverage is sufficiently broad. Worldwide use of this vaccine has resulted in the eradication of wild-type poliovirus in the United States and throughout the Americas. (The last case of indigenous poliovirus infection in the Americas occurred in Peru in 1991.) Poliovirus is in the process of being eradicated in other parts of the world, although it is still endemic in areas of Africa and Asia

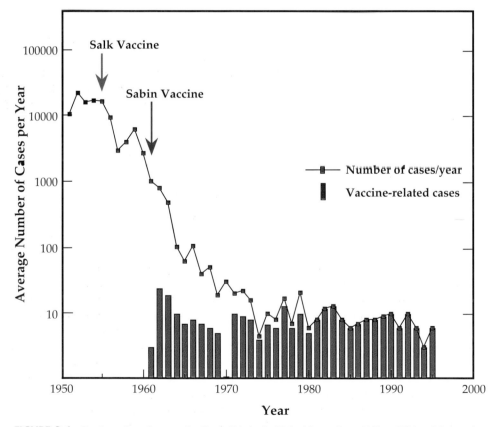

FIGURE 3.4   Total number of cases of poliomyelitis in the United States from 1951 to 1994 and the number of vaccine-related cases after the introduction of the live virus Sabin vaccine. [Data from Nathanson *et al.* (1996, p. 556) and from *Morbidity and Mortality Weekly Report (MMWR)*, Vol. 46, p. 79 (1997).]

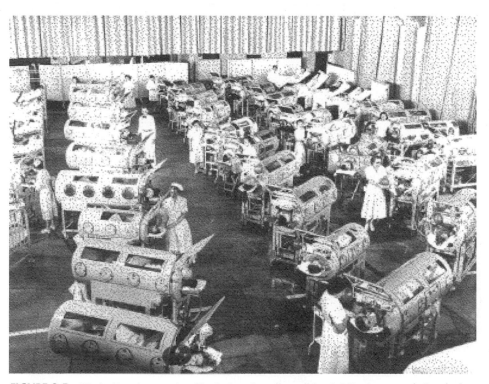

FIGURE 3.5 Ward of iron lungs and rocking beds at the poliomyelitis rehabilitation center in Rancho Los Amigos, California. [From Halstead (1998) with the permission of the author and the publisher.]

(Fig. 3.6). The World Health Organization (WHO) had announced a campaign to eradicate poliovirus worldwide by the year 2000, and as shown in Fig. 3.6B in 2000 wild poliovirus transmission was only present in 33 countries worldwide.

Although the virus has been eradicated from developed countries, there is a large cohort of people infected in the 1950s who are or were paralyzed. Many paralyzed poliomyelitis patients were ultimately able to resume almost normal activities. Through a process of axonal sprouting and reenervation of muscles by the motor neurons that survived the infection, many learned to walk and use their previously paralyzed limbs. However, a syndrome called post-polio syndrome has emerged to plague a significant fraction, perhaps 40%, of the survivors of paralytic poliomyelitis. This syndrome appears 30–40 years after polio infection and is characterized by fatigue, pain, and weakness. The weakness may be severe enough to require the use of a wheelchair. The syndrome results from the degeneration of motor neurons, but the reasons for the degeneration are not clear. The favored hypothesis is that it is the result of overuse of the surviving motor neurons, which are forced to do the work of many. A second possibility is that the surviving neurons were damaged by the original poliovirus infection and fail prematurely. A third possibility is that poliovirus persists in neurons and is somehow reactivated, even in the presence of anti-polio antibody. In model studies using Sindbis virus infection of mice, it has been found that the virus can persist in neurons in a latent state for at least 1–2 years, but there is no evidence that poliovirus might similarly persist in humans for 40 years. Other possible explanations for the failure of motor neurons in post-polio syndrome have also been suggested. Fortunately, paralytic poliomyelitis and its sequelae will soon be a thing of the past.

Other Enteroviruses

Sixty-two other human enteroviruses are currently recognized. Many of these have been known for 50 years, but it is only more recently that the association of many of these viruses with significant human illness has been shown.

Poliovirus was first shown to be a filterable virus in 1908. However, early experiments could only be conducted in monkeys, because the virus will only infect primates. Thus, the amount of information that could be obtained was limited, but such studies eventually showed that more than one poliovirus serotype existed. The development of methods for the cultivation of viruses in cell culture in the 1940s made it possible to screen human stool samples in an effort to type poliovirus isolates, which was necessary if a vaccine was to be produced. Such screening resulted in the discovery of other enteroviruses as well. The study of virology in the United States owes much to the campaign to

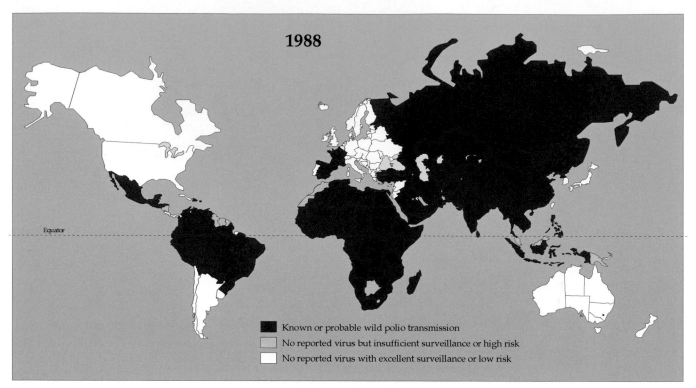

1988

Known or probable wild polio transmission
No reported virus but insufficient surveillance or high risk
No reported virus with excellent surveillance or low risk

(A)

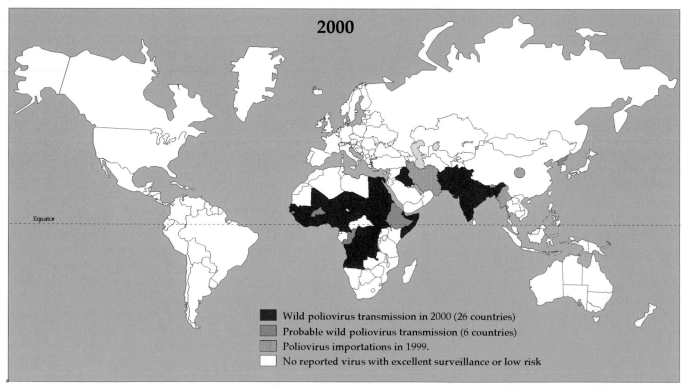

2000

Wild poliovirus transmission in 2000 (26 countries)
Probable wild poliovirus transmission (6 countries)
Poliovirus importations in 1999.
No reported virus with excellent surveillance or low risk

(B)

FIGURE 3.6 Maps showing the worldwide distribution of wild poliovirus and the effects of the global eradication efforts. (A) Wild poliovirus transmission in 1988. [From the WHO web site, http://www.who.int/gpv-surv/graphics/NY_graphics/global_polio_98.htm.] (B) Wild poliovirus transmission as of March 13, 2000. [From *Morbidity and Mortality Weekly Report (MMWR)*, Vol. 49, No. 16, p. 352 (2000).]

develop a vaccine against poliomyelitis. This campaign generated a great deal of public support, which led to funding through private as well as governmental agencies, and the successful development of a vaccine reinforced this support.

The first of these other enteroviruses to be found were two Coxsackie viruses, found by screening patients in Coxsackie, New York, who were suffering from paralysis during a polio epidemic. Coxsackie viruses will infect mice and are classified into two subgroups, called A and B, which differ in their biological properties in mice. They were simply given serial numbers in the order of their isolation—23 Coxsackie A viruses and 6 Coxsackie B viruses are now recognized. Another series of enteroviruses that were first identified in these early studies were called echoviruses (Enteric Cytopathic Human Orphan virus), because these viruses infected the enteric tract of humans, caused cytopathology in cultured cells, and were orphans, not known to cause disease. Echoviruses were distinguished from Coxsackie viruses by their inability to infect suckling mice. Currently 29 echoviruses are recognized in the genus Enterovirus. The latest human viruses to be isolated are now simply called enteroviruses and given serial numbers. Four such viruses, human enterovirus 68, 69, 70 and 71 are now recognized. Numbering started with 68 because at the time there were thought to be 67 polio, Coxsackie, and echoviruses. However, 5 of these (one Coxsackie A virus and 4 echoviruses) were subsequently found to be misidentified, and one (echovirus 22) is sufficiently distinct that it has been renamed human parechovirus and classified into the genus Parechovirus (Table 3.2).

When first isolated, Coxsackie viruses and echoviruses were not known to cause human disease. However, it has now been established that most enteroviruses do cause disease, and many of them cause significant episodes of serious disease (Table 3.3). Study of disease caused by these viruses has been complicated by the fact that there are so many enteroviruses, of which at least some have multiple strains that may differ in disease-causing potential, and by the fact that serious disease is an uncommon complication of infection by most enteroviruses. This has made it difficult to ascribe any particular disease to infection by any particular virus. However, even though serious disease is an uncommon complication, enteroviral infections are very common, and the total number of cases of disease caused by these viruses is large. These illnesses include very infrequent paralytic disease essentially indistinguishable clinically from that caused by poliovirus; myocarditis and pericarditis (caused especially by the Coxsackie B viruses) that is usually subclinical but can be acute and result in significant cardiac compromise; aseptic meningitis; encephalitis; hepatitis; the common cold (perhaps a quarter of summer colds are due to enteroviruses); diarrheal disease; febrile illnesses; rash; hand-foot-and-mouth disease (a common childhood illness caused by several enteroviruses); and epidemic acute hemorrhagic conjunctivitis (an epidemic disease caused by enterovirus 70 that appeared recently and spread around the world). The Coxsackie B viruses are also associated epidemiologically with juvenile onset diabetes in humans but how (or even whether) they cause diabetes is still unresolved. There are no vaccines for any of these viruses.

TABLE 3.3 Human Clinical Syndromes Associated with Enteroviruses

Clinical syndrome	Poliovirus 3 types	Coxsackievirus 23 A types, 6 B types	Echovirus 29 types	Enterovirus 4 types [68–71]
Paralysis	1, 2, 3	Rarely A7, 9 Rarely B2–5	4, 6, 9, 11, 30 (1, 7, 13, 14, 16, 18, 31)	70, 71
Aseptic meningitis	—	A2, 4, 7, 9, 10	All except 12, 24, 26, 29, 32, 33	—
Pericarditis, myocarditis	—	B1–5	1, 6, 9, 19	—
Encephalitis	—	B1–5	2, 6, 9, 19 (3, 4, 7, 11, 14, 18)	70, 71
Hepatitis	—	A4, 9 B5	—	—
Upper respiratory disease, pneumonia	—	A21, 24 B4, 5	4, 9, 11, 20, 25 (91–3, 6–8, 16, 19)	68
Hand-foot-and-mouth disease	—	A5, 10, 16	—	71
Acute hemorrhagic conjunctivitis	—	A24	—	70
Undifferentiated febrile illness	1, 2, 3	B1–6	—	—

Note: Numbers in the columns are the types exhibiting the symptomatology shown in the left column. Numbers in parentheses are types possibly involved (not proven).

Genus Rhinovirus

The rhinoviruses are the causative agents of about half of human colds, the most characteristic symptom of which is rhinitis (inflammation of the nasal mucous membrane and characterized by a runny nose). One hundred serotypes are currently recognized. These are not cross protective and the result is that we are subject to many rhinovirus colds during our lifetimes. Young children, not having been exposed to rhinoviruses and other viruses that cause colds, contract many colds a year. Adults, having become immune to many of these viruses through hard experience, have fewer colds per year, usually only about one. However, the extent and duration of immunity to a particular rhinovirus induced by infection are not well established. There are so many rhinoviruses, and although rhinoviral disease may be miserable it is not life threatening, that detailed studies on cohorts of people over many years have not been done to establish whether immunity to a particular rhinovirus following infection is long lived. For the same reasons, there are no vaccines for any of these viruses.

Rhinoviruses replicate in the upper respiratory tract and are transmitted by direct person-to-person contact. Coughing and sneezing, common syndromes of rhinovirus infection, help spread the virus to nearby contacts. It is not clear how much of the spread is due to aerosolization of the virus on coughing or sneezing followed by inhalation of the aerosolized virus by a susceptible contact, and how much is due to contact with mucus that contains virus, such as by handshake or contact with contaminated doorknobs, followed by transmission of the virus to the nose or mouth.

It is an interesting and informative historical fact that early attempts to isolate rhinoviruses using standard cell culture techniques were unsuccessful. Most cells in the body are maintained at 37°C at a pH of 7.4, and cells in culture are normally maintained under these conditions. However, cells in the upper respiratory tract are maintained at a lower temperature, about 33°C, because the inhalation of outside air through the upper respiratory tract keeps this area cool, and at a pH significantly less than 7.4 because of the high concentration of CO_2 in expired air. Rhinoviruses replicate well in cultured cells under these altered conditions and appear to require the lower temperature and lower pH for efficient growth. In part because of this, rhinovirus infection is limited to the upper respiratory tract, and rhinoviruses almost never cause lower respiratory tract infections.

Rhinoviruses are also known for other animals. Most rhinoviruses are species specific and infect only a single host animal.

Genus Cardiovirus

The cardiovirus genus consists of several viruses of mice of which encephalomyocarditis virus (EMC) has been exten-

sively studied as a model picornavirus. It is closely related to other picornaviruses (Fig. 3.1) although differing in certain important characteristics. The EMC IRES has proven more useful that the poliovirus IRES in experiments that require polycistronic mRNAs or that express proteins in a cap-independent fashion in vertebrate expression systems. Theiler's virus, another member of this genus, causes demyelinating disease in mice and has been extensively studied as a model for this disease.

Genus Hepatovirus

Hepatitis in Man

Many different viruses, belonging to several virus families, are known to cause hepatitis in man. These different viruses have different modes of transmission and cause illness of different degrees of severity (although all hepatitis is serious) that results from destruction of liver cells caused by growth of these viruses in the liver as a target organ. Hepatitis is characterized by fatigue and other symptoms that result from inadequate liver function, and it may be fatal if sufficient destruction of the liver takes place. A characteristic feature of acute hepatitis is the presence of elevated levels of liver enzymes circulating in the blood that results from the destruction of liver cells. Many cases of hepatitis are accompanied by jaundice (turning yellow) because of the destruction of the liver, which is responsible for clearing bilirubin from the blood.

The viruses whose primary disease syndrome in man is hepatitis, or which are closely related to viruses that cause such hepatitis, have been historically named hepatitis virus followed by a letter, in the order of isolation. Thus we have hepatitis A virus, the first to be isolated, hepatitis B virus, the second, and so forth. Because these viruses belong to a number of different families, confusion can arise because of the similar names even though the viruses are unrelated. For ease of reference, Table 3.4 presents a description of the currently known viruses whose name includes hepatitis. Figure 3.7 shows the incidence of hepatitis in the United States in 1997 caused by hepatitis viruses A, B, and C, which are the most important causes of viral hepatitis in the United States. Identification of which hepatitis virus is responsible for any specific case of hepatitis requires immunological tests or virus isolation, because symptoms are similar.

Hepatitis A virus is a picornavirus and will be considered here. The other viruses will be considered when their respective families are introduced.

Hepatitis A Virus

Hepatitis A virus (HAV) is a causative agent of infectious hepatitis in humans. The virus is worldwide in distribution. Only one serotype is known, but isolates from different

TABLE 3.4   Causative Agents of Viral Hepatitis

Virus	Family/*genus*	Genome type/ *size in kb*	Transmission	Chronicity?	Long-term effects	Annual U.S. cases/*deaths*[a]	Chronic hepatitis (millions of cases) U.S./*World*
Hepatitis A	Picornaviridae/ *Hepatovirus*	ss(+)RNA/ *7.5 kb*	Fecal/oral	Very little	Few if any	25,000/ *90*	*0/0*
Hepatitis B	Hepadnaviridae/ *Orthohepadnovirus*	dsDNA (RT)/[b] *3.2 kb*	Parenteral, sexual, vertical	10% of adults, 90% of neonates	HCC,[c] cirrhosis	10,000/*~900*	*1.25–350*
Hepatitis C	Flaviviridae/ *Hepacivirus*	ss(+)RNA/ *9.4 kb*	Parenteral, sexual, vertical	>50%	HCC,[c] cirrhosis	3500/*~2500*	*3.9–170*
Hepatitis D	*Deltavirus*	ss, circular RNA/ *1.7 kb*	Parenteral (sexual, vertical?)	Yes	Exacerbates symptoms of hepatitis B		*0.07/?*
Hepatitis E	Unassigned (formerly Caliciviridae)	ss(+)RNA/ *7.5 kb*	Fecal/oral	No	Few if any	Rare/*0*	*0/0*
Hepatitis F	?[d]						
Hepatitis G	Flaviviridae/ *(Not assigned to a genus)*	ss(+)RNA/*9.4 kb*	Parenteral, other?	Yes	??	??/*none*	*??*

[a]As reported in Summary of Notifiable Diseases in the United States, 1998 *MMWR*, (Vol. 47, No. 53 1999); These numbers may reflect significant underreporting.

[b]RT is reverse transcriptase. Nucleic acid in virion is partially dsDNA, consisting of a full-length minus-strand DNA of 3.2 kb and an incomplete plus-strand DNA that is variable in length.

[c]HCC, hepatocellular carcinoma.

[d]Isolate from a fulminant case of hepatitis; not further characterized.

areas or different times can be grouped into different genotypes or strains. The most distantly related HAV isolates share about 75% nucleotide sequence identity, but most isolates are much more closely related. HAV is a typical picornavirus but is the outlier of the family (Fig. 3.1). It shares only 28% amino acid identity in its structural proteins with any other picornavirus, whereas most picornaviruses are more closely related to one another.

The number of cases of hepatitis A in the world is estimated to be more than 1.4 million each year. In 1998 in the United States, ~37,000 cases of hepatitis were reported, of which ²/₃ were diagnosed as caused by HAV. The virus is spread through contaminated food and water. Filter-feeding shellfish like oysters are known to concentrate the virus, and consumption of raw shellfish has been the cause of many epidemics of hepatitis A. Infection by HAV usually results in a self-limited illness in which the patient recovers with relatively few sequelae. The illness can be quite serious, even fatal, however, because 90% of the liver tissue can be destroyed by virus infection, and liver function is severely impaired until the liver recovers. The seriousness of disease is age dependent. Very young children suffer little disease but with advancing age infection by the virus becomes more serious. The mortality rate in children younger than 14 is only 0.1%, but HAV infection in people older than 40 results in a fatality rate of 2.1%.

Until recently, the only prophylaxis for HAV was injection of immune gamma globulin, which provided protection from the virus for a few weeks. Recently, however, two inactivated virus vaccines against HAV have been licensed that promise to give long-lived protection. It is anticipated that the introduction of these vaccines will result in a decline in the importance of this disease in the developed world, but it will remain endemic in many areas unless inexpensive vaccines suitable for mass distribution in Third World countries are developed.

Genus Aphthovirus

Foot-and-mouth disease viruses (FMDV) belong to seven currently recognized serotypes. They cause a debilitating disease, foot-and-mouth disease, in cattle and other animals, and are economically important pathogens. FMDV was eliminated from the United States many years ago by the simple expedient of killing all infected animals until such time as the virus was extirpated. The last epidemic in the United States occured in 1929. The virus still circulates in Europe, South America, and other parts of the world, and the U.S. Department of Agriculture maintains strict quarantines in order to prevent the virus from reappearing in this country. In the United States, work with the virus is allowed only on Plum Island in Long Island Sound, in order to prevent its accidental release. Because of its agricultural importance, the molecular biology of the virus has been intensively studied.

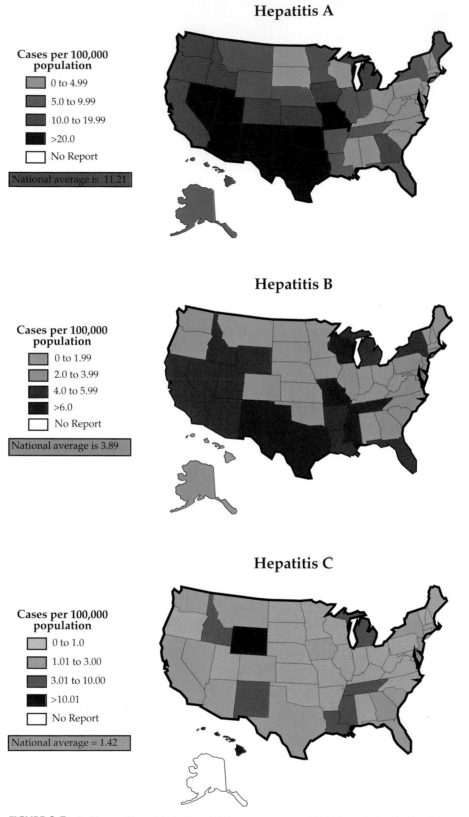

FIGURE 3.7   Incidence of hepatitis A, B, and C (cases per year per 100,000 population) in the United States in 1997, by state. This illustrates the noncoincidence of the three types. [From *MMWR*, Summary of Reportable Diseases (1998).]

In 2001, a large epidemic of FMDV occurred in Western Europe. The epidemic began in February in British sheep and spread to cattle and pigs in Britain and on the continent. By June more than 2000 infected animals had been detected. The epidemic was controlled by restricting the movement of sheep, cattle and swine, and culling of herds in which FMDV was found. Almost 4 million animals were destroyed in this process. The damage to the British cattle industry was particularly distressful because this epidemic occurred only a few years after widespread culling of cattle to control an epidemic of a prion disease called "mad cow disease" (Chapter 7). Beginning in March, many rural areas often visited by tourists were closed to prevent the spread of the virus, and the United States Department of Agriculture was especially vigilant in examining travelers returning from affected countries. British authorities considered vaccinating cattle with commercial vaccines to control the epidemic, but vaccines have not been used to date. Vaccination for FMDV is used in some parts if the world, but is controversial because it is then difficult to distinguish between vaccinated animals and infected animals. Thus, for example, the United States does not allow the importation of beef from areas where vaccination is practiced because it is not possible to rule out the presence of FMDV infection. Although FMDV was thought to have eradicated by June, 2001 and restriction were about to be lifted, 16 new cases of FMDV have been found very recently in Britain.

FAMILY CALICIVIRIDAE

The caliciviruses are nonenveloped viruses with icosahedral symmetry and a diameter of about 30 nm. The name comes from the Latin word for cup or goblet because there are cup-like depressions in the surface of the virion when viewed in the electron microscope. The characteristics of a number of caliciviruses are shown in Table 3.5. Caliciviruses are currently classified into four genera, two of which, the Norwalk-like viruses and the Sapporo-like viruses, contain agents that cause human gastroenteritis. Hepatitis E virus was formerly considered a calicivirus but has been removed from this family recently; it will be considered here for convenience, however, because it is currently unclassified as to family.

The human caliciviruses have been difficult to study and, as a consequence, we know less about them than we do about such well-studied viruses as the picornaviruses. The first calicivirus described was a virus of sea lions (San Miguel sea lion virus, a vesivirus), but caliciviruses of other mammals are now known (feline calicivirus, vesicular exanthema virus of swine, rabbit hemorrhagic disease virus, as well as the two human viruses). What we know of the molecular biology of caliciviruses comes from studies of the nonhuman viruses, but there has not been sufficient interest in these particular viruses to warrant extensive studies. The complete sequences of a number of the human viruses have recently been obtained, which has greatly expanded our knowledge of these viruses, but details of the molecular biology of their replication are still lacking.

TABLE 3.5   Caliciviridae

Genus/members	Virus name abbreviation	Usual host(s)	Transmission	Disease	World distribution
Vesivirus					
Swine vesicular exanthema	VESV	Swine	Oral, contact	Fever, lesions on snout and feet (flippers)	United States[a]
San Miguel sea lion	SMSLV	Pinnipeds			California
Feline calicivirus	FCV	Cats	Contact	Rhinitis, pneumonia, fever	Worldwide
Lagovirus					
Rabbit hemorrhagic disease	RHDV	Rabbits	Water-borne, oral/fecal	Hemorrhages	China, Europe, Australia[b]
Norwalk-like Viruses					
Norwalk	NV	Humans	Water-borne, oral/fecal	Epidemic gastroenteritis	Worldwide
Southampton					
Many other small round structured viruses of humans					
Sapporo-like Viruses					
Sapporo/82	SV	Humans	Water-borne, oral/fecal	Epidemic gastroenteritis	Worldwide
Manchester					
Parkville					

[a]VESV was extirpated from swine by 1956; VESV and SMSLV are now thought to be the same virus.
[b]RHDV was inadvertently introduced into Australia.

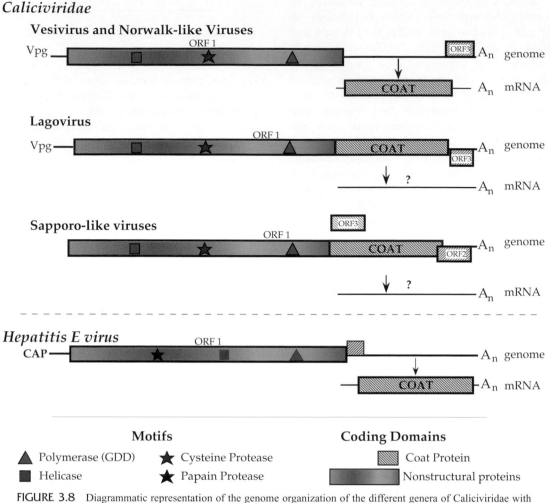

Motifs

▲ Polymerase (GDD)   ★ Cysteine Protease

■ Helicase   ★ Papain Protease

Coding Domains

Coat Protein

Nonstructural proteins

FIGURE 3.8 Diagrammatic representation of the genome organization of the different genera of Caliciviridae with that of hepatitis E virus shown for comparison. Notice that the coat protein is encoded in ORF2 of vesiviruses, Norwalk-like viruses, and hepatitis E virus, but as the C-terminal portion of ORF1 in Lago- and Sapporo-like viruses. [Adapted from Numata *et al.* (1997) and Schlauder *et al.* (1998).]

The caliciviruses are distant relatives of the picornaviruses. The (+)RNA genome of about 8 kb has a 5'-terminal VPg and a 3'-terminal poly(A), as do the picornaviruses, and calicivirus proteins share sequence identity with the picornavirus 2C^{ATPase}, 3D^{pol}, and 3C^{pro}. Unlike the picornaviruses, however, most of the caliciviruses produce (at least) one subgenomic mRNA. The genomes of most contain three ORFs, as illustrated in Fig. 3.8. ORF 1, translated from the genomic RNA, encodes a polyprotein that contains the domains related to picornaviral 2C^{ATPase}, 3D^{pol}, and 3C^{pro}. This polyprotein is presumably processed by 3C^{pro}. In vesiviruses and the Norwalk-like viruses the subgenomic mRNA is translated into the structural protein(s) of the virus. There is evidence that, in at least some caliciviruses, this capsid protein is produced as a precursor that is cleaved. A third ORF is found in the 3'-terminal region of the genome (or upstream of the coat protein gene in hepatitis E

virus). There is no firm evidence that this ORF is translated, although it seems unlikely that an unused ORF would be maintained in a virus with limited coding capacity.

Norwalk Group of Viruses

A number of human viruses causing gastroenteritis, the Norwalk group, have recently been classified as human caliciviruses. There is no cultured cell line in which to propagate these viruses, nor even an animal model in which to grow the virus, and studies have relied on human volunteers for their propagation. This has severely limited the amount of information (and material) that can be obtained. Thus, for example, it is assumed that the Norwalk viruses possess a 5'-terminal VPg as does San Miguel sea lion virus, but to date it has not been possible to prove this. However, because of the power of modern gene cloning technology, the entire genome of Norwalk has been sequenced, starting from stools of experimentally infected human volunteers. These sequences tell us that Norwalk virus is related to animal caliciviruses, and studies are under way that express parts of the genome in various systems in order to study the functions of the different regions of the genome.

Several isolates of the Norwalk group of viruses have been studied, all of which share more than 50% sequence identity and which are named after the location where they were first isolated. These include Norwalk virus, the Hawaii agent, the Snow Mountain agent, and Southampton virus. These viruses are extraordinarily infectious. In one epidemic investigated by the Centers for Disease Control (CDC, now renamed the Centers for Disease Control and Prevention) and local health authorities, a baker preparing food for a wedding was ill with gastroenteritis. After using the toilet, he washed his hands thoroughly before handling food, but his hands were still contaminated with virus, perhaps under the fingernails. He used his hands to stir a very large pot of icing used to glaze cakes and doughnuts that were distributed at the wedding reception, and managed to contaminate the icing with virus. Every guest at the reception who ate as much as a single doughnut contracted gastroenteritis.

The Norwalk viruses regularly cause epidemics of gastroenteritis. The incubation period is short (24 hr on average) and the course of disease is also short (1–2 days). Following recovery, immunity is established to the virus, but the duration of immunity appears to be fairly short (perhaps one or a few years). Studies of immunity are made difficult by the finding that some fraction of the human population seems to be resistant to any particular Norwalk virus studied, perhaps because of a lack of receptors for the virus, and by the fact that there are so many viruses that cause gastroenteritis.

The Norwalk group of viruses is worldwide in its distribution but epidemiological studies of these viruses are limited. Some outbreaks have been associated with the consumption of raw oysters. Other outbreaks have been due to improper handling of food, as described above. They are considered to be a common cause of gastroenteritis in humans.

Rabbit Hemorrhagic Disease Virus

The genome organization of rabbit hemorrhagic disease virus is of interest because it seems to represent an intermediate state between caliciviruses and picornaviruses, in which the capsid protein is fused to the ORF encoding the

FIGURE 3.9 Worldwide incidence of epidemic and endemic hepatitis E. In addition, serosurveys indicate that 1–2% of blood donors in the United States and Western Europe have detectable IgG antibodies to hepatitis E. [Adapted from Fields *et al.* (1996, p. 2838).]

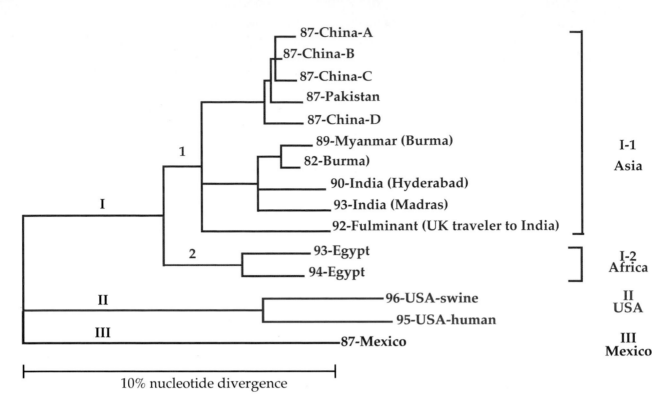

FIGURE 3.10 Phylogenetic tree of HEV isolates based on the complete sequences of ORF2 (1983nt), encoding the coat protein. Isolates are listed by year and location of isolation. Branch lengths are proportional to the evolutionary distance between sequences. Roman numerals are used to denote genotypes (<85% nucleotide sequence identity), and Arabic numbers denote subgenotypes (<92.5% nucleotide sequence identity). The genetic groupings of the HEV strains reflect the geographic relationships of the places from which they were isolated, showing that the viruses are geographically isolated and do not circulate over wide areas. [Adapted from Fig. 1 in Tsarev *et al.* (1999).]

replicase genes. There are reports that the capsid protein can be translated as a polyprotein together with the replicase genes, but there are other reports that a subgenomic mRNA is produced, whose function is unknown.

This rabbit calicivirus was being studied by Australian scientists in a laboratory on an island offshore of the mainland when it suddenly appeared on the mainland and began to spread rapidly. The rapid spread is thought to have been aided by local farmers who were trying to control the rabbit population, a topic to which we will return in Chapter 6.

Hepatitis E Virus

Hepatitis E virus (HEV) was classified by the ICTV as a calicivirus until very recently, but has now been removed from this family and is currently unclassified. Its genome organization and size resemble that of caliciviruses (Fig. 3.8), but it probably represents yet another family of SRVs (small round viruses). It may have a 5'-terminal VPg like the caliciviruses, but its RNA genome has motifs suggesting that the virus encodes a capping enzyme, which would imply that the viral RNA is capped rather than

having a VPg. The virus-encoded protease, whose sequence has been deduced from the sequence of the viral genome, appears to be a papain-like protease rather than a 3C^{pro}-like enzyme, another difference from the caliciviruses.

HEV is one of several viruses that cause human hepatitis, in this case epidemic water-borne hepatitis. It is found in Asia, Africa, Southern Europe, and Mexico, where it causes thousands of cases of hepatitis each year (Fig. 3.9). A dendrogram of various geographical isolates of HEV illustrates that the New World strain has diverged significantly from the Asian isolates (Fig. 3.10). Thus, there is little or no circulation of virus between different geographic regions. The disease is severe but the fatality rate is low (<1%), with the prominent exception that the fatality rate in pregnant women can be 20%. There is no vaccine or treatment for the virus at present.

FAMILY ASTROVIRIDAE

Astroviruses constitute a recently described family of animal viruses. Some of these are human viruses that cause

TABLE 3.6   Astroviridae

Genus/members	Virus name abbreviation	Usual host(s)	Transmission host(s)	Disease	World distribution
Astrovirus					
Human astrovirus (5)	HAstV	Humans	Water-borne, oral/fecal	Gastroenteritis	Worldwide
Bovine astrovirus (2)	BAstV	Cattle	Water-borne, oral/fecal	Gastroenteritis	Worldwide
Duck hepatitis virus type 2	DAstV	Ducks	Water-borne, oral/fecal	Fatal hepatitis	Worldwide
Ovine astrovirus 1	OAstV	Sheep	Water-borne, oral/fecal	Gastroenteritis	Worldwide
Porcine astrovirus 1	PAstV	Swine	Water-borne, oral/fecal	Gastroenteritis	Worldwide

gastroenteritis, but astroviruses for cattle, pigs, sheep, and ducks are also known (Table 3.6). The name comes from the Greek word for star, from star-like structures on the surface of the virion. Unlike the human caliciviruses, the human astroviruses will grow in cultured cells, and progress on understanding their molecular biology has been more rapid. They are small viruses, 30 nm, with icosahedral symmetry, and a genome of 7 kb.

The replicase proteins of astroviruses are translated from the genomic RNA as two polyproteins (Fig. 3.11). The smaller translation product (1a) terminates at a stop codon; ribosomal frameshifting (Chapter 1) at a retrovirus-like slippery sequence allows read-through into a second reading frame (1b) to produce a longer polyprotein (1ab). A subgenomic mRNA is translated into the capsid protein of the virus.

Analysis of the sequence of astrovirus genomes suggests that they contain a serine protease (with serine in the active site) and an RNA polymerase related to other viral RNA polymerases, but there is no evidence for a helicase or a capping enzyme. It has been suggested that astroviruses contain a VPg, consistent with the lack of a capping enzyme, but no domain encoding a VPg has been identified.

There may be yet other small viruses that cause gastroenteritis in humans. Virus particles have been seen in stools of humans suffering from gastroenteritis that have not as yet been characterized and are referred to simply as SRVs. Some of these may be members of families not yet characterized. A summary of virus families known to contain viruses that cause gastroenteritis and acute diarrhea in humans and other vertebrates is shown in Table 3.7.

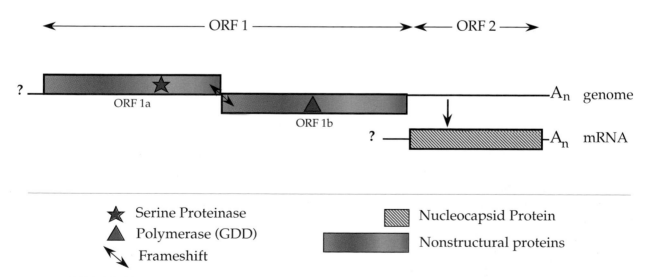

FIGURE 3.11   Genome organization of a human astrovirus. ORF 1a is in a different reading frame than ORF 1b. Ribosomal frameshifting (arrows) during translation is required to produce the ORF 1b protein.

TABLE 3.7   Viruses Causing Acute Diarrhea

Family	Virus[a]	Nucleic acid	Host range
Caliciviridae	Sapporo, Norwalk, feline calicivirus	ss(+)RNA	**Humans**, cattle, swine, chickens, dogs, cats
Astroviridae	Numerous astroviruses	ss(+) RNA	**Humans**, cattle, swine, cats, dogs, birds
Coronaviridae	PEDV, TGEV, and others	ss(+)RNA	Swine, cattle, foals, mice, rabbits, dogs, cats, turkeys, (humans?)[b]
	Numerous toroviruses	ss(+)RNA	Cattle, horses (goats, sheep, swine, rabbits, mice, humans?)
Flaviviridae	Pestivirus BVDV	ss(+)RNA	Cattle
Picornaviridae	Aichi, human parechovirus 1	ss(+)RNA	**Humans**
Paramyxoviridae	Canine distemper	ss(−)RNA	Dogs
	Newcastle disease		Chickens, fowl
Adenoviridae	Human Ad40,41	dsDNA	**Humans**
Reoviridae	Rotavirus	dsRNA	
	Group A		All mammals and birds
	Group B		Swine, cattle, sheep, rodents, **humans**
	Group C		Swine, ferrets, **humans**
	Group D, F, G		Birds
	Group E		Swine
Parvoviridae	Numerous parvoviruses	ssDNA	Cattle, cats, dogs, mink (humans?)

Source: Adapted from Granoff and Webster (1999, p. 442).

[a]*Abbreviations:* PEDV, porcine epidemic diarrhea virus; TGEV, transmissible gastroenteritis virus; BVDV, bovine viral diarrhea virus.

[b]The role of the listed viruses in causing diarrhea has not been proven for species listed in parentheses.

FAMILY TOGAVIRIDAE

The family Togaviridae contains two genera, genus *Alphavirus* and genus *Rubivirus*. The family name comes from the Latin word for cloak, and the name was given to them because they are enveloped. The 26 alphaviruses have a (+)RNA genome of about 12 kb, whereas rubella virus, the only member of the *Rubivirus* genus, has a genome of 10 kb. The genomes of alphaviruses and of rubella virus are organized in a similar fashion, as illustrated in Fig. 3.12. The virions of the two groups are also roughly similar in size (70 nm for alphaviruses, 50 nm for rubiviruses) and structure (icosahedral nucleocapsids surrounded by a lipoprotein envelope). Structures of alphaviruses are illustrated in Figs. 2.5, 2.14, and 2.21. However, although the two genera exhibit similarities, they are only distantly related. As a historical footnote, the flaviviruses, described after the togaviruses, were once classified as a genus within the Togaviridae. When sequences of flaviviruses were determined, however, they were found to be unrelated to alphaviruses and were placed into a new family.

Genus Alphavirus

The alphaviruses have a worldwide distribution. They get their name from the Greek letter α, because they were once known as the Group A arboviruses. Many cause important illnesses in humans, and information for a representative selection of these viruses is presented in Table 3.8. Because of their importance as disease agents and aided by the fact that alphaviruses grow well in cultured cells, this group of viruses has been well studied. The genomes of many of them have been sequenced in their entirety. All members of the genus are closely related and share extensive amino acid sequence identity. A dendrogram that illustrates their relationships is shown in Fig. 3.13. This dendrogram illustrates the interesting fact that during evolution of alphaviruses, there was a singular recombination event between Eastern equine encephalitis virus and a Sindbis-like virus to produce Western equine encephalitis virus. Recombination events in which the recombinant virus persists and prospers appear to be rare, but there is much evidence that recombination has played an important role in the evolution of viruses. The

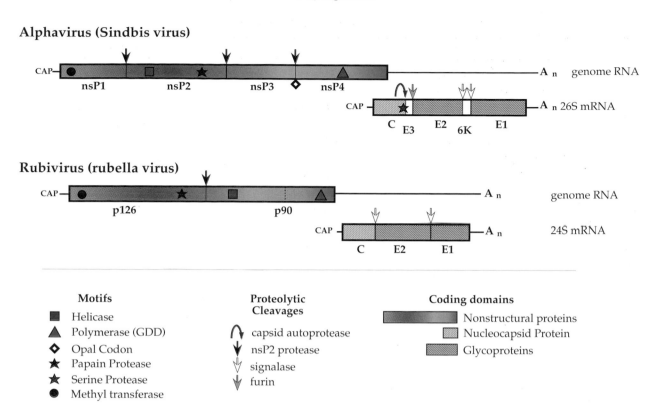

FIGURE 3.12 Genome organizations of the Togaviridae. A number of protein motifs are indicated, as well as the enzymes responsible for the proteolytic cleavages. The opal codon shown between nsP3 and nsP4 is leaky; read-through produces small amounts of nsP4. [Adapted from Strauss and Strauss (1994, Fig. 34).]

TABLE 3.8   Togaviridae

Genus/members	Virus name abbreviation	Usual hosts(s)	Transmission	Disease	World distribution
Alphavirus					
Sindbis	SINV	Mammals[a], birds	Mosquito-borne	Arthralgia, rash, fever	Old World
Semliki Forest	SFV	Mammals[a]	Mosquito-borne	Arthralgia, fever	Africa
Ross River, Barmah Forest	RRV, BFV	Mammals[a]	Mosquito-borne	Polyarthritis, fever, rash	Australasia
Ft. Morgan, Buggy Creek	FMV, —	Birds	Vectored by swallow bug	?	North America
Chikungunya, O'Nyong-nyong	CHIKV, ONNV	Humans	Mosquito-borne	Arthralgia, fever	Africa
Mayaro	MAYV	Mammals[a]	Mosquito-borne	Arthralgia, fever	South America
Eastern, Western, Venezuelan equine encephalitis	EEEV, WEEV, VEEV	Horses, birds, humans	Mosquito-borne	Encephalitis	Americas
Rubivirus					
Rubella	RUBV	Humans	No arthropod vector	Rash, congenital abnormalities	Americas, Europe

[a]Humans can be infected by these viruses, but humans are not the primary mammalian reservoir.

A.   Nonstructural proteins nsP2-nsP3   **B.**   Virion glycoproteins E2-E1

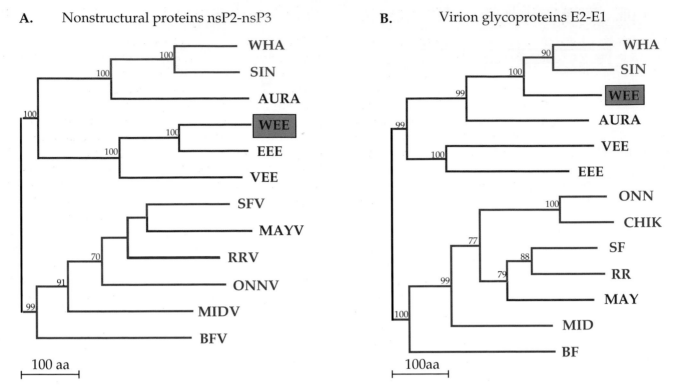

FIGURE 3.13   Phylogenetic trees of the alphaviruses, constructed using maximum parsimony and bootstrap analysis. The Old World viruses are shown in blue, the New World viruses in orange. The vertical distances are arbitrary, but the lengths of the horizontal branches indicate the number of amino acid substitutions along the branch and some bootstrap values are indicated at the nodes. (A) A tree derived from the amino acid sequences of nonstructural proteins nsP2 and the conserved domain of nsP3, comprising about 1100 amino acids. (B) A tree derived from the sequences of the virion envelope glycoproteins E2 and E1, or about 850 amino acids. The trees are very similar in architecture with the exception of the position of WEE (boxed), whose nonstructural proteins are most closely related to EEE, but whose structural proteins resemble SIN, indicating that a recombinational event has taken place during the evolution of WEE. Most virus abbreviations are found in Table 3.9; AURAV, Aura; MIDV, Middelburg; WHAV, Whataroa. (These trees were kindly provided by Scott Weaver.)

dendrogram also illustrates the fact that there have only been a limited number of transfers of alphaviruses between the Americas and the Old World. In fact, three transfers between the Americas and the Old World are sufficient to explain the dendrogram, and the majority of members of the three major lineages are restricted to either the Americas or to the Old World. This contrasts with most virus families, where individual viruses are often worldwide in distribution and evidence is abundant for the mixing of lineages between these two hemispheres.

Expression of the Genome

The alphavirus genome, which is capped and polyadenylated, is translated into a nonstructural polyprotein that is cleaved into four polypeptides by a viral protease (Fig. 3.12). Activities present in these proteins include a capping activity in the N-terminal protein (nsP1), helicase and papain-like protease activities in nsP2, and RNA poly-

merase in nsP4. A viral encoded capping activity appears to be required to cap the viral mRNAs (the genomic RNA and a subgenomic RNA) because replication occurs in the cytoplasm and the virus does not have access to cellular capping enzymes. The RNA helicase is needed to unwind the RNAs during replication and the protease to process the precursor polyprotein. The RNA polymerase is needed to synthesize viral RNA. All four nonstructural proteins are required to synthesize the viral RNA. Replication of the RNA and synthesis of a subgenomic mRNA follow the pattern illustrated schematically in Figure 1.9B.

Studies of the viral nonstructural protease have shown that the cleavages that process the polyprotein control viral RNA replication. During or shortly after translation, the full-length polyprotein precursor (called P1234) cleaves itself in *cis* to produce P123 and nsP4. These form an RNA synthetase, probably together with cellular proteins, that can make complementary (−)RNA from the genomic RNA template, but which cannot make (+)RNA efficiently.

Subsequent cleavage of P123 in *trans,* between nsP1 and nsP2, gives rise to a synthetase that can make both (+)RNA and (–)RNA. A second cleavage between nsP2 and nsP3 gives rise to a synthetase that can make only (+)RNA. Thus, (–)RNA templates are made early, but as infection proceeds and the concentration of protease builds up, *trans* cleavage occurs and (–)RNA synthesis is shut down (Fig. 3.14). This control mechanism probably evolved to make the infection process more efficient, but it also has the effect that the infected cell becomes resistant to superinfection by the same or a related virus because no (–)RNA templates can be made. The resistance to superinfection by the same or related viruses is called superinfection exclusion or homologous interference.

The rate of cleavage of the early synthetase that makes (–)RNA to convert it to one that can make (+)RNA is thought to be controlled in part by a leaky stop codon between nsP3 and nsP4 in most alphaviruses (Fig. 3.12). Termination at this codon produces P123, which can act in *trans* as a protease but cannot act as a synthetase because it lacks the nsP4 RNA polymerase. In addition to a more rapid buildup of protease that accelerates the rate of conversion to (+)RNA synthesis, this additional P123 and its cleavage products may serve to accelerate the rate of RNA synthesis. There is evidence from genetic studies that domains in nsP1 and nsP2, among others, are required for the recognition of viral promoters and the initiation of RNA synthesis, and additional helicase activity could speed up the rate of RNA synthesis.

During infection by alphaviruses, a subgenomic mRNA is produced that serves as the message for the production of the structural proteins of the virus, which consist of a capsid protein and two glycoproteins. The 4.1-kb subgenomic RNA is transcribed by the viral replicase from the (–)RNA template using an internal promoter of 24 nucleotides. The structural proteins are translated as a polyprotein and cleaved by a combination of viral and cellular enzymes. The capsid protein is itself a serine autoprotease that cleaves itself from the N terminus of the nascent polyprotein. It has a fold that is similar to that of chymotrypsin (Fig. 2.15), suggesting that it was derived from a cellular serine protease during evolution of the virus. After release of the capsid protein, N-terminal signal sequences and internal signal sequences in the glycoprotein polyprotein precursor lead to its insertion into the endoplasmic reticulum. This precursor is cleaved by signalase, a cellular enzyme that resides in the lumen of the endoplasmic reticulum, to produce glycoprotein PE2 (a precursor to glycoprotein E2), 6K (a small hydrophobic peptide located between E2 and E1), and glycoprotein E1. PE2 and E1 quickly form a heterodimer. During transport of the heterodimer to the cell plasma membrane, PE2 is cleaved by another cellular enzyme, furin or a furin-like enzyme, to form E2 and E3. E3 is a small glycoprotein that in most alphaviruses is lost into the culture fluid. Cleavage by furin is required to activate the virus for infection of a cell.

Viral Promoters

The replication of alphaviral RNA requires the recognition of specific promoters in the viral RNA by the viral RNA synthetase. These promoters act in *cis*, that is, they must be present in the RNA to be used as a template, and both viral and cellular proteins may be involved in the recognition of these promoters. The best understood of these promoters is that for the production of the subgenomic mRNA for the structural proteins. The basal promoter consists of 24 nucleotides, of which 19 are upstream of the transcription start site and 5 are copied into the subgenomic RNA. The subgenomic promoter can be placed in front of any RNA sequence and the viral synthetase will use it to synthesize a subgenomic mRNA. This property of the promoter has made alphaviruses useful as expression vectors (Chapter 9).

The promoters for synthesis of full-length genomic RNA and the antigenomic RNA template are less well understood. A sequence element at the 3′ end of the (+)RNA genome is required for (–)RNA synthesis, and an element at the 3′ end of the minus strand, which can form a stem-loop structure, is required for genomic RNA synthesis from the antigenomic template. There is also a requirement for

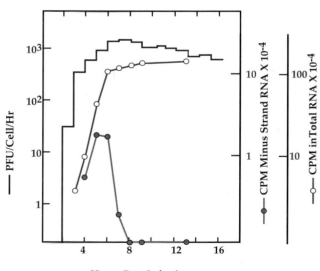

FIGURE 3.14 Growth curve of Sindbis virus infection in chicken cells at 30°C. At the times shown, the medium was harvested and replaced with fresh medium, and the titer was determined. The red line shows release of infectious virus into the culture fluid. For determining the rate of RNA synthesis, cells infected as for virus assay were pulsed with radioactive uridine for 1 hr at the times shown. Monolayers were harvested and incorporation into total RNA (green line) and minus strand RNA (blue line) was determined. [Adapted from Strauss and Strauss (1994, Fig. 5).]

an element of 51 nucleotides present in the sequence of the nsP1 gene for efficient RNA replication.

Assembly of Progeny Virions

The structure of alphaviruses has been described in Chapter 2. Virions mature when a preassembled nucleocapsid consisting of the genomic RNA and 240 copies of capsid protein buds through the cell plasma membrane to acquire a lipoprotein envelope containing 240 copies of the E1–E2 heterodimer (Fig. 2.21). The assembled virion has icosahedral symmetry and a diameter of 70 nm (Figs. 2.5 and 2.14).

Alphaviruses Are Arboviruses

The alphaviruses are arboviruses (*ar*thropod-*bo*rne animal viruses) and were once referred to as the Group A arboviruses. In nature, they alternate between replication in arthropod vectors, usually mosquitoes, and higher vertebrates. A mosquito may become infected on taking a blood meal from a viremic vertebrate, which can have 10^8 or more infectious particles per milliliter of blood. The infection in the mosquito, which is asymptomatic and lifelong, begins in the midgut and spreads to the salivary glands, as illustrated in Fig. 3.15. After infection of the salivary glands, the mosquito can transmit the virus to a new vertebrate host when it next takes a blood meal. Infection in the vertebrate begins in the tissues surrounding the bite or in regional lymph nodes, but then spreads to other organs. The infection is usually

self- limited and the vertebrate is capable of infecting mosquitoes for only a brief time, e.g. after viremia is established but before an immune response limits circulating virus. The necessity to alternate between two such different hosts has constrained the evolution of arboviruses— changes that adapt the virus to one host or that are neutral in one host are often deleterious in the alternate host. Thus, the evolutionary pressures on arboviruses are different from those on viruses such as poliovirus, which infects only primates.

Different alphaviruses infect different spectra of mosquitoes and vertebrates in nature. It is useful to distinguish between reservoir hosts in which the virus is maintained in nature and dead-end hosts in which infection normally does not lead to continuity of the infection cycle. We can also distinguish between enzootic cycles, in which the virus is continuously maintained in nature and which may or may not result in disease in the enzootic host, and epizootic cycles, in which the virus breaks out and causes epidemics of disease that may die out with time. Two types of natural transmission cycles are illustrated in Fig. 3.16. In Fig. 3.16A, a simple transmission cycle is illustrated, such as that of urban yellow fever infection of humans (see the section on flaviviruses below). In this cycle, humans are the only vertebrate hosts and the virus alternates between infection of a human and infection of the mosquito vector *Aedes aegypti*. Figure 3.16B illustrates a complex transmission pattern, using as an example the transmission of Eastern equine encephalitis virus in North America. This virus has a vertebrate reservoir consisting primarily of migratory songbirds and is transmitted by the mosquito, *Culiseta melanura*, a

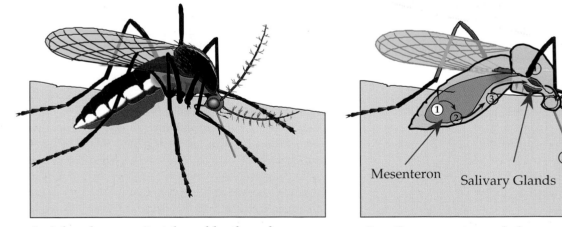

A. A female mosquito takes a blood meal

B. Cutaway view of the mosquito showing steps in the replication and transmission of an arbovirus.

FIGURE 3.15 Sequential steps necessary for a mosquito to transmit an arbovirus. (1) A female mosquito ingests an infectious blood meal and virus enters the mesenteron. (2) Virus infects and multiplies in mesenteronal epithelial cells. (3) Virus is released across the basal membrane of the epithelial cells and replicates in other tissues. (4) Virus infects salivary glands. (5) Virus is released from the epithelial cells of the salivary glands and is transmitted in the saliva during feeding. [From Monath (1988, p. 91).]

A. Simple Transmission Cycle

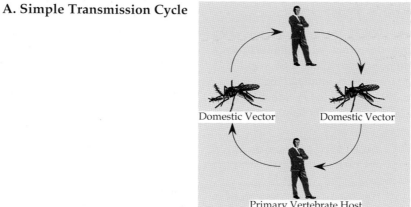

B. Complex transmission cycle

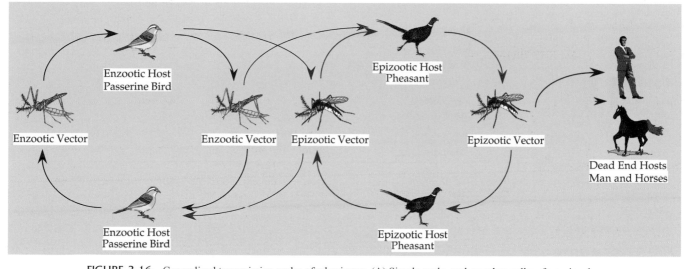

FIGURE 3.16   Generalized transmission cycles of arboviruses. (A) Simple cycle, such as urban yellow fever, involving a single vector (*Aedes aegypti* mosquitoes) and a single vertebrate host (man). (B) Complex cycle, such as that for Eastern equine encephalitis, where the virus is maintained in an enzootic host (passerine birds) with an enzootic vector (*Culiseta melanura*), but can enter an epizootic vector (another insect) and be transmitted to epizootic hosts, and tangentially to dead-end hosts like man. An intermediate type of cycle is illustrated in Fig. 3.48 for Colorado tick fever. [From Monath (1988, p. 129).]

common inhabitant of freshwater swamps in eastern North America. However, the virus is capable of infecting other mosquitoes and has even been isolated from naturally infected chicken mites. It can also infect mammals, including humans. On occasion, the virus breaks out of its enzootic cycle to cause epidemics of disease in pheasants, transmitted and maintained by an epizootic vector mosquito. Either enzootic or epizootic vectors are capable of infecting humans or domestic animals if they invade the areas in which these mosquitoes are present, but these hosts are usually dead-end hosts and do not further spread the virus.

Most alphaviruses are capable of infecting both mammals and birds, and the nature of the vertebrate reservoir depends on the virus or even the strain of virus, which may differ by geographic location. Thus, for example,

Sindbis virus in nature is normally maintained in birds, which are its usual vertebrate reservoir. However, the virus is capable of infecting mammals, including humans, and has also been isolated from amphibians and reptiles. Numerous species of mosquitoes form its insect reservoir, but it has also been isolated from other hematophogous arthropods, including mites. In contrast, Ross River virus is maintained in small marsupial mammals in Australia and does not appear to use birds as hosts.

Arboviruses that are transmitted by mosquitoes, ticks, sandflies, or other blood-sucking arthropods are known from several families of viruses. A selection of arboviruses from three virus families, most of which cause human disease, are listed in Table 3.9, together with the diseases they cause.

Encephalitic Alphaviruses

Eastern (EEE), Western (WEE), and Venezuelan equine encephalitis (VEE) viruses, three New World alphaviruses, cause fatal encephalitis in horses. WEE and EEE regularly cause encephalitis in humans as well, although the number of cases is small (Table 3.9 and Fig. 3.17). Most human infections by WEE or EEE are inapparent or result in febrile illness that is usually mild. Encephalitis, which occurs in only 1 of 1000 humans infected by WEE and at a somewhat higher rate for EEE, is a complication that arises when the virus manages to pass the blood–brain barrier and infects neurons in the CNS. When encephalitis does develop, it is fatal about half the time in EEE infection, and survivors have neurological deficits. The encephalitis produced by WEE is less severe, but it is fatal in 5–10% of

TABLE 3.9   Representative Arboviruses That Cause Human Disease

		Predominant disease manifestations[a]			
	Nonfatal systemic febrile illness	Encephalitis		Hemorrhagic fever (HF)	
Family/virus		Frequency[b]	% Mortality[c]	Frequency[b]	% Mortality[d]
Togaviridae					
Chikungunya	Most cases, E			Rare, E	Rare
Mayaro	Most cases, E				
O'Nyong-nyong	Most cases, E				
Ross River	Most cases, E				
Sindbis	Most cases, E				
EEE	Most cases	Rare	50–70		
VEE	Most cases, E	Rare	0.1–20[e]		
WEE	Most cases	Rare	5–10		
Flaviviridae					
Dengue (1–4)	Most cases, E			Rare, E	3–12
West NileE	NonE	Rare			
Japanese encephalitis		<1%	30–40[f]		
Kyasanur Forest		NonE	5	NonE	5
Murray Valley		E	20–70		
Rocio	E	13			
St. Louis encephalitis		E	4–20		
Tick-borne encephalitis					
Eastern		Rare	30		
Central European		Rare	1–10		
Omsk hemorrhagic fever				NonE	1–2
Yellow fever				Most cases	5–20
Bunyaviridae					
Bunyamwera	NonE				
Germiston	NonE				
Sandfly fever	E				
Rift Valley fever	E			NonE	1–5
California encephalitis		NonE	1		
Crimean hemorrhagic fever				NonE	15–20

[a]In addition to the disease manifestations listed, most viruses in this table can cause mild febrile illness; some viruses are endemic (nonE), but others cause occasional outbreaks or epidemics (E).
[b]Frequency relates to the relative number of cases exhibiting encephalitis or HF relative to the total number of infections. This number can be difficult to estimate, since only the most seriously ill (for example, hospital patients in an epidemic) may be reported as infections.
[c]Percent mortality is the percent of those *with encephalitis* who succumb.
[d]Percent mortality is the percent of those *with HF* who succumb.
[e]Mortality in children is at the high end of the range given.
[f]Mortality generally lower in children.

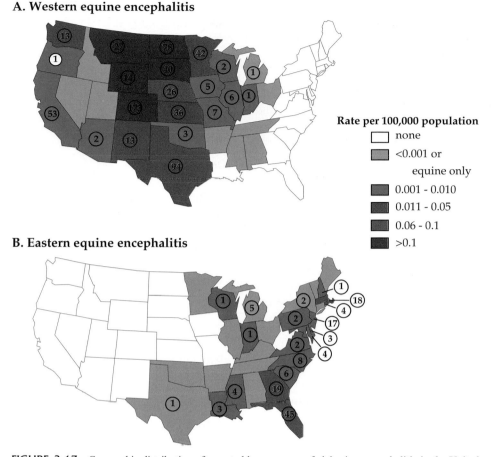

A. Western equine encephalitis

B. Eastern equine encephalitis

Rate per 100,000 population

- none
- <0.001 or equine only
- 0.001 - 0.010
- 0.011 - 0.05
- 0.06 - 0.1
- >0.1

FIGURE 3.17 Geographic distribution of reported human cases of alphavirus encephalitis in the United States from 1964 to 1993. Colors indicate the rate in cases per 100,000 population by state and the actual numbers of cases are shown. Note there were two human cases of WEE in 1994, but none have been reported since then. Reported cases of EEE since 1993 (no state locations given) are 1 in 1994, 4 in 1995, 5 in 1996, and 14 in 1997 (of which 12 were in the South, 1 in New England, and 1 in the upper Midwest). [Adapted from Fields *et al.* (1996, p. 875, using data from Tsai, 1991) and additional data from *MMWR*, Summary of Notifiable Diseases in the United States for 1998. *MMWR*, **47**, No. 53 (1999) pp. 23–25.]

cases. The tendency to develop encephalitis after infection is age dependent, and young humans are much more susceptible than adults. These findings correlate with laboratory experiments in mice, in which many different alphaviruses cause age-dependent encephalitis.

WEE and EEE are both present in the United States and cause small outbreaks of encephalitis, often only a few cases per year (Fig. 3.17). They are less of a problem than formerly, probably because the widespread use of insect repellents and the adoption of air conditioning and window screens has resulted in fewer bites by mosquitoes, especially night flying mosquitoes, and because fewer horses are used in farming. Of particular interest is the fact that no case of WEE encephalitis has been reported since 1994. No vaccines for these viruses are available for widespread use, although experimental vaccines exist that

are given to laboratory personnel who work with the viruses. In Central and South America, WEE and EEE are more of a problem because the horse is still in widespread use as a farm animal and window screens are less common.

VEE is also a major problem in Latin America because of its ability to cause fatal encephalitis in horses. The virus is maintained in an enzootic cycle involving small mammals, but epizootic outbreaks in horses occur that lead to widespread epidemic disease. VEE infection often results in a febrile illness in man that can be debilitating. It rarely causes fatal infection or encephalitis in humans, however, although an outbreak in 1995 resulted in a number of cases of encephalitis in children. Both inactivated virus vaccines and attenuated virus vaccines are in widespread use for the immunization of horses against VEE.

Alphaviruses That Cause Polyarthritis

Several alphaviruses cause epidemic polyarthritis in man. The best known of these is Ross River virus, which is widespread in Australia, but strains of Sindbis virus found in Northern Europe (called Ockelbo or Karelian fever virus) and Mayaro virus of South America also cause polyarthritis. The arthritis is characterized by pain, often accompanied by frank swelling, in the small joints of the hands and feet and in the knees. Arthritis (joint inflammation, from *arth* = joint and *itis* = inflammation) or arthralgia (joint pain, from *arth* and *algia* = pain) can last for almost a year with relapses of severe arthritis being common during this period.

Of particular interest was a wide-ranging epidemic of Ross River polyarthritis that swept through the South Pacific some years ago. The epidemic began when a single viremic traveler from Australia landed in Fiji. The epidemic began near the airport and eventually spread throughout the island. From there it jumped to other islands having air contact with Fiji. During this epidemic, it is believed that humans were the primary or only vertebrate host, with the disease being transmitted from mosquito to human to mosquito without the intervention of another animal reservoir (the cycle illustrated in Fig. 3.16A). This epidemic was explosive (most of the people on the affected islands contracted the disease) but eventually burned itself out because humans had become immune and the virus failed to establish an endemic cycle in other animals in the region.

Other Alphaviruses

Other alphaviruses cause disease characterized by headache, fever, rash, and arthralgia, and only a few examples will be cited here. The prototype virus is Sindbis virus, named after the town of Sindbis, Egypt, where it was first isolated in 1953 from mosquitoes. This virus has been widely studied as a model for alphavirus replication. Most Sindbis strains cause no human illness but some strains cause disease characterized by fever, rash, and arthralgia, and others, as described above, cause polyarthritis. Semliki Forest virus, named after the Semliki Forest in Uganda, has also been extensively characterized as a model system. Most strains cause no human illness, but strains from central Africa cause a disease characterized by exceptionally severe headache, fever, and rash. One known case of fatal human encephalitis caused by this virus occurred in a laboratory worker. Chikungunya virus causes large epidemics of arboviral illness in Africa and Asia whose most characteristic symptom, in addition to rash, is a very painful arthralgia and, frequently, arthritis. In urban settings the virus is transmitted by *A. aegypti*, the only known case of epidemic alphavirus transmission by this urban mosquito, and the virus may have been more widely distributed previously. O'Nyong-nyong virus is an African virus, closely related to Chikungunya, that caused an epidemic,

starting in the late 1950s, resulting in 2 million cases of disease similar to that caused by Chikungunya. The virus subsequently almost totally disappeared, with only sporadic cases reported, until 1996, when another, smaller epidemic occurred in south-central Uganda. In these epidemics the virus was transmitted by *Anopheles funestrus* and *Anopheles gambiae*, mosquitoes that are major vectors of malaria, and these epidemics represent the only known cases of epidemic transmission of an alphavirus by Anopheline mosquitoes.

Rubella Virus

Rubella virus is less well understood than alphaviruses because it grows very poorly in cultured cells and its genome possesses an extraordinarily high GC content (70%), which retarded efforts to sequence and express the viral genome. The complete sequences of several strains are now known and detailed molecular studies are under way. The genome is approximately 10 kb in size and is expressed similarly to the alphavirus genome: The genomic RNA is translated into a polyprotein cleaved by a papain-like protease into two pieces, and a subgenomic mRNA is translated into structural proteins consisting of a capsid protein and two envelope glycoproteins (Fig. 3.12).

Rubella virus infects only humans and there is no other reservoir. Infection is by person-to-person contact, primarily through aerosols. It causes a relatively benign illness, sometimes called German measles, with a characteristic rash and is (or was) one of the typical childhood diseases. However, infection of a pregnant woman in the first trimester of pregnancy can have devastating effects on the developing fetus. The virus sets up a long-lived infection in the fetus that often causes developmental abnormalities resulting in children with severe handicaps (congenital rubella syndrome). An attenuated virus vaccine against rubella has been developed that is now routinely given to children as part of mumps–measles–rubella vaccination (MMR). Since the vaccine was introduced, there has been a drastic reduction in the number of cases of rubella in the United States (Fig. 3.18) and other developed countries.

Because the postnatal disease caused by rubella virus is trivial, the purpose of the rubella vaccine is to protect against future birth defects rather than to protect the individual vaccinated. This raises interesting ethical questions about its use. In some societies, only females were vaccinated, since they would want to protect their future children from the effects of rubella. However, because males remained susceptible to the virus, it continued to circulate in the population and rubella-caused birth defects continued to occur. To protect against this, the only solution is to vaccinate the entire population so as to eliminate the virus from the society.

The rubella vaccine has now been in use for many years and is generally safe and effective when given to children.

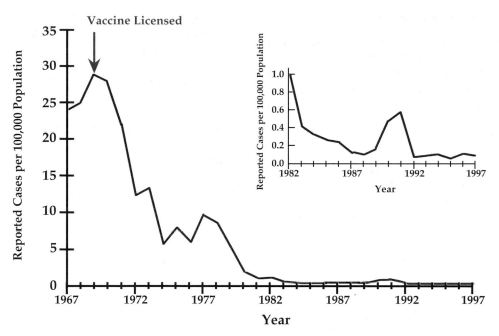

FIGURE 3.18   Incidence of rubella virus, by year, in the United States. [From Summary of Notifiable Diseases in the United States for 1997 (*MMWR*), Vol. 46, No. 54, (1998).]

The present vaccine has a high incidence of side effects in adults, however, especially arthralgia and arthritis, and is seldom administered to adults. Thus, an adult woman who becomes pregnant but is not immune to rubella is at risk in the event of an epidemic, and the need exists to improve the vaccine. Current efforts to understand the molecular biology of the virus in more detail will hopefully lead to the development of a better vaccine.

FAMILY FLAVIVIRIDAE

The Flaviviridae are so-called from the prototype virus of the genus *Flavivirus*, yellow fever virus, *flavus* being the Latin word for yellow. The Flaviviridae are divided into three genera, the genus *Flavivirus*, the genus *Pestivirus*, and the genus *Hepacivirus*. A partial listing of viruses in the three genera is given in Table 3.10. In the discussion below, the term flavivirus refers only to members of the genus *Flavivirus*, unless otherwise specified.

The genome organizations of members of the three genera are shown in Fig. 3.19. The genomes of the three genera are similar in size (11 kb for flaviviruses, 12.5 kb for pestiviruses, 9.4 kb for hepaciviruses) and organization. These viruses, like the picornaviruses, have a genome that contains only a single ORF. This ORF is translated into a long polyprotein that is processed by cleavage into 10 or more polypeptides. Processing of the precursor polyprotein is complicated, as is clear from the figure. Cleavage is effected by a combination of one or two viral encoded proteases and two

or more cellular proteases. The structural proteins are encoded in the 5′-terminal region of the genome (like picornaviruses). However, all members of the Flaviviridae are enveloped, unlike the picornaviruses, and the structural proteins consist of a nucleocapsid protein and two or three envelope glycoproteins. Cellular proteases make the cleavages that separate the glycoproteins, but the cleavages in the nonstructural region of the polyprotein, which is required for RNA replication, are made by one or two virus-encoded proteases. Even so, cellular signalase makes at least one of the cleavages in the nonstructural domain of flaviviruses. The cleavage pathways in this genus are described in detail below.

All members of the Flaviviridae encode a serine protease with a catalytic triad consisting of serine, histidine, and aspartic acid. The protease resides in the nonstructural region called NS3, just upstream of a helicase. The crystal structure of the hepatitis C virus (HCV) protease has been solved to atomic resolution and it possesses a fold similar to chymotrypsin, as is the case for other viral serine proteases whose structures have been solved. The enzyme is interesting in that a second polypeptide is required for activity, NS2B in flaviviruses and NS4A in HCV. From the atomic structure of the HCV protease complexed with the region of NS4A required for activity, it is clear that NS4A forms an integral part of the folded protease. Thus, it is puzzling that the protease consists of two cleaved products rather than one continuous polypeptide chain.

Flaviviruses encode only the NS3 protease, but pestiviruses and hepaciviruses also encode a second protease. In HCV, the second protease is known to be a metallopro-

TABLE 3.10   Flaviviridae

Genus/members	Virus name abbreviation	Usual hosts(s)	Transmission	Disease	World distribution
Flavivirus					
Dengue fever (types 1–4)	DENV	Humans	Mosquito-borne	Dengue fever, shock, hemorrhage	Worldwide
Yellow fever	YFV	Primates[a]	Mosquito-borne	Hemorrhage, liver destruction	Africa, Americas
Japanese encephalitis	JEV	Mammals[a], especially swine	Mosquito-borne	Encephalitis	Widespread in Asia
St. Louis encephalitis	SLEV	Mammals[a], birds	Mosquito-borne	Encephalitis	North America
Murray Valley encephalitis	MVEV	Mammals[a], birds	Mosquito-borne	Encephalitis	Australia
Tick-borne encephalitis	TBEV	Mammals[a]	Tick-borne	Encephalitis	Europe, Asia
Pestivirus					
Classical swine fever	CSFV	Swine	Contact	Fever, acute gastroenteritis	Europe, Americas
Bovine viral diarrhea	BVDV	Cattle	Contact	Usually none[b]	Worldwide
Hepacivirus					
Hepatitis C	HCV	Humans	Parenteral, transfusion	Hepatitis, liver cancer	Worldwide

[a]Including humans.
[b]Calves infected *in utero* develop persistent infections that can lead to mucosal disease.

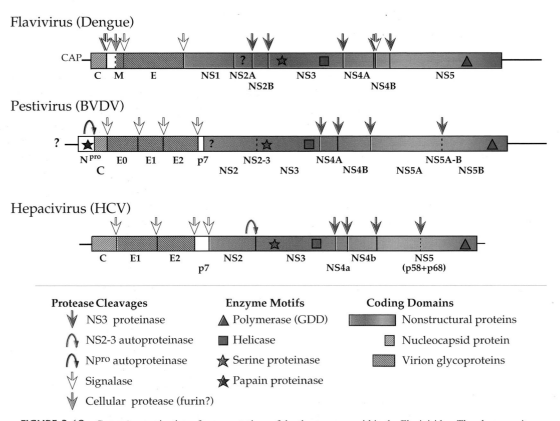

FIGURE 3.19   Genome organization of representatives of the three genera within the Flaviviridae. The cleavage sites indicated with dashed lines have not been precisely localized. [Data for this figure came from Chambers *et al.* (1990), Bartenschlager and Lohmann (2000), and Meyers and Thiel (1996).]

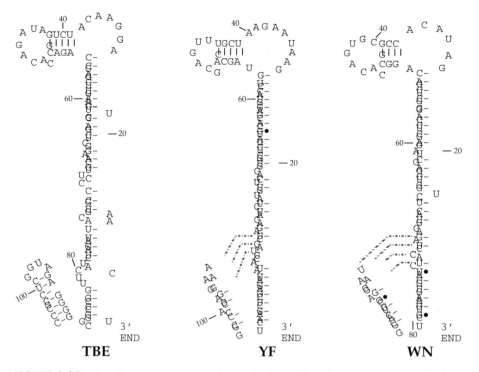

FIGURE 3.20 Two-dimensional structures that can be formed from the sequences at the 3′ ends of flavivirus RNAs. Examples are shown for three flaviviruses: tick-borne encephalitis virus (TBE), yellow fever virus (YF), and West Nile virus (WN). Although there is little sequence conservation, the structures are very similar and have similar predicted thermodynamic stabilities. Nucleotides are numbered from the 3′ terminus. It is thought that YF and WN form a pseudo-knot by interactions between the small 5′ loop and a disordered region of the main stem, as indicated with the dashed lines. [Adapted from Shi *et al.* (1996) and Rauscher *et al.* (1997).]

tease. This is the only example known of a metalloprotease in viruses.

Flaviviruses have capped genomes whose translation is cap-dependent. In contrast, the hepacivirus and pestivirus genomes are not capped and have an IRES in the 5′ nontranslated region. Members of Flaviviridae do not have a poly(A) tail at the 3′ end of the RNA. A stable stem-loop structure present at the 3′ end of the Flavivirus genome is illustrated in Fig. 3.20. This structure is required for replication of the genomic RNA and for its stability. No nucleotide or amino acid sequence identity can be detected between members of different genera except for isolated motifs that are signatures of various enzymatic functions.

Viruses in the family are enveloped. They mature at intracytoplasmic membranes rather than at the plasma membrane.

Genus Flavivirus

There are about 70 known flaviviruses, of which a representative sample is listed in Tables 3.9 and 3.10. All members of the genus are closely related, as illustrated by the dendrogram in Fig. 3.21. They share significant amino acid sequence identity in their proteins, which results in serolog-

ical cross reactivity. Historically, members of this genus were assigned to it on the basis of these cross reactions. Most are arthropod-borne, and they were once referred to as Group B arboviruses. They can be divided into three major groups based on the vector utilized: the mosquito-borne group (which includes yellow fever, the dengue complex, and the Japanese encephalitis complex), the tick-borne encephalitis group (the TBE complex), and a group that lacks an arthropod vector. The last are of limited medical importance and will not be considered here. Notice that in the phylogenetic tree in Fig. 3.21, the tick-borne viruses and the mosquito-borne viruses belong to different lineages. Thus, the viruses are adapted to a tick vector or to a mosquito vector, and interchange of vectors does not occur.

Expression of the Viral Genome

The genome organization of a typical flavivirus is illustrated in Fig. 3.19. The processing of the long polyprotein produced from the genome is complicated and is illustrated in Fig. 3.22 as an example of complex processing events that can occur in viral polyproteins associated with lipid bilayers. The nucleocapsid protein is 5′ terminal in the genome and is removed from the precursor polyprotein by

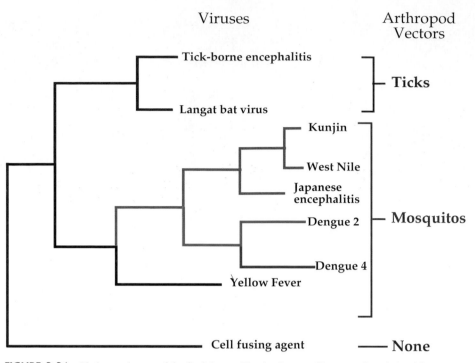

FIGURE 3.21 Phylogenetic tree of the flaviviruses. The dendrogram illustrates the relationships among flaviviruses, based on the nucleotide sequences encoding the NS5 protein. The cell-fusing agent is a distantly related flavivirus that was found as a contaminant in mosquito cell cultures. Its natural host range and distribution are unknown. [From Marin *et al.* (1995) reported in Strauss and Strauss (1996, p. 115).]

the viral NS2B/NS3 protease. Two envelope proteins, prM (precursor to M) and E (envelope), follow. Both are anchored in the endoplasmic reticulum by C-terminal membrane-spanning domains and are usually, but not always, glycoproteins. A series of internal signal sequences is responsible for the multiple insertion events required to insert prM, E, and the following protein, NS1 into the endoplasmic reticulum. After separation of these three proteins by signalase, prM and E form a heterodimer. prM is cleaved to M by furin during transport of the heterodimer or during virus assembly. After cleavage, E forms a homodimer. These events are shown schematically in Fig. 3.23. If cleavage of prM does not occur, an immature form of the virion is produced that is not infectious.

Following the E protein is NS1 (NS for nonstructural). NS1 is a glycoprotein and is required for RNA replication (how is an interesting question since it is external to the cell). Cleavage at its C terminus is by an unknown cellular protease. Although lacking a C-terminal anchor, some fraction of it remains cell associated. Next are two hydrophobic polypeptides called NS2A and NS2B. These proteins are cleaved by the viral NS2B/NS3 protease. They are associated with membranes and may serve to anchor parts of the replication machinery to internal membranes in the cell. NS2B forms a complex with NS3 that activates the serine protease, which cleaves many bonds in the polyprotein. It

also has at least two other activities—the middle domain of NS3 is a helicase, required for RNA replication, and the C-terminal domain has RNA triphosphatase activity, an activity that is required for the capping of the viral genome.

NS4A and NS4B are hydrophobic polypeptides that are associated with membranes. They may function in assembly of the viral replicase on intracellular membranes. Both the viral NS2B/NS3 protease and cellular signalase are required to produce the final cleaved products.

NS5 is the viral RNA polymerase. It also has methyltransferase activity required for capping of the viral genome. Thus, the capping activity is divided between proteins NS3 and NS5. NS5 appears to be a soluble cytoplasmic protein that associates with membranes through association with other viral peptides.

RNA replication is associated with the nuclear membrane. The composition of the replicase complex is not understood but is assumed to consist of many (most?) of the viral nonstructural proteins with associated cellular proteins.

Structure of the Virion

The nucleocapsid is thought to be icosahedral in symmetry, perhaps having a triangulation number of 3. The structure of E has been solved to atomic resolution for tick-borne encephalitis virus (Fig. 3.24). Unlike the glycopro-

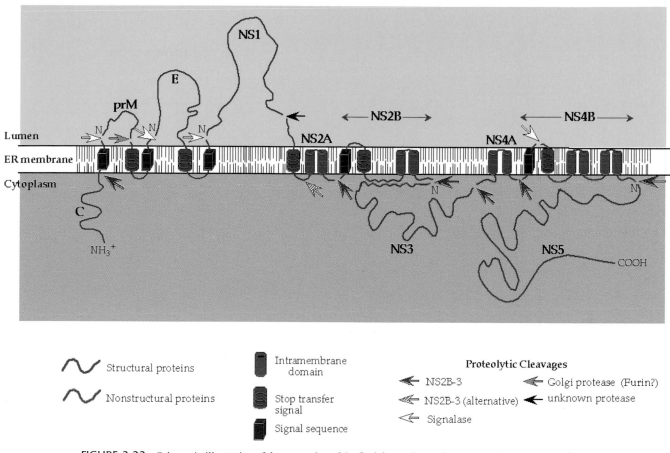

FIGURE 3.22 Schematic illustration of the processing of the flavivirus polyprotein precursor into the structural proteins (blue) and the nonstructural proteins (green). As described in the text, some of the cleavages are cotranslational, while others are delayed. The proteases responsible for the various cleavages are colored as shown in the key and the amino termini of the major proteins are labeled "N." The cellular enzyme that cleaves following NS1, the first nonstructural protein, has not been identified. An alternative site of cleavage within NS2A is shown that might lead to an anchored form of NS1. NS2B and NS3 form a complex that involves the central 40 amino acids of NS2B and is required for expression of the proteolytic activity of NS3. This interaction also ties NS3 to the membrane. The orientation of NS4A and NS4B within the membrane has not been determined, but this model is consistent with the sequences of these peptides. [Redrawn from Strauss and Strauss (1996).]

teins of other enveloped viruses, the homodimer lies flat along the membrane rather than projecting upward as a spike. Thus, the surface of the flavivirus lacks projecting spikes and is relatively smooth (Fig. 2.5). This also has the effect that the diameter of the flavivirion (about 50 nm) is less than that of many enveloped viruses. For example, the alphaviruses, which are otherwise fairly comparable to flaviviruses, have a diameter of 70 nm). Very recent studies have shown that E of flaviviruses and E1 of alphaviruses have the same structure, and thus were derived from the same ancestral protein. Although the protein structures are the same, they are used in somewhat different ways to construct the protein shell external to the lipid bilayer, as is clear from a comparison of the cryoelectron reconstructions of alphaviruses and flaviviruses in Fig. 2.5.

Flaviviruses mature at intracellular membranes. Budding figures have been described only rarely and assembly may

be associated with the complex processing of the viral polyprotein.

Diseases Caused by Flaviviruses

Many flaviviruses are important pathogens of humans. Different viruses may cause encephalitis, hemorrhagic fever with shock, fulminant liver failure, or disease characterized by fever and rash. Several important viruses and their diseases are listed in Tables 3.9 and 3.10. Many of these viruses are individually described below.

Yellow Fever Virus

The type flavivirus is yellow fever virus (YFV), once greatly feared and still capable of causing large epidemics. The virus is viscerotropic in primates, the only natural hosts for it. The growth of the virus in the liver, a major target

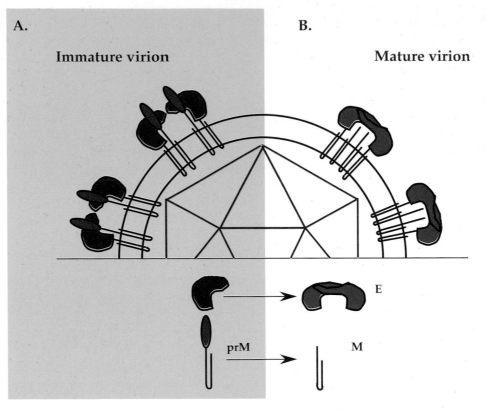

FIGURE 3.23 Schematic representation of the configuration of membrane proteins of mature and immature flavivirions, with the heterodimers of prM and E on the left and homodimers of E following cleavage of prM on the right. [Adapted from Heinz *et al.* (1994).]

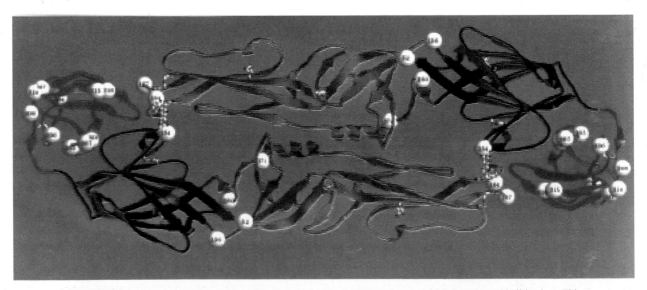

FIGURE 3.24 Crystal structure (ribbon diagram) of the dimer of the E protein of tick-borne encephalitis virus. This is a top view, looking down onto the surface of the virion. Numbered sites are those in which mutations alter the virulence of the virus. [From Rey *et al.* (1995).]

organ, causes the major symptoms of disease and the symptoms from which the name of the virus derives, jaundice following destruction of liver cells. The virus also replicates in other organs, such as kidney and heart, and causes hemorrhaging. Illness is accompanied by high fever. Death occurs in 20–50% of serious infections, usually on days 7–10 of illness and usually as a result of extensive liver necrosis.

YFV is present today in Africa and Latin America. It originated in Africa and spread to the Americas with European colonization and the introduction of slaves. The virus is maintained in two different cycles. In an endemic or sylvan cycle, it is maintained in *Aedes africanus* and other *Aedes* mosquitoes in Africa and in *Haemogogus* mosquitoes in the Americas. Monkeys form the vertebrate reservoir. In this cycle, forest workers and other humans who enter deep forests are at risk. Infection of humans can lead to the establishment of an epidemic or urban cycle in which the virus is transmitted by the mosquito *A. aegypti* and man is the vertebrate reservoir. In this cycle, all urban dwellers are at risk. *Aedes aegypti* is a commensal of man, breeding around human habitation. It is widespread in the warmer regions of the world, including the southern United States, Central America and the Caribbean, large regions of South America, sub-Saharan Africa, the Indian subcontinent, southeast Asia, Indonesia, and northern Australia.

In the 1800s, YFV was continuously epidemic in the Caribbean region, where it had a pronounced influence on the development and settlement of the Americas by the Europeans. Caucasians and Native Americans are very sensitive to yellow fever, usually suffering a serious illness with a high death rate. Black Africans, who were brought as slaves to the New World to replace Native American slaves who had died in large numbers from European diseases, in general suffer less severe disease following yellow fever infection, presumably having been selected for partial resistence by millenia of coexistence with the virus. Their relative resistance to yellow fever resulted in the importation of even more black slaves into yellow fever zones. The high death rate among French soldiers sent to the Caribbean region to control black slaves was probably responsible for the decision by Napoleon to abandon the Louisiana territory by selling it to the United States. The high death rate among French engineers and workers in the 1880s under de Lesseps, who had previously supervised the construction of the Suez Canal, led to the abandonment of the attempt by the French to build a canal through Panama. The Panama Canal through Panama was built by the United States only after yellow fever was controlled.

From its focus in the Caribbean, yellow fever regularly spread to port cities in the southern and southeastern United States and as far north as Philadelphia, New York, and Boston. Epidemic yellow fever even reached London. The virus also spread up the Mississippi river from New Orleans. The virus was transported from its focus in the Caribbean by ships, which carried freshwater in which mosquitoes could breed. If there was yellow fever on the ship, the disease was maintained and could be transmitted by the mosquitoes or by infected individuals to ports at which the ships called. Yellow fever epidemics could afflict most of the population of a city and result in death rates of 20% or more of the city's original population.

One telling account of an epidemic in Norfolk, Virginia, in 1855 is described in the report of a committee of physicians established to examine the causes of this epidemic. Quarantine procedures to prevent the introduction of yellow fever were often thwarted by captains who concealed the presence of the disease to avoid a lengthy quarantine, even going to the extreme of secretly burying crew members who died while in quarantine. On June 6, 1855, the steamer *Ben Franklin* arrived from St. Thomas and anchored at the quarantine ground. The health officer, Dr. Gordon, visited the ship and was told that there was no disease on the ship. After 13 days in quarantine the ship was allowed to dock and yellow fever soon appeared in Norfolk. The first cases were crew and passengers from the ship. A number of early cases among the citizens of the town were ascribed to the ship passing within a half mile of their homes, and it is possible that infected mosquitoes were blown ashore. The disease then spread in all directions at a uniform rate of about 40 yards per day until it encompassed the whole city. The epidemic peaked at the end of August and died out after October. During the epidemic, an estimated 10,000 cases of yellow fever occurred in a population of 16,000, and 2000 died of the disease. The report established two other facts about the disease: Persons who had had yellow fever previously were immune, and the epidemic was not spread by person-to-person contact.

At the turn of the century, there was much debate as to the mechanism by which yellow fever spread. The Department of the Army sent an expedition, under the command of Walter Reed, to Cuba, recently acquired by the United States from Spain, to study the disease. The commission undertook to test the thesis that the virus was transmitted by mosquitoes, using themselves as human volunteers. Mosquitoes were allowed to feed on yellow fever patients and then on volunteers. At first there was a lack of understanding about the fact that mosquitoes are infected only by feeding on patients early in their disease, before an effective immune response arises, and about the necessity for an extrinsic incubation period in the mosquito, during which the virus establishes an infection in the salivary glands, before it can transmit the virus. Ultimately, however, the investigation team did succeed in proving mosquito transmission and one member of the commission, Dr. Jesse Lazear, died of it. Fortunately, his was the only death recorded in these experiments. It is of note that in the days before the introduction of a vaccine, most researchers who studied yellow fever in the field or in the laboratory ultimately contracted the disease and many of them died.

With the discovery that the virus was mosquito-borne, the U.S. Army began a campaign in Havana to eliminate mosquito breeding places by eliminating sources of water around human habitation. It was (and still is) common for drinking water to be stored around houses throughout Latin America in large pots that served as excellent breeding places for *A. aegypti*. The campaign succeeded in breaking the mosquito transmission cycle and yellow fever as an epidemic agent disappeared from Havana within months. This approach was later exported to other areas with great success, including Panama. It was at first believed that yellow fever could be eradicated, but the discovery of the endemic cycle of yellow fever dispelled this idea. Forest workers who cut down trees and brought the mosquitoes down from the upper canopy, where they transmit the disease to monkeys, were particularly at risk. Once infected, a person is able to bring the disease back to town where it can get into the *A. aegypti* population and start an urban epidemic.

In the late 1920s, yellow fever virus was successfully propagated in Rhesus monkeys, in which it causes a lethal disease and in which it can be experimentally passed from monkey to monkey. One such strain was derived from an infected human named Asibi. Theiler and Smith passed the Asibi strain of yellow fever in chicken cells, and after approximately 100 passages, it was found that the resulting virus was no longer virulent for Rhesus monkeys. After additional passages, this virus, called 17D, was ultimately used as a live virus vaccine in man and has proven to be one of the best and most efficacious vaccines ever developed. The vaccine virus causes very few side reactions and is essentially 100% effective in providing long-lasting protection against yellow fever. This vaccine is routinely given to travelers to regions where yellow fever is endemic and is used to control the spread of epidemic yellow fever in Latin America and, with less success, in Africa. The success of this vaccine has served as a model for the development of other live virus vaccines, namely, passing the virus in cultured cells from a non-native host.

Although not as wide ranging as previously, yellow fever continues to cause epidemics in Africa and South America as illustrated in Fig. 3.25. On an annual basis, 50–300 cases

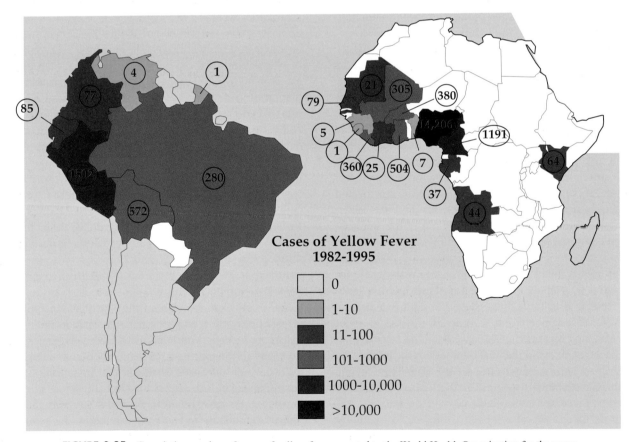

FIGURE 3.25 Cumulative number of cases of yellow fever reported to the World Health Organization for the years 1982 through 1995, by country. It is suspected that cases in Africa may be underreported by at least a factor of 10. Immunization coverage in Africa has remained low and the disease has recently emerged after more than a decade, with cases reported in Kenya (1992), Ghana (1993), Gabon (1994), and Liberia (1995). [From Fields *et al.* (1996, p. 1014) and the WHO web site: http://ww.who.int/vaccines-diseases/diseases/Yellow_fever.htm.]

are officially reported in South America and up to 5000 cases in Africa, but these figures are significantly under-reported. Between 1986 and 1991, annual outbreaks of yellow fever occurred in Nigeria that probably resulted in hundreds of thousands of cases. Epidemics of yellow fever has also occurred in Peru and Bolivia in recent years. There was one imported case of yellow fever in the United States in 1996, in which an American who visited the jungles of Brazil along the Amazon River without being immunized returned to the United States with yellow fever and died of the disease. Because of the endemic cycle in which monkeys are the reservoir, it is probably impossible to erad-icate the virus as has been done with smallpox and as is planned for poliovirus and measles virus.

Dengue Viruses

The four dengue viruses have recently undergone a dra-matic expansion in range and cause tens to hundreds of mil-lions of case of dengue fever in humans each year. Uncomplicated dengue fever is characterized by headache, fever, rash, myalgia (muscle pain, from *myo* = muscle and *algia* = pain), bone pain, and prostration. The disease may be mild or it may be extremely painful, but it is almost never fatal. However, the virus can cause illnesses charac-terized by hemorrhage (dengue hemorrhagic fever or DHF) or shock (dengue shock syndrome or DSS), which have mortality rates of several percent. Up to 250,000 cases of DHF and DSS are recorded each year, most of them in Southeast Asia, and DHF and DSS are a leading cause of mortality in children in southeast Asia. It is hypothesized that DHF and DSS are caused by immune enhancement in which infection by one serotype of dengue virus expands the population of cells that can be infected by a second serotype. In this model, infection with one serotype predis-poses a person to shock or hemorrhage on infection with a second serotype when the second infection occurs within a limited time period, usually 1–2 years. For this reason, the development of vaccines against dengue has progressed slowly, because of the possibility that immunizing against one serotype might put a person at risk for a more serious illness. Current efforts in Thailand are directed toward developing a quadrivalent attenuated virus vaccine that would immunize against all four serotypes simultaneously. U.S. scientists are independently attempting to develop vaccines for the viruses, based either on attenuated dengue viruses or on the development of chimeric flaviviruses that express dengue envelope antigens in a yellow fever vaccine background.

Dengue viruses are maintained in *A. aegypti* in urban settings in most of the world, but also in *Aedes albopictus* in Asia, and man is the primary vertebrate reservoir. Forest cycles have been documented in Africa and Madagascar in which the vertebrate reservoir is monkeys and other mos-quito species maintain the virus. Such a forest cycle does not exist in the Americas.

Dengue viruses, which have been continuously active over large areas of Asia and the Pacific region for a long time, have recently expanded their range in the Americas, as illustrated in Fig. 3.26. The viruses may have caused large epidemics in the Americas, including the United States, in the 1800s and into the early 1900s. However, it is impossible to determine with certainty from descrip-tions of the disease written at the time whether dengue was the causative agent of these epidemics or whether other viruses that cause similar illnesses might have been responsible. Dengue almost died out in the Americas largely because of efforts to control *A. aegypti*. In the mid-1900s, a serious effort was made in the Americas to eradicate *A. aegypti* from large regions, in order to control viral diseases spread by these mosquitoes. These efforts succeeded in eliminating the mosquito from large areas of Central and South America, as illustrated in Fig. 3.26A. However, by 1970 these efforts were aban-doned because of the expense involved and the detrimen-tal effects of DDT on the environment, and the mosquito reestablished itself over most of the region. The reintro-duction of multiple dengue strains into the Americas from foci in Asia after the reestablishment of *A. aegypti* resulted in the outbreak of huge epidemics of dengue fever (Fig. 3.26B). Furthermore, as multiple strains have become epidemic, the incidence of DHF and DSS has started to rise, so that dengue diseases now constitute a major plague throughout Central and South America and the Caribbean region.

Japanese Encephalitis Virus and Related Viruses

The Japanese encephalitis (JE) complex of flaviviruses includes a number of related viruses, many of which cause encephalitis. In addition to JE, these include St. Louis encephalitis (SLE), Murray Valley encephalitis (MVE), Kunjin, and West Nile viruses. The close relationships of these viruses are illustrated in Fig. 3.27. Notice in this den-drogram that the the viruses do not always group by place of isolation. Kunjin virus, found in Australia and the South Pacific, is more closely related to West Nile virus, found in Africa and Europe, than it is to Murray Valley encephalitis virus, an Australian virus. Thus, circulation of these viruses has been widespread, as is also true for the dengue viruses described in the preceding section.

JE virus is distributed throughout Asia, including Japan, India, Southeast Asia, Indonesia, the Philippines, and Borneo (Fig. 3.28). Reported cases of JE encephalitis average 35,000 per year with 10,000 deaths, but the disease is greatly underreported. Only 1 JE virus infection in 200 or 300 results in encephalitis, with children and the elderly being at higher risk. The fatality rate following JE

A. *Aedes Aegypti* in the New World

B. Dengue fever and dengue hemorrhagic fever(DHF) in the New World

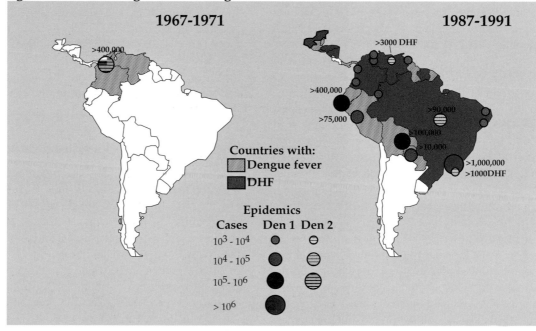

FIGURE 3.26 Changing distribution of dengue virus, DHF, and the vector for dengue in the New World. (A) Distribution of the vector mosquito *A. aegypti* in the Americas in 1970 and 1994. *Aedes aegypti* spread rapidly during the 1970s and 1980s due to the collapse of mosquito control programs and urbanization. (B) Increase and spread of dengue fever and DHF, and the introduction of multiple dengue serotypes between 1967 and 1991. Size of circles indicates the size of the epidemics. Although data shown are primarily for dengue 1 and dengue 2, dengue 3 and dengue 4 have also been active recently in the Americas. [Redrawn from Fields *et al.* (1996, pp. 1001 and 1021, respectively).]

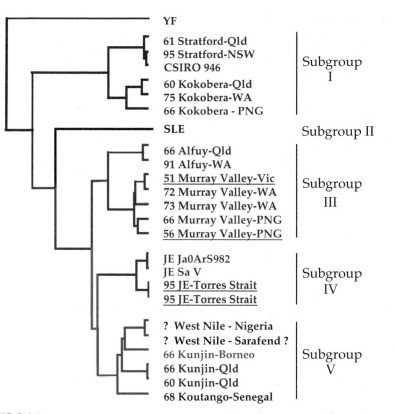

FIGURE 3.27 Phylogenetic tree of JE subgroup flaviviruses, based on nucleotide identity in the NS5 gene and the 3'NTR. Virus names in green are Australasia isolates; blue-green are PNG viruses; blue are from Japan, orange are from Africa. Underlined names are isolates from humans; the remainder are mosquito isolates. Isolates are named with year of isolation, name, and location of isolation where known. Yellow fever virus (YF) from Africa is shown as an outgroup. JE, Japanese encephalitis virus; Qld, Queensland; WA, Western Australia; NSW, New South Wales; VIC, Victoria; PNG, Papua New Guinea. [Adapted from Poidinger et al. (1996).]

encephalitis is 2–40% in different outbreaks, but 45–70% of survivors have neurological sequelae. In endemic areas, virtually all people have been infected by the time they reach adulthood. The vertebrate reservoirs for the virus are birds and pigs. Domestic pigs are particularly important amplifying hosts for human disease because they are found in proximity to their human owners. Various species of *Culex* mosquitoes transmit the virus. During peak transmission seasons, up to 1% of *Culex* mosquitoes around human habitations may be virus infected. Travelers to endemic regions have a probability of about 10^{-4}/week of contracting JE, and 24 cases of JE encephalitis in travelers were reported between 1978 and 1992. Inactivated virus vaccines are in use in different regions of Asia. The Japanese have long used such a vaccine to eliminate JE encephalitis from their population, and the Chinese have recently developed a vaccine that is being used in China and Thailand. The Japanese vaccine is also available in the United States for travelers to endemic regions. Of considerable interest is the finding that JE virus infection may reactivate in mice after the immune system first damps it out. Reactivation in other

animals may also occur and could be important for persistence of the virus in nature.

MVE virus and Kunjin virus are Australian viruses that are closely related to JE virus. They cause encephalitis in man, but the number of cases is small. Birds are the primary vertebrate reservoir, and epidemics of MVE have been associated with wet years when the mosquito population expands and nomadic waterfowl invade regions that are normally too dry to support them. *Culex annulirostris* is the primary vector for MVE.

SLE virus is a North American virus that belongs to the JE complex and that causes regular epidemics of encephalitis in the United States. The virus is widely distributed and cases of SLE encephalitis have been recorded in every state, with the majority of cases occurring in the Mississippi River valley, Texas, California, and Florida. Data for the years 1964–1993 are shown in Fig. 3.29. In the epidemic year 1975 there were 1815 cases of SLE encephalitis officially reported in the United States, but in nonepidemic years there may be fewer that 50 cases. The most recent epidemic occurred in 1990 in Florida with 223 cases and 11 deaths. The case fatal-

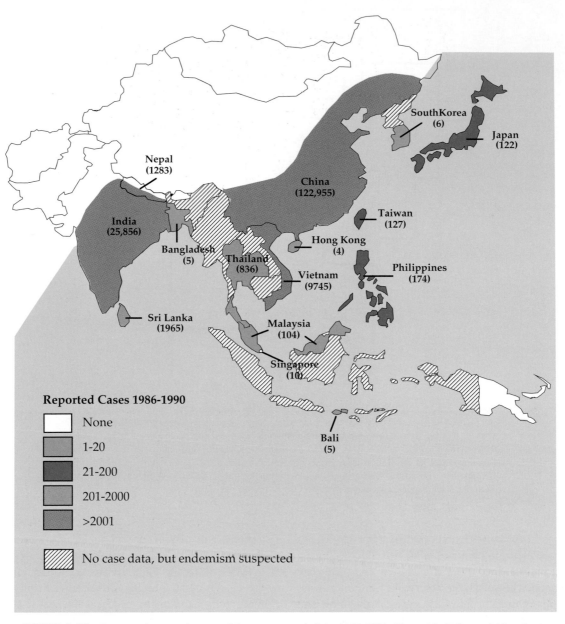

FIGURE 3.28   Range and reported cases of Japanese encephalitis, 1986–1990. [From *Morbidity and Mortality Weekly Report (MMWR)*, Vol. 42, RR-1, Fig. 1, p. 2 (1993).]

ity rate is about 7% overall, but is higher in the elderly. Most infections by SLE are inapparent, as is the case for many encephalitis viruses. The ratio of inapparent to clinical infection is age dependent and varies from 800 to 1 in children to 85 to 1 in the elderly. The virus is transmitted by *Culex* mosquitoes, and the primary vertebrate reservoirs are wild birds.

Another member of this complex that has caused recent headlines in the United States is West Nile virus. Known primarily from Africa, the virus has recently caused large epidemics of disease in Eastern Europe, where outbreaks have been associated with disruptions resulting from changes in government and civil strife. In the summer of 1999, West Nile appeared in North America for the first time. There were 62 human cases of West Nile disease in the New York City area, of whom 7 died of encephalitis. Numerous birds also died, including exotics in zoos as well as native birds. With the end of the mosquito season the epidemic died out, but surveys indicate that the virus is still present in overwintering mosquitoes and may have become established. By August, 2001 West-Nile virus had been reported from 15 states along the Atlantic seabound and it is feared that West Nile will continue to spread.

FIGURE 3.29 Distribution of cases of St. Louis encephalitis occurring between 1964 and 1993, shown by state. The large number of cases in Florida includes the most recent U. S. epidemic, which occurred in 1990, during which Florida reported 223 cases and 11 deaths. [From Fields *et al.* (1996, p. 981) and Summary of Notifiable Diseases in the United States for 1996, *MMWR*, Vol. 45, No. 53, (1997).]

Tick-Borne Encephalitis Viruses

The tick-borne encephalitis (TBE) viruses are important pathogens of Europe and Asia, and there is also a representative in North America. The viruses include Central European encephalitis (CEE), louping ill, Russian spring–summer encephalitis (RSSE), Kyasanur Forest disease, Omsk hemorrhagic fever, and Powassan viruses. Members of the TBE complex form a distinct group within the flaviviruses (Fig. 3.21), but share 40% amino acid sequence identity with the mosquito-borne flaviviruses, showing their close relationship to other flaviviruses. TBE viruses are transmitted by *Ixodes* ticks and can cause a fatal encephalitis in humans. An inactivated virus vaccine is widely used in Central Europe to protect people exposed to ticks. Even so, several thousand cases of TBE encephalitis occur each year. The case fatality rate is 1–2%, with 10–20% of survivors exhibiting neurological sequelae in the milder CEE form. However, the fatality rate is 20% with 30–60% of survivors having sequelae in the RSSE form. RSSE, and perhaps other TBE viruses, can also be contracted by drinking raw goat's milk and possibly other forms of raw milk. The virus has a tendency to set up persistent infection in experimental animals and possibly in humans as well. Although *Ixodes* ticks are the primary vector, *Dermacentor* ticks and ticks of other genera are also capable of transmitting the virus. The distributions of two species of *Ixodes* ticks that are important vectors of TBE are shown in Fig. 3.30.

Powassan virus is a member of the complex found in North America and in Russia. In North America, 20 cases of Powassan encephalitis have been reported since 1958.

Genus Pestivirus

The pestiviruses consist of three closely related viruses that are important pathogens of domestic animals. These are bovine viral diarrhea virus (BVDV), classical swine fever virus (CSFV) (also called hog cholera virus), and border disease virus of sheep (BDV). These three viruses share more than 70% amino acid sequence identity and exhibit extensive serological cross reactivity. Their genome organization is similar to those of other viruses in the family (Fig. 3.19).

BVDV exhibits an important and interesting disease syndrome in cattle. Animals infected as adults by BVDV may exhibit no disease or may have symptoms that include diarrhea, but they recover uneventfully. However, when a pregnant cow is infected by the virus, infection of the fetus may cause the fetus to become immunologically tolerant to the virus, resulting in a chronic infection that lasts for the life of the animal. Such *in utero* infection may lead to developmental abnormalities or runting in the calf, and may render the calf sensitive to infection by other microorganisms, all of which have serious economic effects. A more interesting effect of the chronic infection, however, is the development in some animals of fatal mucosal disease at the age of 1–2 years; once symptoms appear, the animal dies within weeks. Animals that die of this disease are found to be infected by two types of BVDV. One is the normal wild-type virus, which is noncytopathic in cultured cells. The second type of BVDV is a new strain that is cytopathic in cultured cells. The cytopathic BVDV strain is derived from the wild-type strain by recombination, which occurs during the persistent infection. Several different cytopathic BVDV

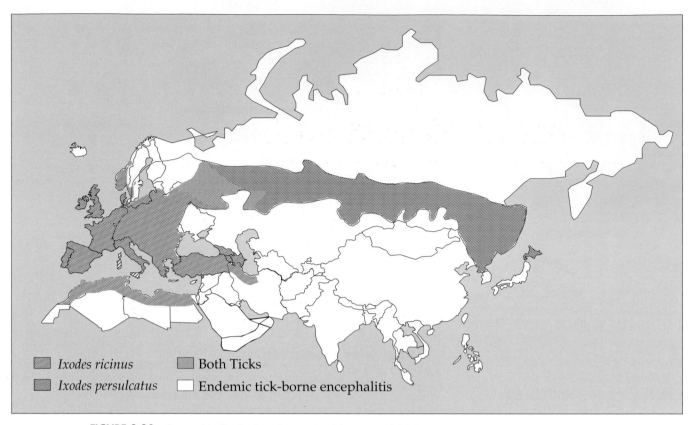

FIGURE 3.30   Geographic distribution of two major tick vectors of tick-borne encephalitis. Also shown is the major region in which TBE is endemic. [Adapted from Porterfield (1995, p. 207).]

strains have been sequenced, and they all have in common that NS2–3 (formerly called p125) is cleaved to produce NS3 (also called p80). Evidently it is the production of NS3 that renders the virus cytolytic in culture and causes lethal mucosal disease in cattle. As illustrated in Fig. 3.31, the cleavage to produce NS3 can be induced in several different ways. In at least three cytopathic BVDV strains, cellular ubiquitin sequences were inserted (in different ways) within the sequence encoding this protein, such that a cellular enzyme that cleaves specifically after ubiquitin cleaves the BVDV polyprotein to produce NS3. Another mechanism to produce NS3 was the insertion of the BVDV N^{pro} autoprotease immediately upstream of the NS3 sequence. Why the production of NS3 should have the effects that it does is a mystery whose solution will tell us more about the interactions of viruses with their hosts.

CSFV is epidemic in pig populations and causes serious illness, with different isolates differing in their virulence. Infection of pregnant sows can lead to abortion or to birth of persistently infected piglets, which soon die. BDV also can cause congenital infection, which can lead to abortion or to birth of animals that display a number of defects.

Genus Hepacivirus

Hepatitis C virus (HCV) forms a third genus in the Flaviviridae. The genome organization is similar to those of the other members of the family (Fig. 3.19). The genome is slightly smaller than those of the other two genera. HCV is a causative agent of blood-borne hepatitis in man. In the United States, HCV was once spread primarily through transfusion of contaminated blood, but the development of a diagnostic screen for the virus has virtually eliminated this source of infection in the developed world. However, the virus continues to be transmitted through the sharing of needles by drug users, and there are additional mechanisms of transmission that are not well understood. The virus is worldwide in distribution, as illustrated in Fig. 3.32. It has been estimated that ~3% of the world's population is chronically infected by the virus. The highest infection rate found was among Egyptian blood donors, where up to 19% were seropositive for HCV.

Infection with HCV can be extremely serious. The initial infection may cause no disease or may result in hepatitis accompanied by jaundice; fulminant liver failure is rare.

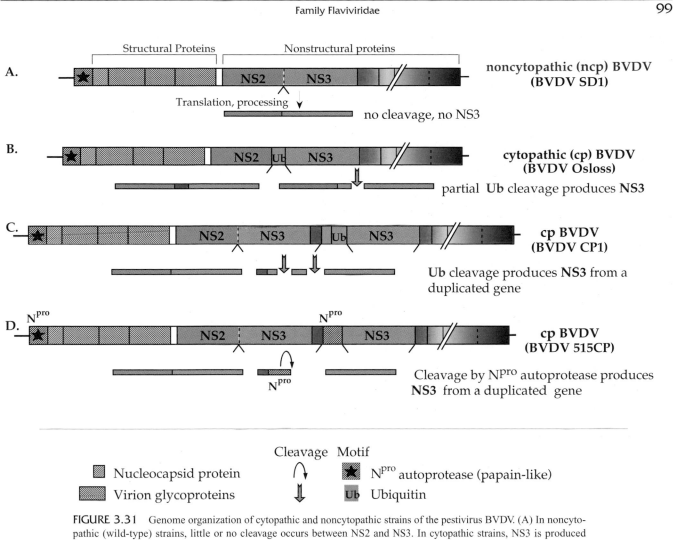

FIGURE 3.31 Genome organization of cytopathic and noncytopathic strains of the pestivirus BVDV. (A) In noncytopathic (wild-type) strains, little or no cleavage occurs between NS2 and NS3. In cytopathic strains, NS3 is produced either by an upstream insertion of ubiquitin (see B), insertion of multiple ubiquitin sequences plus duplication of NS3 sequences (see C), or duplication of the N^pro proteinase and insertion immediately upstream of a duplicated NS3 (see D). [Data for this figure came from Meyers and Thiel (1996).]

However, in the majority of cases the infection becomes chronic. This chronic infection is well tolerated by some, but it leads to cirrhosis in some fraction of infected individuals. End-stage liver disease is associated with multiple organ complications, and patients with HCV are at risk for development of hepatocellular carcinoma (HCC). The only treatment currently available for chronic HCV infection is a course of interferon therapy, which is often poorly tolerated. This is helpful in over half of treated individuals, but most relapse on withdrawal of interferon, and HCV is a common indication for liver transplantation. HCV is thus a medical problem of worldwide dimensions.

There is no cell culture system in which the HCV can be grown. Chimpanzees can be infected with the virus, but no other animal. This has limited the ability to study the molecular biology of the virus or even to determine its structure. Most of what we know of its molecular biology has been obtained from studies in which DNA copies of parts of the

HCV genome have been inserted into expression vectors. Nevertheless, the identification of cDNA clones containing segments of the viral genome, and the use of these clones to obtain the entire sequence of the HCV genome, has been a recent advance that has led to rapid increases in our understanding of this important viral pathogen. A full-length cDNA clone of the virus has been constructed, from which virus can be recovered by injection of RNA transcribed from the clone into the liver of a chimp. This validates the sequence of the clone, including the sequence of the ends, and makes possible many experiments to study the biology of the virus.

Viruses related to HCV, called GB viruses (from the initials of a surgeon with hepatitis from which they were first isolated), have been known for some time, but only recently have they been sequenced. GBV-A and GBV-B viruses have a genome organization very similar to that of HCV, but share little amino acid sequence identity with HCV or with

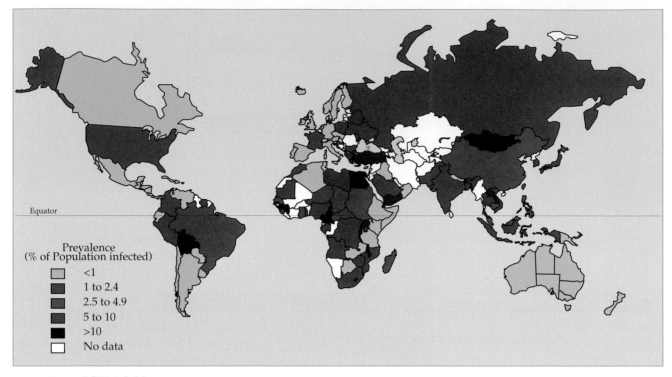

FIGURE 3.32 Worldwide prevalence of hepatitis C as of June 1999, based on published data. [From the World Health Organization (2000).]

each other. They may eventually be classified as two new genera within the Flaviviridae, more closely related to genus Hepacivirus than to genus Flavivirus or genus Pestivirus. A third virus, GBV-C, also called hepatitis G virus or HGV, is related to GBV-A. These three viruses appear to be widely distributed and establish chronic infections. Whether they are associated with human illness, and in particular with hepatitis in humans, is not yet clear.

FAMILY CORONAVIRIDAE

The name *coronavirus* comes from the Latin word meaning crown, from the appearance of the array of spikes around the enveloped virion. The family is composed of a number of RNA-containing animal viruses currently classified into two genera, the genus *Coronavirus* (whose members will here be called coronaviruses) and the genus *Torovirus* (whose members will be referred to as toroviruses). A representative listing of viruses in the two genera is found in Table 3.11. The family has recently been classified together with the Arteriviridae (described below) in the Order *Nidovirale*s, after the Latin word *nido* meaning nest, because they produce a nested set of mRNAs. Coronaviruses are somewhat larger in size (120–160 nm) than the toroviruses (120–140 nm) and have a larger genome (about 30 kb compared to 20 kb). In contrast to

other (+)RNA viruses, the nucleocapsids of Coronaviridae are constructed using helical symmetry. The coronaviruses have a helical nucleocapsid 10–20 nm in diameter, whereas the toroviruses have a tubular nucleocapsid that appears toroidal in shape in the virion. The coronavirus virion is roughly spherical, whereas the torovirus virion is disk shaped or rod shaped. The viruses mature by budding through intracytoplasmic membranes. The coronaviruses have been well studied, whereas the toroviruses, which are comprised of one pathogen of horses, one pathogen of cattle, a presumptive human torovirus, and a possible torovirus of swine, have only recently been described.

Genus Coronavirus

The coronaviruses have the largest RNA genome known. The genome size of RNA viruses is thought to be limited by the mutation rate of RNA. Because there is no proofreading during RNA synthesis, an inherent mistake frequency results on the order of 10^{-4}. Thus, error-free replication of an RNA genome becomes impossible once the genome becomes too large. The 30-kb genome of coronaviruses may represent this upper limit. It is also possible that because the coronaviruses undergo highfrequency recombination, as described below, they may be able to accommodate these large genomes because recombination offers a possible mechanism for correcting defective genomes.

TABLE 3.11   Coronaviridae

Genus/members	Virus name abbreviation	Usual host(s)	Transmission	Disease	World distribution
Coronavirus					
Infectious bronchitis	IBV	Birds	Mechanical, oral/fecal	Bronchitis	Worldwide
Human coronavirus	HCoV	Humans	Aerosols	Common cold	Americas, Europe
Murine hepatitis	MHV	Mice	Aerosols, contact	Gastroenteritis, hepatitis	Laboratory mouse colonies worldwide
Transmissable gasteroenteritis	TGEV	Swine	Contact	Gastroenteritis	US, Europe
Torovirus					
Berne (equine torovirus)	EqTV	Horses	Oral/fecal	Diarrhea	Europe, Americas
Breda (bovine torovirus)	BoTV	Cattle	Oral/fecal	Diarrhea	?
Human torovirus	HuTV	Humans	?	Diarrhea	?

Replication and Expression of the Genome

The coronavirus genome is, as in the case of all plus-strand RNA viruses, a messenger, and the naked RNA is infectious. The organization of the genome of avian infectious bronchitis virus (IBV) is shown in Fig. 3.33 as an example for the genus. The RNA, which is capped and polyadenylated, is translated into two polyproteins required for the replication of the viral RNA and the production of subgenomic mRNAs. The first polyprotein terminates at a stop codon about 12 kb from the 5′ end of the RNA. Ribosomal frameshifting occurs frequently, however, and in the shifted frame, translation continues to the end of the RNA replicase encoding region at ~22 kb. The resulting polyproteins are cleaved by virus-encoded proteases, as illustrated in Fig. 3.33B. All coronaviruses possess at least two proteases, one papain-like and the other serine-like, and some encode a third protease. Processing is complicated, as indicated in the figure.

The members of the *Nidovirales* produce a nested set of subgenomic mRNAs, which are also capped and polyadenylated. The number produced depends on the virus. Each subgenomic RNA is a messenger for a different protein, and usually only a single protein is translated from each messenger (translated from the 5′ ORF in the mRNA). Coronaviruses produce five to seven subgenomic RNAs. The five subgenomic mRNAs of IBV and the proteins translated from them are illustrated in Fig. 3.33A. Three of the subgenomic mRNAs are translated into the major structural proteins in the virion. The other two are translated into small proteins of unknown function.

The mechanism by which these subgenomic RNAs are produced has been difficult to determine. The first mechanism proposed was primer-directed synthesis using the (−)RNA template. In this model, a primer of about 60 nucleotides is transcribed from the 3′ end of the template, which is therefore identical to the 5′ end of the genomic RNA. The primer is proposed to dissociate from the template and to be used by the viral RNA synthetase to reinitiate synthesis at any of the several subgenomic promoters in the (−)RNA template. Evidence for this model includes the fact that each subgenomic RNA has at its 5′ end the same 60 nucleotides that are present at the 5′ end of the genomic RNA, and that there is a short sequence element present at the beginning of each gene that could act as an acceptor for the primer. A more recent model, however, proposes that the bulk of the subgenomic mRNAs are produced by independent replication of these RNAs as replicons. Such replication is thought to be possible because the mRNAs contain both the 5′ and 3′ sequences present in the genomic RNA, and therefore possess the promoters required for replication. Evidence for this model includes the fact that both-plus sense and minus-sense subgenomic RNAs are present in infected cells. In this modified model, the subgenomic RNAs are first produced by synthesis from either the genomic RNA or the (−)RNA antigenome. If synthesis is from the (−)RNA, leader-primed synthesis would be used as described above. If synthesis is from the genome, synthesis would initiate at the 3′ end and then jump to the 5′ leader at one of the junctions between the genes. At the current time, a model featuring synthesis from the plus strand is favored. Once produced, the subgenomic RNAs begin independent replication.

Coronaviruses undergo high-frequency recombination in which up to 10% of the progeny may be recombinant. It is proposed that the mechanism for generation of the subgenomic RNAs, which requires the polymerase to stop at defined sites and then reinitiate synthesis at defined promoters, may allow the formation of perfect recombinants at high frequency.

Envelope Glycoproteins

All coronaviruses possess at least two envelope glycoproteins, a spike protein (S) and a membrane protein (M).

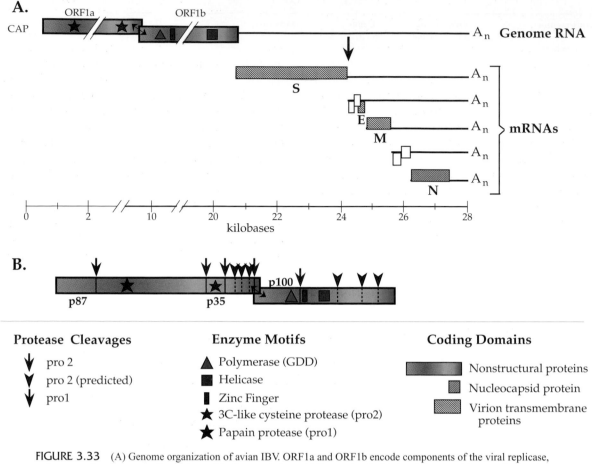

FIGURE 3.33 (A) Genome organization of avian IBV. ORF1a and ORF1b encode components of the viral replicase, and are translated as two polyproteins, with ribosomal frameshifting at the double-headed arrow. The remaining viral components are encoded in a nested set of mRNAs. The shaded proteins are polypeptides found in virions. Open boxes are ORFs of unknown function. E is a minor virion component, but essential for virus assembly. (B) Proteolytic processing of the IBV ORF1ab polyprotein. Motifs of papain-like proteases (pro1), 3C-like cysteine protease (pro2), RNA polymerase (GDD), zinc finger, and helicase are indicated with various symbols. Arrows at known cleavage sites are color coded according to the protease responsible. Green arrowheads are predicted cleavage sites for pro2. [Adapted from de Vries *et al.* (1997).]

The spike protein, present in the virion as multimers, probably trimers, possesses the receptor binding activity, the major neutralizing epitopes, and the fusion activity of the virion. It is this protein that gives coronaviruses their characteristic corona around the envelope. The M protein is an integral membrane protein that spans the lipid bilayer three times and has only a small fraction of its mass exposed outside the bilayer. Some coronaviruses also have a third glycoprotein, a hemagglutinin-esterase (H-E). Remarkably, this protein appears to be homologous to the H-E of influenza C virus (described in the next chapter). It appears that recombination between a coronavirus and an influenza C virus occurred that led to exchange of this protein. Because only some coronaviruses possess H-E, whereas all influenza C viruses possess it, the simplest hypothesis is that H-E was an influenza C protein that was acquired by a coronavirus. Presumably, this acquisition was maintained because it extended the host range of the coronavirus by allowing it to infect cells that lack a receptor recognized by the S protein.

In addition to these well-defined glycoproteins, a small protein of about 100 residues has been identified as a component of at least some coronaviruses. This protein, called sM for small membrane protein, is translated from a different mRNA than S or M, as illustrated in Fig. 3.33A.

Diseases Caused by Coronaviruses

Coronaviruses are responsible for about 25% of human colds and are spread by a respiratory route. Unlike rhinoviruses, they cause not only upper respiratory tract infections but sometimes lower respiratory tract infections as well, which are more serious. Some coronaviruses may cause gastroenteritis in humans, because there have been reports of coronaviruses in the stools of people suffering from gastroenteritis.

Coronaviruses for many other animals are known, including mice, chickens, pigs, and cats. Diseases associated with various coronaviruses in lower animals include respiratory disease, gastroenteritis, hepatitis, and a syndrome similar to multiple sclerosis of humans, as well as other illnesses. Because the human coronaviruses have been difficult to cultivate in cultured cells, they have been less well studied in laboratory settings than those infecting other animals. Mouse hepatitis virus has been particularly well studied as a model for the genus. Thus, much of what is known of the molecular biology of their replication comes from model studies with nonhuman coronaviruses.

FAMILY ARTERIVIRIDAE

The family Arteriviridae contains four viruses, which are listed in Table 3.12. There are no known human viruses in the family, but it is of interest because it represents an intermediate between the coronaviruses and other enveloped (+)RNA viruses. The genome of equine arterivirus is illustrated in Fig. 3.34. The arteriviruses have a 13-kb genome that is very similar in organization and expression strategy to that of coronaviruses. The virion (60 nm) is enveloped, as are the coronaviruses, but the nucleocapsid is icosahedral rather than helical. The arteriviruses could have arisen by the acquisition of new structural proteins by a coronavirus (or vice versa). The existence of this family, which appears to be a coronavirus with structural proteins that lead to icosahedral symmetry rather than helical symmetry, illustrates a problem for taxonomy. The ICTV has classified

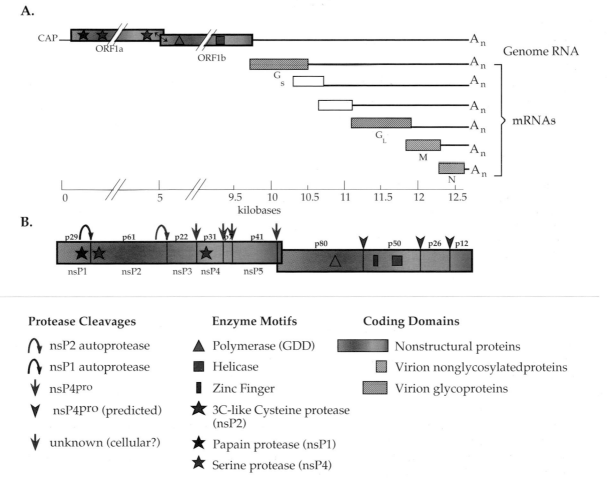

FIGURE 3.34 (A) Genome organization of an arterivirus, equine arteritis virus. ORF1a and ORF1b encode components of the viral replicase and are translated as a polyprotein with ribosomal frameshifting at the arrow. The remaining viral components are encoded in a nested set of mRNAs. The shaded proteins are polypeptides found in virions. (B) Proteolytic processing of the equine arteritis virus ORF1ab polyprotein. Positions of motifs of proteases, polymerase, zinc finger, and helicase are indicated with various symbols. Arrows are color coded to indicate cleavage by the corresponding protease. Arrowheads are predicted cleavages. Blue arrowhead is a cleavage site possibly cleaved by a cellular protease. [From de Vries et al. (1997), with permission, and den Boon et al. (1991).]

TABLE 3.12　Arteriviridae

Genus/members	Virus name abbreviation	Usual hosts(s)	Transmission	Disease	World distribution
Arterivirus					
Equine arteritis	EAV	Horses	Aerosols, contact	Fever, necrosis of arteries, abortion	Worldwide
Porcine reproductive and respiratory syndrome	PRRSV	Pigs	Oral/fecal?	Infertility, respiratory distress	?
Lactic dehydrogenase-elevating	LDV	Mice	Biting	?	?
Simian hemorrhagic fever	SHFV	Monkeys	Biting	Hemorrhagic fever	?

these viruses as a distinct family, but created the order Nidovirales to indicate their relation to the coronaviruses.

The four arteriviruses are lactate dehydrogenase-elevating virus of mice (LDV), equine arteritis virus (EAV), simian hemorrhagic fever virus (SHFV), and porcine reproductive and respiratory syndrome virus (PRRSV). The primary target cells in their respective hosts are macrophages, and all are associated with persistent, long-term infections. LDV causes a lifelong infection of mice that requires special care to detect. EAV causes epizootics of subclinical or mild respiratory diseases in adult horses. Infection can lead to abortions in pregnant mares, and infection of young horses causes a more serious illness. The virus persists for long periods, and in stallions the virus may be secreted in semen for the life of the animal.

PRRSV causes respiratory distress in pigs of all ages and abortions and stillbirths in pregnant sows. SHFV is an African virus that causes persistent, inapparent infections in African monkeys. When introduced into colonies of Asian monkeys, however it causes fatal hemorrhagic fever.

THE PLUS-STRAND RNA VIRUSES OF PLANTS

Most plant viruses possess (+)RNA as their genome. Some have as their genome a single RNA molecule and produce subgenomic mRNAs, whereas in others the viral genome is divided into two or three or more segments. In plant viruses in which the genome is present in more than one segment, each segment is packaged separately into different particles and infection requires the introduction into the same cell of at least one of each genome segment. Many (+)RNA plant viruses are rod shaped, formed using helical symmetry (e.g., tobacco mosaic virus, Fig. 2.2), while others are icosahedral (e.g., the comovirus cowpea mosaic virus, Figs. 2.5 and 2.7). No (+)RNA plant viruses are enveloped. Many of these viruses are major agricultural pathogens responsible for a great deal of crop damage worldwide. Although important as plant pathogens, plant

viruses will not be covered here except for a description of the genomes of certain families that are of particular interest because of what they tell us about the evolution of viruses.

Several families of (+)RNA plant viruses share sequence homology with one another and with the alphaviruses. This collection of viruses, sometimes referred to as the Sindbis superfamily or the alphavirus superfamily, includes the alphaviruses, the tobamoviruses, the bromoviruses, and other families of plant viruses. The genomes of the tobamovirus tobacco mosaic virus (TMV), the bromovirus brome mosaic virus (BMV), and the alphavirus Sindbis virus are compared in Fig. 3.35. The genome of TMV is one molecule of (+)RNA, and two subgenomic RNAs are produced. The genome of BMV consists of three molecules of (+)RNA, and one subgenomic RNA is made. The alphaviruses have been described. Notice that a characteristic of this superfamily is that all viruses in it produce at least one subgenomic mRNA. The members of this superfamily all share three proteins (or protein domains) with demonstrable sequence homology, as indicated in the figure. These three are a viral RNA polymerase, a helicase, and a capping enzyme (characterized by methyltransferase activity). In the case of the alphaviruses and the tobamoviruses, all three domains are found on one genome segment and read-through is required to translate the polymerase. In the bromoviruses, the capping enzyme and the helicase are encoded on one segment, but the polymerase is encoded on a different segment. It is interesting that the alphaviruses encode a protease to separate the three domains from one another, but the plant viruses do not. While these three shared proteins have clearly diverged from a common ancestral source, other domains within the nonstructural proteins are different from family to family. The alphavirus protease and nsP3 are not shared with the plant viruses, while the plant viruses possess movement proteins that are not shared with the alphaviruses. The structural proteins of the different families are also distinct. These observations clearly point to the occurrence of extensive recombinational events during the evolution of this

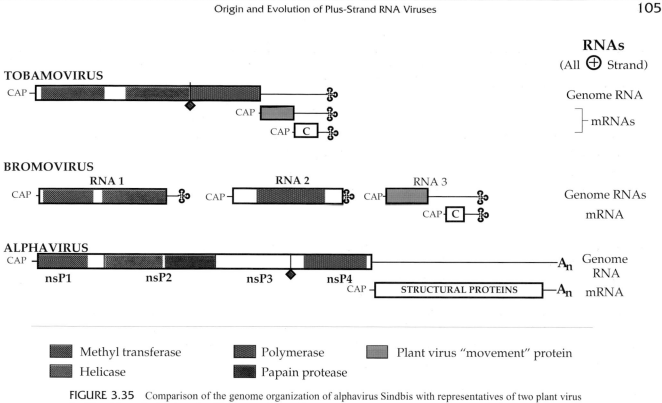

FIGURE 3.35 Comparison of the genome organization of alphavirus Sindbis with representatives of two plant virus families. Three shaded domains illustrate regions of low but significant sequence homology, which extend over hundreds of amino acids, within the methyl transferase, helicase, and polymerase proteins. The blue diamond is a leaky termination codon that is read through to produce the downstream blue-shaded domains in the tobamoviruses and the alphaviruses. C is the coat protein. The plant viruses have no module corresponding to the protease in nsP2 nor to protein nsP3. The alphaviruses have no domain corresponding to the "movement" protein of plant viruses. [Adapted from Strauss and Strauss (1994, Fig. 35).]

group of viruses from a common ancestral source. Recombination has brought together new combinations of genes appropriate to the different lifestyles of the various members of the superfamily.

Similar considerations pertain to two families of plant viruses (the Comoviridae and the Potyviridae) and the animal picornaviruses, which are all related to one another and are sometimes referred to as the picornavirus superfamily. The Comoviridae have a bipartite genome, whereas the Potyviridae and the Picornaviridae have a single molecule of RNA as their genome. A characteristic of this superfamily is that no subgenomic RNAs are produced. The genome organizations of two members of the Comoviridae that belong to different genera, tomato black ring virus (genus Nepovirus) and cowpea mosaic virus (genus Comovirus), are compared with that of picornavirus poliovirus in Fig. 3.36. The members of this superfamily have demonstrable homologies in their RNA polymerases, 2C helicases, and $3C^{pro}$ proteases. Further, the RNA genomes have a 5′ VPg and are polyadenylated. Proteases, VPg's, and poly(A) are very unusual in plant viruses, found only in members of this superfamily. It is clear that these viruses are all related to one another, and that multiple recombination events have taken place to give rise to the current families.

ORIGIN AND EVOLUTION OF PLUS-STRAND RNA VIRUSES

A reasonable hypothesis for the origin of the RNA viruses is that they began as an mRNA that encoded an RNA polymerase. The acquisition of an origin of replication that allowed the mRNA itself to be replicated by its encoded product would give rise to a self-replicating RNA and could have represented the first step in the development of a virus. Subsequent recombination with an mRNA encoding an RNA-binding protein that could be modeled into a capsid would give rise to a very simple virus. This protovirus could then evolve through continued mutation and recombination into something more complex.

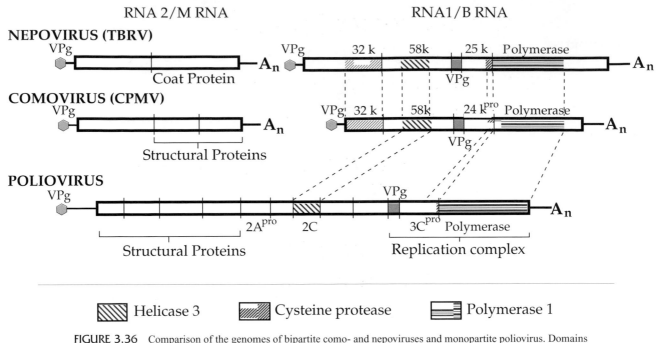

FIGURE 3.36 Comparison of the genomes of bipartite como- and nepoviruses and monopartite poliovirus. Domains in the helicase, polymerase, and protease that share sequence homology over long stretches of amino acids are identified with differently colored patterns. The related 32k proteins of como- and nepoviruses have no counterpart in poliovirus, but are related to the "movement" proteins of other plant viruses. TBRV, tomato black ring virus; CPMV, cowpea mosaic virus. The structural proteins of the three viruses show no sequence similarity. [Adapted from Strauss and Strauss (1997, Fig. 2.12).]

Examples of the importance of recombination in the evolution of RNA viruses have been discussed. Computer-aided studies that have attempted to align the amino acid sequences of the proteins of different (+)RNA viruses have suggested that all these viruses share core functions that have common ancestral origins. These results are summarized in Fig. 3.37. All RNA viruses possess an RNA polymerase and these all appear to have derived from a common ancestral source. However, three lineages of RNA polymerases can be distinguished that probably diverged from one another early in the evolution of RNA viruses. Most RNA viruses also possess an RNA helicase that is required to unwind the RNA during replication. These helicases also appear to have diverged from a single source, but three lineages can be distinguished here as well. A third shared function in those RNA viruses with capped mRNAs is a methyltransferase gene (an activity required for capping), and two methyltransferase lineages can be distinguished. Finally there are the viral proteases that process polyproteins. The two distinct types of proteases with independent origins are the proteases derived from serine proteases (which may possess serine or cysteine at the active site) and the papain-like proteins. The different lineages of these four core activities have been reassorted in various ways during the evolution of the RNA viruses, as shown in the figure.

The second mechanism for divergence among viruses is mutation. Lack of proofreading in RNA replication means that the mistake frequency during replication is very high, on the order of 10^{-4}. Most mistakes are deleterious and do not persist in the population. However, because the mistake frequency is so high, many different sequences can be tried rapidly because of the rapid replication rate of viruses. The net result is that the rate of sequence divergence in RNA viruses is very high, up to 10^6-fold faster than their eukaryotic hosts. Three studies of the rate of sequence divergence in RNA viruses are illustrated in Fig. 3.38. In these studies, regions of the genomes of viruses isolated over a period of many years were compared. The rates of sequence divergence in a picornavirus and in influenza virus (Chapter 4) were found to be 0.5–1% per year. Changes in third codon positions, which are usually silent, occur more rapidly than changes in first or second codon positions, which usually result in an amino acid substitution. In alphaviruses, which alternate between insect and vertebrate hosts, the rate of divergence was significantly less, 0.03% per year, because changes that might be neutral or positively selected in one host are often deleterious in the other host. One of the

and plants. Several viruses are important pathogens of man.

Other dsRNA viruses are known in addition to the family Reoviridae. They include the birnaviruses, icosahedral viruses 60 nm in diameter containing two genome segments of dsRNA totaling about 7 kb, which infect chickens, fish, and arthropods; the totiviruses of fungi and protozoa, which have only one RNA segment; and the partiviruses and hypoviruses of fungi and plants. Many of these viruses have only recently been described. For this reason as well as the fact that no human pathogens are known among these viruses, they will not be considered further .

Overview of the Family Reoviridae

The nine genera of Reoviridae are listed in Tables 3.13 and 3.14 together with a partial listing of viruses in each genus, their hosts and modes of transmission, the diseases they cause, and their distributions. As a group, Reoviridae have a wide host range. Members of the genera *Orthoreovirus*, *Rotavirus*, *Orbivirus*, and *Coltivirus* infect humans as well as other vertebrates, aquareoviruses infect fish, cypoviruses infect arthropods, and members of three genera infect plants. The four genera that contain human pathogens are compared in Table 3.15.

TABLE 3.13   Reoviridae of Vertebrates

Genus/members	Virus name abbreviation	Usual host(s)	Transmission or vector	Disease	World distribution
Orthoreoviruses					
Nonfusogenic					
Mammalian reoviruses types 1, 2, 3	MRV	Humans, cattle, sheep, swine	Oral/fecal	Gastroenteritis, respiratory disease	Worldwide
Fusogenic					
Nelson Bay virus	NBV	Flying foxes (bats)			Australia
Baboon reovirus	BRV	Monkeys	Oral/fecal	Wide range of symptoms from inapparent to lethal	?
Avian reoviruses (5 serotypes)	ARV	Birds	Oral/fecal		Worldwide
Reptilian reoviruses	?	Snakes			?
Orbiviruses					
African horse sickness (10 serotypes)	AHSV	Equines	*Culicoides* (midges)	Cardiopulmonary disease	Africa
Bluetongue (24 serotypes)	BTV	Sheep, cattle	*Culicoides* (midges)	Rhinitis, stomatitis	Africa, Australia, Asia, Americas
Changuinola	CGLV	Humans	Phlebotomines	Fever	Panama
Kemerovo serogroup					
Kemerovo	?	Humans			
Great Island	GIV	Seabirds	Ticks	Fever, encephalitis	E. Europe, United States
Chenuda	CNUV				
Wad Medani	WMV	Domestic animals			
Coltiviruses					
Colorado tick fever	CTFV	Humans	Ticks	Fever, encephalitis	North America, Europe
Rotaviruses					
Group A	RV-A	Humans, animals	Oral/fecal	Infant diarrhea	Worldwide
Group B	RV-B	Humans, animals	Oral/fecal	Epidemic adult diarrhea	Primarily China
Group C	RV-C	Humans, animals	Oral/fecal	Clinical significance unknown	
Groups D, E, F		Mammals, birds	Oral/fecal		
Aquareoviruses					
Five serogroups		Fish	?	?	?

TABLE 3.14   Reoviridae of Plants and Insects

Genus/members	Usual hosts(s)	Transmission or vector	Disease	World distribution
Cypoviruses				
Cytoplasmic polyhedrosis viruses	Arthropods	Ingestion	Diarrhea, starvation due to changes in the gut	Worldwide
Fijivirus				
Fiji disease virus	Plants	Delphacid leafhoppers		Australasia, Asia, South America, Northern Europe
Phytoreovirus				
Wound tumor virus	Plants	Cicadellid leafhoppers		
Rice dwarf virus	Rice	Cicadellid leafhoppers	Stunting	Southeast Asia, China, Japan, Korea
Oryzavirus				
Rice ragged stunt virus	Plants (*Gramineae*)	Planthoppers		

Many of the Reoviridae are transmitted by arthropod vectors (Tables 3.13 and 3.14). The orbiviruses are transmitted by gnats or midges of the genus *Culicoides*, phlebotomine flies, or ticks, and members of this genus are true arboviruses, as are the coltiviruses, transmitted by ticks. The members of the three genera of plant reoviruses are also transmitted by arthropods and are effectively plant arboviruses (although the term arbovirus is usually reserved for vertebrate viruses that are transmitted by arthropods). Thus, members of six of the nine genera of Reoviridae possess the ability to replicate in arthropods, of which one genus contains viruses that replicate exclusively in insects and five genera contains viruses that have an alternate vertebrate or plant host.

Members of the family Reoviridae replicate in the cytoplasm. The virion is icosahedral ($T = 13$), 60–80 nm in diameter, and double or triple shelled (the shells consist of protein). The structures of viruses belonging to three genera of the Reoviridae were shown in Chapter 2: reovirus (genus *Orthoreovirus*) in Figs. 2.1 and 2.5; rotavirus (genus *Rotavirus*) in Fig. 2.5; and bluetongue virus (genus *Orbivirus*) in Fig. 2.11 (in this case of the core of the virion). The members of the genus *Orthoreovirus* have been the best studied and have served as a model system for the family, but because of the medical importance of the members of the genus *Rotavirus* and the veterinary importance of the members of the genus *Orbivirus*, these viruses have recently come under increased scrutiny.

TABLE 3.15   Comparison of Orthoreovirus, Orbivirus, Rotavirus, and Coltivirus

Characteristic	Orthoreovirus	Orbivirus	Rotavirus	Coltivirus
Segments	10	10	11	12
Size of genome	23.5 kb	19.2 kb	18.6 kb	28.5 kb
Type virus	Reovirus type 3	Bluetongue-1	Simian rotavirus SA11	Colorado tick fever
Portal of entry	Oral	Skin	Oral	Skin
Tissue tropism	Intestinal tract, upper respiratory tract	Hemopoietic	Intestinal tract	Hemopoietic and muscle
Vector	None	Culicoid flies, ticks, mosquitoes, phlebotomines	None	Ticks, mosquitoes
Human disease	Upper respiratory infections, infant enteritis	See Table 3.17	Diarrhea, particularly in children <5 years old	See Table 3.17

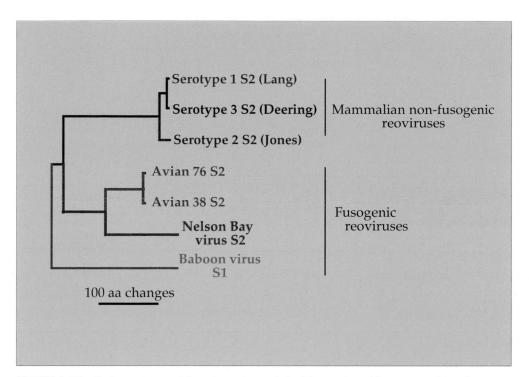

FIGURE 3.39   Phylogenetic tree of the orthoreoviruses, derived from amino acid sequences of the σ2 core proteins, which are encoded in most viruses by the S2 RNA segment. A virtually indistinguishable tree is given by the sNS sequences, and one with only slightly longer arms, but the same topology, by the outer capsid protein sequences. The scale bar indicates the length of the arm for 100 changes. [Adapted from Duncan (1999).]

Genus Orthoreovirus

The genus *Orthoreovirus* (the "true" reoviruses, to distinguish the genus from the family) contains three viruses that infect many mammals, including humans, referred to as the mammalian reoviruses types 1, 2, and 3. They were originally named from the first initials of the words *r*espiratory *e*nteric *o*rphan virus—they grow in the respiratory tract and in the enteric tract but were orphans, not known to cause human illness. The viruses are widespread and the majority of humans have antibodies against all three serotypes by the time they are adults. Most infections do not result in symptomatic disease or result in only mild symptoms. Studies with human volunteers have shown that some individuals develop a mild disease characterized by headache, pharyngitis, sneezing, rhinorrhea, cough, and malaise.

These three orthoreoviruses are virtually ubiquitous viruses of mammals. They have been isolated from many different mammals as well as from sources such as river water and untreated sewage. There is little evidence for host range specificity among these mammalian viruses. However, reovirus infection of lower mammals is sometimes associated with more serious illness than infection of humans.

Another branch of the orthoreovirus genus consists of the fusogenic reoviruses and includes a number of serotypes of avian reoviruses as well as a baboon reovirus. In general, the avian viruses do not grow in mammalian cells or must be adapted to mammalian cells before they will grow, and thus have a host range distinct from that of the mammalian viruses. The relationships among these viruses are illustrated by the tree in Fig. 3.39. Notice that the three mammalian viruses group closely together and are distinct from the other reoviruses, including the baboon virus.

Reovirus serotype 3 has been extensively studied as a model for the members of the orthoreovirus genus. In the following discussion, in which aspects of the genome organization, replication, and structure of orthoreoviruses are described, specific details refer to reovirus ST3. These details are summarized in Table 3.16.

The Genome of Orthoreoviruses

The genomes of orthoreoviruses consist of 10 segments of dsRNA (Fig. 3.40). The genome segments range in size from 3.9 to 1.2 kb and sum to 23.5 kb for reovirus ST3 (Table 3.16). Twelve proteins are produced of which eight are components of the virion, four in the outer shell and four in the inner shell.

TABLE 3.16   Characteristics of the Proteins Encoded by the 10 Genome Segments of
Orthoreovirus Serotype 3[a]

Genome segment	Protein product	Length (nt)	5'NT[b] (nt)	ORF (aa)	3'NT[b] (nt)	Function of protein	Location/molecules of protein per virion
L1	λ3	3854	18	1267	35	Catalytic subunit of polymerase/transcriptase	Core/12
L2	λ2	3916	13	1289	36	Guanylyl transferase	Turrets on core/60
L3	λ1	3896	13	1233	184	NT binding motif, Zn finger	Core/120
M1	μ2	2304	13	736	83	Putative polymerase component	Minor core component/12
M2	μ1–μ1C	2203	29	708	50	Major structural protein	Outer capsid shell/600
M3	μNS (μNSC)	2235	18	719	60	Binds ssRNA	Nonstructural
S1	σ1	1416	12	455	39	HA, neut Ag[c], cell attachment protein	Capsid/36
	σ1NS			120	—	Unknown	Nonstructural
S2	σ2	1331	18	418	59	Structural protein	Core/240
S3	σNS	1198	27	366	73	Binds ssRNA	Nonstructural
S4	σ3	1196	32	365	69	Major structural protein	Outer capsid shell/600

[a]Information in this table is from Joklik and Roner (1996).
[b]5'NT, nucleotides at the 5' terminus of the RNA segment that are not translated into protein; 3'NT, nucleotides at the 3' terminus of the RNA that are not translated.
[c]HA, hemagglutinin; neut Ag, contains epitopes recognized by neutralizing antibodies.

Entry of Orthoreoviruses into the Cell

After attachment to receptors, orthoreoviruses are internalized into endosomes. In endosomes or in lysosomes, proteolysis of two proteins in the outer shell, σ3 and μ1C, produces what has been termed an ISVP (infectious subviral particle or intermediate subviral particle). In this process, μ1C is cleaved to produce two fragments, δ and a small C terminal ϕ, whereas σ3 is degraded. These cleavages are illustrated schematically in Fig. 3.41. They can be blocked by agents that prevent acidification of endosomes, which demonstrates the importance of the endosomes in the process. ISVPs can also be produced by treating virions with proteases *in vitro*.

The ISVP is capable of breeching the endosomal membrane and gaining entry into the cytoplasm. The process of penetration may involve μ1N, whose N terminus is myristoylated and lipophilic. When ISVPs are produced by treatment of virions with proteases *in vitro*, they are capable of penetrating into the cell by way of the plasma membrane, which confirms the importance of proteolytic processing for the activation of domains required for penetration.

Synthesis of mRNAs

On penetration of ISVPs into the cytoplasm, they are converted into cores by further loss of δ, ϕ, and the σ1 fiber, and the rearrangement of the protein λ2 (Fig. 3.41). Core particles are transcriptionally active: Each of the 10 segments of dsRNA within them is used to synthesize an mRNA molecule that is the same length as the plus strand of the genome. These mRNAs are capped but not polyadenylated. All enzymatic activities required for the initiation of RNA synthesis, capping, and elongation of the

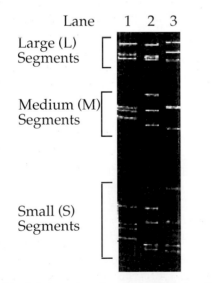

FIGURE 3.40   Gel electrophoresis of the orthoreovirus RNA genome segments, showing the variation of the segment size with serotype. Lane 1, reovirus serotype 2 (Jones); lane 2, reovirus serotype 1 (Lang); lane 3, serotype 3 (Deering). The segments cluster into three groups: 3L (large), 3M (medium), and 4S (small). [From Fields *et al.* (1996, p. 1559).]

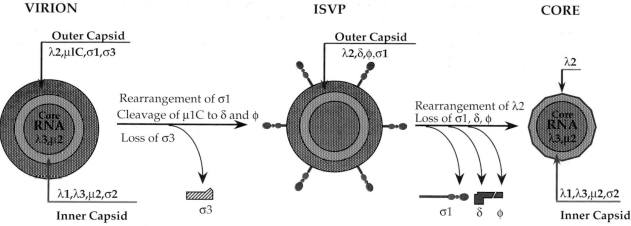

FIGURE 3.41 Schematic of the structure of a reovirion and subviral particles derived from it. At left, a cross section through the particle shows the two protein shells (in yellow and blue) surrounding the RNA (green). In the middle, the ISVP is shown, after the loss of σ3, the cleavage of μ1C to δ and φ, and the extension of σ1. The right illustration shows the core after the loss of σ1, δ, and φ and the rearrangement of λ2. [Redrawn from Niebert and Fields (1995); reprinted in Fields *et al.* (1996, p. 1562).]

product are present in the core, and occur *in vitro* if cores are supplied with appropriate substrates. Synthesis of mRNA is conservative: The newly synthesized mRNAs are extruded from the core and both strands of the parental dsRNA remain within the core. Extrusion is an active process. Electron microscopic studies have suggested that the mRNAs are extruded from the 12 vertices of the icosahedral structure. The enzymatic activities are organized about these 12 fivefold axes, and it has been suggested that there is an independent transcription unit for each genome segment, consistent with the fact that no member of the family Reoviridae has more than 12 genome segments.

Translation of Proteins

The 10 mRNAs are translated into 12 proteins (Table 3.16). For 8 mRNAs, only one reading frame is used and only one protein is produced. For the mRNA produced from segment M3, only one reading frame is used but two different in-frame AUGs are used to initiate translation. Thus, two proteins (μNS and μNSC) are produced from this segment that differ only in that the longer version has an N-terminal extension. The mRNA from segment S1 is translated using two different, out-of-frame AUGs, however, so that two different proteins (σ1 and σ1NS) are produced. The various mRNAs are translated with widely different efficiencies so that different amounts of the 12 proteins are produced.

Of the proteins produced, eight are components of the virion and four are nonstructural, present only within the infected cell. Proteins λ1 and σ2, present in 120 and 240 copies, respectively, form the shell of the core. Proteins λ3 and μ2 are present in 12 copies within the core, at the 12

fivefold axes. Protein λ3 is the catalytic subunit of the RNA polymerase, and μ2 is believed to be a component of this enzyme complex.

Pentamers of protein λ2, present in 60 copies, form "turrets" at the 12 fivefold axes of the core through which the mRNAs are extruded. This protein has guanylyl transferase activity and is a component of the complex that caps the mRNAs. Protein μ1 and its cleavage products, μ1N and μ1C, together with protein σ3, both present in 600 copies, form the outer shell of the virion. Protein σ1 is the cell attachment protein on the surface of the virion. Trimers of this protein are located at the 12 fivefold axes. Interestingly, a complete complement of σ1 trimers is not present in all virions. Virions contain from 0 to 12 trimers, with the median number of trimers being 7. Virions devoid of σ1 are not infectious, but virions containing one or more trimers are infectious.

Four proteins, μNS, μNSC, σ1NS, and σNS, are nonstructural. μNS, μNSC, and σNS are RNA-binding proteins and probably participate in virion assembly. The function of σ1NS is not known; it is found in the nucleus of infected cells.

Assembly of Progeny Virions

The mRNAs serve as intermediates in the replication of reoviruses. Following synthesis and release from the core, mRNAs quickly become associated with proteins μNS, σNS, and σ3 to form single-strand-RNA-containing complexes. These complexes contain only a single mRNA molecule associated with 10–30 protein molecules. All complexes contain μNS, but only half contain σ3 and one-quarter contain σNS. Complexes containing dsRNA appear later, which contain the three proteins named above but also contain protein λ2 and the RNA polymerase. Significantly, all 10 dsRNA genome segments are present in equimolar quantities, suggesting that the selection and assortment of the 10 genome segments into progeny virions is associated with the conversion of (+)RNA into double-stranded RNA. Because the particle-to-infectious virus ratio is almost one, the assembly process is clearly precise.

During maturation, protein μ1 undergoes a cleavage to produce two fragments called μ1N (the N-terminal fragment) and μ1C (the C-terminal fragment). Fragment μ1N is small (4 kDa) and myristoylated, and this process bears a striking similarity to the cleavage event that occurs during the maturation of picornaviruses.

Genus Rotavirus

Rotaviruses are viruses of higher vertebrates and are very widely distributed. They cause gastroenteritis in their various hosts and many different serotypes are known. They have been isolated from monkeys, cattle, dogs, cats, pigs, sheep, horses, chickens, and turkeys, as well as from humans. Viruses isolated from different animals exhibit extensive serological cross-reactivity, but limited ability to replicate in other hosts. As one example, rotaviruses isolated from monkeys and cows can infect humans but cause much milder symptoms than human rotaviruses, and have been examined for use as vaccines.

Structure of Rotaviruses

Rotaviruses contain 11 genome segments that sum to 18.5 kb (Table 3.15). The replication of the virus and the overall structure of the virion are similar to those of the reoviruses, but with some exceptions. The virion is distinguishable from that of reoviruses in the electron microscope (Fig. 2.5): It resembles a wheel with spokes (*rota* = wheel in Latin). The assembly of rotaviruses differs in one important detail from that of reoviruses. Subviral particles assembled in the cytoplasm bud through the endoplasmic reticulum, acquiring an envelope that is subsequently lost. During this maturation process, the outer capsid layer of the virion is acquired.

Activation of infectivity of the virion requires cleavage of one of the major outer capsid proteins, a process that normally occurs in the enteric tract. Uncleaved rotavirus will bind to cells but cannot penetrate. The protein is cleaved at a hydrophobic sequence that is postulated to possess fusion activity. This process is analogous to the process that occurs in enveloped viruses such as influenza virus, where precursor glycoproteins must be cleaved to activate fusion activities.

The Human Rotaviruses

The rotaviruses cause diarrhea, primarily in newborns and the young, and the human rotaviruses are the single most important cause of severe diarrheal diseases of infants and young children. In one study in the United States, nonbacterial infectious gastroenteritis was found to be the second most common disease in humans, accounting for 16% of illnesses over a 10-year period and occurring on average 1.5 times per person per year. Severe diarrhea can result in dehydration that can be fatal if fluids are not replaced. In the developing world where hospitalization is not readily available, rotaviruses are a major cause of infant mortality. It has been estimated that in Asia, Africa, and Latin America there are more than 1 billion cases of diarrhea each year with 2–3 million deaths. The majority of deaths occur in children less that 5 years of age, where 1–4% of diarrheal episodes are fatal. About half of all cases of severe diarrheal disease in both developed countries and in Third World countries are caused by rotaviruses, as illustrated in Fig. 3.42. The global distribution of deaths caused by rotaviruses is illustrated in Fig. 3.43.

Epidemics of rotaviral disease occur in the winter. This is illustrated in Fig. 3.44 for epidemics in Melbourne, Australia, and in Washington, D.C. In southern Australia, epidemics peak in the June–September time frame (their winter), while epidemics peak in the January–March time frame in the eastern United States. In the United States, epidemics peak in November and December in the warmer southwestern states and move to progressively later months as they spread to cooler states north and east (Fig. 3.45). As is the case for seasonality in epidemics of disease caused by other viruses, the reasons for the association of rotaviral outbreaks with cool weather are poorly understood.

Because of the seriousness of rotaviral disease on a worldwide basis, there has been an active effort to develop vaccines against rotaviruses. Such vaccines have been directed at newborns and infants, in whom the problem is most acute. Vaccine development and interpretation of results from clinical trials have been complicated by several factors, including the possible presence of maternal antibodies in newborns, the fact that rotaviral infections often do not produce absolute and lasting immunity to subsequent reinfection, and the fact that there are multiple serotypes of rotaviruses that infect humans. There are four major serotypes of rotaviruses that cause widespread and serious infections of humans, and thus recent efforts have been directed toward developing a quadrivalent vaccine that would protect against all four serotypes. Furthermore, the objective of vaccination has now been defined as the prevention of severe rotaviral disease in very young children rather than prevention of all rotaviral disease.

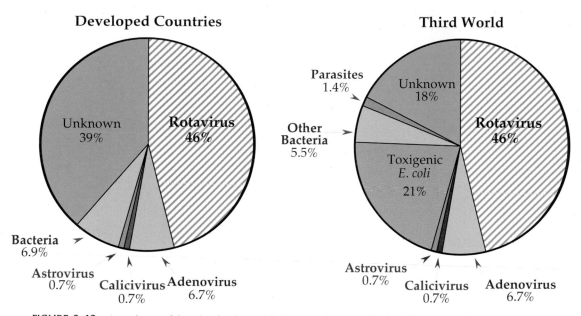

FIGURE 3.42   An estimate of the role of various etiologic agents in severe diarrheal disease requiring hospitalization in infants and young children in developed countries and in the Third World. [Adapted from Kapikian (1993).]

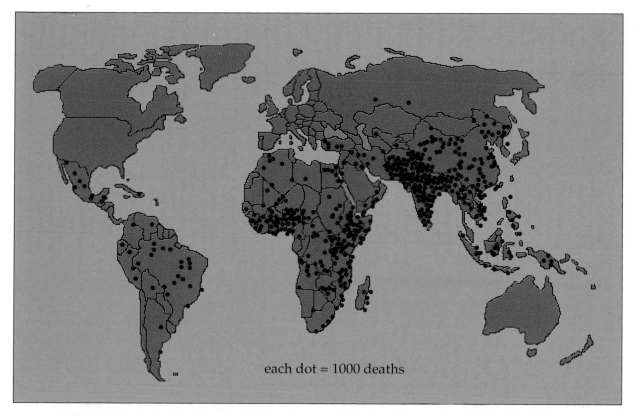

FIGURE 3.43   Estimated global distribution of the almost 1 million annual deaths due to rotavirus diarrhea. [Redrawn from that found in Glass *et al.* (1997).]

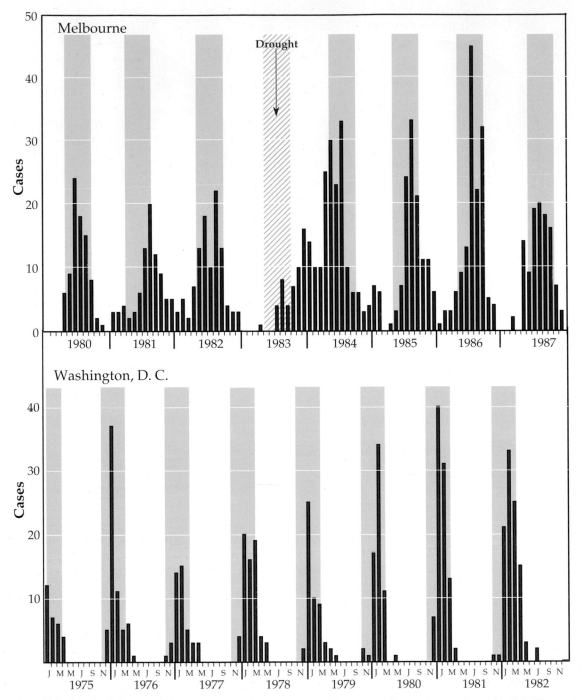

FIGURE 3.44 Monthly rotavirus infections over a multiyear period. Seasonal fluctuations are clear, with the cooler months shaded in blue. Upper panel shows hospitalizations for gastroenteritis in Melbourne, Australia. The lower panel shows hospitalizations in a children's hospital in Washington, D. C. Note the anolmalous pattern for Melbourne in 1983, a year in which there was a severe drought. [Data from Barnes *et al.* (1998) and Brandt *et al.* (1983).]

With these objectives, the most successful approach to date has been to use a rotavirus from another animal species, the Rhesus rotavirus, as a human vaccine. The Rhesus rotavirus replicates well enough in humans to elicit an immunizing response but does not cause serious illness, and the use of this virus as a human vaccine has been compared to the use of cowpox virus by Jenner to immunize against smallpox (Chapter 6). The Rhesus rotavirus will successfully immunize people against only one of the four major rotavirus serotypes, and not against the others. However, rotaviruses

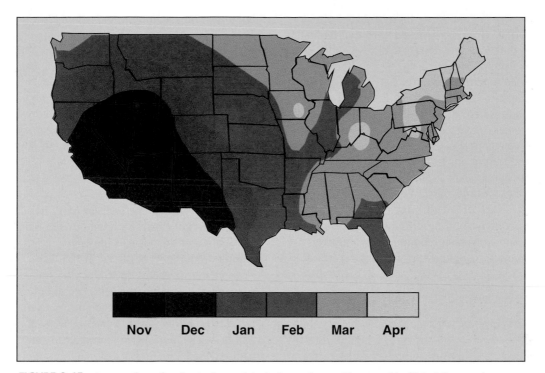

FIGURE 3.45 Average time of peak rotavirus activity in the contiguous 48 states of the United States, using cumulative data from July 1991 to June 1997. This contour plot was derived using the median value for time of peak activity reported by each regional or state diagnostic laboratory. [The surveillance system and analytic methods used to create this map are described in greater detail in Török *et al.* (1997).]

contain 11 genome segments and the different segments readily undergo reassortment when tissue culture cells are infected by more than one strain, that is, genome segments can be interchanged in progeny viruses. This property was used to isolate reassorted Rhesus rotaviruses in which all of the segments were derived from Rhesus rotavirus except for the segments encoding the surface proteins, which were derived from human viruses of the other three major serotypes. Clinical trials showed that a quadrivalent vaccine based on Rhesus rotavirus and three reassorted Rhesus rotaviruses, so that all four serotypes important for humans were present, was successful at preventing severe rotaviral disease in newborns and the very young. This vaccine was licensed in 1998 for general use in the United States. It was anticipated that this vaccine would be useful not only in the developing world where rotavirus infection causes many deaths in infants, but also in the developed world where rotaviral infections lead to many cases of diarrhea each year that require hospitilization or visits to physicians. However, on widespread use in the United States, it was found that a small number of infants developed the bowel obstruction called intussusception after immunization. This obstruction sometimes clears spontaneously but can require a fairly benign treatment by medical personnel. In a small minority of cases, surgery is required to correct the defect. The vaccine was withdrawn and its future remains in doubt.

The withdrawal of the rotavirus vaccine raises interesting legal and moral questions, and illustrates the difficulties associated with developing and introducing vaccines. Roughly 1 out of 2000 infants develop intussusception in the first 2 years of life. At least some of these cases may be due to natural infection with rotaviruses. Although it seems clear that vaccination with the rotaviral vaccine triggers intussusception in a small fraction of vaccinees, it is not known whether vaccinated infants are more likely to develop the obstruction in the first 2 years of life than are nonvaccinated infants. It is even possible that the vaccine is actually protective in that fewer vaccinated infants will ultimately develop intussusception than nonvaccinated infants. However, because of legal issues and the ethical dilemma of giving a problematical vaccine for a disease that is not life threatening in developed countries, it seems unlikely that data to address this issue will be forthcoming, at least in the near future. However, if the vaccine is not used in the United States or other developed countries because of concerns about its safety, it is unlikely that developing nations will adopt it, despite the fact that the vaccine would undoubtly save hundreds of thousands of infants from fatal rotaviral infection if it were widely used in such countries.

Other approaches have been pursued for the development of a vaccine. One such approach is the isolation of attenuated human rotavirus strains that are cold sensitive or temperature sensitive and therefore less virulent. Following isolation, reassortant viruses can be selected that retain the genes rendering the virus cold sensitive but which have the surface antigens from the different serotypes. Whether such a vaccine would be superior to the Rhesus rotavirus vaccine is unknown. Another possible approach is the use of empty particles as subunit vaccines. Virus-like particles assemble spontaneously when rotavirus structural proteins are produced in large concentrations using baculovirus vectors. Particles containing from one to four structural proteins are formed when the proteins are expressed in the proper combinations, and these empty particles have been useful in studies of the structure of rotaviruses. Whether a vaccine that uses these findings could be developed that was inexpensive enough and convenient enough for widespread use in developing countries is unknown.

Genus Orbivirus

The genus *Orbivirus* is widely distributed. It contains more than 100 serotypes of viruses grouped into 14 serogroups or species. Orbiviruses contain 10 segments of dsRNA totaling 19 kb in an icosahedral virion 86 nm in diameter. Unlike reoviruses and rotaviruses, orbiviruses are arboviruses, transmitted by biting flies, mosquitoes, or ticks, and able to replicate in the vector as well as in the vertebrate host. Among the vectors known for various orbiviruses are insects of the genus *Culicoides* (midges and gnats), phle-

botomine flies, culicine and anopheline mosquitoes, and *Ixodes* ticks. Members of five serogroups, including African horse sickness, bluetongue, and epizootic hemorrhagic disease viruses, cause disease in domestic animals, and members of four other serogroups cause human disease. Human disease caused by naturally acquired orbiviruses may require hospitalization but is normally not life threatening. However, the animal pathogens may cause serious illness in domestic or wild animals (Table 3.17).

Veterinary Pathogens

African horse sickness virus (AHF) has caused many epidemics of fatal illness in horses in sub-Saharan Africa. The distribution of this virus is illustrated in Fig. 3.46C. The first recorded epidemic occurred in the Cape Colony in 1719. Thereafter, disastrous epidemics occurred every 20 years or so up until this century, when vaccines became available. The virus appears to have been endemic, presumably in the zebra, but wherever horses were introduced, epizootic AHF was sure to follow. The mortality in introduced horses is close to 90%, and AHF had a major impact on agriculture, exploration, and conquest in Africa. The military had to operate without cavalry and the early explorers often walked rather than rode. The virus represents an example of an endemic virus that appears to cause little disease in its native host (zebra), but which causes very serious illness when transmitted to a non-native host (horses).

TABLE 3.17   Representative Orbiviruses and Coltiviruses Causing Disease in Humans and Domestic Animals

Genus	Serogroup	Hosts	Vector	Disease syndromes	Distribution
Orbivirus					
	African horse sickness	Horse, dog, zebra	*Culicoides* (midges)	Cardiopulmonary disease, hemorrhagic fever	Africa, Asia
	Bluetongue	Sheep, cow, goat, deer	*Culicoides* (midges)	Fever, frothing at mouth, shock	Africa, Asia, Australia[a], Americas
	Epizootic hemorrhagic disease	Deer	*Culicoides* (midges)	Similar to bluetongue	Americas, Australia, Africa
	Palyam	Cow	*Culicoides* (midges)	Abortion	South Africa, Japan
	Orungo	**Humans**	Mosquitoes	Febrile illness	Africa
	Changuinola	**Humans**	Phlebotomines	Febrile illness	Panama
	Kemerovo	**Humans**	Ticks	Febrile illness, encephalitis	Russia, Eastern Europe
Coltivirus					
	Colorado tick fever	**Humans**	Ticks	Febrile illness, encephalitis, hemorrhagic fever	North America
	Eyach	**Humans**	Ticks	Encephalitis?	Europe, China

[a] Australian isolates of bluetongue virus are not pathogenic.

A. Bluetongue virus (BTV)

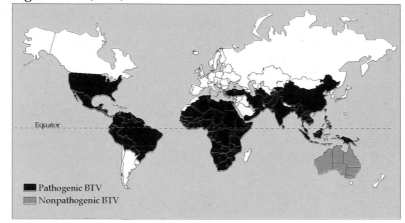

B. Hemorrhagic disease of deer (IHDD), Ibaraki virus, and the Kemerovo group viruses

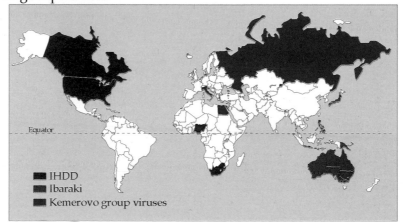

C. African Horse Sickness (AHSV)

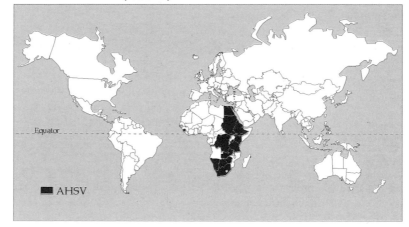

FIGURE 3.46 Geographical distribution of orbiviruses that cause disease in animals and humans. [From Fields *et al.* (1996, p. 1736).]

Another important veterinary pathogen is bluetongue virus, which causes a serious disease in sheep. Bluetongue virus is very widely distributed (Fig. 3.46A) and causes widespread disease, but, interestingly, the strains of bluetongue virus in Australia are not pathogenic. Also widespread are viruses that cause epizootic hemorrhagic disease in animals (Fig. 3.46B), including outbreaks in deer in North America.

Persistence of Orbiviruses on Erythrocytes

At least some orbiviruses have evolved an unusual way of persisting in nature as arboviruses. In tropical areas of the world, arthropods may be continuously active and viruses associated with such arthropods may persist by continuous passage between the vertebrate and the invertebrate host. However, in many areas of the world, including some tropical areas, the vector activity may become low or nonexistent during some periods, such as during the dry season or during the winter. Arboviruses must be able to survive such periods of low vector activity, and different arboviruses have solved this problem in different ways. Most have evolved ways to persist in the invertebrate host during periods of inactivity. Many can be passed transovarily in the arthropod host, others can survive in diapausing insects. Orbiviruses, in contrast, have evolved a mechanism to persist in the vertebrate host for long periods. Viral infection of vertebrates is normally cleared rapidly by the immune system so that viremia of sufficient titer to infect a new arthropod taking a blood meal normally lasts only a few days. However, bluetongue virus becomes associated with red blood cells by binding to glycophorins on the surface of the cell, where it persists in indentations in the membrane in a nonreplicating state protected from the immune system. The virus evidently can remain attached and viable for the life of the erythrocyte. When an arthropod takes a blood meal, the virus bound to erythrocytes is able to initiate infection of the arthropod. Because erythrocytes have an average lifetime of 160 days, the virus can persist in a viable state in the vertebrate host for many months.

Genus Coltivirus

The coltiviruses possess 12 genome segments summing to 26 kb (Table 3.15). The virion is icosahedral and 80 nm in diameter. Like the orbiviruses, the coltiviruses are arboviruses, transmitted by ticks. Four viruses are recognized, found in North America, Europe, and China, and all infect humans. The prototype virus is Colorado tick fever virus (CTF), from which the genus gets its name (*Colorado tick fever*). CTF is present in the western United States and is transmitted by *Dermacentor andersoni* ticks (Fig. 3.47). The life cycle of the virus is illustrated in Fig. 3.48. CTF

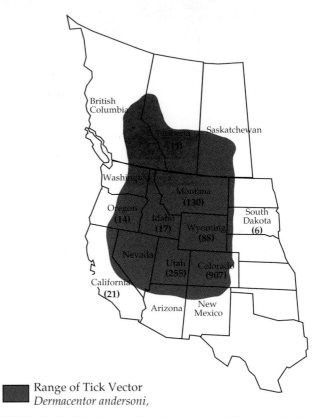

■ Range of Tick Vector
Dermacentor andersoni,

FIGURE 3.47   Distribution of the primary vector of Colorado tick fever, *D. andersoni*, shown in color, and the number of diagnosed human cases of Colorado tick fever in various states between 1980 and 1988. [From Tsai (1991); reprinted in Fields *et al.* (1996, p. 1753).]

infects many mammals, including humans, but the natural cycle of transmission involves primarily small mammals and larval and nymphal ticks. Adult ticks may transmit the virus to humans and large mammals outside the normal transmission cycle. Transovarial transmission does not occur in ticks and larval ticks are normally infected by feeding on small mammals that are viremic. In humans, CTF causes an illness characterized by fever, myalgia, chills, headache, and malaise, and 20% of cases require hospitalization. The acute illness lasts 5–10 days. Recovery may be uneventful or convalescence may be prolonged for several weeks. CNS infection or hemorrhagic fever may occur, almost always in children. Fatalities are rare (<0.1%).

Persistence of CTF in Erythrocytes

CTF has evolved a way to persist in the vertebrate host that resembles that used by bluetongue virus. CTF infects bone marrow cells early after infection, including erythrocyte precursors. The virus remains within the erythrocyte after it matures, in a nonreplicating state, and appears to persist for the life of the erythrocyte. When a tick takes a blood meal, it

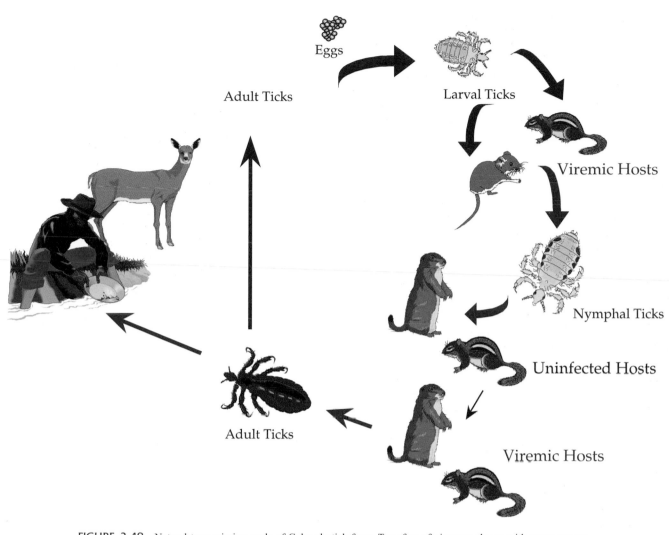

FIGURE 3.48 Natural transmission cycle of Colorado tick fever. Transfers of virus are shown with green arrows. Larval ticks feed on small mammals that can remain viremic for long periods of time and then transmit virus to other small mammals. Adult ticks, although not important for maintaining the virus in nature, may then bite nonreservoir hosts such as deer or man. Adult ticks lay eggs to produce the next generation of larval ticks, but no transovarial transmission of virus occurs. [Adapted from Bowen (1988, Fig. 4).]

can be infected by the virus within the erythrocyte. The differences in the mechanisms used by CTF and bluetongue virus to persist in the blood of vertebrates are thought to reflect the different vectors used. *Culicoides* flies, the vector of bluetongue, take smaller blood meals and digest the blood meal in a different way than ticks, the vector of CTF.

Comparison of the Reoviridae with Other RNA Viruses

Some aspects of the replication cycle of the Reoviridae resemble those of single-strand (+)RNA viruses and other aspects resemble those of single-strand (−)RNA viruses (Chapter 4). Replication proceeds through a plus- strand intermediate that exists, at least transiently, free in the cytoplasm, a characteristic of (+)RNA viruses. Reoviruses could have originated from the plus-strand viruses through the acquisition of an RNA synthetase that is packaged as part of the virion. In contrast, packaging of the RNA synthetase in the virion, which is necessary to begin the infection process, and the retention of the genome in the entering subviral particle, from which it is never released, are features that are shared with the minus-strand RNA viruses. Packaging of the RNA synthetase makes it feasible for a virus of vertebrates to have its genome in multiple segments, a feature of reoviruses and of many of the (−)RNA viruses. Segmented genomes allow reassortment to occur during mixed infection, and reassortment is known to occur in nature. The acquisition of novel genome segments from related virus strains has occurred frequently during reovirus evolution and has been advantageous for the survival of the virus in nature.

FURTHER READING

General

Porterfield, J. S., ed. (1995). "Exotic Viral Infections," Kass Handbook of Infectious Diseases. Chapman & Hall Medical, London.

Nathanson, N., Ahmed, R., Gonzalez-Scarano, F., *et al.*, eds. (1996). "Viral Pathogenesis." Lippincott-Raven, Philadelphia.

Picornaviridae

Halstead, L. S. (1998). Post-polio syndrome. *Sci. Am.* **278**; April; pp. 42–47.

Xiang, W., Paul, A. V., and Wimmer, E. (1997). RNA signals in entero- and rhinovirus genome replication. *Semin. Virol.* **8**; 256–273.

Hyypiä, T., Hovi, T., Knowles, N. J. *et al.* (1997). Classification of enteroviruses based on molecular and biological properties. *J. Gen. Virol.* **78**; 1–11.

Gould, T. (1995). "A Summer Plague: Polio and Its Survivors." Yale University Press, New Haven.

Caliciviridae

Thiel, H. J., and Konig, M. (1999). Caliciviruses: An overview. *Vet. Microbiol.* **69**; 55–62.

Hepatitis E

Mast, E. E., and Krawczynski, K. (1996). Hepatitis E: An overview. *Ann. Rev. Med.* **47**; 257–266.

Tsarev, S. A., Binn, L. N., Gomatos, P. J., *et al.* (1999). Phylogenetic analysis of hepatitis E virus isolates from Egypt. *Med. Virol.* **57**; 68–74.

Astroviridae

Glass, R. I., Noel, J. Mitchell, D., *et al.* (1996). The changing epidemiology of astrovirus-associated gastroenteritis: A review. *Archi. Virol.* pp. 287–300.

Hart, C. A. and Cunliffe, N. A. (1999). Viral gastroenteritis. *Curr. Opin. Infect. Dis.* **12**; 447–457.

Togaviridae

Monath, T. P., ed. (1988). "The Arboviruses: Epidemiology and Ecology." CRC Press, Inc., Boca Raton, FL.

Calisher, C. H. (1994). Medically important arboviruses in the United States and Canada. *Microbiol. Rev.* **7**; 89–116.

Strauss, J. H., and Strauss, E. G. (1994). The alphaviruses: Gene expression, replication, and evolution. *Microbiol. Rev.* **58**; 491–562.

Strauss, J. H., and Strauss, E. G. (1997). Recombination in alphaviruses. *Semin Virol.* **8**; 85–94.

Flaviviridae

Chambers, T. J., Hahn, C. S., Galler, R., *et al.* (1990). Flavivirus genome organization, expression, and replication. *Ann. Rev. Microbiol.* **44**; 649–688.

Rey, F. A., Heinz, F. X., Mandl, C. W., *et al.* (1995). The envelope glycoprotein from tick borne encephalitis virus at 2Å resolution. *Nature London* **375**; 291–298.

Strauss, J. H., and Strauss, E. G. (1996). Current advances with yellow fever. *In* "Microbe Hunters—Then and Now." (H. Koprowski and M.B.A Oldstone, eds.). Medi-Ed Press, Bloomington, IL, pp. 113–137.

Meyers, G., and Thiel, H.-J. (1996). Molecular characterization of pestiviruses. *Adv. Virus Res.* **47**; 53–118.

Rigau-Pérez, J. G., Clark, G. G., Gubler, D. J., *et al.* (1998). Dengue and dengue haemorrhagic fever. *Lancet* **352**; 971–977.

Di Bisceglie, A. M. (1999). The unmet challenges of hepatitis C. *Sci. Am.* **28**; 80–85 OCT.

Bartenschlager, R., and Lohmann, V., (2000). Replication of Hepatitis C virus. *J. Gen. Virol.* **81**; 1631–1648.

Coronaviridae and Arteriviridae

de Vries, A. A. F., Horzinek, M. C., Rottier, P. J. M., *et al.* (1997). The genome organization of the Nidovirales: Similarities and differences between arteri-, toro-, and coronaviruses. *Semin. Virol.* **8**; 33–47.

Reoviridae

Joklik, W. K., and Roner, M. R. (1996). Molecular recognition in the assembly of the segmented reovirus genome. *Prog. Nucl. Acid Res.* **53**; 249–281.

Lee, P. W. K., and Gilmore, R. (1998). Reovirus cell attachment protein σ1: Structure-function relationships and biogenesis. *Curr. Top. Microbiol. Immunol.* **233**; 137–153.

Saragovi, H. U., Rebai, N., Roux, E., *et al.* (1998). Signal transduction and antiproliferative function of the mammalian receptor for type 3 reovirus. *Curr. Top. Microbiol. Immunol.* **233**; 155–166.

Evolution

Koonin, E. V., and Dolja, V. V. (1993). Evolution and taxomony of positive-strand RNA viruses —implications of comparative analysis of amino-acid sequences. *Crit. Rev. Biochem. and Mol. Biol.* **28**; 375–430.

4

Minus-Strand RNA Viruses

INTRODUCTION

Seven families of viruses contain minus-strand RNA [(–)RNA], also called negative-strand RNA, as their genome. These are listed in Table 4.1. Included in the table are the names of the genera belonging to these families and the hosts infected by these viruses. Six of the families are known to contain members that cause epidemics of serious human illness. Diseases caused by these viruses include influenza (Orthomyxoviridae), mumps and measles (Paramyxoviridae), rabies (Rhabdoviridae), encephalitis (several members of the Bunyaviridae), upper and lower respiratory tract disease (numerous viruses in the Paramyxoviridae), and hemorrhagic fever (many viruses belonging to the Bunyaviridae, the Arenaviridae, and the Filoviridae), as well as other diseases. Bornavirus, the sole representative of the Bornaviridae, also infects humans and may cause neurological illness, but proof of causality is lacking. Many of the (–)RNA viruses presently infect virtually the entire human population at some point in time (for example, respiratory syncytial virus, influenza virus), whereas others did so before the introduction of vaccines against them (for example, measles virus and mumps virus). These viruses are thus responsible for a very large number of cases of human illness. The diseases caused by such widespread viruses are usually serious but have a low (although not insignificant) fatality rate. In contrast, some (–)RNA viruses, such as rabies and Ebola viruses, cause illnesses with high fatality rates but (fortunately) infect only a small fraction of the human population. The (–)RNA viruses are major causes of human suffering, and all seven families and the viruses that belong to these families will be described here.

OVERVIEW OF THE MINUS-STRAND RNA VIRUSES

Viruses belonging to four families of (–)RNA viruses, the Paramyxoviridae, the Rhabdoviridae, the Filoviridae, and the Bornaviridae, contain a nonsegmented RNA genome having similar organization. They are grouped into the order *Mononegavirales* (*mono* because the genome is in one piece, *nega* for negative-strand RNA). This was the first order to be recognized by the International Committee on Taxonomy of Viruses and still is one of only three orders currently recognized. Viruses belonging to the other three families, the Arenaviridae, Bunyaviridae, and Orthomyxoviridae, possess segmented genomes with two, three, and seven or eight segments, respectively. Regardless of whether the genome is one RNA molecule or is segmented, the genomes of all (–)RNA viruses encode a similar constellation of genes, as illustrated in Fig. 4.1. In the Mononegavirales, the order of genes along the genome is conserved among the viruses (although the number of genes may differ). In the viruses with segmented genomes, the genes can be ordered in the same way if the segments are aligned as shown. In addition, many features of virion structure and of replication pathways are shared among the (–)RNA viruses.

Structure of the Virions

All (–)RNA viruses are enveloped and have helical nucleocapsids. The different families encode either one or two glycoproteins (called G in most of the families but called HA, NA, F, or HN in some, after hemagglutinating, neuraminidase, or fusion properties). These glycoproteins are present in the viral envelope. In most cases, cleavages

TABLE 4.1   Minus-Strand RNA Viruses

Family/genus	Genome size (kb)	Type virus[a]	Host(s)[b]	Transmission
Mononegavirales (nonsegmented)				
Rhabdoviridae	13–16			
Vesiculovirus		VSIV	Vertebrates	Some arthropod-borne
Lyssavirus		Rabies	Vertebrates	Contact with saliva
Ephemerovirus		BEFV	Cattle	Arthropod-borne
Novirhabdovirus		IHNV	Fish	—
Two genera of plant viruses			Plants	Arthropod-borne
Filoviridae	13			
"Marburg-like" viruses		Marburg	Vertebrates	—
"Ebola-like" viruses		Zaire Ebola virus	Vertebrates	—
Paramyxoviridae	16–20			
Respirovirus		Sendai	Vertebrates	Airborne
Morbillivirus		Measles	Vertebrates	Airborne
Rubulavirus		Mumps	Vertebrates	Airborne
Megamyxovirus[c]		Hendra	Vertebrates	Airborne
Pneumovirus		HRSV	Vertebrates	Airborne
Metapneumovirus		TRTV	Turkeys	Airborne
Bornaviridae	~9			
Bornavirus		BDV	Vertebrates	
Segmented negative-strand RNA viruses				
Orthomyxoviridae	13 in 8 segments			
Influenzavirus A		Influenza A	Vertebrates	Airborne
Influenzavirus B		Influenza B	Vertebrates	Airborne
Influenzavirus C		Influenza C	Vertebrates	Airborne
Thogotovirus		Thogoto	Vertebrates	Arthropod-borne
Bunyaviridae	11–20 in 3 segments			
Bunyavirus		Bunyamwera	Vertebrates	Mosquito-borne
Hantavirus		Hantaan	Vertebrates	Feces/urine/saliva
Nairovirus		Dugbe	Vertebrates	Tick-borne
Phlebovirus		Rift Valley fever	Vertebrates	Arthopod-borne
Tospovirus		TSWV	Plants	Thrips
Arenaviridae	10–14 in 2 segments			
Arenavirus		LCMV	Vertebrates	Urine/saliva

[a]*Abbreviations*: VSIV, vesicular stomatitis Indiana virus; BEFV, bovine ephemeral fever virus; IHNV, infectious hematopoietic necrosis virus; HRSV, human respiratory syncytial virus; TRTV, turkey rhinotracheitis virus; BDV, Borna disease virus; TSWV, tomato spotted wilt virus; LCMV, lymphocytic choriomeningitis virus.
[b]In all cases, vertebrates include humans as hosts.
[c]Megamyxovirus is a provisional genus at present.

are required to produce the mature glycoproteins, such as cleavage to release a signal peptide, cleavage to separate two glycoproteins produced as a common precursor, or cleavage to activate viral infectivity. The glycoproteins project from the lipid bilayer as spikes that are visible in the electron microscope (see, e.g., Fig. 2.18D).

All (–)RNA viruses have a single major nucleocapsid protein (called N) that encapsidates the virion RNA to form the helical nucleocapsid. Also present in the nucleocapsid is a phosphorylated protein that is required for RNA synthesis, variously called P (for phosphoprotein) or NS (for nonstructural protein because it was not originally known to be

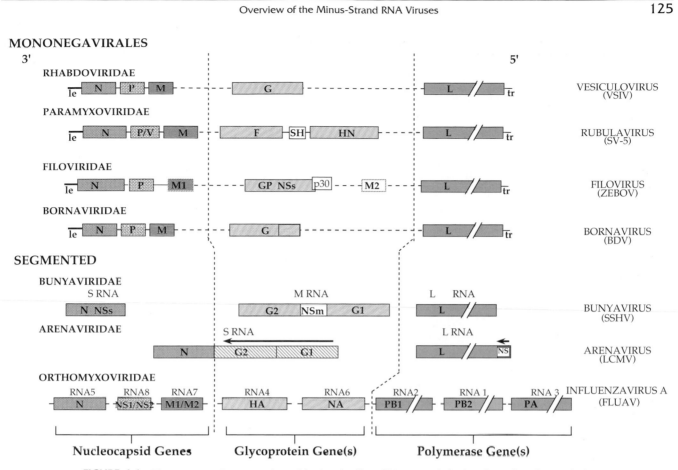

MONONEGAVIRALES

FIGURE 4.1 The genomes of representatives of the four families of Mononegavirales have been aligned to maximize functional similarity between gene products. The genome organization of a member of the Bunyavirus genus of the Bunyaviridae, of an arenavirus, and of influenzavirus A are also shown. The individual gene segments of the members of the three families with segmented genomes have been aligned according to similarity of function with the Mononegavirales above. Gene expression strategies for the other genera of Bunyaviridae vary (see Fig. 4.19). Abbreviations of virus names are as follows: VSIV, vesicular stomatitis Indiana virus; SV-5, simian virus 5; ZEBOV, Zaire Ebola virus; BDV, Borna disease virus; SSHV, snowshoe hare virus; LCMV, lymphocytic choriomeningitis virus; FLUAV, influenzavirus A. The gene products are abbreviated as follows: le is a leader sequence; N is the nucleoprotein; P is the phosphoprotein; M (M1, M2) are matrix proteins; G (G1, G2) are membrane glycoproteins; F is the fusion glycoprotein; HN is the hemagglutinin-neuraminidase glycoprotein; L is the RNA polymerase, NA is the neuraminidase glycoprotein; HA is the hemagglutinin glycoprotein; NS (NS2, SH, NSs, NSm) are minor proteins of unknown function; PB1, PB2, and PA are components of the influenza RNA polymerase, tr is the trailer sequence. Within a given genome, the genes are drawn approximately to scale. mRNAs for most genes would be synthesized left to right; however, an arrow over a gene means that it is in the opposite orientation (ambisense genes). [Redrawn from Strauss *et al.* (1996, Fig. 5).]

a component of the virion), as well as a few molecules of an RNA-dependent RNA polymerase. The polymerase is a large, multifunctional protein called L in most families but is present as three proteins in the Orthomyxoviridae. L and P form a core polymerase that replicates the viral genome and synthesizes mRNAs.

A matrix protein (M) is present in all of the viruses except the bunyaviruses and the arenaviruses. M underlies the lipid bilayer where it interacts with the nucleocapsid. M also inhibits host transcription and shuts down viral RNA synthesis prior to packaging.

The (–)RNA virions are heterogeneous to a greater or lesser extent. Members of five families often appear

roughly spherical in the electron microscope. The example of influenza virus is shown in Figs. 2.1 and 2.18, and the paramyxovirus Sendai virus is shown in Fig. 2.18. The compositions of these virions are not rigorously fixed and some variability in the ratios of the different components, particularly in the glycoprotein content, is present. The rhabdoviruses are bullet shaped or bacilliform and appear more regular (Fig. 2.19), but even here variations in the composition of the glycoproteins in the envelope can occur. The filoviruses are filamentous (Fig. 2.19). Influenza viruses and human respiratory syncytial virus also produce filamentous forms as well as round virions. In fact, clinical isolates of these viruses are predominantly filamentous.

Synthesis of mRNAs

For all (−)RNA viruses, the first event in infection is the synthesis of mRNAs from the minus-strand genome by the RNA polymerase present in the nucleocapsid. Because this polymerase is necessary for the production of the mRNAs, and because the proteins translated from the mRNAs are required for replication of the genome, the naked genomes of (−)RNA viruses are not infectious, nor are complementary RNA copies of the genomes. It has been possible, nonetheless, to rescue virus from cDNA clones of viral genomes by using special tricks, as described in Chapter 9.

Multiple mRNAs are produced from minus-strand genomes. By definition, each region of the genome from which an independent mRNA is synthesized is called a gene. In (−)RNA viruses with segmented genomes, it is obvious that multiple mRNAs are produced (the number of mRNAs produced actually exceeds the number of segments, as described below). In the Mononegavirales, multiple mRNAs arise from the use of a single polymerase entry site at the 3′ end of the genome. The polymerase then recognizes conserved start and stop signals at the beginning and end of each gene to generate discrete mRNAs. Synthesis of mRNAs is controlled by the location of the gene relative to the single polymerase entry site, because mRNA synthesis is obligatorily sequential and attenuation occurs at each gene junction.

Most of the mRNAs are translated into a single protein, but a few of the genes produce mRNAs that are translated into more than one product. Multiple products can be produced from the same gene by the use of alternative translation initiation codons during translation of an mRNA; by the introduction of nontemplated nucleotides during mRNA synthesis, which results in a shift in the reading frame; or by splicing of an mRNA. The (−)RNA viruses do not produce polyproteins and virus-encoded proteases are unknown among them. Most of the glycoproteins that they encode are produced as precursors, however, that are processed by cellular enzymes. In no case are the mRNAs exact complements of virion RNAs. This is obvious in the case of the Mononegavirales, where as many as 7–10 mRNAs are produced from a single long genomic RNA, but is also true of the segmented (−)RNA viruses. The mRNAs lack cis-active sequences required for encapsidation and replication that are present near the ends of the antigenome or antigenome segments.

Replication of the Genome

Replication of the (−)RNA genome requires the production of a complementary copy of the genome, called an antigenome or virus complementary RNA (vcRNA), which is distinct from the mRNAs (schematically illustrated in Figs. 1.9C and D). Neither the genomic (−)RNA nor the antigenomic template produced during replication is ever free in the cytoplasm. Instead, replication of the genome, as well as the synthesis of mRNAs, takes place in nucleocapsids (sometimes referred to as ribonucleoprotein or RNP), which always contain the phosphoprotein and the polymerase as well as N and the viral RNA. Replication can only occur in the presence of ongoing protein synthesis to produce the new proteins required to encapsidate the genome or antigenome. The mRNAs can be synthesized in the absence of viral protein synthesis and lack encapsidation signals, so that they are released into the cytoplasm where they can associate with ribosomes and be translated. Thus, early after infection, mRNAs are synthesized. After translation of the mRNAs, which leads to production of sufficient amounts of viral proteins, a switch to the production of antigenomes for use as templates occurs, followed by production of genomic RNA from the antigenomic templates.

The genomes (or genome segments) of all (−)RNA viruses have sequences at the ends that are complementary. In the bunyaviruses, the RNA forms panhandles, circular structures that are visible in the electron microscope. In other viruses, circles have not been seen but may form transiently during replication. It is possible that these complementary sequences exist to promote cyclization of the RNA, which may be required for replication of the genome or synthesis of mRNAs. However, it seems more likely that the promoter at the 3′ end of the genomic RNA that is recognized by the viral RNA polymerase for the production of antigenomes is the same, at least in part, as the promoter at the 3′ end of the antigenomic RNA that is used to initiate the production of genomic RNA. In this event, the sequences at the two ends of the genome or antigenome that encompass these promoters would be complementary.

Host Range of the (−)RNA Viruses

All seven families contain members that infect higher vertebrates, including humans. For five of the families, only vertebrate hosts are known. The rhabdoviruses and bunyaviruses, however, have a broader host range. Some are arboviruses that replicate in an arthropod vector as well as in a vertebrate host, and others infect only insects. In addition, some genera of rhabdoviruses and bunyaviruses consist of plant viruses. Some of these are transmitted to the plants by insect vectors in which the viruses also replicate.

FAMILY RHABDOVIRIDAE

The genome organization of the rhabdoviruses is the simplest of the (−)RNA viruses and it is useful to begin our coverage with this group. The genome is a single piece of minus-strand RNA 11–15 kb in size. In most viruses, the genome has five genes, which result in the production of six or seven proteins, five of which are present in the virion. The animal rhabdoviruses are bullet shaped, approximately 200 nm long and 75 nm in diameter (Fig. 2.19), whereas some of the plant viruses are bacilliform, being rounded at both ends. The rhabdoviruses infect mammals, birds, fish, insects, and plants, and are presently divided into six genera. A listing of these genera and a representative sample of the viruses in each genus, together with several characteristics of each virus, are shown in Table 4.2.

Members of three genera infect mammals, namely, the vesiculoviruses (type virus: vesicular stomatitis Indiana virus), lyssaviruses (type virus: rabies virus), and ephemeroviruses (type virus: bovine ephemeral fever virus). The novirhabdoviruses infect fish, and the cytorhabdoviruses and nucleorhabdoviruses infect plants. Some or all of the members of four genera are transmitted by arthropods (Table 4.2). In addition, a large number of rhabdoviruses have not been assigned to a genus. The animal rhabdoviruses replicate in the cytoplasm, but certain of the plant rhabdoviruses may replicate in the nucleus.

TABLE 4.2   Rhabdoviridae

Genus/members[a]	Virus name abbreviation	Usual host(s)	Transmission/vector?	Disease	World distribution
Vesiculovirus					
Vesicular stomatitis Indiana	VSIV	Humans, horses, ruminants, swine	Airborne, insects?	Vesicles on tongue and lips	Americas
Chandipura virus	CHPV	Mammals including humans	Sandflies	Febrile illness	India, Asia?
Piry	PIRYV	Mice, humans	Sandflies	Febrile illness	Brazil
Lyssavirus					
Rabies	RABV	Humans, dogs, skunks, foxes, raccoons	Infectious saliva	Malaise, then delirium, then coma and death	Worldwide, except Australia, Papua New Guinea, and Antarctica
Bat lyssaviruses	ABLV, EBLV, LBV[b]	Bats, humans	Infectious saliva	Like rabies	Europe, Africa, Australia
Mokola		Humans, dogs, cats, shrews	?	Like rabies	Africa
Ephemerovirus					
Bovine ephemeral fever	BEFV	Cattle, water buffalo	Hematophagous arthropods	Fever, anorexia	Africa, Asia, Australia
Adelaide River virus	ARV	Cattle	?	?[c]	Australia
Berrimah virus	BRMV	Cattle	?	?[c]	Australia
Novirhabdovirus					
Infectious hematopoietic necrosis (and other fish viruses)	IHNV	Salmonid fish	?	Hemorrhage	Pacific Northwest of North America
Cytorhabdovirus					
Lettuce necrotic yellows		Plants	Aphids		
Northern cereal mosaic		Plants	Leafhopper		
Strawberry crinkle		Plants	Aphid		
Nucleorhabdovirus					
Potato yellow dwarf		Plants	Leafhopper		
Maize mosaic		Plants	Leafhopper		
Sonchus yellow net		Plants	Aphid		

[a]Representative members of each genus are shown, and the first virus listed is the type species.
[b]ABLV, Australian bat lyssavirus; EBLV, European bat lyssavirus; LBV, Lagos bat virus.
[c]Not known as animal pathogens, but will infect healthy sentinel cattle.

Genus Vesiculovirus

Vesicular stomatitis virus (VSV) has been extensively studied and serves as a model for the replication of (−)RNA viruses in general and rhabdoviruses in particular. Its genome size is 11,161 nt. The genome is neither capped nor polyadenylated, consistent with the fact that it is minus-strand RNA. Two serotypes exist, New Jersey (VSNJV) and Indiana (VSIV).

Synthesis of mRNAs

The VSV nucleocapsid has about 1200 copies of N protein as its major structural component, but also contains about 500 molecules of P and 50 copies of L. It can synthesize RNA, and P, L, and N are all required for this activity. The organization of the genome and the production of five mRNAs from it are illustrated in Fig. 4.2. There is a single polymerase entry site at the 3′ end of the genome, and production of mRNAs is obligatorily sequential. Synthesis begins at the exact 3′ end of the genome and a leader RNA of 48 nucleotides is first synthesized. The leader is released and synthesis of the first mRNA, that for N, is initiated. The RNA polymerase complex has capping activity, and the mRNA is capped during or shortly after initiation. At the end of the gene for N, the transcriptase reaches a conserved sequence AUACU$_7$, where it begins to stutter and produces a poly(A) tract at the 3′ end of the mRNA. The polymerase complex will not terminate or stutter unless the conserved AUAC is present immediately upstream of the U$_7$ tract, and the sequence AUACU$_7$ is therefore a consensus termination-polyadenylation signal. The capped and polyadenylated mRNA for N is terminated and released, the transcriptase skips the next two nucleotides, which are referred to as the intergenic sequence, and initiates synthesis of the second mRNA, that for P, at the conserved gene start signal UUGUC. Following synthesis of this mRNA, the polymerase again stutters at the oligo(U) tract in the AUACU$_7$ signal to produce a poly(A) tract, releases the capped and polyadenylated mRNA, skips the next two nucleotides, and begins synthesis of the third gene, that for M. The process continues in this way through the fourth gene (the G protein) and the fifth gene (the L protein, L for large because it comprises about 60% of the genome). In this way, five capped and polyadenylated mRNAs are produced. In VSV, the intergenic sequence is always two nucleotides. After releasing the L mRNA, the polymerase complex terminates synthesis some 50 nucleotides before the 5′ end of the genome is reached.

mRNA synthesis always begins by initiating synthesis of the leader at the 3′ end and mRNAs are always produced in strict sequential order. Initiation at each downstream gene requires the termination of the preceding upstream gene. However, during synthesis of mRNAs, attenuation occurs

A. Location of intergenic sequences of VSV (a rhabdovirus), and detailed view of the M/G intergenic region

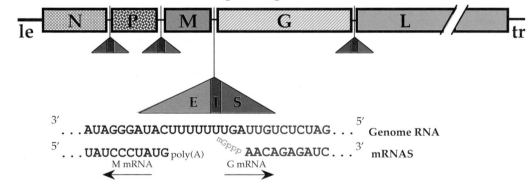

B. Genomic sequences at other intergenic regions in the VSV genome

       3′   5′

N/P   . . . CGAUGUAUACUUUUUUUGAUUGUCUAUAG . . .

P/M   . . . CAUCUGAUACUUUUUUUCAUUGUCUAUAG . . .

G/L   . . . UUAAAAAUACUUUUUUUGAUUGUCGUUAG . . .

FIGURE 4.2 (A) Schematic diagram of the VSV genome, where "le" is the leader sequence and "tr" is the trailer sequence. The five genes N, P, M, G, and L were defined in the legend to Fig. 4.1 and are described in more detail in the text. The positions of the conserved regulatory sequences at the gene boundaries are shown by the triangles. Each of these intergenic sequences is composed of E (end), I (intergenic), and S (start) domains. (B) Sequences in VSV at the other three gene boundaries. [Data for this figure came from Rose and Schubert (1987, Fig. 3).]

at each initiation step so that there is a gradient in the amounts of mRNAs produced. N mRNA is produced in highest abundance and L mRNA in least abundance. This attenuation appears to be important for regulation of the virus life cycle, so that the mRNAs for proteins needed in most abundance are produced in most abundance. Reorganization of the genome to change the order of genes gives rise to viable virus, but the yield of such virus during an infection cycle in cultured cells is less.

The mRNAs for N, M, G, and L are each translated into a single protein. That for P is translated into three proteins. The major translation product of this mRNA is P, which is produced using an initiation codon near the 5′ end of the mRNA. Initiation of translation also occurs at two downstream AUGs. These two downstream AUGs are in frame with one another but in a different reading frame from P. Use of these alternative AUGs leads to the synthesis of short proteins of 55 and 65 amino acids (of which the shorter protein is a truncated version of the longer one). The functions of these small proteins are not known.

Replication of the Genome

Synthesis of viral proteins, in particular of the N protein, allows the enzymatic activity present in the genomic nucleocapsid to switch from synthesis of messengers to replication of the genome. Replication requires producing a full-length antigenomic template. This RNA is encapsidated into plus-strand RNP, containing N, P, and L, during its synthesis. It has been postulated that it is the immediate encapsidation of the RNA that leads to the replication mode rather than to synthesis of mRNAs. A requirement for immediate encapsidation in order to replicate the genome means that ongoing synthesis of N is required for replication. The M protein also appears to regulate RNA synthesis. In the replication mode, the polymerase complex ignores all of the initiation, termination, and polyadenylation signals utilized to produce mRNAs, and instead produces a perfect complementary copy of the genome. The antigenomic RNA can be copied by the polymerase activity in the (+)RNP to produce more genomic RNA. This also requires that the RNA be immediately encapsidated. The new genomic RNP can be used to amplify the replication of viral RNA or, later in infection, can bud to produce progeny virions.

Maturation of Virus

The G protein has a 16-residue N-terminal signal sequence that leads to its insertion into the endoplasmic reticulum during translation. The signal is removed by signalase. The resulting 495-residue protein is anchored near the C terminus by a 20-residue transmembrane anchor, with the 29 C-terminal residues forming a cytoplasmic domain (i.e., it is a type 1 integral membrane protein). G is glycosylated on two asparagine residues and transported to the plasma membrane, where progeny viruses are formed by budding (Fig. 2.19D). The M protein appears to form an adaptor between the glycoprotein present in the plasma membrane and the nucleocapsids assembled inside the cell. M also acts to repress RNA synthesis by the viral nucleocapsid. The G protein contains the fusion activity and receptor recognition activities of the virus, and it is the only protein present on the surface of the virion. It is present in the virion as trimers that form spikes visible in the electron microscope (Fig. 2.19C).

Vesiculovirus Diseases

The two major serotypes of VSV are Indiana and New Jersey. They generally cause nonfatal but economically important disease in cattle, pigs, and horses. The name of the virus comes from the vesicles that it induces on the tongue and lips. Human infection is common in rural areas where VSV is endemic in domestic animals; 25–90% of farmers in such areas may have anti-VSV antibodies, showing past infection by the virus. Human infection is largely asymptomatic or associated with a mild febrile illness, sometimes accompanied by herpes-like lesions in the mouth or on the lips or nose. The virus can also replicate in mosquitoes and other arthropods, and has been isolated from mosquitoes during VSV epidemics. The epidemiologic importance of mosquitoes or other hematophagous arthropods in transmission of the virus is not clear.

More than 20 other vesiculoviruses are known. Chandipura virus is widespread in India, where it causes a febrile illness in humans. It has been isolated from sandflies and may be an arbovirus. The novirhabdoviruses infect salmon and other fish and are responsible for economic losses in fish farming operations.

Genus Lyssavirus

The rhabdovirus of greatest medical interest is rabies virus, which belongs to the genus *Lyssavirus*. Rabies is a uniformly fatal disease of man and of other mammals, and has been known since the 23rd century B.C. Rabies virus is present in the saliva of a rabid animal and is transmitted by its bite. Infection begins in tissues surrounding the site of the bite. Without treatment the virus may be transmitted to the brain, where replication of the virus leads to the disease called rabies. It is believed that the virus enters neurons by using acetylcholine receptors as a receptor, followed by transport up the axon until it reaches the cell body. The probability that rabies will develop following the bite of a rabid animal depends on the location of the bite, the species doing the biting, and the virus strain. In the absence of treatment, bites on the face and head result in rabies in

40–80% of cases, whereas bites on the legs result in rabies in 0–10% of cases. The incubation period to development of symptomatic rabies can vary from less than a week to several years. Once the virus reaches the brain, it spreads from there to a variety of organs. To be transmitted, it must spread to the salivary glands. Infection of neurons in the brain may result in behavioral changes that cause the animal to become belligerent and bite other animals, so that the virus present in salivary fluid is transmitted. In humans, the disease may be paralytic or may result in nonspecific neurologic symptoms including anxiety, agitation, and delirium. Biting behavior is not a consequence of rabies-induced neurologic disease in humans, and human-to-human transmission does not occur. Two to 7 days after symptoms of rabies appear, coma and death ensue. Only three cases of humans recovering from symptomatic rabies have been recorded.

For centuries, the saliva of a rabid dog was thought to be the source of rabies infection, but it was only in 1804 that Zinke succeeded in transmitting rabies from it. In the late 1800s, Pasteur adapted rabies virus to laboratory animals and developed the concept of protective vaccination against rabies. The dessicated spinal cords from rabies-infected rabbits became the first rabies vaccine. On July 6, 1885, this vaccine was used to immunize Joseph Miester, who had been bitten 14 times by a rabid dog. Because of the multiplicity of bites, he would almost surely have died, but the Pasteur vaccine saved him. A vaccine grown in nervous system tissue and inactivated by phenol rather than drying was the accepted rabies vaccine for decades. In the 1960s, a safer inactivated virus vaccine derived from virus grown in cultured human cells was introduced. The rabies vaccine is unique in that it is normally given after exposure to the virus, in conjunction with anti-rabies antiserum. This is possible because there is a window of time following the bite of a rabid animal before rabies develops, during which a protective immune response can be induced. Veterinarians and wildlife workers who are potentially exposed to rabid animals, as well as biologists who work with rabies virus in the laboratory, are immunized prophylactically, but the protective immune response can be of short duration and immunity must be tested at regular intervals.

In the United States, Canada, and Western Europe, where vaccination of domestic dogs is widely practiced, wild animals such as raccoons and skunks maintain the virus and transmit it to man or his domestic animals. Figure 4.3 shows the decline in number of cases of rabies in dogs and man in the United States since the 1940s and the increase in rabies in wild animals. Figure 4.4 illustrates the explosive spread of rabies in raccoons on the Eastern seaboard in the last 20 years. In other parts of the world, where licensing and immunization of pets is not required, domestic dogs continue to be the principal vectors that transmit rabies to humans. Rabies remains a significant global health problem. More than one million people annually undergo antirabies treatment following exposure to the virus, and

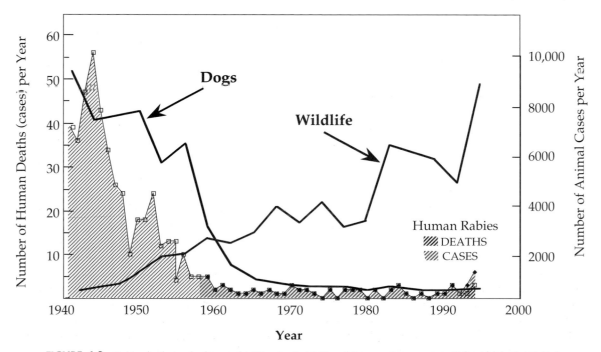

FIGURE 4.3   Rabies in domestic dogs and wild animals (right scale) versus human cases (left scale) in the United States 1940–1995. Note that untreated rabies in humans is uniformly fatal. [Data from Smith *et al.* (1995) and Summaries of Notifiable Diseases in the United States for 1996. *MMWR* **45**; No. 53 (1997).]

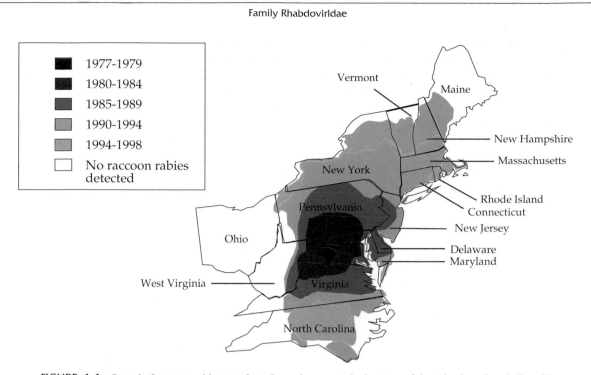

FIGURE 4.4 Spread of raccoon rabies over 3- to 5-year increments in the states of the Atlantic seaboard. Over 20 years the virus has spread from a small focal area in northern Virginia to encompass much of the entire region from southern Maine to North Carolina. [From *MMWR* **45**; p. 1119 (1997).]

50,000 people die of rabies each year. Efforts to control rabies in wildlife in the United States and Western Europe have met with some success. These efforts involve vaccinating wildlife with attenuated rabies virus or with recombinant vaccinia viruses that express the rabies G protein, using bait containing one of these viruses that is often dispersed by airplane.

The perpetuation of rabies in nature is somewhat of a mystery because of the fact that it can be maintained only in rabid animals who die quickly of the infection. How is it that the virus manages to persist? One possibility arises from recent findings that rabies virus can establish a latent infection in humans. Five cases have been documented in which people did not develop symptoms for 7 or more years after infection with the virus. In at least some of these cases, progression to rabies appeared to be triggered by hormonal changes during puberty. If the virus can establish a latent infection in other animals that is later followed by reemergence of the virus and its transmission to new susceptibles, this could serve as a reservoir of the virus.

Bats may also be an important reservoir of rabies virus. Rabies virus infection of bats seems to take longer to kill the animal, during which time the virus may be transmissible through the bite of an infected bat or through aerosols from infected bat feces or saliva spray. However, rabies virus in bats is distinguishable from rabies virus strains in other wildlife by nucleotide sequence analysis. Thus,

mixing of bat rabies and rabies in other wildlife is infrequent. Bats can transmit rabies to humans, and cases of human rabies transmitted by bats in the United States have been documented. In fact, in the United States in recent years, cases of human rabies resulting from infection with bat-associated rabies virus have been more numerous than cases resulting from infection by bites of other rabid wildlife. In many cases of bat-associated rabies, the mechanism by which the virus was transmitted to the human is not known, because no exposure to bats, rabid or otherwise, could be shown.

Australia was long believed to be completely free of rabies. However, it has recently been found that many Australian bats, including large fugivorous bats known as flying foxes that are present in vast numbers in northern Australia, carry a virus known as Australian bat lyssavirus. Two cases of fatal human rabies that were caused by infection with this bat virus have occurred in the last few years. No rabies has been found in other animals, presumably because there is no efficient mechanism for transmission of the virus among other mammals present in the continent. However, the disease could potentially spread to dogs and cats that have been introduced into Australia over the years.

Other bat lyssaviruses are also known, as is a virus called Mokola virus (Table 4.2). All lyssaviruses can cause rabies-like illness in man and other mammals, but the importance of lyssaviruses other than rabies virus as human disease agents is probably limited.

The genome organization of rabies virus is identical to that of VSV and the replication pathways are the same. However, very little sequence identity exists between the genomes of the viruses belonging to the two genera.

Genus Ephemerovirus

Bovine ephemeral fever virus is an arbovirus that causes economically important disease in domestic bovines in tropical areas of the Old World. This virus, as well as other members of the genus, are not known to be human disease agents.

FAMILY PARAMYXOVIRIDAE

The family Paramyxoviridae has six genera. These genera are shown in Table 4.3, together with representative viruses in each genus. The relationships among the genera are illustrated in the tree shown in Fig. 4.5. Each genus represents a distinct lineage. Furthermore, *Respirovirus*, *Morbillivirus*, *Megamyxovirus*, and *Rubulavirus* are more closely related to one another than to *Pneumovirus* and *Metapneumovirus*, and the family is divided into two subfamilies, Paramyxovirinae and Pneumovirinae. Many of the viruses belonging to this family are very important human pathogens. Some, such as measles virus and mumps virus,

TABLE 4.3   Paramyxoviridae

Genus/members[a]	Virus name abbreviation	Usual host(s)	Transmission	Disease	World distribution
Paramyxovirinae					
Respirovirus					
Human parainfluenza 1, 3	HPIV-1,3	Humans	Airborne	Respiratory disease	Worldwide
Bovine parainfluenza 3	BPIV-3	Cattle, sheep	Airborne	Respiratory disease	Worldwide
Sendai	SeV	Mice	Airborne	Respiratory disease	Worldwide
Rubulavirus					
Mumps	MuV	Humans	Airborne	Parotitis, orchitis, meningitis	Worldwide
Newcastle disease, avian paramyxoviruses 2–9	NDV	Gallinaceous birds	Airborne	Respiratory distress, diarrhea	Worldwide
Human parainfluenza 2, 4a, 4b	HPIV-2,4	Humans	Airborne	Respiratory disease	Worldwide
Simian virus 5	SV-5	Monkeys, canines	Airborne	Respiratory disease	Worldwide
Morbillivirus					
Measles	MeV	Humans, monkeys	Airborne	Fever, rash, SSPE[b], immune suppression	Worldwide
Rinderpest	RPV	Cattle, swine	Airborne	Gastroenteritis	Worldwide
Distemper	CDV, PDV[b]	Dogs, marine mammals	Airborne	Immune suppression, gastroenteritis, CNS disease	Worldwide
Megamyxovirus					
Hendra (equine morbillivirus)	??	Humans, equines, *Pteropus* fruit bats	Body fluids?	Respiratory disease, encephalitis	Australia
Nipah	??	Humans, swine, cats, dogs	Body fluids?	Respiratory disease, encephalitis	Malaysia, Singapore
Pneumovirinae					
Pneumovirus					
Human respiratory syncytial	HRSV	Humans	Airborne	Respiratory disease	Worldwide
Bovine respiratory syncytial	BRSV	Cattle	Airborne	Respiratory disease	Worldwide
Pneumonia virus of mice	PVM	Mice	Airborne	Respiratory disease	Worldwide
Metapneumovirus					
Turkey rhinotracheitis	TRTV	Turkeys	Airborne	Respiratory disease	Worldwide

[a]Representative members of each genus are shown, and the first virus listed is the type species.
[b]*Abbreviations:* SSPE, subacute sclerosing panencephalitis; CDV, canine distemper virus; PDV, phocine distemper virus.

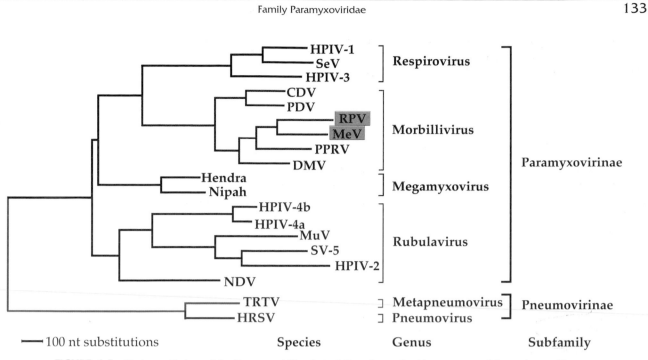

FIGURE 4.5 Phylogenetic tree of the Paramyxoviridae derived from the nucleotide sequences of the nucleocapsid protein gene. Most of the virus abbreviations are found in Table 4.3. CDV, canine distemper; PDV, phocine distemper; PRRV, peste-des-petits-ruminants; DMV, dolphin morbillivirus. Notice that the closest relative of measles (MeV, boxed), a human virus, is rinderpest (RPV, boxed), a virus of cattle and pigs. [Adapted from Chua *et al.* (2000).]

have been known for a long time—the infectious diseases caused by these viruses were known to the ancients. At the other extreme, Hendra virus, first described as an "equine morbillivirus of Australia," and the related Nipah virus of Southeast Asia have been known for less than a decade. These new viruses have been classified into a new genus, Megamyxovirus.

Replication of the Paramyxoviridae

The genome organizations of five viruses representing five genera of the paramyxoviruses are shown in Fig. 4.6. The paramyxovirus genome is larger than that of the rhabdoviruses, 15–20 kb, and encodes more proteins, 8–11 or more. Most paramyxoviruses possess six genes of which one, the P gene, uses more than one reading frame to encode multiple proteins. Rubulaviruses possess a seventh gene, encoding a protein called SH, and pneumoviruses possess an even larger constellation of genes, 10 in number. It is possible that still other genes are hidden within some of these large genomes. For example, the SH gene, encoding a very small protein, was discovered only recently.

The N or NP (= N in rhabdoviruses), P, M, and L genes serve the same functions as their counterparts in rhabdoviruses. G of rhabdoviruses is replaced by two glycoproteins in paramyxoviruses, one called F and the other H or HN or G, depending on the virus. The order of genes in the paramyxoviruses is the same as in the rhabdoviruses, and

the genome of the ancestral paramyxoviruses could have arisen from that of a rhabdovirus by insertion of extra genes (or vice versa by deletion of genes).

Virus replication occurs in the cytoplasm. Like the rhabdoviruses, paramyxovirus mRNAs are transcribed sequentially beginning at the 3' end of the genome and the mechanisms to produce these mRNAs are similar to those employed by rhabdoviruses. A leader is first transcribed, poly(A) tracts are added by stuttering at oligo(U) stretches at the end of each gene, intergenic nucleotides are skipped by the polymerase during synthesis of mRNAs, and attenuation of mRNA synthesis occurs at each junction. The intergenic nucleotides are variable among paramyxoviruses, however. They are GAA or GGG for some viruses, but are variable in sequence and in length, from 1 to 60 nucleotides, for others. The mechanisms by which the virus switches from synthesis of mRNAs to replication of the genome are the same as those used by the rhabdoviruses.

The Viral Glycoproteins

Paramyxovirus virions are 150–350 nm in diameter and contain a helical nucleocapsid that is 8–12 nm in diameter. Virions are usually round but pleomorphic, and are produced by budding from the plasma membrane (Figs. 2.18C and 2.21B). The virion size differs even within a single species and the composition of the virion is not as well

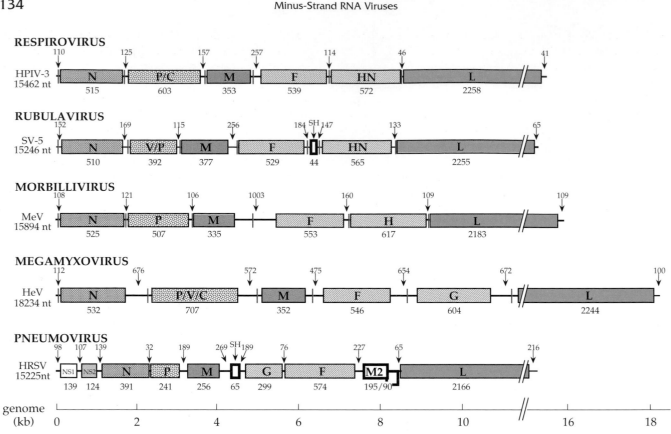

FIGURE 4.6 Genome organizations of five genera of the Paramyxoviridae. The genome is shown 3' to 5' for the minus-strand RNA. For the top four genera, each gene begins with the vertical pink bar marking the intergenic sequence. The untranscribed intergenic sequences of respiroviruses, morbilliviruses, and megamyxoviruses are 3 nt in length. Those of rubulaviruses and pneumoviruses vary in length from 1 to more than 60 nt. The boxes are the ORFs encoding the nucleocapsid (N), the P/V/C complex, the matrix protein (M), the fusion protein (F), the second glycoprotein (G, HN, or H), and the polymerase (L). Numbers above the arrows are the total number of nucleotides between the ORFs; numbers below the boxes are the number of amino acids in the protein. HPIV-3, human parainfluenza virus 3; SV-5, simian virus 5; MeV, measles virus; HeV, Hendra virus; HRSV, human respiratory syncytial virus.

defined as for some enveloped viruses. There are two glycoproteins on the surface. One is a fusion protein that is required for the fusion of the viral membrane with the cell plasma membrane. Paramyxoviruses fuse with the plasma membrane, not with endosomal membranes, and fusion does not require exposure to low pH. The fusion protein is produced as a precursor called F_0. When first synthesized, F_0 has an N-terminal signal sequence that results in its insertion into the endoplasmic reticulum during translation. The signal sequence is removed by signal peptidase and the resulting type 1 integral membrane protein is anchored by a membrane-spanning region near the C terminus. F_0 is cleaved either by cellular furin within the cell or by other cellular enzymes after release of the virion from the cell, depending on its sequence. Cleavage is required for the virus to be infectious and the cleavage products, F_1 (the N-terminal part of the precursor) and F_2 (the C-terminal part which is anchored in the membrane), remain covalently linked through a disulfide bond. The fusion domain consists of the N-terminal 20 amino acids of F_2, but this domain is not fusogenic until cleavage of F_0 has occurred. Those strains whose F_0 can be cleaved intracellularly by furin (which recognizes the sequence RXRR or RXKR) are often more virulent than strains that require cleavage of F_0 by proteases after the release of (noninfectious) virions from the cell. F oligomerizes to form trimers which are visible as spikes on the surface of the virion.

The second glycoprotein is called the hemagglutinin-neuraminidase (HN), the hemagglutinin (H), or simply G, depending on the virus. This protein is a type 2 integral membrane protein. The signal sequence at the N terminus is not removed but instead serves as the transmembrane anchor for the protein, so that it has its N terminus inside and its C terminus outside. This protein contains the receptor-binding activity of the virus. For many paramyxoviruses, the receptor is sialic acid (*N*-acetylneuraminic acid) bound to protein or lipids. Because this receptor is also present on red blood cells, these viruses can cause red blood cells to clump or agglutinate, a process called hemagglutination (*heme* = the red compound in red blood cells that binds oxygen). Many viruses hemagglutinate, but the majority of these, including many paramyxoviruses, use receptors other than sialic acid.

In paramyxoviruses that use sialic acid as a receptor, this second glycoprotein is also a neuraminidase, in which case it is called HN. Neuraminidase removes sialic acid from potential receptors and from virus glycoproteins. By removing sialic acid from the virus glycoproteins and from the cell surface, it prevents released virus from aggregating with itself or sticking to infected cells. It also increases the probability that the virus will successfully initiate infection of a suitable animal. Mucus, which lines the respiratory tract where the virus begins infection, contains sialic acid and might otherwise bind virus, preventing its entry into cells, if the virus could not destroy these receptors.

In paramyxoviruses that do not hemagglutinate, the second glycoprotein is simply called G, for glycoprotein. The second glycoprotein, best studied in the case of the HN of some paramyxoviruses, oligomerizes to form tetrameric spikes on the surface of the virion.

Some paramyxoviruses belonging to the genera Rubulavirus and Pneumovirus contain a third protein that may be a component of the envelope. This small (44–64 residues) type 2 integral membrane protein is called SH or 1A. It is glycosylated in the pneumovirus respiratory syncytial virus but not in the rubulaviruses SV5 and mumps. The function of this protein, or even whether the SH proteins of pneumoviruses and rubulaviruses serve the same function, is not known.

The P Gene

Synthesis and translation of the P gene of paramyxoviruses belonging to the subfamily Paramyxovirinae, sometimes called the P/V or P/C/V gene, are remarkable, as illustrated in Fig. 4.7. The P gene, or its equivalent, of most (–)RNA viruses is translated into more than one gene product, as was described above for the rhabdoviruses and as will be described below for other viruses, but the translation strategies used by some paramyxoviruses result in maximal use of the potential information contained within this gene. In some paramyxoviruses, alternative AUG start codons are used to produce two different proteins translated from different reading frames, similar to what occurs in the rhabdoviruses. A second strategy used by paramyxoviruses is to add nontemplated nucleotides to the mRNA during synthesis in order to shift the reading frame downstream of the added nucleotides. The ultimate use of the paramyxovirus P gene occurs in some viruses in which all three reading frames are translated by using one or both of these strategies to produce four or more proteins.

In respiroviruses, morbilliviruses, and megamyxoviruses, translation of P mRNA can start at one of two different AUGs that are in different reading frames. One of the two proteins produced is called C and the other P (Fig. 4.7). In addition, during transcription of P mRNA in most members of the Paramyxovirinae, nontemplated G residues are added at a specific site in the gene. In measles or Sendai viruses, addition of one G shifts the reading frame after this point to produce a new protein called V, which is rich in cysteine residues. Thus P and V share their N-terminal sequence but diverge after the site where the extra G is added. In the case of parainfluenza virus 3, addition of one nontemplated G leads to the production of mRNA for V, but addition of two leads to mRNA translated into a protein called D. In parainfluenza virus 3, then, all three reading frames are used over a considerable span of the P gene to produce four different proteins. In the rubulaviruses, the V protein is translated from the unmodified transcript, and production of mRNA for P requires addition of two nontemplated G residues. In mumps virus, addition of 4 G residues also occurs to produce a third protein.

During translation of these P mRNAs, the situation becomes even more complicated. In some viruses, multiple in-frame start codons are used to initiate translation of C. Thus, different forms of C are produced that are variously truncated at their N terminus. The start codons used include not only AUG but also ACG and GUG! The many different protein products produced from the P gene have not been fully characterized, and perhaps not yet fully enumerated, and the various functions of this wealth of proteins have as yet to be determined.

Addition of the nontemplated G residues is thought to involve a mechanism similar to the stuttering that produces a poly(A) tract opposite a string of U's in the template. The nontemplated G's are always added at a specific, unique place in the genome characterized by a string of C's. There must be some signal within the genome that is recognized by the viral polymerase for the addition of the extra G's, similar to the case for the addition of the poly(A) tract at the end of mRNAs.

Genus Respirovirus

The genus Respirovirus contains several parainfluenza viruses (abbreviated PIVs) and Sendai virus (from Sendai, Japan, where it was isolated; also called mouse PIV-1) (Table 4.3). The two human respiroviruses, HPIV-1 and HPIV-3, cause a respiratory illness similar to that caused by influenza virus and utilize sialic acid as a receptor, as does influenza. They were once grouped with influenza virus as myxoviruses (*myxo* from mucus because the viruses attach to mucus, which contains sialic acid). When they were separated from influenza virus into a distinct family, they were called parainfluenza viruses and the family was named Paramyxoviridae.

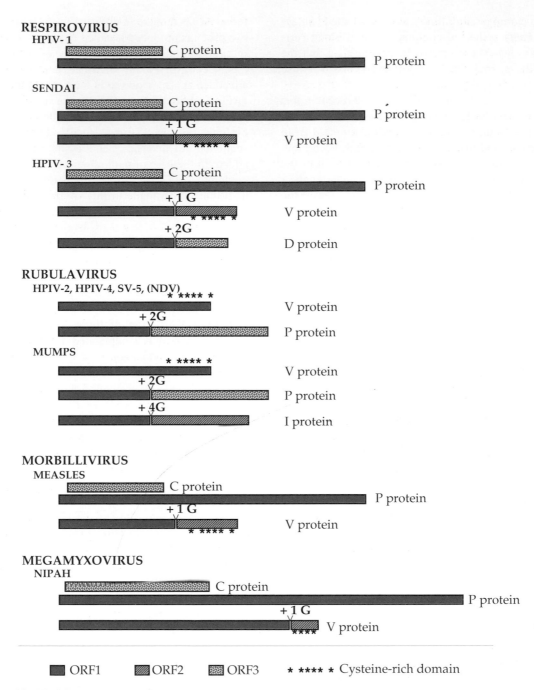

RESPIROVIRUS
  HPIV- 1
    C protein
    P protein

  SENDAI
    C protein
    P protein
    +1 G
    V protein
    * **** *

  HPIV- 3
    C protein
    P protein
    +1 G
    V protein
    * **** *
    +2 G
    D protein

RUBULAVIRUS
  HPIV-2, HPIV-4, SV-5, (NDV)
    * **** *
    V protein
    + 2 G
    P protein

  MUMPS
    * **** *
    V protein
    + 2 G
    P protein
    + 4 G
    I protein

MORBILLIVIRUS
  MEASLES
    C protein
    P protein
    +1 G
    V protein
    * **** *

MEGAMYXOVIRUS
  NIPAH
    C protein
    P protein
    +1 G
    V protein
    ****

    ■ ORF1   ▨ ORF2   ▧ ORF3   * **** * Cysteine-rich domain

FIGURE 4.7   Translation strategy of the P gene of paramyxoviruses. In most paramyxoviruses, nontemplated nucleotides are inserted during synthesis of the PmRNA to shift the translation frame. Alternative translation start codons are also used. The result is the production of up to four proteins from this one gene. [Adapted from Strauss and Strauss (1991, Fig. 1) and Chua et al. (2000).]

The respiratory tract infections caused by HPIV-1 and HPIV-3 may be limited to the upper respiratory tract, causing colds, or may also involve the lower respiratory tract, causing bronchopneumonia, bronchiolitis, or bronchitis. These viruses are widespread around the world and are an important cause of lower respiratory tract disease in young children. Serological studies have shown that most children are infected by HPIV-3 by 2–4 years of age, and that the incidence of infection can be as high as 67 out of 100 children per year during the first 2 years of life (that is, reinfections are common). Thus, immunity is incomplete and the viruses continue to reinfect older children and adults. However, subsequent infections are normally less severe and there is a reduction in the incidence of lower respiratory tract disease (which is more serious than infection of the upper respiratory tract). The viruses, as is common for respiratory tract infections, are spread by respiratory droplets.

Attempts to develop vaccines against the HPIVs have not met with success. Because of incomplete immunity produced by natural infections, the primary purpose of a vaccine would be to decrease the severity of natural infection by the virus. Even so, results to date have been disappointing. Inactivated virus vaccines developed for HPIV-1 and -3, as well as for HPIV-2, a rubulavirus, were antigenic but failed to induce resistance to the viruses. This could have resulted from failure to develop IgA following a parenterally administered vaccine (Chapter 8), and attempts to develop effective vaccines are continuing.

Genus Rubulavirus

Mumps Virus

The genus *Rubulavirus* gets its name from an old name for mumps, which is the disease produced in humans by mumps virus. The only natural host for mumps virus is humans and the virus is transmitted from person to person by contact. The disease has been known since the 5th century B.C. The incubation period, that is, the period of time between infection by the virus and the development of symptoms, is about 18 days. During the last 7 days of the incubation period, a person sheds virus and is capable of infecting others. Infection of children is usually not serious, but mumps virus infection can cause serious illness, particularly in adults. Infection begins in the upper respiratory tract but becomes systemic with the virus infecting many organs, where it replicates in epithelial cells. It is best known for infection of the parotid salivary glands leading to painful swelling of these glands. More serious disease can result from the replication of the virus in other organs, however. The central nervous system (CNS) is a common target for the virus and 0.5–2.3% cases of mumps encephalitis are fatal. Infection of the pancreas can occur, and it has been suggested that mumps may be associated with sudden onset insulin-dependent diabetes. The heart is sometimes infected, resulting in myocarditis. Infection of the testes in adult males can lead to orchitis and, in rare cases, to sterility. Infection of the fetus can result in spontaneous abortion.

At one time, mumps was one of the common childhood diseases that was contracted by almost everyone. It is now controlled in developed countries by an effective attenuated virus vaccine that was selected by passage of the virus in embryonated eggs. This mumps vaccine is given as part of the MMR (measles–mumps– rubella) combination vaccine. The dramatic decline in cases of mumps in the United States after introduction of this vaccine is shown in Fig. 4.8. Because mumps is exclusively a human virus that induces effective immunity following infection, and infection of an individual requires direct contact with a person actively shedding the virus, the virus requires a population of about 200,000 people to sustain it. Such a population density was first attained 4000 or 5000 years ago, before which mumps could not have existed, at least in its current form.

Other Rubulaviruses

Other human rubulaviruses include HPIV-2 and HPIV-4. They are named human parainfluenza viruses because the disease they cause is similar to that caused by HPIV-1 and HPIV-3. However, they are genetically related to the rubulaviruses rather than to the respiroviruses (Fig. 4.5). Other members of the rubulavirus genus infect many mammals and birds. One of the most intensively studied rubulaviruses (studied as a model system for replication of members of the family) has been SV-5 (simian virus 5). During the development of the polio vaccine, Rhesus monkey kidney cells were used for replication of poliovirus in culture, and these cultures were often contaminated with monkey viruses. These simian viruses (SVs) were simply numbered as they were isolated, and any particular SV may be totally unrelated to any other. Two of the most widely studied are SV-5 and SV-40, which are not related to one another: SV-5 is an RNA-containing paramyxovirus and SV-40 is a DNA-containing polyomavirus (Chapter 6).

The avian viruses include nine serologically distinct paramyxoviruses. These viruses form a distinct lineage, but clearly group with the rubulaviruses (Fig. 4.5). APMV-1 is also known as Newcastle disease virus (NDV), which causes a highly contagious and fatal disease of birds. NDV has serious economic consequences because it infects chickens, among other avian hosts. When epidemics break out, and they do with some regularity, many birds die, causing economic losses. Quarantines are placed on the movement of birds during epidemics in an effort to curtail the spread of the virus, which could have further economic consequences.

Genus Morbillivirus

The genus *Morbillivirus* contains measles virus as well as a number of nonhuman pathogens that include rinderpest virus, which infects cattle and pigs, and distemper viruses of dogs, dolphins, and porpoises. The relationships among these viruses are illustrated in Fig. 4.5.

Measles

Measles virus causes serious illness in man. Infection begins in the upper respiratory tract but becomes systemic, and many organs become infected. Lymphoid organs and

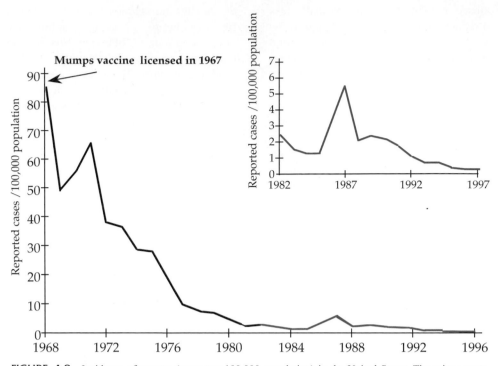

FIGURE 4.8 Incidence of mumps (cases per 100,000 population) in the United States. The minor resurgence of mumps cases in the late 1980s is thought to be due to a pool of susceptible teenagers and young adults who were not aggressively immunized during the first decade after the introduction of the vaccine. In 1998 there were a total of 666 cases of mumps in the United States, the lowest number ever reported for one year (<0.25 cases per 100,000). [From Summary of Notifiable Diseases in the United States for 1996. *MMWR* **45**; No. 53, p. 45 (1997) and the comparable summary for 1998.]

tissues are prominent sites of viral replication, and one consequence of virus infection is immune suppression that lasts for some weeks. Immune suppression can result in secondary infections that may be serious, even life threatening, and interference with immune function is a major cause of measles mortality. Measles also has uncommon neurological complications, including encephalomyelitis and subacute sclerosing panencephalitis (SSPE). In SSPE, the virus sets up a persistent but modified infection in the brain in which M protein is produced in only low amounts; downregulation of production of M protein appears to be necessary to establish the disease syndrome. Symptoms of SSPE appear several years after measles infection, and the disease progresses slowly but inexorably. Serious complications caused by viral infection of other organs can also occur.

Natural History of Measles Virus

Like mumps, measles is a disease of civilization. The virus is a human virus. Although subhuman primates are infected by the virus and suffer the same disease as man, man is the only reservoir of the virus in nature. Infection requires direct contact with an infected person and recovery from infection results in solid lifelong immunity to the virus. Thus,

a minimum size population is required to maintain the virus, in which the continuing birth of new susceptibles occurs at a rate sufficient to maintain continuous virus infection within the community. The requirement for a minimum sized human population to sustain the virus is illustrated in Fig. 4.9 In this figure, data from 1949–1964 (before tourism became as popular as it is today) are plotted that show the duration of measles epidemics on various islands. In Fig. 4.9A, we see that an island must have a population sufficient to produce about 16,000 surviving newborns a year (population about 500,000) in order to maintain the virus continuously in the population. If the population is smaller, the epidemic burns itself out when all susceptibles have been infected. The island is then free of measles until sufficient newborn susceptibles have been born and measles is once again introduced into the island from outside. The smaller the population, the longer this takes. Note that Guam and Bermuda, with their heavy tourist influx, had measles present more than expected from the curve because the virus is introduced more often, that is, the island population is not truly isolated. Figure 4.9B illustrates that the more densely packed the population, the more readily the virus spreads and therefore the sooner the epidemic burns itself out. The islands shown in this panel all have about the same population, but when the population is

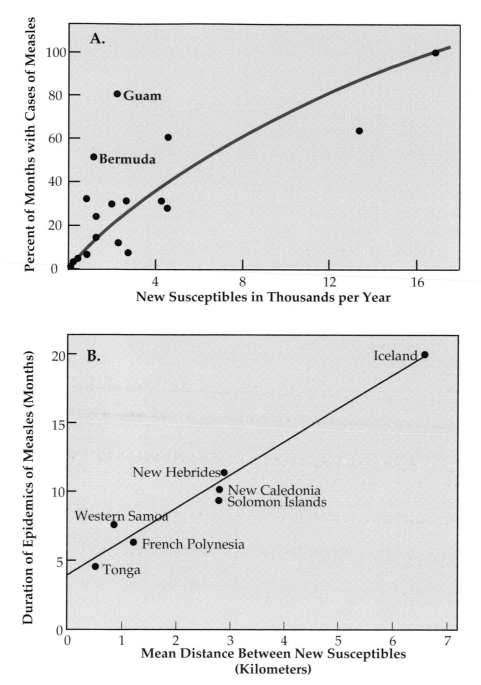

FIGURE 4.9 Effect of population size and density on the epidemiology of measles. (A) Percent of months with measles (true endemicity = 100%) in island populations as a function of number of new susceptibles per year. Measles periodically fades out in isolated populations of less that 500,000 (= approximately 15,000 new susceptibles per year). Each dot represents a different island population. Note that Guam (with a transient military population) and Bermuda (with a steady influx of tourists) do not fall on the curve, because they are not truly isolated populations. Vaccination for measles can reduce the number of new susceptibles, even in large urban populations, below the number needed to sustain transmission. [Data from Black (1966).] (B) Relation between the average duration of measles epidemics and the dispersion of populations in isolated islands. All of the islands shown have about the same population, sufficient to introduce 2000–4000 new susceptible children per year. The abscissa is which represents the average distance between infants added to the population each year. Population input is defined as births minus infant mortality.

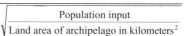

compressed into a smaller area, such as in Tonga, person-to-person spread is more efficient and epidemics do not last as long as when the population is dispersed over a larger area, such as in Iceland.

The study of measles epidemics on islands first demonstrated that lifelong immunity arises following infection by measles. After an epidemic of measles in the Faeroe Islands in 1781, the islands were free of measles until the virus was again introduced in 1846 by a Danish visitor. In the 1846 epidemic, 77% of the population of the islands contracted measles, but no one over 65 years of age came down with the disease.

The requirement for a minimum sized population to maintain the virus means that even though measles virus is extraordinarily infectious, the virus could not have existed until perhaps 5000 years ago when human population density became sufficient to support it. At about this time, large population centers arose in the Fertile Crescent, a region of the Middle East encompassing parts of modern Iraq, Syria, Jordan, Israel, Lebanon, and Turkey, which included the upper Tigris and Euphrates rivers and whose climate was conducive to primitive agriculture. These population centers were associated with the cultivation of food plants and the domestication of animals, including bovines. Measles virus is most closely related to rinderpest virus (Fig. 4.5), which infects cattle and swine. An obvious hypothesis is that the close contact between humans and their domesticated animals allowed rinderpest virus, or perhaps another virus of domestic animals, to jump to humans and evolve to become specific for humans. Subsequent coexistence of the virus with its human host led to the present situation where infection results in significant but relatively low morbidity and mortality.

Introduction of Measles into the Americas and Island Populations

Epidemics of measles were undoubtedly widespread in the Old World following the appearance of measles, although it is difficult now to ascertain the causes of epidemics that occurred thousands of years ago. However, it is clear that measles was widespread in Europe at the time the Europeans began their explorations of the Americas and of the many isolated island communities around the world, and Europeans carried measles with them as they traveled. Introduction of measles virus into virgin populations resulted in very high mortality. Mortality was 26% in Fiji islanders when measles was introduced in 1875, for example. It has been estimated that 56 million people in the Americas died of Old World diseases following European exploration of the New World, and measles and smallpox (Chapter 6) were significant contributors to these deaths. The introduction of measles and smallpox by the Spaniards facilitated the conquest of the Americas by them, and the subsequent depopulation of Central and South America

allowed the Spaniards to remain dominant. It has been suggested that the depopulation of the Americas caused by these diseases led the Europeans to introduce Africans as slaves to replace Native Americans being used as slaves.

The very high mortality caused by the virus in naive populations, which contrasts with the low mortality in Europeans, was probably due to two causes. Europeans and other Old World peoples have been continuously exposed to measles for millenia and have been selected for resistance to measles. The people in the Americas had never experienced measles infection, however. A second factor that led to high mortality rates was the introduction of measles into a virgin population, in which not only young children but also all of the adults were susceptible, meaning that the entire population became seriously ill simultaneously. This surely disrupted the ability of the society to maintain itself because there was no one healthy enough to care for the sick.

Vaccination against Measles

At one time, measles virus was epidemic throughout the world and caused one of the childhood illnesses contracted by almost everyone. Because of the extraordinary infectiousness of the virus, very few people escaped infection by it. In the United States, there were about 4 million cases of measles a year, of which about 50,000 required hospitalization and 500 were fatal. There were 4000 cases of measles encephalitis each year, with many patients suffering permanent sequelae. In addition, some fraction of children infected as infants went on to develop SSPE, which is a progressive neurological disease that results in death within about 3 years of the appearance of symptoms. Throughout the world, an estimated 2.5 million children died annually of measles.

Because measles was a widespread and serious disease, attempts to develop a vaccine began at about the same time as attempts to develop a poliovirus vaccine. One vaccine used in the United States from 1963 to 1967 consisted of inactivated measles virus. The vaccine was poorly protective and recipients of this vaccine exposed to measles sometimes developed a more serious form of measles, called "atypical measles," characterized by higher and more prolonged fever, severe skin lesions, and pneumonitis (inflammation of the lungs, from *pneumon* = lung and *itis* = inflammation). The increased severity may have resulted from an unbalanced immune response primed by the formalin inactivated virus or to a lack of local immunity in the respiratory tract (see Chapter 8). Similar problems occurred following vaccination with inactivated respiratory syncytial virus, a paramyxovirus described below.

An attenuated measles virus vaccine, now given as part of the MMR combination vaccine, produced much more satisfactory results. The live virus vaccine gives solid protection from disease caused by the virulent virus and has largely controlled the virus within the United States (Fig. 4.10).

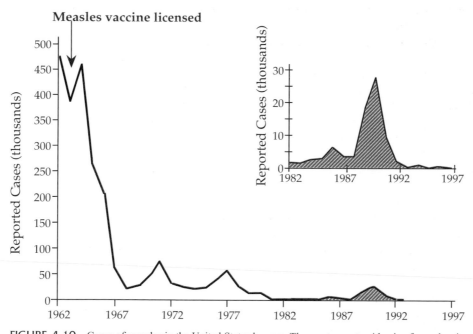

FIGURE 4.10 Cases of measles in the United States by year. The most recent epidemic of measles, in 1989–1991, was probably due to vaccine failure and waning immunity in children immunized earlier with only a single dose of vaccine. In 1998, a total of 100 cases of measles was reported, a 27% decrease from the previous record low of 138 cases in 1997. Of these 100 cases, 71% came from international importation. [From Summary of Notifiable Diseases in the United States for 1998. *MMWR* **47**; No. 53, p. 48 (1999).]

Following introduction of the vaccine, the number of cases dropped dramatically. The virus continued to circulate among nonimmunized individuals, however, and thousands of cases per year still occurred, sometimes associated with epidemics of more than 50,000 cases. As vaccine coverage became more effective, cases dropped to new lows, but another epidemic in 1989–1991 caused about 50,000 cases. This epidemic occurred in young immunized adults as well as in young children who had not been immunized. Some of the cases in young adults were due to vaccine failures (about 5% of humans immunized with a single dose of the measles vaccine fail to develop immunity to measles), but other cases appear to have been due to waning immunity. Thus, immunity induced by the vaccine is probably not lifelong, in contrast to natural infection by wild-type measles virus. The guidelines now call for a second immunization on entry to elementary or middle school. This not only boosts immunity in individuals whose immunity is waning, but also usually leads to immunity in those who did not develop immunity after the first dose. In addition, some colleges require immunization on entry. With these changes, the number of cases of measles in the U.S. was only 100 in 1998.

Molecular genotyping has increased our understanding of the few cases of measles that occur annually in the United States today (Fig. 4.11). In 1988–1992, all the reported isolates of measles virus were subgroup 2, the indigenous North American strain. However, by 1994–1995, all outbreaks were caused by one of four other subgroups that are endemic in other parts of the world. Thus, these outbreaks were initiated by viremic visitors from Asia and Europe. One notable outbreak is thought to have been initiated by a single visitor to Las Vegas and resulted in small epidemics in five states.

After control of measles in the United States and other developed countries, the virus remained epidemic in many parts of the developing world. Control has recently been established throughout most of the Americas, but measles remains a serious pathogen in other parts of the developing world. The World Health Organization has initiated a campaign to eliminate measles during the next decade. Because the virus infects only humans in nature, it should be possible to eradicate it using the same techniques that were used for smallpox and that are being used for poliovirus. A major problem with measles, however, has been the inability to effectively immunize young infants against the disease before they become naturally infected by the virus. Newborns are protected from infection for 6–12 months by maternal antibodies, and a live vaccine does not take while they are thus protected. In many societies, measles is so pervasive that very shortly after the infant becomes susceptible to infection, infection by wild-type virus occurs, thus keeping wild-type virus in circulation. As described in Chapter 8, attempts to overcome maternal immunity by

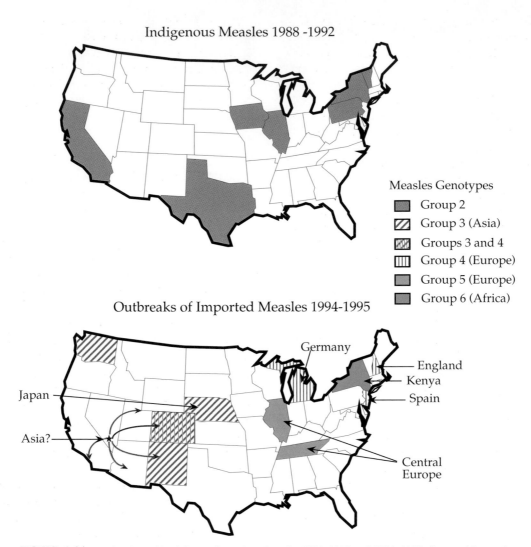

FIGURE 4.11 Molecular epidemiology of measles virus for 1988–1992 and 1994–1995. States with measles isolates are shaded to indicate the genotype found. In the lower panel, arrows indicate the source of imported viruses. Orange arrows show spread from a single index case in Las Vegas. [From Rota *et al.* (1995).]

increasing the dose of the vaccine virus have not led to satisfactory results.

Measles Neuraminidase

The receptor for measles virus is a protein called CD46 that is expressed on the surface of human and monkey cells (Chapter 1), which is bound by the measles H protein. In view of the fact that sialic acid was not the measles receptor, the apparent lack of neuraminidase activity in the H protein was not surprising. Recent studies of the morbillivirus H protein have shown that it is related in structure to the HN protein of paramyxoviruses, however, and that the related rinderpest and peste-des-petits-ruminants viruses possess neuraminidase activity. This neuraminidase activity differs in its specificity from that exhibited by the respiroviruses and the orthomyxoviruses, explaining why it had not been observed previously. Presumably this activity is also found in the measles H protein, and the function of this enzyme in the measles life cycle remains to be determined.

Emerging Paramyxoviruses: Hendra Virus and Nipah Virus

In 1994–1995, three cases of human illness occurred in Queensland, Australia, that were associated with two outbreaks of respiratory disease in horses. Two of the humans died, one of encephalitis and the other of respiratory disease, as did 13 of the infected horses. The disease was found to be caused by an unknown virus belonging to the Paramyxoviridae, and it was first described as an equine morbillivirus. Analysis of sera from healthy humans and horses in the areas failed to detect the presence of antibody, and analysis of 5264 sera from a variety of wild animals trapped in the areas also failed to detect antibody. However, further study revealed that a significant fraction of flying foxes had antibodies to the virus. Experimental infection of flying foxes and other animals showed that flying foxes were readily infected by the virus and developed antibodies to the virus, but suffered no clinical symptoms of disease. Experimentally infected horses became quite ill, however, and some died. The virus did not spread to noninfected contacts, so that it is not readily transmissible by casual contact. Thus, there is good evidence that flying foxes are the reservoir species and from this reservoir the virus occasionally spreads to horses or other mammals. The human cases are thought to have been contracted through close association with infected horses. Sequence analysis of the virus has shown that it belongs to the Paramyxoviridae but is not particularly closely related to the morbilliviruses (Fig. 4.5), and so its initial description as an equine morbillivirus was wrong on both counts. It is now referred to as Hendra virus, after the town in Queensland where the first recognized cases occurred. It is classified as the type member of a new genus, Megamyxovirus (because its genome is so much larger than other Paromyxoviridae), in the subfamily Paramyxovirinae (Table 4.3).

At this point Hendra virus, and the genus it represents, was an oddity, associated with only three cases of human disease, and it was assumed to only rarely infect humans. The two outbreaks in Australia were separated by 1000 km, however, so that Hendra virus was clearly widespread. Then, in late 1998 and early 1999, an outbreak of 258 cases of human encephalitis occurred in Malaysia and Singapore that had a 40% mortality rate. The disease was associated with pigs and it was first thought that it was due to infection by Japanese encephalitis (JE) virus (see Chapter 3 on the importance of pigs as amplifying hosts for this virus). The Malaysian government vaccinated 2.4 million pigs against JE virus, but when this did not slow the epidemic, it was decided to slaughter pigs in an attempt to reduce the incidence of disease. In March 1999, with the assistance of the Centers for Disease Control and Prevention, the virus responsible for the epidemic was identified as a Hendra-like virus, a virus related to but distinct from Hendra virus. In this epidemic, it is thought that the human cases were contracted from pigs, which also showed symptoms of respiratory disease associated with neurological symptoms. The virus responsible has been called Nipah virus, after the village in Malaysia where the disease first appeared, and it is classified as a second member of the genus Megamyxovirus. Like Hendra virus, the reservoir of Nipah virus is flying foxes. It is possible that there are other, as yet uncharacterized, viruses of flying foxes and other bats that may be potential human pathogens.

Hendra virus and Nipah virus represent emerging pathogens. They are previously unknown viruses that are causing serious disease over widely separated geographic areas. It is not clear whether new strains of virus are involved that have recently acquired the ability to infect humans and domestic animals, or whether the disease in humans was previously less prevalent and not recognized as distinct from the many cases of human encephalitis that occur each year from infection by a variety of viruses.

Genus Pneumovirus

The genus Pneumovirus, subfamily Pneumovirinae, contains the respiratory syncytial viruses (RSVs). RSVs are known for cattle, mice, sheep, goats, and turkeys, as well as man. The genome of RSV is more complex than other Paramyxoviridae, having more genes (Fig. 4.6). The polymerase gene of RSV is more closely related to those of the filoviruses than to those of the Paramyxovirinae, making classification of these viruses problematical.

Human RSV is the most important cause of pneumonia in infants and children worldwide. Half of hospital admissions in the United States in January and February of infants less that 2 years old are due to infection by RSV. Infants are normally infected at 6 weeks to 9 months of age. Infection begins as an upper respiratory tract infection that progresses to the lower respiratory tract in 25–40% of primary infections. Immunity following infection is incomplete and reinfection is common in children and adults, but reinfection tends to produce less severe disease. Symptoms can include bronchitis and pharyngitis (*itis* = inflammation, so inflammation of the mucous membranes of the bronchi or pharynx), rhinorrhea (runny nose), cough, headache, fatigue, and fever. Pneumonia (inflammation of the lungs in which the air sacs become filled with exudate) can result, particularly in infants or the elderly. RSV infection is particularly serious in the immunocompromised. As one example, individuals of any age undergoing bone marrow transplantation have a 90% mortality rate if infected by RSV.

No vaccine is available at the current time for RSV. Because of the widespread prevalence of infection by the virus and the severity of the disease it causes, especially in infants, efforts are ongoing to develop a vaccine that would provide protection against disease or that would at least protect against severe disease. A clinical trial with an inactivated virus vaccine in a group of children some years ago gave disastrous results, however. Not only did the inoculation with the candidate vaccine fail to protect the children against subsequent infection by RSV, but it was found that when infected the vaccinated group suffered a much higher proportion of serious illnesses such as viral pneumonia than did the control group. Thus, immunization potentiated illness, possibly because of an unbalanced immune response. This result has impeded efforts to develop a vaccine and made it clear that a better understanding of the interaction of the virus with the immune system is important.

Viruses and Respiratory Disease

Many viruses belonging to several different families have now been described that cause respiratory disease, and more viruses belonging to other families will be described below. For comparative purposes, an overview of viruses that cause respiratory disease is shown in Table 4.4. This table is not meant to be comprehensive and includes only a sampling of viruses. Furthermore, some of the viruses in the table, such as measles virus, are better known for disease other than respiratory disease. However, the table makes clear that a large number of viruses can infect the respiratory tract and cause illness.

FAMILY FILOVIRIDAE

Table 4.5 lists the known filoviruses, which are classified into two genera, the Marburg-like viruses and the Ebola-like viruses. The filovirus genome is 19 kb in size and contains seven genes, which result in the production of seven or eight proteins following infection (Fig. 4.1). The molecular biology of filoviruses is not well understood, in part because most known filoviruses are severe human pathogens that must be handled under biosafety level 4 conditions. The genomes of

Ebola virus and Marburg virus have both been sequenced, and they have a genome organization similar to that of other members of the Mononegavirales. Their sequences suggest that they are most closely related to the pneumoviruses, and they are assumed to replicate in a manner similar to that for the rhabdoviruses and paramyxoviruses.

The filovirus virion is enveloped, as is the case for all minus-strand viruses, but rather than being spherical, the virion is long and thread-like (whence the name *filo* as in filament). The infectious virion is thought to be 800–1000 nm in length and 80 nm in diameter (Fig. 2.19E), but preparations examined in the electron microscope are pleomorphic and oddly shaped, often appearing as circles or the number 6 but sometimes branched (Figs. 2.19F and G). There is one glycoprotein (called GP) in the envelope, present as homotrimers, that is both *N*- and *O*-glycosylated and has a molecular weight of 120–170 kDa. GP mRNA is edited in Ebola virus, but not in Marburg virus, to produce an mRNA for a second protein called secreted glycoprotein. Its function is unknown but one speculation is that it interferes with the host immune system.

The filoviruses first came to the attention of science in 1967 when outbreaks of hemorrhagic fever occurred in Marburg and Frankfurt, Germany, and in Belgrade, Yugoslavia. The cause was a virus subsequently named Marburg that was present in African green monkeys whose

TABLE 4.4 Viruses Causing Respiratory Disease

Family	Virus[a]	Nucleic acid	Host range	Disease(s)
Orthomyxoviridae	Influenza	ss(−)RNA	**Humans**, birds, horses, swine	Rhinitis, pharyngitis, croup, bronchitis, pneumonia
Paramyxoviridae	RSV	ss(−)RNA	**Humans**, cattle	Rhinitis, pharyngitis, croup, bronchitis, pneumonia
	Canine distemper		Dogs	Bronchitis, pneumonia
	NDV		Birds	Respiratory distress
	Human parainfluenza		**Humans**	Rhinitis, pharyngitis, croup, bronchitis, pneumonia
	Measles		**Humans**	Pneumonia
Picornaviridae	Rhinoviruses	ss(+)RNA	**Humans**	Common cold (rhinitis), pharyngitis
	Coxsackie A		**Humans**	Rhinitis, pharyngitis
Caliciviridae	Feline calicivirus	ss(+)RNA	Cats	Rhinitis, tracheitis, pneumonia
Coronaviridae	HCoV	ss(+)RNA	**Humans**	Rhinitis
	IBV		Fowl	Bronchitis
Adenoviridae	Human Ad40, 41	dsDNA	**Humans**	Rhinitis, phryngitis, pneumonia
	CLTV		Dogs	Pharyngitis, tracheitis, bronchitis, and bronchopneumonia
Herpesviridae	Cytomegalovirus	dsDNA	**Humans**	Pharyngitis, pneumonia
	Herpes simplex, EBV, varicella		**Humans**	Pharyngitis, pneumonia
	Various alphaherpesvirinae		Cattle, cats, horses, chickens	Rhinotracheitis

Source: Adapted from Granoff and Webster (1999, pp. 1493, 1494).
[a]RSV, respiratory syncytial virus; NDV, Newcastle disease virus; HCoV, human coronavirus; IBV, infectious bronchitis virus; CLTV, canine laryngotracheitis; EBV, Epstein–Barr virus.

TABLE 4.5   Filoviridae

Genus/members	Virus name abbreviation	Usual host(s)[a]	Transmission	Disease	World distribution
Marburg-like viruses					
Marburg	MARV	Humans	Contact with blood or other body fluids	Severe hemorrhagic disease	Africa
Ebola-like viruses					
Sudan Ebola	SEBOV				
Zaire Ebola	ZEBOV	Humans	Contact with blood or other body fluids	Severe hemorrhagic disease	Africa
Cote d'Ivoire Ebola	CIEBOV				
Reston Ebola	REBOV	Cynomolgus monkeys	?	Severe hemorrhagic disease in monkeys, attenuated in man	Philippines

[a]Natural reservoirs unknown.

kidneys were being processed for cell culture production (for use in preparing poliovirus vaccine). Twenty-five laboratory workers were infected and six secondary cases resulted; of these 31 infected people, 7 died. The monkeys in the shipment, which originated in Uganda, also died. Subsequent studies with the virus isolated during the outbreak showed that it caused lethal illness in African green monkeys following experimental infection. The natural history of the virus remains a mystery. There were 3 cases of Marburg in South Africa in 1975 (the source of infection was probably Zimbabwe) with one death, 2 cases in Kenya in 1980 (infection probably in Uganda), one case in Kenya in 1987, and an outbreak of 99 cases in Zaire (now Congo) in 1998–2000. The locations of these outbreaks are shown on the map in Fig. 4.12. The fatality rate in these various outbreaks has averaged about 30%. There must be a reservoir of the virus somewhere in Africa from which it can spread to man or to monkeys, but the identity of this reservoir is unknown.

The second filovirus known is Ebola virus. Ebola virus was first isolated during a 1976 epidemic of severe hemorrhagic fever in Zaire and Sudan and named for a river in the region. During this epidemic, the >600 cases resulted in 430 deaths. Asymptomatic infection appears to be rare. One case of Ebola occurred in 1977, and in 1979 there were 34 cases with 22 deaths in the Sudan. In this latter epidemic, an index case was brought to the hospital and the virus spread to four people there, who then spread it to their families. After this, Ebola disease in Africa disappeared until 1994. In late 1994, a Swiss ethologist working in the Ivory Coast performed necropsies on chimps. She contracted Ebola but survived, and a new strain of Ebola was isolated from her blood. Then, in May 1995, there was an epidemic in Kikwit, Zaire, that resulted in at least 315 cases with >75% mortality. This was followed by several deaths in western Africa that resulted from consumption of a monkey that had died of Ebola. Recently there has been a prolonged series of smaller outbreaks in Gabon from 1995 through 1997. In 2000, Ebola appeared in Uganda for the first time and caused an epidemic of more than 425 cases as of the latest tally. A map showing these various filoviral outbreaks is shown in Fig. 4.12.

The natural reservoir of Ebola virus in Africa also remains a mystery, despite extensive effort by the Centers for Disease Control (CDC) and other health care agencies to identify it. It is clear that monkeys can be infected by the virus and spread it to humans, but how the monkeys contract it or how human epidemics get started when monkeys are not implicated is not known.

Ebola virus has recently caused epidemics of hemorrhagic fever in monkeys imported from the Philippines. The first epidemic occurred in Reston, Virginia, near Washington, D.C. in 1989. The deaths were at first attributed to simian hemorrhagic fever virus (SHFV), but investigation by the U.S. Army Medical Research Institute for Infectious Diseases and the CDC found that both SHFV and Ebola virus were present in the monkeys. Believing that the community was at risk for Ebola, the army team quickly decided to euthanize the monkeys and decontaminate the facility. Follow-up studies showed that four animal handlers at the facility had been infected by the virus but had suffered no illness. Thus the strain of Ebola present in the Reston monkeys, called Ebola-Reston to distinguish it from Ebola-Zaire, seems to be nonpathogenic for man although it remains pathogenic for monkeys. The story of the Reston incident was recounted in a book called *The Hot Zone* by Richard Preston. Since this first epidemic of Ebola-Reston, further outbreaks have happened in Reston and in an animal facility in Texas. Nucleotide sequencing has shown that Ebola-Reston is closely related to Ebola

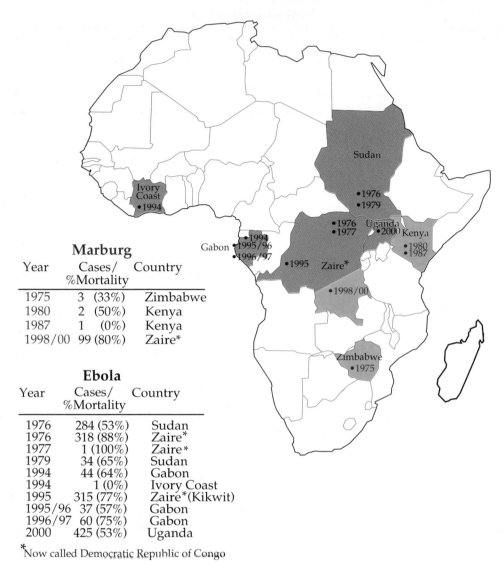

Marburg

Year	Cases/ %Mortality	Country
1975	3 (33%)	Zimbabwe
1980	2 (50%)	Kenya
1987	1 (0%)	Kenya
1998/00	99 (80%)	Zaire*

Ebola

Year	Cases/ %Mortality	Country
1976	284 (53%)	Sudan
1976	318 (88%)	Zaire*
1977	1 (100%)	Zaire*
1979	34 (65%)	Sudan
1994	44 (64%)	Gabon
1994	1 (0%)	Ivory Coast
1995	315 (77%)	Zaire*(Kikwit)
1995/96	37 (57%)	Gabon
1996/97	60 (75%)	Gabon
2000	425 (53%)	Uganda

*Now called Democratic Republic of Congo

FIGURE 4.12 Map of Africa showing the different filovirus outbreaks. [Data from Porterfield (1995, p. 320) and later data from Georges-Courbot *et al.* (1997), Peters and Kahn (1999), and news bulletins from the World Health Organization (2000) at their web site: http://www.who.int/disease-outbreak-news/

from the Sudan epidemics. The reason it is attenuated in man is still unexplained, and high containment is used for studies of Ebola-Reston in the laboratory. The natural history of the virus is unknown, but it is thought that it might be a Philippine virus or an Asian virus, rather than an African virus.

Because of the dramatic symptoms caused by Ebola virus, as well as by other hemorrhagic fever viruses, and the high fatality rate following infection, it has been the subject of discussion in the popular press and has appeared in a number of works of nonfiction as well as fiction. To date, the virus has caused only a limited number of human cases, but there is always the fear that if the virus were to

adapt to man in a way that allowed for easier transmission, it could become a very big problem.

FAMILY BORNAVIRIDAE

Borna disease virus is a recently characterized virus that has been assigned to a new family in the order Mononegavirales (Table 4.6). It has a nonsegmented, minus-sense genome of 8.9 kb, containing five genes, that in general organization resembles that of other members of the Mononegavirales (Fig. 4.1). However, the virus replicates in the nucleus, not in the cytoplasm. Splicing of mRNAs

TABLE 4.6   Bornaviridae

Genus/members	Virus name abbreviation	Usual host(s)	Transmission	Disease	World distribution
Bornavirus					
Borna disease	BDV	Horses, sheep, other mammals	??	Encephalopathy, fatal paralysis	Europe, possibly worldwide

occurs to form an incompletely characterized set of mRNAs from the five genes.

Borna disease virus appears to have a very wide host range. It was originally described as a pathogen of sheep and horses in Germany, but is now known to infect a wide variety of warm-blooded vertebrates, birds as well as mammals. Its natural host range may be all warm-blooded vertebrates, including man. Its geographic distribution is probably worldwide. Its mode of transmission is not known.

In nonhumans, the virus establishes an infection characterized by neurotropism and low production of virus. Infection is not cleared despite an immune response to the virus. Infection may be asymptomatic or may result in disease characterized by movement and behavioral abnormalities. Naturally infected horses exhibiting such abnormalities usually recover, but the disease may progress to paralysis and death. Experimentally infected rats and primates also exhibit behavioral abnormalities. Because of these effects on other animals, several recent studies have tried to determine if the virus is associated with neurological disease in man, in particular with schizophrenia. Serological surveys have found that psychiatric patients are more likely to have antibodies to bornavirus than normal controls. Surveys to examine for the presence of viral RNA in peripheral blood mononuclear cells (PMBCs) are even more suggestive: psychiatric patients, including schizophrenics, show a positivity rate of up to 66% in some surveys, compared to <5% in normal controls. Furthermore, very small amounts of virus-specific RNA have been isolated from postmortem brain samples from patients suffering from schizophrenia and bipolar disorder, but not from normal individu-

als or patients suffering from other neurological disorders. Interestingly, a recent study found that two patients hospitalized for severe depression exhibited a rise in bornavirus antigen in PMBCs during the course of the disease, which fell to very low levels on recovery. Whether these different associations are indicative of causality remains to be determined, but it is conceivable that the virus causes recurrent episodes of depression on reactivation of a latent infection.

FAMILY ORTHOMYXOVIRIDAE

The family Orthomyxoviridae (*ortho* = normal or correct) contains the genera *Influenzavirus A*, which contains influenza virus A; *Influenzavirus B*, which contains influenza virus B; and *Influenzavirus C*, which contains influenza virus C (Table 4.7). Thogoto virus, a tick-borne virus of mammals, forms a fourth genus, *Thogoto-like viruses*. Influenza viruses A and B are closely related, but influenza A infects a wide spectrum of birds and mammals including humans, whereas influenza B infects only humans. Influenza C is more divergent. Eight segments of (−)RNA, totaling about 14 kb, comprise the genomes of influenza A and B viruses (Fig. 4.1) whereas influenza C has only 7 segments. Influenza viruses use sialic acid as a receptor, but the form used by influenza A and B viruses diff~ from that used by influ~~za C~~ encoded by the viruses to ~~ ingly different. All three ~~ cause disease, but influen~~

TABLE 4.7   Orthomyxoviridae

Genus/members	Virus name abbreviation	Usual host(s)	Transmission	
Influenzavirus A				
Influenza A	FLUAV	Humans, birds, swine	Airborne	Resp
Influenzavirus B				
Influenza B	FLUBV	Humans	Airborne	Respi
Influenzavirus C				
Influenza C	FLUCV	Humans	Airborne	Respir
Thogoto-like viruses				
Thogoto virus	THOV	Mammals	Tick-borne	

human pathogen because it causes very large, recurrent epidemics with significant mortality. Influenza A has therefore been the most intensively studied and has been the focus of efforts to control influenza in humans.

Proteins Encoded by the Influenza Viruses

The proteins encoded in the different gene segments of influenza A and influenza C viruses are described in Table 4.8. Influenza A produces 10 proteins from its eight genome segments, and most of these proteins have analogues in other (–)RNA viruses (Fig. 4.1). The matrix protein, M1, and the nucleocapsid protein, NP, perform functions similar to those of M (when present) and N of other (–)RNA viruses. The three proteins encoded in the three largest segments of influenza, called PB2, PB1, and PA (B or A refers to a basic or acidic pK), possess the RNA polymerase activities encoded in the L protein and the P protein of other (–)RNA viruses. Influenza A and B have two surface glycoproteins, called HA and NA, but influenza C has only one, called HEF. These glycoproteins have the receptor binding, fusion, and receptor-destroying activities present in surface glycoproteins of (–)RNA viruses.

Two proteins, called NS1 and NS2 (NS for nonstructural), are produced from RNA segment 8. NS1 is produced from the unspliced mRNA (replication occurs in the nucleus). It binds to RNAs in the nucleus, including cellular pre-mRNAs, cellular snRNAs which are involved in splic-

ing, and dsRNA. Its activities inhibit the transport of cellular mRNAs from the nucleus and promote the synthesis of influenza mRNA. NS1 also regulates splicing of influenza mRNAs and their transport from the nucleus to the cytosol. Another function of NS1 is to interfere with the interferon pathway (Chapter 8), perhaps by binding dsRNA. Influenza virus lacking NS1 is very sensitive to interferon, whereas the wild-type virus is resistant. NS2 is produced from a spliced mRNA. It interacts with M1 attached to influenza RNP and promotes the transport of the RNP to the cytoplasm. It is present in small quantities in the virion and so is not truly nonstructural.

Protein M2 is produced from a spliced mRNA from segment 7. It forms ion channels in membranes, probably as a tetramer, that allow passage of H^+ ions. During transport of HA to the cell surface, the presence of M2 in the membrane of the transport vesicle causes the pH within the vesicle to equilibrate with that in the cytosol. This prevents low pH activation of the fusion activity of HA during transport, because transport vesicles are otherwise acidic. M2 is also present in virions and is required for the disassembly of the virus and for the activation of the RNA polymerase activity. To become active, the polymerase in the interior of the virus must be exposed to low pH. Influenza virus enters the cell in endosomes, which are progressively acidified. The acidic pH not only triggers a conformational change in HA that results in fusion of the viral membrane with the endosomal membrane, but

TABLE 4.8   Genome Segments of Influenza Viruses

Influenza A					Influenza C			
RNA segment	Length (nt)	Encoded protein		Function	RNA segment	Length (nt)	Encoded protein	
		Name	(aa)				Name	(aa)
1	2341	PB2	759	Cap recognition, RNA synthesis	1	2365	PB2	774
2	2341	PB1	757	RNA synthesis	2	2363	PB1	754
3	2233	PA	716	RNA synthesis	3	2183	PA	709
4	2073	HA	566	Hemagglutinin, fusion, major surface antigen, sialic acid binding; HEF of FLUCV also has esterase activity	4	2073	HEF	655
5	1565	NP	498	Nucleocapsid protein	5	1809	NP	565
6	1413	NA	454	Neuraminidase				
7	1027	M1	252	Matrix protein	6	1180	—	—
	Spliced	M2	97	Ion channel ??		Spliced	M1	242
						Internal initiation	CM2	139
8	934	NS1	230	Nonstructural protein	7	934	NS1	286
	Spliced	NS2	121	Nuclear export protein ??		Spliced	NS2	122

Source: Adapted in part from Fields *et al.* (1996, Table 2, p. 1355).

it also activates the RNA polymerase of the virion through the activity of M2. M2 is the target of the drug amantadine, one of the relatively few drugs that are effective against a viral disease. Amantadine binds M2 and prevents it from acting as an ion channel, which prevents the activation of the polymerase. When taken early during infection, amantidine ameliorates the symptoms of influenza.

Influenza Glycoproteins

Comparison of the glycoproteins of influenza A virus and the paramyxovirus SV-5 is of interest. In both influenza A virus and SV-5, one of the glycoproteins is type 1 (N terminus out) and one is type 2 (C terminus out). In both cases, the type 1 glycoprotein is produced as a precursor that must be cleaved to activate the fusion activity required for entry into cells. The type 1 glycoprotein of influenza A has fusion and receptor-binding (hemagglutinating) activities and is called the hemagglutinin or HA. The precursor is called HA_0 and the cleaved products are called HA_1 and HA_2 (which remain covalently linked by a disulfide bond after cleavage of the peptide bond). The SV-5 type 1 glycoprotein has only fusion activity and is called F. As described above, it is produced as a precursor F_0 which is cleaved to F_1 and F_2. The receptor bound by both influenza A virus and by SV-5 for entry into cells is sialic acid. The type 2 glycoprotein of influenza has neuraminidase activity and is called the neuraminidase or NA. It removes sialic acid from glycoproteins for the same reasons as described for the paramyxoviruses that use sialic acid as a receptor. The type 2 glycoprotein of SV-5 has both neuraminidase activity and receptor-binding (hemagglutinating) activities and is called HN.

Influenza HA is present as a trimer on the surface of the virus (as is F of SV-5). The trimeric spike has a long stalk and a head containing the sialic acid binding sites. As described in Chapter 1, exposure to acid pH in endosomes produces a dramatic rearrangement of the spike in which the fusion peptide, which forms the N terminus of HA_2, is moved over a distance of more than 10 nm to the tip of the spike (Fig. 1.5). Here it inserts into the target membrane and somehow promotes fusion of the viral membrane with the target membrane. NA is present as a tetramer (as is HN of SV-5), and forms a spike that is distinguishable in the electron microscope from the HA spike.

There is only one surface glycoprotein in influenza C, the hemagglutinin-esterase-fusion protein (HEF). Influenza C virus has, therefore, one fewer gene segments than influenza A. HEF has receptor-binding (hemagglutination), fusion, and receptor-destroying activities. The receptor is sialic acid, but the activity that destroys the receptor is an esterase activity. The esterase does not remove sialic acid from proteins as does NA of influenza A. Instead it removes the 9-O-acetyl group from 9–O-acetyl-N-acetylneuraminic acid, the receptor used by influenza C, and the virus does not bind to the deacylated sialic acid.

Replication of Influenza RNA and Synthesis of mRNAs

Synthesis of influenza virus RNAs occurs in the nucleus, rather than in the cytoplasm as for most RNA viruses. This makes possible the differential splicing observed for two of the mRNAs. Following infection by the virus, the viral RNPs are transported to the nucleus and mRNA synthesis begins. During synthesis of mRNA, influenza engages in a process called "cap-snatching." Capped cellular pre-mRNAs present in the nucleus are bound by NS1, and the 5′-terminal 10–13 nucleotides, containing the 5′ cap, are removed by PB2. This oligonucleotide is used to prime synthesis of mRNA from the influenza genome segments, as illustrated in Fig. 4.13. Once initiated, other aspects of mRNA synthesis resemble those that occur in rhabdo- and paramyxoviruses. Synthesis continues to near the end of the genome segment, where an oligo(U) stretch is encountered. Here the enzyme stutters to produce a poly(A) tail on the messenger and then releases it. In addition to its role as a primer, using a cap derived from cellular mRNA relieves the virus of the necessity of encoding enzymes required for capping and ensures that the virus mRNA has a cap suitable for the cell in which it is replicating. This mechanism also results in interference with the synthesis and transport of host mRNAs. Furthermore, because the mRNAs have a different 5′ end and lack the 3′ end of the antigenomic RNA, they lack promoters required for replication and packaging and are therefore dedicated mRNAs.

Each genome segment gives rise to one primary mRNA species. However, two of these can be spliced, and both the unspliced and spliced RNAs serve as messengers. Thus, two mRNAs are formed from each of two of the segments, and in total, 10 mRNAs are formed and 10 proteins are produced. The formation of the two mRNAs from segment 7 and their translation into proteins is illustrated schematically in Fig. 4.14.

When sufficient amounts of viral proteins have been synthesized and transported to the nucleus, viral RNA replication begins. Replication requires encapsidation of progeny genomic and antigenomic RNAs as described for other (–)RNA viruses, and the mechanisms that lead to a switch between synthesis of mRNAs and replication are thought to be similar to those that occur in rhabdoviruses and paramyxoviruses. During replication, the viral genome is copied into a faithful antigenomic RNA (vcRNA)

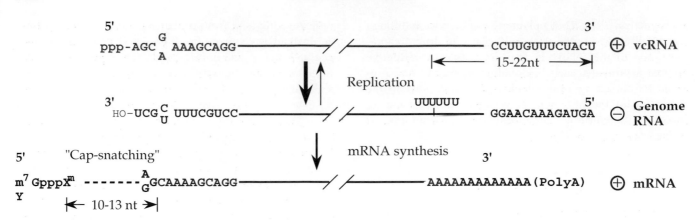

FIGURE 4.13 Relationship between genome RNAs, mRNAs, and vcRNAs of influenza virus. Synthesis of mRNAs in the cell nucleus requires a primer of 10–13 nucleotides derived from cellular pre-mRNAs by "cap-snatching," and mRNAs terminate with a poly(A) tail. Those portions of the mRNA that are not complementary to the genome RNA are shown in red. In contrast, vcRNAs are exact complements of the genomic minus strands. [Adapted from Strauss and Strauss (1997).]

(Fig. 4.13), which is a perfect complement of the genome and serves as a template for production of genomic RNA.

Assembly of Progeny Virions

Influenza virus matures by budding of nucleocapsids through the cell plasma membrane (Fig. 2.21C). The virion is often spherical, averaging 100 nm in diameter, but virus preparations are often pleomorphic, and filamentous forms may be present. During assembly, the eight genome segments are reassorted in progeny virions if the cell is infected

with more than one strain of influenza. Reassortment to produce viruses with mixed genomes is very efficient—the segments are almost randomly reassorted to give all possible combinations of genome segments in the progeny virions. This process is analogous to the reassortment of chromosomes that takes place during sexual reproduction in diploid organisms.

Budding must result in the packaging of the 8 different genomic segments that constitute the viral genome into one virus particle if it is to be infectious. Reoviruses (Chapter 3) have an assembly mechanism whereby the 10–12 differ-

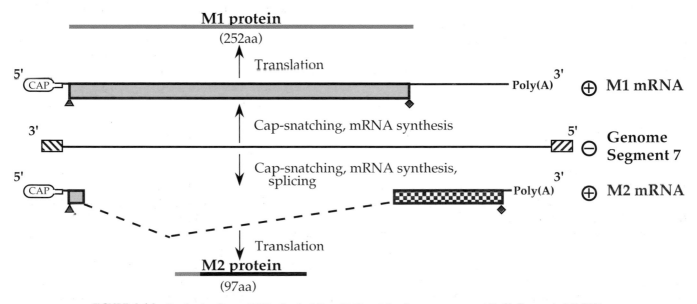

FIGURE 4.14 Synthesis of two mRNAs for the M1 and M2 proteins from gene segment 7 of influenza A. M1 RNA is translated from ORF1 (open box). M2 RNA starts identically, but after the splice it is translated in ORF2 (checked box). Both proteins are found in infected cells. The AUG initiation codon is shown as a triangle; termination codons are shown as filled diamonds. Patterned boxes at the end of the genome RNA are self-complementary sequences not present in the mRNAs that could form panhandles.

ent segments are recognized and assorted so that each virus particle has one each of the different segments. This is probably not true for influenza virus, however. Instead, the virus appears to package more than 8 segments, probably about 10, that are randomly chosen from the intracellular pool. Random packaging of 10 segments would result by chance in about 3% of the virions having at least 1 each of the 8 different genome segments.

Influenza A Virus

Natural History of Influenza Virus

Influenza A virus infects a wide variety of birds and mammals. A phylogenic tree that shows the relationship of viruses isolated from many different animals is shown in Fig. 4.15. In this tree, the sequences of NP are used, whose sequence is less variable than those of the surface glycoproteins. Notice that the human isolates are found in lineage III and are most closely related to certain isolates from pigs. Other pig isolates have an NP protein from lineage V.

Influenza viruses are characterized by their two major surface antigens, HA and NA. There are 15 different HA subtypes (numbered 1 to 15). HAs in different subtypes differ by more than 10% in sequence. There are also 9 different NA subtypes (numbered 1 to 9). Only a limited number of these HAs and NAs have been isolated from viruses infecting humans. The first influenza virus isolated, in 1933, was called H1N1 (i.e., HA type 1 and NA type 1). The virus isolated in the epidemic of 1957 had a different subtype of both HA and NA and was called H2N2. The epidemic of 1968 was caused by H3N2 virus.

The major reservoirs of influenza A in nature are wild ducks and other waterfowl such as gulls, terns, and shearwaters. Influenza replicates in the lung and in the gut of birds and the infection is normally asymptomatic (but epidemics of fatal influenza have occurred in turkeys and chickens). Ducks can excrete virus in feces for weeks, infecting other ducks via contaminated water, and a significant fraction of ducks may become infected by the virus in this process. Migratory ducks then spread the virus around the world, normally in a north–south direction. Viruses containing all 15 subtypes of HA and all 9 subtypes of NA have been isolated from waterfowl (a representative sampling of these is shown in Fig. 4.15).

The gene segments of influenza A virus reassort readily during mixed infection, and viruses with new combinations of genes arise frequently. Newly arising reassortants can cause major epidemics of influenza when introduced into humans. However, not all combinations of genes give rise to viruses that replicate in humans and are capable of epidemic spread. Only three subtypes of HA (HA1, 2, and 3) and two subtypes of NA (NA1 and 2) have been isolated from epidemic strains of human influenza virus. Similarly,

only certain types of the other segments are compatible with infection of and epidemic spread in humans. For example, the nucleocapsid gene has diverged into five lineages, but only one of these lineages, lineage III of Fig. 4.15, is present in viruses isolated from humans. Importantly, lineage III is also found in viruses isolated from pigs. It is thought that reassortment can result in the introduction of a new HA or NA gene into a human virus. The new HA and/or NA gives rise to a virus with different antigenic properties that may cause a new epidemic, a process called antigenic shift. One possible scenario is that pigs serve as intermediates ("mixing vessels") in the recombination process, because pigs are infectable by both avian and human viruses and reassortment could occur in this host.

It is noteworthy that influenza A virus causes devastating epidemics of disease in humans, but human infection is not required for the maintenance of the virus in nature. The reservoir of the virus is ducks and other birds, and viral infection is usually asymptomatic in these hosts.

Epidemics of Influenza

Influenza A virus causes a serious human illness, influenza. It is perhaps confusing and unfortunate that the term *flu* is often used to describe any respiratory tract infection (and at times even infections of the gastrointestinal tract), even those that are fairly mild. The symptoms of true influenza are usually more severe than those resulting from other respiratory tract infections and include fever, headache, prostration, and significant muscle aches and pains (myalgias) that last for 3–6 days. Weakness and cough can last 1–2 weeks more. The fever can be high (39–40°C is not uncommon in adults and can be higher, especially in children). The morbidity that accompanies the disease can cause the patient to remain bedridden for a week or longer. In young children, the high fever can result in Reye's syndrome, an encephalopathy that may be fatal. The probability of contracting Reye's syndrome is higher if aspirin is administered to control the fever.

Lower respiratory tract infection can also occur following influenza infection and result in viral pneumonia. Invasion of the damaged lungs by pathogenic bacteria may follow and result in secondary bacterial pneumonia. Influenza can be fatal, usually as a result of the accompanying pneumonia, especially in the very young (whose immune system is not fully developed) and in the elderly (whose immune system may be waning). Before the advent of antibiotics, bacterial pneumonia killed many following severe bouts of influenza, but even today influenza remains a serious killer. The annual death rate in the United States from influenza A in people over 65 is 1 per 2200, and in an epidemic year the death rate may be 1 in 300 (that is, 1 of

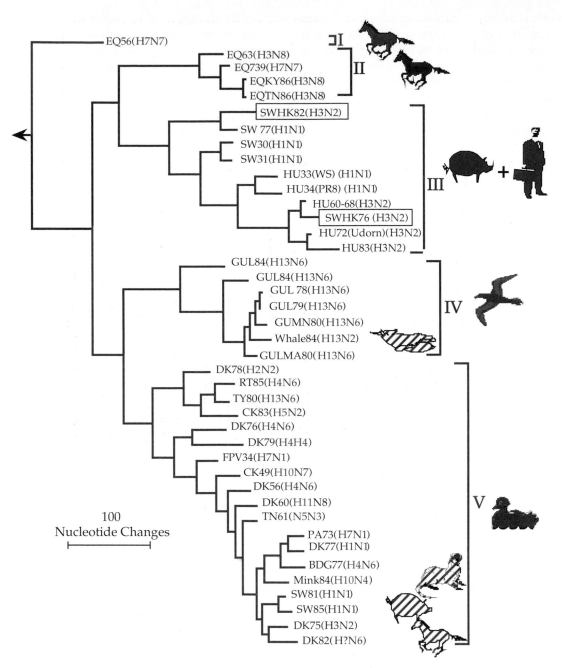

FIGURE 4.15 Phylogenetic tree of the influenza A viruses determined from the nucleotide sequence of the nucleo-protein (NP) genes. Horizontal distance is proportional to the minimum number of nucleotide differences to join nodes. Vertical spacing is for branches and labels. The NP sequences divide the viruses into five species-specific lineages. Diagonally hatched animals indicate viruses in a lineage that have been transmitted to other hosts. The genotype of the hemagglutinin (H) and the neuraminidase (N) is indicated for each isolate. In general, the names indicate only the species and the year, but human isolates have been given their traditional names (PR8, Udorn, etc.). Species abbreviations are as follows: EQ, horse; SW, pig; HU, human; GUL, seagull; DK, ducks (including mallard and grey teal); RT, ruddy turnstone; TY, turkey; CK, chicken; FPV, fowl plague virus; TN, tern; PA, parrot; BDG, budgerigar. The disparity between the colors of the branches and the colors of the H and N genotypes illustrates the amount of reassortment that has taken place. In particular the two boxed isolates are clear reassortants. [Adapted from Gorman *et al.* (1990).]

every 300 people over the age of 65 die of influenza during the epidemic). The excess mortality in the elderly caused by influenza epidemics is illustrated in Fig. 4.16, in which the different strains of influenza A or of influenza B responsible for the epidemics are indicated. Note the large increase in mortality during the epidemic of 1957 caused

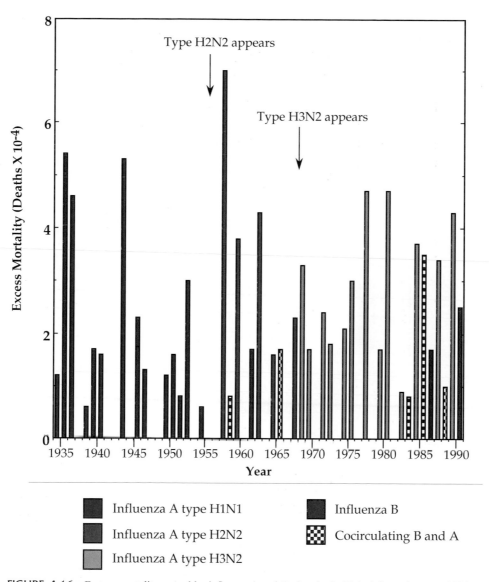

FIGURE 4.16 Excess mortality caused by influenza A and B virus in the United States between 1934 and 1990. "1935" refers to the winter of 1934–1935. Excess mortality due to the three dominant subtypes of influenza A and influenza B are indicated by the colors shown in the key. Cross-hatched bars are excess mortality in years when both A and B viruses circulated. In 1955 and 1965, type H2N2 circulated with B, in 1983 and 1988 type H1N1 circulated with B, and in 1985 H3N2 circulated with B. Note that the excess mortality of 70,000 in 1957 meant that 0.3% of the U.S. population that was over 65 years old died of influenza in that year. [Redrawn from Fields *et al.* (1996, p. 1421).]

by the appearance of influenza type H2N2, which replaced the H1N1 strain that circulated previously. This figure also illustrates that although influenza A is the most serious cause of mortality in the elderly among the influenza viruses, in some years influenza B is more of a problem than influenza A.

Although the very young and the elderly are normally at the most risk from influenza, the influenza pandemic of 1918–1919 was unusual in that mortality was high in healthy young adults. The age distributions of people dying of

influenza and the related pneumonia are compared for the years 1917 and 1918 in Fig. 4.17. The much higher death rates in the young and the elderly in 1917, the normal pattern, is apparent. The dramatic increase in the death rate in the 20- to 29-year-old group in 1918, in which people of this age were more likely to die than the old and the young, is striking. The epidemic caused by this extremely virulent virus spread around the world over a period of about a year and ultimately infected an estimated 20% of the world's population. The overall mortality was perhaps 2% but in some

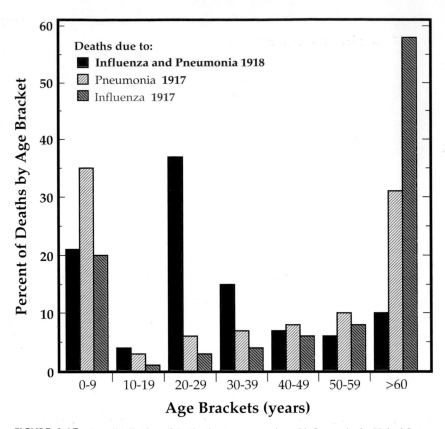

FIGURE 4.17 Age distribution of deaths due to pneumonia and influenza in the United States in 1917 and 1918. Age at death of patients has been divided into seven intervals of 10 years each. The percent of deaths due to pneumonia in 1917, due to influenza in 1917, and due to the combined effects of pneumonia and influenza during the great epidemic year 1918 which fall into each age bracket are shown. The epidemic shows the atypical preponderance of deaths in the 20–29 and 30–39 year old brackets during the 1918 epidemic. [Data from Crosby (1989).]

regions of the world, for example, regions of Central America and certain islands in the Pacific, 10–20% of the entire population died in the epidemic. In some remote Alaskan villages, more than 70% of all adults died. Estimates of the final death toll worldwide vary widely, from 20 to 100 million, but was high enough that overall life expectancy was notably reduced (Fig. 4.18). The death toll exceeded that produced by World War 1, which was ongoing at the time. In fact, 80% of deaths in the U.S. Army during World War I resulted from influenza, and it is thought that the final collapse of the German army in 1918 may have been precipitated by widespread influenza in the troops. The surgeon general of the United States had expressed the hope that WWI would be the first war in which more U.S. soldiers died of war injuries than died of disease, but this hope was shattered by the influenza epidemic. Descriptions of the epidemic with a focus on its effects on U.S. society are found in the books *Flu*, by G. Kolata, and *America's Forgotten Pandemic*, by A. W. Crosby.

The reasons for the extreme virulence of the 1918 virus, and why healthy young people were more likely to die, remains a mystery. But the devastation caused by this virus

raises continuing concern that a strain of influenza of equal virulence might appear and again cause immense suffering worldwide. Such concerns are heightened by the continual appearances of new strains of influenza virus and the fact that a strain of the virus epidemic in 1933 (the H1N1 strain) reappeared essentially unchanged 20 years later and caused a new epidemic. Thus, it is possible that the 1918 strain itself might reemerge. The pandemic of 1918 occurred before influenza virus could be isolated and it has not been possible to study the virus in the laboratory using modern tools. However, the sequences of the HA and NA gene of the 1918 virus have been obtained recently in a feat that demonstrates the power of modern molecular biology. Samples of preserved lung tissue taken at autopsy from two U.S. soldiers who died of influenza in 1918 were found to contain detectable influenza RNA, albeit in fragmented condition. Reverse transcriptase–polymerase chain reaction technology was used to obtain sequences from HA and NA that could be used to reconstruct the complete sequences of these genes. A third source of influenza RNA came from an Alaskan victim of the 1918 influenza who had been

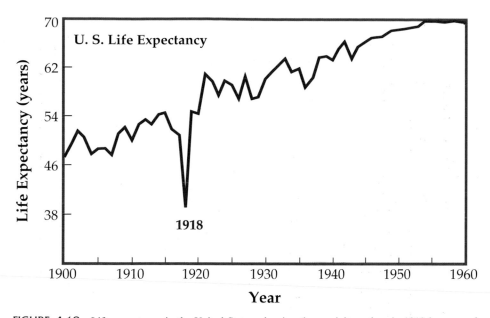

FIGURE 4.18   Life expectancy in the United States, showing the precipitous drop in 1918 because of deaths due to the "Spanish flu." This drop interrupted an otherwise fairly uniform increase in life expectancy that resulted from better health care, sanitation, and living conditions. [Adapted from *ASM News* (July 1999).]

buried in permafrost, and whose body was sufficiently well preserved that lung samples containing (fragmented) viral RNA were obtained. The sequences from these three victims were almost identical and showed that the virus belonged to strain H1N1.

Antigenic Shift and Drift

Immunity to influenza A virus following infection is long lived but may not be complete and is subtype specific and even strain specific. The continuing appearance of new strains that arise from antigenic drift and of new subtypes that arise from antigenic shift lead to continuing epidemics. Normally, two or three strains of influenza A are in the human population at any one time. Spread from person to person is by respiratory droplets, requiring close proximity, but people travel extensively and new strains of the virus speed around the globe as they arise. Antigenic drift is the process by which mutations accumulate in the virus genome, usually because of immune selection, that result in the development of new strains of the virus. These new strains are partially resistant to the immunity induced by infection with previous strains of virus. After several years of drift, the strain may be sufficiently distinct to cause disease in a person previously infected, but the illness is usually less severe because of partial immunity to the new strain. However, every 10–20 years or so, a new virus arises by reassortment in which one or both of the segments encoding the surface glycoproteins of the virus are replaced by segments from another source, a process called antigenic shift. The reassortants that cause the biggest problems are those belonging to a new subtype (as is illustrated by Fig. 4.16). Such a new subtype often causes pandemics in the human population in which a significant fraction of people become infected, because there is little immunity to the virus carrying these new surface antigens in a large fraction of the population. Such events happened in 1957 (subtype H2N2) and 1968 (subtype H3N2), and presumably were responsible for the pandemic of 1918 (H1N1).

H1N1 virus, which had disappeared with the appearance of the H2N2 epidemic strain in 1957, suddenly reappeared in 1977. This H1N1 virus, which first appeared in northern China in May 1977 and was called the Russian flu, was virtually identical to influenza virus isolated from an epidemic in humans in 1950. It circulated in young people who had not been exposed to H1N1 virus. Because it was virtually unchanged despite 27 years having elapsed, it seems unlikely that it arose again *de novo*. It has been hypothesized that it had been preserved in a frozen state, perhaps in a laboratory freezer.

Vaccination against Influenza A Virus

Because of the seriousness of influenza disease, especially in the elderly, attempts are made each year to vaccinate the population at risk. Because of drift and shift, the vaccine must be reformulated every year to reflect the viruses currently circulating in the human population. The viruses included in the vaccine are those that are circulating in late spring, because these viruses are usually those that will cause epidemics the following winter. Inactivated virus vaccines are currently used, but attempts to devise attenuated virus vaccines that could be developed more rapidly and require smaller doses of vaccine are promising. One such approach is to develop an attenuated strain of a human influenza virus and each year introduce into it the HAs and NAs of the epidemic strains by reassortment. Because the attenuation of the virus results from changes in other genome segments, the recombinant strain is also attenuated.

Continuing surveillance of influenza strains in nature is required in order to reformulate the vaccines each year. This surveillance also serves to watch for the possible appearance of another killer strain of influenza. An episode that occurred during the Ford administration, however, illustrates the difficulties of identifying such a strain and reacting in time. In 1976, a young soldier at Fort Dix died of influenza and others became seriously ill. Tests showed that most of the soldiers were suffering from the A/Victoria strain of influenza that was epidemic in the United States at the time or from adenovirus infection. However, the soldier who died and three other soldiers who were ill were infected with an influenza strain that was epidemic in pigs, referred to as swine flu. The swine flu virus was closely related to the 1918 pandemic virus, and is thought to have been introduced into pigs in 1918 from humans and to have continued to circulate in pigs after it had died out in humans. Could it be possible that the 1918 virus had reappeared as an epidemic virus in humans? The decision was made by President Ford, in consulation with leading scientists, to begin a crash program to develop a vaccine against swine flu and to begin to immunize the American population. It was thought, with some justification, that to wait for an epidemic to begin before an immunization program was undertaken would mean that it would be too late to be effective, given the speed with which influenza epidemics spread. Forty million Americans were immunized against swine flu. No epidemic of swine flu developed, however, and litigation began. The pharmaceutical companies had been reluctant to participate in the program, pointing out that at any one time a certain fraction of Americans would develop encephalitis or rheumatoid arthritis or any one of hundreds of other diseases. If disease developed in proximity to receiving a new and relatively untested vaccine, a lawsuit would certainly follow and the potential damages were enormous. The program could only advance when Congress agreed to indemnify the pharmaceutical houses. Although the vaccine was never conclusively shown to cause disease, litigation went on for years and substantial damages were paid out. In retrospect it is easy to criticize the program as an overreaction, but what would have been the reaction if nothing had been done and an influenza epidemic developed that resulted in 50–100 million Americans

becoming seriously ill with 1–2 million deaths? Given the state of knowledge at the time, many leaders felt there was no choice.

A more recent scare occurred when 18 people in Hong Kong became seriously ill from influenza in 1997 and 6 died. The culprit was an avian influenza (H5N1) that was epidemic in birds imported from China for food. Avian viruses do not normally infect people, and there was fear that an avian virus had made the jump to humans and might cause an epidemic of lethal influenza. The Hong Kong authorities destroyed 1.6 million domestic birds in order to eradicate the epidemic in birds. No human-to-human transmission took place and no epidemic in humans developed.

Influenza B and C Viruses

Influenza B virus infects only humans and no animal reservoir of the virus exists. It also causes influenza in humans but there exists only one subtype, and antigenic shift does not occur. Antigenic drift does occur, and the virus can cause epidemics of serious illness that result in increased mortality among the elderly, as shown in Fig. 4.16. However, wide-ranging pandemics do not occur and the virus is therefore not as much of a problem as influenza A. Less attention has accordingly been given to the control of this virus. Influenza C is not a serious human pathogen and has been less well studied.

FAMILY BUNYAVIRIDAE

The family Bunyaviridae contains more than 300 viruses grouped into five genera. A representative sampling of these viruses is shown in Table 4.9. Members of four genera, *Bunyavirus*, *Nairovirus*, *Phlebovirus*, and *Hantavirus*, infect vertebrates and contain important human pathogens, whereas viruses belonging to the genus *Tospovirus* infect plants. The human pathogens in the family variously cause hemorrhagic fever, a pulmonary syndrome that can be fatal, encephalitis, or milder febrile illnesses, as shown in the table. Some of these pathogens were listed in Table 3.8, which contains a partial listing of arboviruses that cause disease in humans. All members of the Bunyaviridae except the hantaviruses are transmitted to their vertebrate or plant hosts by arthropods. The hantaviruses, in contrast, are associated with rodents and are transmitted to man by aerosolized excreta from infected rodents. In the description below, the term bunyavirus refers to any member of the family unless indicated otherwise.

Replication of the Bunyaviridae

Genome Organization

The genomes of viruses belonging to 5 genera of the Bunyaviridae are illustrated in Fig. 4.19. All bunyavirus

TABLE 4.9 Bunyaviridae

Genus/members[a]	Virus name abbreviation	Usual host(s)	Transmission/ vector	Disease in humans	World distribution
Bunyavirus (~150 types)					
Bunyamwera	BUNV	Rodents, rabbits	*Aedes* mosquitoes	Febrile illness	Worldwide
La Crosse	LACV	Humans, rodents	*Aedes triseriatis*	Encephalitis	Midwest United States
Snowshoe hare	SSHV	Lagomorphs	Mosquitoes (*Culiseta* and *Aedes*)	Rarely infects humans	Northern United States
California encephalitis	CEV	Rodents, rabbits	*Aedes melanimon, A. dorsalis*	Encephalitis (rare)	Western United States, Canada
Jamestown Canyon	JCV	White-tailed deer	*Aedes species, C. inornata*	Rarely infects humans	North America
Hantavirus					
Hantaan	HTNV	*Apodemus agrarius*	Feces, urine, saliva	Hemorrhagic fever	Worldwide
Seoul	SEOV	*Rattus* species	Feces, urine, saliva	Hemorrhagic fever	Eastern Asia Eastern Europe
Prospect Hill	PHV	*Microtus pennsylvanicus*		None?	United States
Sin Nombre	SNV	*Peromyscus*	Feces, urine, saliva	Pulmonary syndrome	Western United States, Canada
Nairovirus					
Dugbe	DUGV	Sheep, goats	Tick-borne		Africa
Nairobi sheep disease	NSDV	Sheep, goats	Tick-borne		Africa
Crimean-Congo hemorrhagic fever	C-CHFV	Humans, cattle, sheep, goats	Tick-borne	Hemorrhagic fever	Africa, Eurasia
Phlebovirus (~50 types)					
Sandfly fever Sicilian	SFSV	Humans	Phlebotomous flies	Nonfatal febrile illness	Mediterranean
Rift Valley fever	RVFV	Sheep, humans, cattle, goats	Mosquitoes, also contact, aerosols	Hemorrhagic fever	Africa
Uukuniemi	UUKV	Birds	Tick-borne	??	Finland
Tospovirus					
Tomato spotted wilt	TSWV	Plants	Thrips	None	Australia, Northern Hemisphere

[a]Representative members of each genus are shown.

genomes consist of three segments of RNA, referred to as S(mall), M(edium), and L(arge), that together total about 12 kb. The S segment encodes the nucleocapsid protein, M the two surface glycoproteins, and L the polymerase protein. In addition, viruses belonging to three of the genera illustrated encode two nonstructural proteins, NS_s in segment S and NS_m in segment M.

Replication of bunyavirus genomes and the synthesis of mRNAs take place in the cytoplasm. Like influenza viruses, these viruses engage in cap-snatching in order to prime mRNA synthesis. In bunyaviruses, however, the caps are captured from cytoplasmic mRNAs rather than from nuclear pre-mRNAs. Synthesis of mRNAs is assumed to occur by processes similar to those used by influenza virus, but the termination of an mRNA does not appear to be precise and no poly(A) is added to the 3′ end. Thus, the mRNAs are capped but not polyadenylated. The switch to replication is assumed to use the same mechanisms as used by other (−)RNA viruses.

Expression of Proteins Encoded in S

The S segment of bunyaviruses encodes one or two proteins (Fig. 4.19). In the *Hantavirus* and *Nairovirus* genera, S encodes only N. In the other genera, S encodes both N and NS_s, using one of two different mechanisms. In the Bunyavirus genus, the two proteins are translated from a single mRNA using two different start codons in different reading frames. The coding region for NS_s is completely contained within that for N. In the phleboviruses and

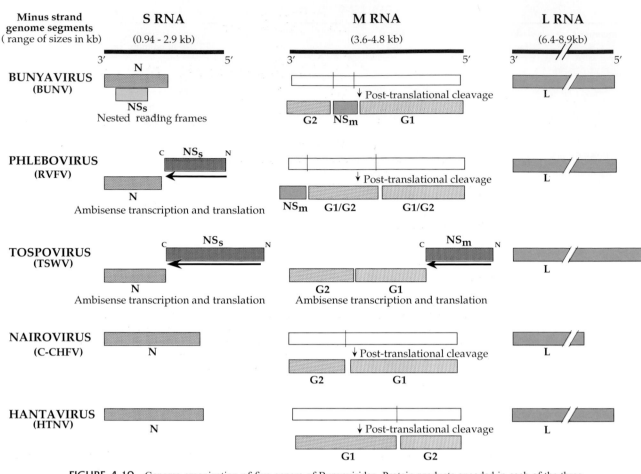

FIGURE 4.19 Genome organization of five genera of Bunyaviridae. Protein products encoded in each of the three genome segments and the various strategies used to produce these proteins are shown. Unless otherwise noted the mRNA (not shown) would extend 5′ to 3′ from left to right, and the protein product is shown N terminal to C terminal in the same direction. The products are illustrated roughly to scale. Color coding for functionality is the same as in Fig. 4.6. Patterned blocks are virion structural proteins; white boxes are precursor proteins. Arrows indicate the direction of synthesis and translation of ambisense mRNAs (magenta). Virus abbreviations are as follows: BUNV, Bunyamwera; RVFV, Rift Valley fever; TSWV, tomato spotted wilt; C-CHFV, Crimean-Congo hemorrhagic fever; HTNV, Hantaan. [From Fields *et al.* (1996, Tables 2 and 3, pp. 1475 and 1476).]

tospoviruses, however, an ambisense coding strategy is used for the two proteins (*ambi* = both). In this strategy, the two genes encoded in a genomic segment are linked tail to tail so that they are in different polarities, as illustrated in Fig. 4.20. The gene for N is present at the 3′ end of the genome S segment in the minus-sense orientation, and synthesis of the mRNA for N occurs from the genome segment. Expression of this gene occurs early because its mRNA is synthesized from the entering genome by the polymerase activity present in viral nucleocapsids. The gene for NS$_s$ is plus sense within the genome, but the genomic RNA does not serve as mRNA. Instead, an mRNA for NS$_s$ is synthesized from the antigenomic RNA. Thus, NS$_s$ is expressed late because its mRNA can only be made after replication of the incoming genomic RNA to produce the antigenomic RNA. Termination of either mRNA occurs at a secondary structure between the genes for N and NS,

which appears to cause the polymerase to fall off and release the mRNA.

Expression of Proteins Encoded in M

Two glycoproteins, usually called G1 and G2, are translated from mRNA made from M (Fig. 4.19). They are thought to be produced as a polyprotein that is cleaved by a cellular protease to separate the two glycoproteins, analogous to what happens in some of the (+)RNA viruses that have envelopes (e.g., coronaviruses and flaviviruses). However, there is also evidence that reinitiation to produce the second glycoprotein can occur, at least in some viruses.

The M segments of hantaviruses and nairoviruses encode only the two glycoproteins, but in the other three genera M encodes a third protein called NS$_m$ (Fig. 4.19). In phle-

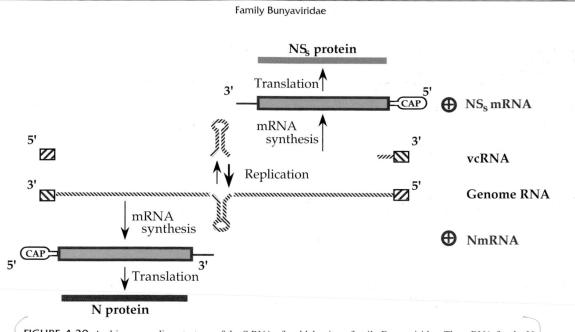

FIGURE 4.20 Ambisence coding strategy of the S RNA of a phlebovirus, family Bunyaviridae. The mRNA for the N protein is synthesized from the S genome segment using primers derived by cap-snatching (similar to the mechanism for influenza mRNA priming in Fig. 4.13) from cytoplasmic host mRNAs. The mRNA for the NSs protein is formed in the same way, but with vcRNA as the template. Diagonally striped boxes are the self-complementary termini. The loops in the middle of the viral genomic and antigenomic RNAs indicate a secondary structure in the RNAs, which terminates synthesis of the mRNAs. No poly(A) is added to the 3' terminus of the mRNA's.

boviruses and members of the genus *Bunyavirus*, NSm forms part of the polyprotein translated from the single mRNA produced from M. NSm is formed during post-translational processing of the polyprotein. In tospoviruses, an ambisense strategy is used to encode NSm and the translation strategy is the same as that shown in Fig. 4.20.

There is no matrix protein in bunyaviruses. Budding occurs from intracellular membranes and there is assumed to be a direct interaction between the glycoproteins and the nucleocapsid protein. The virion is spherical, averaging 100 nm in diameter. The three nucleocapsids are circular when isolated from the virion.

Genus Bunyavirus

There are more than 100 members of the genus *Bunyavirus*, which together have a worldwide distribution. These viruses are mosquito-borne and are true arboviruses, replicating in the mosquito vector as well as in vertebrates. The viruses of greatest medical interest are those belonging to the California encephalitis group, of which La Crosse virus is the best known. La Crosse virus was named for the town of La Crosse, Wisconsin, where it was first identified as the causative agent of encephalitis, primarily in children. About 100 cases per year of encephalitis are caused by La Crosse virus, concentrated in the Midwest.

Mortality is low (0.3%) but 10% of patients suffer neurological sequelae. No vaccine exists for the virus and control measures have involved control of the mosquito vector. The principal vector of La Crosse is *Aedes triseriatus*. This mosquito breeds in tree holes, but abandoned tires filled with rainwater constitute an important breeding area for it close to human habitation. Such abandoned tires serve as a beautiful incubator for the development of mosquito larvae. Efforts to eliminate this source of mosquitoes, as well as the institution of other mosquito control measures, resulted in a reduction in the number of cases of disease.

Abandoned tires are important in the transmission of other arboviruses as well. Old tires are abundant in Puerto Rico, for example, and contribute to the endemic transmission of dengue virus, all four serotypes of which are present on the island. Old tires have also been responsible for the introduction of *Aedes albopictus*, the so-called Asian tiger mosquito that is the vector of dengue virus in Asia, into the United States. Loads of old tires that were brought from Asia to Houston for recycling contained eggs or larvae of the mosquito. After its introduction into the Houston area, this mosquito spread over large areas of the United States and there is fear that it might become an efficient vector of arboviral disease in this country.

Genus Phlebovirus

More than 50 phleboviruses are known. All are arboviruses transmitted by mosquitoes, phlebotomine flies, or ticks. The most important of these is Rift Valley fever virus, an African virus that was first isolated in 1930 in the Rift Valley of East Africa. The virus is transmitted by mosquitoes and causes hemorrhagic fever in humans. It also causes disease in domestic animals, and many widespread epidemics in cattle, sheep, and man have occurred over the years in Africa. In 1977–1978, for example, an epizootic in Egypt infected 25–50% of cattle and sheep in some areas, and 200,000 human cases resulted in at least 600 deaths. A recent large epidemic in East Africa in 1997–1998 was associated with the heaviest rainfall in 35 years, 60–100 times normal in some areas. Losses of 70% of sheep and goats and 20–30% of cattle and camels were reported, and there were hundreds of cases of human hemorrhagic fever. Contact with livestock was statistically associated with acute infection with Rift Valley fever virus, indicating that during epidemics contact transmission becomes important as a means of spread to humans. Laboratory-acquired cases contracted through aerosols are also known.

Sandfly fever virus is transmitted by phlebotomine flies and causes an acute, nonfatal influenza-like disease in man. It is found in the Mediterranean area, North Africa, and southwest Asia. Related viruses are found in South America.

Genus Nairovirus

The nairoviruses, named for Nairobi sheep disease, are tick-borne. Crimean-Congo hemorrhagic fever virus is the most important virus in terms of human disease. It is found from southern Africa through eastern Europe and the Middle East to western China. Infection of humans is relatively rare, but disease caused by the virus can be serious, with a high mortality rate. Transmission to hospital personnel treating infected patients has occurred.

Genus Hantavirus

Hantaviruses cause serious human disease, including hemorrhagic fevers and hantavirus pulmonary syndrome. Unlike other members of the Bunyaviridae, they are not arboviruses. The hantaviruses are associated with rodents, which form their natural reservoir, and are transmitted to man through contact with aerosolized urine or feces from infected rodents. Each hantavirus establishes persistent infections in one particular species of rodent and is maintained in nature in this way. Man is not an important host for the virus and does not contribute to its maintenance in nature. Related to this is the fact that the viruses do not cause serious disease in their rodent hosts, but many cause quite serious illness in man.

An evolutionary tree of hantaviruses is shown in Fig. 4.21. The rodent hosts for the viruses are also indicated. The viruses assort by host rather than by geographical proximity. All of the viruses whose hosts belong to the order *Murinae* group together, as do those that use rodents in the order *Arvicolinae* and those that use rodents in the order *Sigmodontinae*. As one example, consider Prospect Hill virus and New York virus, both found in the northeastern United States. Prospect Hill virus is associated with rodents of the genus *Microtus*, order *Arvicolinae*, and is more closely related to Puumala virus of Europe, which uses *Clethrionomys glareolus*, order *Arvicolinae*, than it is to New York virus. New York virus is associated with rodents in the genus *Peromyscus*, order *Sigmodontinae* and is closely related to Sin Nombre virus of the southwestern United States, which is associated with *Peromyscus maniculatus*. The fact that the evolutionary tree of the hantaviruses resembles that of their rodent hosts rather than being based on geographical proximity is evidence that they have coevolved with their rodent hosts over a very long period of time.

The first of the hantaviruses to be identified was the causative agent of epidemic hemorrhagic fever with renal syndrome, now called Korean hemorrhagic fever, that occurred in U.S. troops during the Korean war. The virus was called Hantaan virus after a river in the area where it was isolated. In Korea, Hantaan virus is associated with the field mouse *Apodemus agrarius*. The virus also occurs in eastern Europe and China, where it is associated with *Apodemus flavicollis* and causes a disease similar to Korean hemorrhagic fever (Fig. 4.22).

Viruses related to Hantaan virus have now been isolated from all over the world, including the United States. Puumula virus occurs in Western Europe (Fig. 4.22) and causes a disease characterized by acute fever with renal involvement. Seoul virus, first identified in Seoul, Korea, is associated with wild urban rats (*Rattus norvegicus*) and has been found all over the world because wild urban rats have been inadvertantly introduced almost everywhere. It causes a mild form of Korean hemorrhagic fever in Seoul but does not cause apparent illness in most other areas where it has been found. The discovery of Seoul virus led to an intensive study of rats in central Baltimore, where it was found that a high percentage of them were infected with Seoul virus and, furthermore, that a substantial fraction of the people living in the slums of downtown Baltimore showed evidence of infection by hantavirus. No disease is known to be associated with this virus, but statistical studies suggest that infection may lead to high blood pressure and, possibly, renal failure. Prospect Hill virus has been isolated from microtine rodents in many states in the northern and eastern United States, including Maryland, but no human disease has been associated with it.

More interesting was the isolation, in May 1993, of a new hantavirus that causes acute respiratory distress in humans that can lead to rapid death, a syndrome called hantavirus

Rodent-borne Hantaviruses   **Rodent hosts**

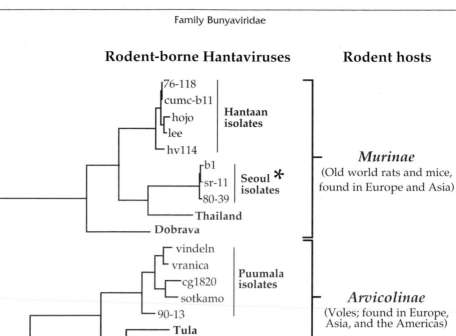

FIGURE 4.21 Phylogenetic tree of rodent-borne hantaviruses derived from the nucleotide sequence of the M RNA segment. This tree illustrates that hantaviruses have coevolved with their rodent hosts for millions of years. However, note (*) that in contrast to other members of this group, Seoul virus, which infects *Rattus norvegicus*, is found worldwide, due to the widespread distribution of these rats. [From Peters (1998b, Fig. 2).]

pulmonary syndrome (HPS) or human acute respiratory disease syndrome (ARDS). The virus was isolated by the CDC in collaboration with local health authorities following an epidemic in the Four Corners area of the southwestern United States that resulted in approximately 25 deaths. The virus is associated with the deer mouse *Peromyscus maniculatus*. It is thought the epidemic may have resulted from an abundance of pine nuts in the area during a good growing year, leading the local people to harvest larger amounts of these than usual and store them in their homes when their normal storage areas became full. The rodent population exploded and invaded people's homes to get to the pine nuts, and it is thought that this more intimate contact between humans and rodents may have led to the epidemic. Since these original observations, studies have shown that this virus or closely related viruses are present in virtually all states within the United States and into Latin America, and that fatalities due to infection by the virus have occurred in many states. One of the cases in California is of interest because the person died more than a year before the Four Corners epidemic; retrospective studies of serum collected from the

patient at the time of his hospitalization showed that he was infected with a hantavirus. This hantavirus is now called Sin Nombre virus, which is Spanish for "without a name." Early suggestions that it be called Four Corners virus or Muerto Canyon virus (after a geographical feature in the area) drew objections from local residents, and eventually the CDC simply named it Sin Nombre (there is a small creek in the area called the Sin Nombre river that serves as justification for the choice of name). As related viruses were isolated in other regions of the Americas, many associated with other rodents in the order Sigmodontinae, they were given names of local features in order to distinguish them. These include New York, Monongahela, Bayou, and Black Creek Canal viruses, all of which have caused HPS in the United States. The number of cases of HPS in the Americas from 1993 to 1998, totaled by country, and the names of the viruses responsible in various areas are shown in Fig. 4.23.

The mortality rate following infection with Sin Nombre virus or its close relatives is greater than 50%. The mortality in the earliest cases was even higher because the pulmonary syndrome results from the rapid extravasation of fluids into

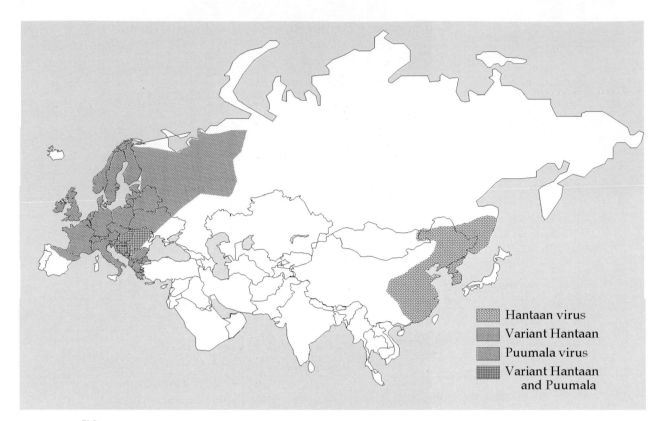

FIGURE 4.22   Map of Eurasia showing the disjunct distribution of different hantaviruses. [From Porterfield (1995, p. 276).]

the lungs, which can result in respiratory death. This loss of fluids from the intravascular compartment also leads to an increase in the hematocrit (the percentage of blood volume occupied by red blood cells). Early attempts to decrease the hematocrit by supplying fluid intraveneously simply exacerbated the pulmonary edema. Even with the best treatment today, however, the mortality rate is still very high.

It is clear that hantaviruses are widely distributed around the world and have been present in their rodent hosts for a very long time. Although many are capable of causing serious illness in man, the number of human cases is fortunately small. However, there is always the fear that one of these viruses might acquire the ability to spread more readily from human to human and thereby become a more serious problem.

FAMILY ARENAVIRIDAE

A listing of representative arenaviruses is found in Table 4.10. The arenaviruses share many features with the hantaviruses. They are associated with rodents and have coevolved with them, as have the hantaviruses. They are transmitted to humans by contact with aerosolized rodent

urine or feces; many cause very serious illness, often hemorrhagic fever, with a high mortality rate. Their genome organization has much in common with the hantaviruses, as described below.

Arenaviruses are typical (–)RNA viruses. The virion is enveloped and approximately spherical, averaging 100 nm in diameter. The nucleocapsid is helical. The arenaviruses bud from the plasma membrane (Fig. 2.21D). They are named from the Latin word for sand (arena) because during budding they often incorporate ribosomes into the virus particle, giving the virions a grainy appearance when examined in the electron microscope. Why this happens is unknown. The ribosomes do not appear to serve a useful function in the virion.

Genome Organization and Expression

The genome organization of an arenavirus is illustrated in Fig. 4.24. Arenavirus genomes consist of two segments of RNA, naturally called L(arge) and S(mall). Both genomic RNAs are ambisense in character. The S segment corresponds to the bunyavirus S and M segments linked tail to tail in an ambisense arrangement (Fig. 4.1). The L segment corresponds to the L segments of bunyaviruses but

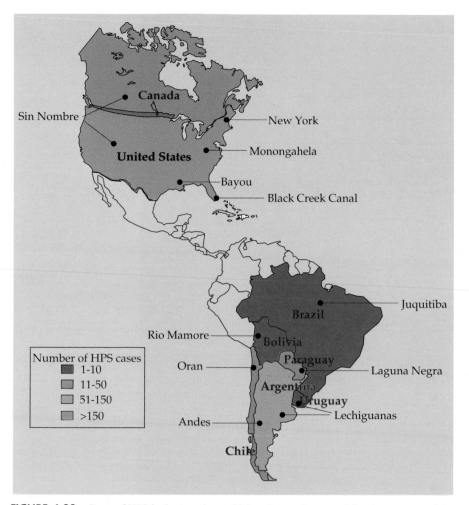

FIGURE 4.23 Cases of HPS in the Americas, with locations and names of the viruses responsible. Case numbers are cumulative totals from the time that HPS was recognized in 1993 in the Four Corners region of Arizona and New Mexico through March 1998. To the current time, cases of HPS have been diagnosed in the United States in 30 states, predominantly in the western U.S., with a few cases as far east as Rhode Island. [Adapted from Peters (1998a,b, Fig. 1 and Table 3).]

with the addition of a second gene, encoding a nonstructural protein called NS or sometimes Z, in an ambisense orientation. Expression of the encoded genes follows an ambisense strategy as described for the some of the bunyaviruses. The mRNA for one gene is synthesized from the genomic RNA and is expressed early, whereas the mRNA for the second gene is synthesized from the antigenomic or vcRNA and is expressed late (Fig. 4.24). As in the bunyaviruses, synthesis of arenavirus mRNA occurs in the cytoplasm using a primer that is snatched from cellular mRNAs, there is a secondary structure in the RNA between the two ambisense genes that causes termination of transcription, and the mRNAs are not polyadenylated.

The genomic S RNA is the template for the synthesis of the mRNA for N, and N is therefore expressed early after infection. Because N is required for the replication of the

viral RNA, as is the case for all (–)RNA viruses, this arrangement is necessary if the virus is to replicate. The mRNA for the glycoproteins G1 and G2 is transcribed from the antigenomic copy of S and is therefore expressed late. The glycoproteins are produced as a polyprotein that is cleaved, presumably by cellular enzymes, in a process that appears to be similar to what happens in the bunyaviruses. Producing the glycoproteins late has the effect of delaying virus assembly. This allows RNA amplification to proceed for an extended period of time before it is attenuated by the incorporation of nucleocapsids into virions.

In the case of the L segment, the mRNA for protein L is produced early by synthesis from the genomic RNA. L is required for RNA replication, and this orientation of the genes is necessary for virus replication. NS mRNA is

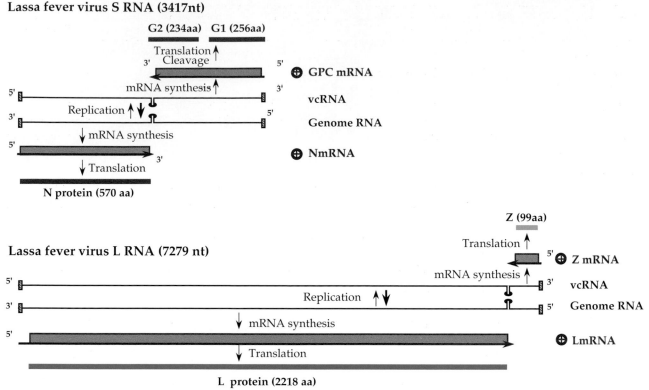

Lassa fever virus S RNA (3417nt)

G2 (234aa) G1 (256aa)

Translation
Cleavage
mRNA synthesis

⊕ GPC mRNA
vcRNA
Genome RNA

Replication

mRNA synthesis

⊕ NmRNA

Translation

N protein (570 aa)

Z (99aa)

Translation
⊕ Z mRNA

mRNA synthesis
vcRNA
Genome RNA

Lassa fever virus L RNA (7279 nt)

Replication

mRNA synthesis

⊕ LmRNA

Translation

L protein (2218 aa)

FIGURE 4.24 Genome organization and replication strategy of an arenavirus, Lassa fever virus. The genome consists of two segments of RNA, L and S. Both segments are expressed using an ambisense strategy. The nucleocapsid protein mRNA is synthesized from the 3′ end of the genomic S RNA, while the GPC mRNA is synthesized from the vc S RNA. A similar strategy occurs with the L segment. In this case, however, more than 95% of the coding capacity is used for the L protein, the RNA-dependent RNA polymerase. Also note that the 5′ nontranslated region of L mRNA is 157 nt, which is unusually long for an arenavirus. The Z protein is a so-called "ring finger protein" that is involved in regulation of transcription and replication. [Drawn from data in Lukashevich *et al.* (1997) and Clegg *et al.* (1990).]

expressed late, after replication of the RNA begins. Its function is not known.

Natural History and Diseases

The natural history of the arenaviruses is very similar to that of the hantaviruses (Table 4.10). They establish a persistent infection in a single rodent host. Many cause hemorrhagic fever in man following transmission by aerosolization of virus excreted in urine or feces. Like the hantaviruses, they appear to have coevolved with their hosts: An evolutionary tree of arenaviruses resembles the tree that describes their rodent hosts, as was true of the hantaviruses. The many similarities in genome organization and expression, the association with a single rodent species, and the nature of the disease caused in man all suggest that the arenaviruses are closely related to the hantaviruses. A reasonable hypothesis is that the arenaviruses arose from the hantaviruses by fusion of the S and M segments to form one segment, which allowed finer control of the virus life cycle.

The arenaviruses can be divided into Old World viruses and New World viruses (Table 4.10). Because of their association with a single rodent species, their geographic range is restricted to that of their host, and rodents have a restricted range. The exceptions are rodents that have been distributed widely by humans, such as the house mouse and the urban rat. Of interest in this regard is the Old World virus lymphocytic choriomeningitis virus (LCMV), which is associated with the house mouse *Mus domesticus* and *Mus musculus.* This virus is widespread in Europe, along with its host, and spread to the Americas with the (inadvertant) introduction of the house mouse by European travelers. LCMV has been intensively studied in the laboratory as a model for the arenaviruses, in part because it is widespread, often being present in colonies of laboratory mice as well as being found in wild mice, and in part because it is less virulent for man than most arenaviruses. LCMV infection of man usually results in mild illness, although serious illness can result with occasional mortality.

TABLE 4.10   Arenaviridae

Genus/members[a]	Virus name abbreviation	Natural host(s)[b]	Transmission	Disease in humans	World distribution
Old World Arenaviruses					
Lymphocytic choriomeningitis	LCMV	*Mus musculus*	Urine, saliva	Aseptic meningitis	Worldwide
Lassa	LASV	*Mastomys* sp.	Urine, saliva	Hemorrhagic fever (HF)	West Africa
Mopeia	MOPV	*Mastomys natalensis*	Urine, saliva	Nonpathogenic	Mozambique, Zimbabwe
Mobala	MOBV	*Praomys* sp.	??	??	Central African Republic
New World Arenaviruses					
Tacaribe complex lineage A					
Tamiami	TAMV	*Sigmodon hispidus*	Urine, saliva	Nonpathogenic?	Florida (US)
Whitewater Arroyo	WWAV	*Neotoma albigula*	Urine, saliva	Three fatal cases of ARDS in California[c]	Western United States
Paraná	PARV	*Oryzomys buccinatus*	Urine, saliva	Nonpathogenic?	Paraguay
Flexal	FLEV	*Oryzomys* sp.	Urine, saliva	Nonpathogenic?	Brazil
Pichinde	PICV	*Oryzomys albigularis*	Urine, saliva	Nonpathogenic?	Colombia
Pirital	PIRV	*Sigmodon alstoni*	Urine, saliva	Nonpathogenic?	Venezuela
Tacaribe complex lineage B					
Guanarito	GTOV	*Zygodontomys brevicauda*	Urine, saliva	Venezuelan HF	Venezuela
Amapari	AMAV	*Oryzomys capito, Neacomys guianae*	Urine, saliva	Unknown pathogenicity	Brazil
Junín	JUNV	*Calomys musculinus*	Urine, saliva	Argentine HF	Argentina
Machupo	MACV	*Calomys callosus*	Urine, saliva	Bolivian HF	Bolivia
Sabiá	SABV	Unknown	???	Isolated from a fatal case, and has caused two severe laboratory infections	Brazil
Tacaribe	TCRV	*Artibeus* spp. bats	? Has been isolated from mosquitoes	Unknown pathogenicity	Trinidad

Source: Adapted from Fields *et al.* (1996, Table 1, p. 1522) and Porterfield (1995, Table 11.1, p. 228).
[a]LCMV is the type virus of the family.
[b]Most of these viruses cause chronic infections in their natural rodent hosts.
[c]ARDS, acute respiratory distress syndrome. Until these cases in 1999–2000, WWAV was not known to cause human illness.

Many arenaviruses cause hemorrhagic fever in man with significant mortality rates (Table 4.10). Lassa virus causes an often fatal illness (mortality rate as high as 60% in some outbreaks) characterized by fever, myalgia, and severe prostration, often accompanied by hemorrhagic or neurological symptoms. Survivors may be deaf because of nerve damage. Lassa fever virus was first isolated in 1969 when a nurse in a rural mission hospital in Nigeria became infected. She was transported to Jos, Nigeria where several health care workers became infected. Serum samples were sent to the United States and a well-known virologist at the Yale Arbovirus Research Unit, Dr. Jordi Casals, became infected with the virus while working with it and became very seriously ill. He eventually recovered but later that same year a technician in another laboratory at Yale became infected with Lassa fever virus and died, whereupon Yale ceased to work with the virus. The containment facilities in 1969 were

not of the quality of those in current use and virologists in those days literally took their lives in their hands when working with dangerous agents. The study of virology owes a great deal to the courage exhibited by these earlier workers.

Lassa virus is endemic to West Africa. The full extent of Lassa disease is not known because most Africans infected by the virus do not seek help and there is little monitoring of the disease. However, estimates range from 100,000 to 300,000 cases per year. The virus has been imported to the United States on at least one occasion, when a resident of Chicago attended the funerals of relatives in Nigeria who had died of Lassa fever. He became infected there. On return to Chicago he began suffering symptoms of Lassa fever but the local hospitals were unable to diagnose the cause of his disease, being unfamiliar with it. He eventually died of Lassa fever, but fortunately there were no secondary cases.

New World arenaviruses include several South American viruses that are very important disease agents because they cause large outbreaks of hemorrhagic fever with high mortality rates. The names of a number of these viruses and the places where they are found are shown in Fig. 4.25. They include Junín virus (causative agent of Argentine hemorrhagic fever), Machupo virus (Bolivian hemorrhagic fever), Guanarito virus (Venezuelan hemorrhagic fever), Sabiá virus (cause of an unnamed disease in Brazil), and Oliveros virus (cause of an unnamed disease in Argentina). The diseases caused by these viruses are often referred to as emerging diseases because the number of human cases has increased with development and expanding populations. The increasing number of cases results from development of the pampas or other areas for farming, bringing humans in closer association with the rodent reservoirs. Furthermore, the storage of grain near human habitation results in an increase in the local rodent population, and plowing of the fields leads to the production of aerosols which may transmit the disease to man. An attenuated virus

vaccine against Junín virus has been developed and is widely used in populations at risk. The vaccine is effective and has reduced dramatically the number of cases of Argentine hemorrhagic fever. No vaccines are in use for the other viruses, however.

Two arenaviruses have been isolated in the United States. They are Whitewater Arroyo virus, present in the Southwest, and Tamiami virus, present in Florida. Neither of these viruses had been known to cause illness in humans until very recently. In 1999–2000, three Californians died following infection by Whitewater Arroyo virus. The disease these three suffered was ARDS, although two also had hemorrhagic manifestations. Thus, like the hantaviruses, the U.S. arenaviruses may cause isolated cases of serious illness.

Agents Causing Hemorrhagic Fevers in Humans

Many viruses, belonging to several different families, have been described that cause hemorrhagic fever in

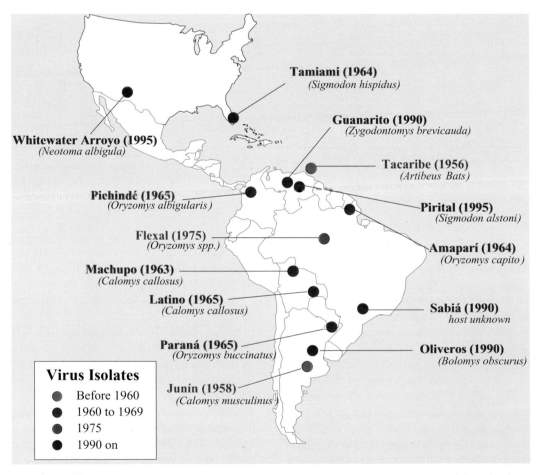

FIGURE 4.25   Arenavirus isolates in the New World. Also shown are the year of first isolation and the rodent host of each virus where known. [Adapted from Peters (1998a, Fig. 1).]

TABLE 4.11 Viruses That Cause Hemorrhagic Fevers in Humans

Virus	Disease[a]	Geographic range	Vector transmission	Case mortality %	Treatment (prevention)
Arenaviridae					
Junin	Argentine HF	Argentine pampas	Infected field rodents, *Calomys musculinus*		Antibody effective, ribavirin probably effective; preventive vaccine exists
Machupo	Bolivian HF	Beni province, Bolivia	Infected field rodents, *Calomys callosus*	15–30	Ribavirin probably effective
Guanarito	Venezuelan HF	Venezuela	Infected field rodents, *Zygodontomys brevicauda*		No data for humans, ribavirin probably effective
Sabiá	HF	Rural areas near Salo, Brazil	Unidentified infected rodents		Intravenous ribavirin effective in one case
Lassa	Lassa fever	West Africa	Infected *Mastomys* rodents	15	Ribavirin effective
Bunyaviridae					
Rift Valley fever	Rift Valley fever	Sub-Saharan Africa	*Aedes* mosquitoes	50	Rapid course; ribavirin or antibody might be effective
Crimean-Congo HF	Crimean-Congo HF	Africa, Middle East, Balkans, Russia, W. China	Tick-borne	15–30	Ribavirin used and probably effective
Hantaan, Seoul, Puumala, and others	HFRS	Worldwide (see Fig. 4.24)	Each virus maintained in a single species of infected rodents	Variable[b]	Ribavirin useful; supportive therapy is mainstay
Sin Nombre and others	HPS, also rare HF	Americas (see Fig. 4.25)	As for viruses causing HFRS	40–50	Rapid course makes specific therapy difficult
Filoviridae					
Marburg, Ebola	Filovirus HF	Africa	Unknown	Marburg: 25 EbolaZ: 30–90	No effective therapy, barrier nursing prevents spread of epidemics
Flaviviridae					
Yellow fever	Yellow fever	Africa, South America	*Aedes* mosquitoes	20	Very effective vaccine
Dengue	DF, DHF, DSS	Tropics and subtropics worldwide	*Aedes* mosquitoes	<1[c]	Supportive therapy useful; vector control
Kyasanur Forest disease	KFD	Mysore State, India	Tick-borne	0.5–9	????
Omsk hemorrhagic fever	OHF	Western Siberia	Poorly understood cycle involves ticks, voles, muskrats??	?	Needs further study

Source: This table includes data from Nathanson *et al.* (1996, Table 32.1, p. 780).

[a] *Abbreviations:* HF, hemorrhagic fever; HFRS, hemorrhagic fever with renal syndrome; HPS, hantavirus pulmonary syndrome; DF, uncomplicated dengue fever; DHF, dengue hemorrhagic fever; DSS, dengue shock syndrome; KFD, Kyasanur Forest disease; OHF, Omsk hemorrhagic fever

[b] Hantaan is 5–15% fatal, whereas Puumala is <1% fatal.

[c] Mortality due to DHF and DSS is 3–12%.

humans. Table 4.11 contains a listing of many of these viruses. These viruses include members of the Arenaviridae, Bunyaviridae, Filoviridae, and Flaviviridae. Many cause severe disease with high mortality, but although the disease is severe, survivors have few sequelae. The dramatic symptom of profuse bleeding has excited the purple prose of many lay authors, best illustrated by recent discussions of Ebola virus, and struck terror in native populations. With the exceptions of yellow fever virus and Junín virus, there are no vaccines, and treatments are primarily supportive, although ribavirin therapy holds some promise for arenavirus disease. Human-to-human transmission is uncommon. Where limited transmission has occurred, it has been by exposure to contaminated blood, or possibly exposure to other bodily fluids, and resulted in limited epidemics for such viruses as Ebola and Machupo virus.

It is said that in some primitive societies, a victim of one of these diseases is placed alone in a hut, with ample food and water left at the doorstep, on the theory that those who will recover will soon be able to feed themselves. If the food is untouched for more than 3 days, the hut is burned. We can hope that in the future more humane, if not more effective, strategies for the control of HF viruses will be found.

EVOLUTION OF MINUS-STRAND RNA VIRUSES

As has been described, all (−)RNA viruses share a number of features. These include virion structure (enveloped viruses with helical nucleocapsids); mechanisms for replicating the genomic RNA (replication within RNP and a requirement for protein synthesis; self-complementarity of the ends of the RNA with its implications for promoter elements involved in replication); mechanisms for synthesis of mRNA (synthesis of leaders or the use of primers for synthesis of mRNA, the presence of intergenic sequences); and the complement of proteins encoded. These similarities make it seem likely that all (−)RNA viruses have diverged from a common ancestor fairly recently on a geological timescale, certainly more recently than the divergence of the extant plus-strand RNA viruses from a common ancestor. The (+)RNA viruses are much more divergent in structure and in the strategies used for replication and expression of the genome, suggesting that they have had a much longer period in which to diverge from one another. Although the complement of proteins encoded is very similar in all (−)RNA viruses, the rate of evolution of RNA viruses is so fast that little sequence identity can be demonstrated between different groups. However, where studies have been performed, evidence for common origin of at least some of these proteins has been shown. As an example, the M proteins of VSV and influenza virus are related and have diverged from a common ancestor. It seems likely that most of the various proteins are related in this way, although it is clear that some viruses have genes that are not represented in all viruses and which presumably arose by recombination events that led to the insertion of new functions, or to deletion events that resulted in a virus with fewer genes, or, probably, to both.

Because the (−)RNA viruses appear to be more recent than the (+)RNA viruses, it is reasonable to postulate that they arose from the (+)RNA viruses. If so, one obvious candidate for the ancestor is a coronavirus. Like the (−)RNA viruses, coronaviruses are enveloped viruses with a helical nucleocapsid that synthesizes RNA leaders and use primers to prime mRNA synthesis, traits in which the coronaviruses differ from other (+)RNA viruses.

If the (−)RNA viruses did arise from the (+)RNA viruses, what traits might account for their success once they arose? One obvious possibility is the ability to synthesize individual mRNAs for each protein needed. This trait also carries with it the necessity to include the RNA synthesis machinery in the virion, but the ability to control the order of synthesis and the translation frequency of the different proteins has obvious advantages for control of the replication cycle. The (−)RNA viruses with segmented genomes also have the ability to undergo ready reassortment, which is clearly advantageous in the Orthomyxoviridae and probably important for all viruses with segmented genomes. (+)RNA viruses that infect animals do not have segmented genomes, except for a few insect viruses with bipartite genomes, for reasons that are not clear. (+)RNA viruses of plants with segmented genomes are common, however. It is perhaps suggestive that the (−)RNA viruses have not been as successful in plants as they have been in animals, and the (−)RNA plant viruses that do exist also replicate in arthropods, which serve as vectors for transmitting the virus to plants.

FURTHER READING

General Reviews

Garoff, H., Hewson, R., and Opstelten, D.-J. (1998). Virus maturation by budding. *Microbiol. Mol. Biol. Rev.* **62**; 1171–1190.
Pringle, C. R., and Easton, A. J. (1997). Monopartite negative strand RNA genomes. *Semin. Virol.* **8**; 49–57.

Rhabdoviridae

Ball, L. A., Pringle, C. R., Flanagan, B., *et al.* (1999). Phenotypic consequences of rearranging the P, M, and G genes of vesicular stomatitis virus. *J. Virol.* **73**; 4705–4712.
Hanlon, C. A., and Rupprecht, C. E. (1998). The reemergence of rabies. *In* "Emerging Infections I" (W. M. Scheld, D. Armstrong, and J. M. Hughes, eds.) pp. 59–80. ASM Press, Washington, DC.
Hooper, P. T., Lunt, R. A., Gould, A. R., *et al.* (1997). A new Lyssavirus—the first endemic rabies-related virus recognized in Australia. *Bull. Inst. Pasteur* (Paris) **95**; 209–218.
Smith, J. S., Orciari, L. A., and Yager, P. A. (1995). Molecular epidemiology of rabies in the United States. *Semin. Virol.* **6**; 387–400.
Winkler, W. G., and Bögel, K. (1992). Control of rabies in wildlife. *Sci. Am.* **266**; 86–92, June.

Filoviridae

Feldmann, H., and Klenk, H.-D. (1996). Marburg and Ebola viruses. *Adv. Virus Res.* **47**; 1–52.
Peters, C. J., and Kahn, A. S. (1999). Filovirus diseases. *Curr. Top. Microbiol. Immunol.* **235**; 85–95.

Bunyaviridae

Peters, C. J. (1998). Hantavirus pulmonary syndrome in the Americas. *In* "Emerging Infections 2" (W. M. Scheld, W. A. Craig, and J. M. Hughes., eds.) pp. 17–64. ASM Press, Washington, DC.

Arenaviridae

Bowen, M. D., Peters, C. J., and Nichol, S. T. (1997). Phylogenetic analysis of the *Arenaviridae*: Patterns of virus evolution and evidence for cospeciation between arenaviruses and their rodent hosts. *Mol. Phylogenet. and Evol.* **8**; 301–316.

Peters, C. J. (1998). Hemorrhagic fevers: how they wax and wane. *In* "Emerging Infections 2" (W. M. Scheld, W. A. Craig, and J. M. Hughes, eds.) pp. 12–25. ASM Press, Washington, DC.

Paramyxoviridae

Black, F. L. (1966). Measles endemicity in insular populations: critical community size and its evolutionary implication. *J. Theor. Biol.* **11**; 207–211.

Chua, K. B., Bellini, W. J., Rota, P. A., *et al.* (2000). Nipah virus: A recently emergent deadly paramyxovirus. *Science* **288**; 1432–1435.

Curran, J., Latorre, P., and Kolakofsky, D. (1998). Translational gymnastics on the Sendai virus P/C mRNA. *Semin. Virol.* **8**; 351–357.

Daszak, P., Cunningham, A. A., and Hyatt, A. D. (2000). Emerging infectious diseases of wildlife-threats to biodiversity and human health. *Science* **287**; 443–449.

Hausmann, S., Garcin, D., Morel, A.-S, *et al.* (1999). Two nucleotides immediately upstream of the essential A_6G_3 slippery sequence modulate the pattern of G insertions during Sendai Virus mRNA editing. *J. Virol.* **73**; 343–351.

Horikami, S. M., and Moyer, S. A. (1995). Structure, transcription, and replication of measles virus. *Semin. Virol.* **8**: 166–175.

Strauss, E. G., and Strauss, J. H. (1991). RNA viruses: Genome structure and evolution. *Curr. Opin. in Genet. Dev.* **1**; 485–493.

Vincent, S., Gerlier, D., and Manié, S. N. (2000). Measles virus assembly within membrane rafts. *J. Virol.* **74**; 9911–9915.

Wang, L.-F., Yu, M., Hansson, E., *et al.* (2000). The exceptionally large genome of Hendra virus: Support for creation of a new genus within the family *Paramyxoviridae*. *J. Virol.* **74**; 9972–9979.

Orthomyxoviridae

Crosby, A. W. (1989). "America's Forgotten Pandemic: The Influenza of 1918." Cambridge University Press, Cambridge, England.

Kolata, G. (1999). "Flu: The Story of the Great Influenza Pandemic of 1918 and the Search for the Virus that Caused It." Farrar, Straus, & Giroux, New York.

Laver, W. G., Bischofberger, N., and Webster, R. G. (1999). Disarming flu viruses. *Sci. Am.* **280**; 78–87 Jan.

Munoz, F. M., Galasso, G. J., Gwaltney, J. M., Jr. *et al.* (2000). Current research on Influenza and other respiratory viruses: 2nd international symposium. *Antiviral Res.* **46**; 91–124.

Neumann, G., Watanabe, T., Ito, H., *et al.* (1999). Generation of influenza A viruses entirely from cloned cDNAs. *Proc. Natl. Acad. Sci. U.S.A.* **96**; 9345–9350.

Ortín, J. (1998). Multiple levels of posttranscriptional regulation of influenza virus gene expression. *Semin. Virol.* **8**; 335–342.

Reid, A. M., Fanning, T. G., Janczewski, T. A., *et al.* (2000). Characterization of the 1918 "Spanish" influenza virus neuraminidase gene. *Proc. Natl. Acad. Sci. U.S.A.* **97**; 6785–6790.

Taubenberger, J. K., Reid, A. H., and Fanning, T. G. (2000). The 1918 influenza virus: A killer comes into view. *Virology* **274**; 214–245.

Bornaviridae

Horikami, S. M., and Moyer, S. A. (1995). Structure, transcription, and replication of measles virus. *Semin.Virol.* **8**; 166–175.

Iwata, Y., Takahashi, K., Peng, X. *et al.* (1998). Detection and sequence analysis of borna disease virus p24 RNA from peripheral blood mononuclear cells of patients with mood disorders or schizophrenia and of blood donors. *J. Virol.* **72**; 10044–10049.

Kohno, T., Goto, T., Takasaki, T. *et al.* (1999). Fine structure and morphogenesis of borna disease virus. *J. Virol.* **73**; 760–766.

Schneemann, A., Schneider, P. A., Lamb, R. A. *et al.* (1995). The remarkable coding strategy of borna disease virus: A new member of the nonsegmented negative strand RNA viruses. *Virology* **210**; 1–8.

5

Viruses Whose Life Cycle Uses
Reverse Transcriptase

INTRODUCTION

Two families of animal viruses utilize reverse transcriptase (RT) in the replication of their genome, the Retroviridae and the Hepadnaviridae. Two floating genera of plant viruses also use RT; their life cycles are more similar to the hepadnaviruses than to the retroviruses. The hepadnaviruses and the plant viruses are sometimes called pararetroviruses, because their life cycle resembles that of the retroviruses. For the viruses that use RT, the genetic information in the genome alternates between being present in RNA and present in DNA. RT, which is encoded in the viral genome, converts the RNA genome of retroviruses, or an RNA copy of the DNA genome of hepadnaviruses, into double-stranded (ds) DNA. In the nucleus of the infected cell, cellular RNA polymerase transcribes the DNA genome of hepadnaviruses, or the DNA copy of the retrovirus genome, to produce the RNA to be reverse transcribed. The retroviruses package this RNA in the virion and are allied to retrotransposons that form a prominent feature of eukaryotic genomes. The hepadnaviruses and the plant viruses reverse transcribe the RNA into DNA during packaging, so that the virion contains DNA. Thus, the replication of the genome of retroviruses can be described as RNA→DNA→RNA, whereas the replication of the genome of hepadnaviruses can be described as DNA→ RNA→DNA. Although the two families differ in the timing of when reverse transcription takes place in their life cycles, this difference may not represent a fundamental distinction between them. Recent studies have indicated that one genus of retroviruses, the spumaviruses, may package DNA in the virion.

It is an interesting feature of reverse transcription that the RNA template is destroyed in the process of conversion to DNA. RT has associated with it an RNase H activity, which specifically degrades the RNA strand of a DNA–RNA hybrid. This activity is essential for the production of a dsDNA copy of the viral RNA by RT. The destruction of the RNA template makes the process of reverse transcription fundamentally different from other mechanisms used for transcription or copying of nucleic acids, in which the template remains intact.

An essential feature of the infection of cells by retroviruses is that the dsDNA copy of the genome is integrated into the host chromosome, where it is called a provirus. Only integrated DNA is stably and efficiently transcribed by the host machinery. Thus, integration is required for productive infection. During infection by hepadnaviruses, however, the viral DNA does not integrate. Instead, it is maintained in the nucleus as a nonreplicating episome. In contrast to retroviruses, hepadnaviral episomal DNA is stably and efficiently transcribed by the host machinery.

The retroviruses have been intensively studied for years because researchers discovered early that avian retroviruses have the ability to induce leukemias and sarcomas in chickens. The study of these viruses led to the discovery of cellular oncogenes, of RT, and of mechanisms that regulate cycling of the animal cell, and several Nobel prizes have been awarded for work with the avian retroviruses (Chapter 1). Although clearly important for our understanding of biology, for many years after their discovery retroviruses were in some ways biological curiosities because no human disease was known to be associated with retroviral infection. This changed with the discovery of human T-cell leukemia viruses, now known as primate T-lymphotropic viruses (PTLV), which cause leukemia in man. More recently, the appearance of human immunodeficiency virus (HIV) and of acquired immunodeficiency syndrome (AIDS) in the human population has dramatically altered our understanding of the disease-causing potential of retroviruses.

The most important hepadnavirus is hepatitis B virus, which is a major cause of hepatitis in man. Like hepatitis C virus, it often establishes a chronic infection that can result in cirrhosis or hepatocellular carcinoma.

FAMILY RETROVIRIDAE

The retroviruses are a very large group of viruses that infect invertebrates as well as vertebrates. Most of what we know about this group of viruses comes from studies of viruses that infect higher vertebrates. Hundreds have been studied and, although considerable divergences exist, they form a well-defined taxon. All are sufficiently similar to be classified as belonging to a single family, the Retroviridae. The family gets it name from the concept that these viruses use retrograde flow of information, from RNA to DNA, whereas the conventional flow of information in living organisms is from DNA to RNA.

The RTs of retroviruses are the most highly conserved elements of these viruses and have been used to study the relationships among them. Figure 5.1 illustrates the rela-

tionships among the retroviruses of higher vertebrates based on the sequences of their RTs. Included in the tree is a lineage of fish viruses, now classified as members of the genus Epsilonretrovirus. The tree is annotated to show where various new genes entered different virus lineages via recombination with the host or with other viruses. Based on these sequence relationships, the retroviruses that infect birds and mammals are classified into six genera, as illustrated in Fig. 5.1 and as listed in Table 5.1. Of these, members of three genera are characterized as simple retroviruses, which encode only the genes *gag*, *pro*, *pol*, and *env* (and sometimes *dut*). The other three genera of retroviruses of higher vertebrates, as well as the fish viruses, encode, in addition, regulatory genes that control their life cycle, and they are called complex retroviruses. Notice that these regulatory genes independently entered the four different lineages of complex retroviruses represented by the four different genera (Fig. 5.1). Thus, recombination to acquire new functions has been an ongoing process in the retroviruses. Also notice that the complex retroviruses do not group together. The epsilonretroviruses are more closely related to the gammaretroviruses, which are simple viruses,

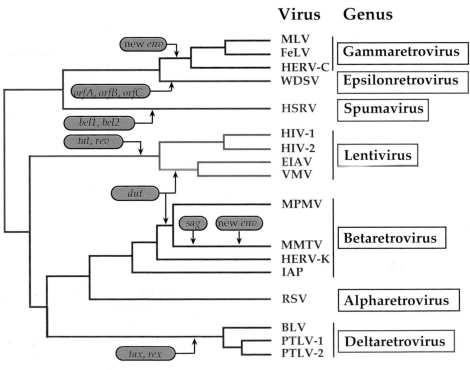

FIGURE 5.1 Phylogenetic tree of the Retroviridae drawn from the amino acid sequences of the reverse transcriptases. The lengths of the branches are proportional to the degree of divergence; the names of the "simple" retrovirus genera are boxed in black, the "complex" genera are boxed in red. The green ovals indicate the acquisition of new genes during the evolution of current extant viruses. Most of the virus name abbreviations are found in Table 5.1; HERV-C and HERV-K are defective retroviruses in the human genome and IAP is a virus-like element in rodent genomes. [Adapted from Coffin *et al.* (1997, Fig. 6, p. 43) and Fields *et al.* (1996, p. 1769).]

TABLE 5.1   Retroviridae

Genus/members[a]	Virus name abbreviation	Host(s)	Related defective viruses	Transmission	Disease	World distribution
Alpharetrovirus (simple)						
Avian leukosis	ALV	Birds	Avian myeloblastosis (AMV) Avian myelocytomatosis (AMCV-29)			Worldwide
Rous sarcoma	RSV	Birds				
Gammaretrovirus (simple)						
Moloney murine leukemia	MLV	Mice	Moloney sarcoma (MoMSV) Harvey murine sarcoma (HaMSV)	T-cell lymphoma		
Feline leukemia	FeLV	Cats, humans	Gardner-Arnstein feline sarcoma (GAFeSV)	T-cell lymphoma, immunodeficiency		
Betaretrovirus (simple)						
Mouse mammary tumor	MMTV	Mice		Vertical, including mothers' milk	Mammary carcinoma, T-cell lymphoma	Worldwide
Mason-Pfizer monkey	MPMV	Monkeys			Unknown	
Deltaretrovirus (complex)						
Bovine leukemia	BLV	Cows			B-cell lymphoma	
Primate T-lymphotrophic[b]	PTLV-1	Humans		Vertical including mothers' milk sexual transmission,	T-cell lymphoma, neurological disorders	
	PTLV-2	Humans		blood	TSP, HAM[c]	
Epsilonretroviruses						
Walleye dermal sarcoma	WDSV	Fish			Benign sarcomas	North America
Lentivirus (complex)						
Human immunodeficiency	HIV-1	Humans		Neonatal infection, sexual transmission,	AIDS	Worldwide
	HIV-2	Humans		blood		
Simian immunodeficiency	SIV	Monkeys			Simian AIDS	Africa
Visna-maedi	VISNA	Sheep			Neurological disease	Northern Europe
Equine infectious anemia	EIAV	Horses			Anemia	Current epidemic in Utah
Spumavirus (complex)						
Chimpanzee foamy	CFV	Monkeys		??	None	
Human spumaretrovirus	HSRV	Humans				

[a]Representative replication competent members are shown; the first listed is the type species of the genus. Alpharetroviruses were formerly "avian type C," betaretroviruses include the "mammalian type B" and the "type D" retroviruses; gammaretroviruses were "mammalian type C," deltaretroviruses were "BLV/HTLV," and epsilonretrovirus is a new genus of piscine viruses.

[b]Primate T-lymphotropic virus 1 (PTLV-1) was formerly known as human T-cell leukemia virus and referred to as HTLV.

[c]TSP, tropical spastic pareparesis; HAM, HTLV-associated myelopathy.

than they are to other complex retroviruses, and the deltaretroviruses, lentiviruses, and spumaviruses are not particularly closely related.

Members of the different genera differ in their structure as visualized in the electron microscope. The simple viruses were formerly classified on the basis of morphology into groups A, B, C, and D. The nucleocapsids of C-type viruses, now classified as alpharetroviruses and gammaretroviruses, assemble during budding, and the nucleocapsid is centrally located in the mature virion. The nucleocapsids of B-type and D-type viruses, now classified as betaretroviruses, assemble before budding and the nucleocapsid is eccentrically located (type B) or bar shaped (type D) in the mature virion (see Figs. 2.1 and 2.17).

Eukaryotic genomes contain a very large number of genetic elements that are related to retroviral genomes. Retrotransposons encode RT and can move around within the genome by a process that uses reverse transcription and insertion, similar to what happens with retroviruses. They are related to retroviruses but have no independent lives as viruses. Other elements in the eukaryotic genome contain additional retrovirus-like genes and appear to have arisen by insertion of retroviral genomes into the germ line at some time in the past. Some of these are still active, capable of giving rise to infectious retroviruses, whereas others are defective. It is clear that this class of elements has been coevolving with the eukaryotes for a long period of time. The integrated copies of retroviruses in the germline constitute a form of fossil record that allows us to trace the lineage of at least some retroviruses for 100 million years or longer. For other viruses, whether RNA or DNA, we can only trace ancestry for much shorter periods of time.

Retroviral Genome

The RNA genome in the retrovirion is diploid, consisting of two copies of a 7- to 10-kb single-stranded (ss) RNA that is capped and polyadenylated. The two copies of the genome are normally identical, but during mixed infection hybrid genomes can result. They are joined near their 5′ ends, and perhaps in other regions as well, by hydrogen bonds.

All retroviruses encode the four genes called *gag*, *pro*, *pol*, and *env*, which are always found in this order in the genome. A simplified illustration of a retroviral genome, its relation to the provirus, and the location of the different gene products in the virion are shown in Fig. 5.2. The name *gag* comes from group-specific *a*ntigen, because the proteins encoded in this gene are more highly conserved, and therefore more widely cross reactive immunologically, than are the envelope proteins of the virion. The peptides derived from the Gag polyprotein, the precursor polyprotein encoded in the *gag* gene, form the capsid of the retrovirion. The *pro* gene encodes a protease (PR) that is required for the processing of Gag, and *pol* encodes three activities, RT, RNase H, and integrase (IN). RNase H forms a separate domain in the RT-RNase H protein, but functions as an integral component of RT. IN is required for the integration of the dsDNA copy of the retroviral genome into the host chromosome. The fourth gene, *env*, encodes the envelope glycoproteins present at the surface of the enveloped retrovirion. The primary product is Env, which is processed by cleavage to form an N-terminal external protein called SU (for surface) and a C-terminal protein that spans the membrane called TM (for transmembrane). The order of genes in the provirus is the same as in the viral genome.

As described above, the genomes of complex retroviruses contain a number of regulatory genes in addition to

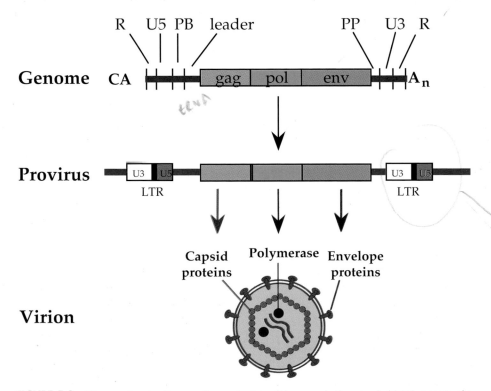

FIGURE 5.2 Diagram showing the overall organization of the genes in the retroviral RNA genome, the comparable organization of the DNA provirus, and the location of the various virus-encoded proteins in the mature virion. The RNA strand is shown in green; the DNA provirus is shown in blue. The open reading frames (ORFs) in the genome are color coded to match their products in the virion. The nontranslated regions of the RNA genome and the long terminal repeats (LTRs) are described in detail in the text. The *pro* gene, located between *gag* and *pol*, is not shown here. [Adapted Goff (1997, Fig. 3.5, p. 143).]

these four basic genes present in all retroviruses. These genes will be described later.

Reverse Transcription of Viral RNA

Most retroviruses penetrate into a cell by fusion with the plasma membrane, but some use the endosomal pathway. Once inside the cell, reverse transcription takes place in a subviral particle to produce a full-length, linear dsDNA copy of the RNA. The composition of this subviral particle is not well defined. It certainly contains RT and its associated activities as well as components derived from Gag, but which Gag components are present and whether cellular proteins form part of this particle are not known.

Reverse transcription begins in the cytoplasm. In most viruses, the full-length dsDNA is produced in the cytoplasm, but in some the finishing touches occur after transfer to the nucleus. After transfer to the nucleus, the viral DNA is integrated into the host chromosome, essentially at random, in a process that requires the activity of IN. In the simple viruses, transfer to the nucleus occurs during cell division, when the nuclear envelope is disassembled. At least some of the complex retroviruses, however, such as HIV, encode proteins that allow the DNA-containing complex to traverse the nuclear membrane. Thus, the simple retroviruses can only productively infect dividing cells, whereas HIV can infect nondividing cells.

Synthesis of First DNA Strand

The process of reverse transcription of the viral genome is illustrated in Fig. 5.3. At the ends of the viral RNA are domains that are essential for production of the dsDNA copy of the RNA genome. These are illustrated schematically in Figs. 5.2 and 5.3, and the sizes of these elements are shown in Table 5.2 for a number of different retroviruses. A direct repeat element called R, 15–230 nt in length depending on the virus, is present at the two ends of the RNA genome. At the 5′ end of the genome, a unique sequence element called U5 (70–220 nt long) is present immediately downstream of R, and at the 3′ end of the genome a unique element called U3 (230–1200 nt) is present immediately upstream of R. On completion of DNA synthesis, these elements give rise to direct repeats present at the ends of the viral DNA, called the long terminal repeats or LTRs, which have the sequence U3–R–U5.

Immediately 3′ of U5 is a primer binding site (PBS), where 18 nucleotides are exactly complementary to the 3′ end of a specific cellular tRNA. Each genomic RNA molecule has bound to it at PBS one molecule of the appropriate tRNA, which is used as a primer for DNA synthesis. The tRNA used depends on the virus, and several different tRNAs are known to be used by different viruses (Table 5.2).

Reverse transcription begins by extending the primer tRNA through U5 and R, and stops when the end of the RNA genome is reached (Fig. 5.3, steps 1 and 2). The cDNA product, called the first strong stop DNA, is then transferred, while still attached to the primer, from the 5′ end of the RNA to the 3′ end ("first jump," Fig. 5.3, steps 3 and 4). It is unknown whether transfer is usually to the 3′ end of the same molecule, to the 3′ end of the second copy of the genome, or randomly to either copy of the RNA in the virus. This transfer uses the repeat element R at the 5′ and 3′ ends—the DNA copy detaches from R at the 5′ end and anneals to R at the 3′ end, so that only one copy of R is present in the DNA transcript. RNase H is presumably important for this. During reverse transcription, the RNA strand is destroyed by RNase H about 18 nt behind the transcription point. The degradation of the RNA strand during transcription of the DNA copy may encourage the jump to the other end. Other components of the particle may also be involved in the jump.

After the jump, reverse transcription resumes until the 5′ end of the RNA template is reached. Note that the RNA strand now ends at PBS because RNase H has degraded the RNA strand of the DNA/RNA hybrid (but not the RNA in PBS because this is an RNA–RNA duplex). This process results in the formation of what is called first- strand DNA or minus-strand DNA, because it is the antimessage sense (step 5).

Synthesis of the Second DNA Strand

The first-strand DNA produced in this way is the template for second-strand (plus-strand) DNA synthesis. The primer for second-strand synthesis is an RNA oligonucleotide, called the polypurine tract or PPT, positioned immediately 5′ of U3. PPT survives RNase H degradation and its 3′ terminus is exact; that is, the cleavages to produce it are precise (Fig. 5.3, step 6). In addition to this precise primer for plus-strand synthesis, which defines the boundary of U3 and thus of the LTR, additional priming sites are used in some retroviruses, such as HIV.

The PPT plus-strand primer is extended through U3, R, U5, and into the region of the primer tRNA, which is still attached to the first-strand DNA. The 18 nucleotides of the tRNA primer that are complementary to the PBS are copied, but further copying is thought to be blocked by a modified nucleotide in the tRNA that cannot be copied. The tRNA primer is then removed by RNase H (step 7) and a second jump occurs. In the second jump, the nascent second-strand DNA is transferred to the other end of the template, using the PBS sequence, which is now present at the 3′ ends of both strands (second jump, step 8). Synthesis of both strands of DNA resumes and it becomes full length and double stranded. The resulting dsDNA is cleaned up,

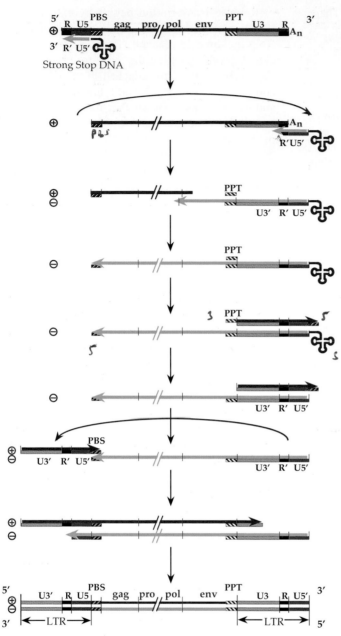

1) Primer t-RNA anneals to PBS sequence in genome RNA

2) tRNA is extended to form DNA copy of the 5′ end of the genomic RNA

3) RNase removes hybridized RNA (R and U5)

4) First Jump. DNA hybridizes to remaining RNA R sequence at 3′ end

5) DNA minus strand extended and completed; most RNA removed.

6) Plus strand DNA primes at poly-purine track (PPT) downstream of *env* gene. 5′ end of plus strand DNA is synthesized

7) RNase H degrades tRNA and PPT.

8) **Second Jump** Plus strand DNA binds to the primer binding sequence (PBS) near the 3′ end of minus strand DNA.

9) Both strands extended and completed to give double stranded DNA with duplicated LTRs in the same orientation at both ends.

FIGURE 5.3 Mechanism of retroviral DNA synthesis (reverse transcription). Green lines are RNA, light blue lines are minus-strand DNA, and the dark blue line in the last three steps is plus-strand DNA. Features within the LTRs (U3, R, U5), as well as PBS and PPT, are indicated by colored bars beneath the lines designating the nucleic acids, to clarify what is present at each step. [Adapted from Fields *et al.* (1996, p. 1792), Goff (1997, Fig. 3.6, p. 145) and Coffin *et al.* (1997, Fig. 2, p. 123).]

probably by cellular enzymes. In the full-length, linear ds copy of the genome, the LTR sequence U3–R–U5 has been formed at both ends of the DNA genome (Fig. 5.3, step 9).

It is of interest that RT, like DNA polymerases, requires a primer to synthesize DNA, but unlike most DNA polymerases it can copy either DNA or RNA, given the appropriate primers. Thus, the enzyme differs from the RNA replicases of RNA viruses, which use other mechanisms for the initiation of RNA replication.

Why Is the Genome Diploid?

Why the retroviral genome is diploid is not clear. In other instances where reverse transcription occurs, such as

TABLE 5.2   Terminal Regions of Retrovirus Genomes

Genus	Prototype virus	Approximate sizes in bases of terminal elements			Primer tRNA used
		U3	R	U5	
Alpharetrovirus	RSV, ALV	230[a]	20	80	Trp
Betaretrovirus	MMTV	1200[b]	15	120	Lys-3
Gammaretrovirus	MLV	450	70	80	Pro/Gln
Deltaretrovirus	HTLV-1	350	230	220	Pro
Epsilonretrovirus	WDSV	440	80	70	His
Lentivirus	HIV-1	450	100	80	Lys-1,2,3
Spumavirus	HRSV	910	190	160	Lys-1,2

Source: Adapted from Coffin *et al.* (1997, Table 2, p. 38).
[a]Includes *v-src* gene.
[b]U3 contains *sag* gene.

in the hepadnaviruses and the retrotransposons, the RNA to be copied is not diploid. Furthermore, *in vitro* studies have shown that retroviral RT can use a single copy of the RNA genome to produce a full-length dsDNA. Thus, two copies of the genome are not essential. However, it is possible that during infection the process is more efficient if the RT can go back and forth between the two copies, and this resulted in selection for a diploid genome in retroviruses. The process of reverse transcription to produce the viral dsDNA is complex with the multiple jumps required, and the diploid genome could be organized in such a way as to make these jumps more efficient. Other possible advantages of a diploid genome that could have resulted in selective pressure for diploidy are the possibility of overcoming at least some damage in the RNA by switching templates, and the fact that switching templates results in recombination. Recombination does occur frequently in retroviruses, and it is clear from many studies of virus evolution that recombination is important in their evolution.

Integration

After the appearance of the full-length dsDNA genome in the nucleus, the viral integrase catalyzes its insertion into a host cell chromosome by means of a single recombinational event. This process is illustrated in Fig. 5.4. Insertion is essentially random within the host genome. The first event in integration is the removal of two nucleotides from the 3′ end of both strands of the viral DNA. The next two nucleotides are always AC, from 3′ to 5′. The 3′ OH of the now 3′-terminal A residue is used to attack an internucleotide phosphate in the host DNA. Attack is coordinated so that both ends of the viral DNA are inserted at once. The spacing between the two insertion points depends on the viral integrase and is 4, 5, or 6 nucleotides in different viruses. The ends of the inserted structure are then cleaned up, probably by host

enzymes. Insertion results in the loss of the two terminal nucleotides of the viral DNA. In addition, a duplication of the 4, 5, or 6 nucleotides of the host that lie between the two insertion points is produced, and these duplicated nucleotides flank the two ends of the viral genome.

The integrated form of the virus, the provirus, is stable. No mechanism for precise excision of the provirus is known, and integration is essentially irreversible. Most retroviruses do not kill the host cell, and the provirus behaves as a simple Mendelian gene that is transmitted to all daughter cells. It is obvious that insertion of such a provirus into the germ line would lead to its transfer to progeny organisms, which appears to have occurred many times in the past.

Transcription of RNA in Simple Retroviruses

The LTR of the integrated provirus has all of the signals required for transcription of the provirus by cellular RNA polymerase II. The transcription signals, almost all found in U3, include TATA boxes as well as an array of binding sites for cellular transcription factors that are optimal for the cell in which the particular retrovirus primarily replicates. Figure 5.5 gives examples of the binding sites for transcription factors in several different retroviruses. This figure illustrates that the constellation of binding sites is complex and is different in different viruses that replicate in different tissues. Only the upstream (5′) LTR is used to initiate transcription—or at least efficient transcription.

Transcription initiates precisely at the 5′ end of R, as indicated by the arrow in Fig. 5.5, and proceeds through the proviral genome and into the 3′ flanking sequences. The transcript is capped by host cell capping enzymes. There is a poly(A) addition signal (AAUAAA) in the RNA, which is usually present about 20–30 nucleotides upstream of the end

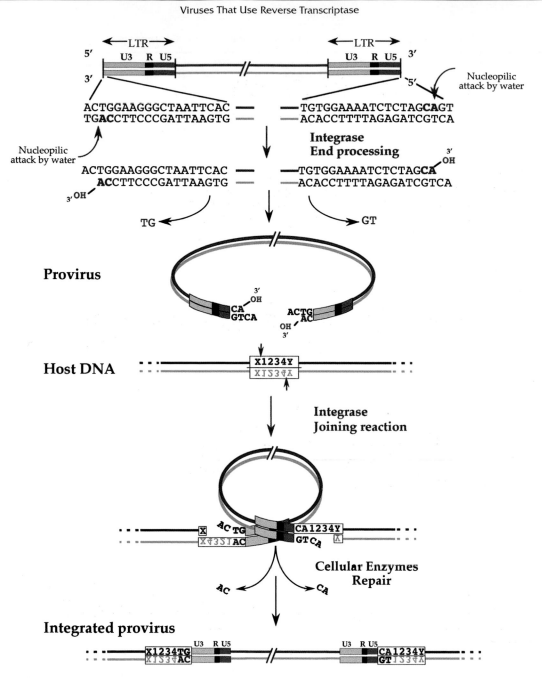

FIGURE 5.4 Steps in the integration of retroviral DNA into the host genome. First, the termini of the blunt-ended viral DNA are attacked by the integrase, and two bases (blue) adjacent to a highly conserved CA dinucleotide (red) are lost by nucleophilic attack by a water molecule, leaving recessed 3′ OH ends. Next, the host DNA is cleaved at the target sequence and the 3′ OH ends of the viral DNA are joined by the integrase to the 5′ phosphates on the host DNA. Overhang removal and gap and nick repair by cellular enzymes complete the integration reaction. [From Hindmarsh and Leis (1999), Goff (1997, Fig. 3.10, p. 145), and Collins *et al.* (1997, Fig. 8, p. 185).]

of the viral genomic RNA sequence (i.e., upstream of the 3′ end of R). Host cell polyadenylation machinery recognizes the AAUAAA signal, cleaves the RNA transcript precisely at the R-U5 boundary, and polyadenylates the RNA. Thus the processed RNA transcript is identical to the genomic RNA and a round-trip from RNA to DNA to RNA has been made.

In many viruses, the polyadenylation signal is present in U3. Thus, there is only one copy of this signal in the RNA transcript, near the 3′ end. However, in some viruses the signal is present in R and is present in both the 5′ and 3′ regions of the transcript. Interestingly, only the 3′ copy of the signal is active for cleavage and polyadenylation of the transcript.

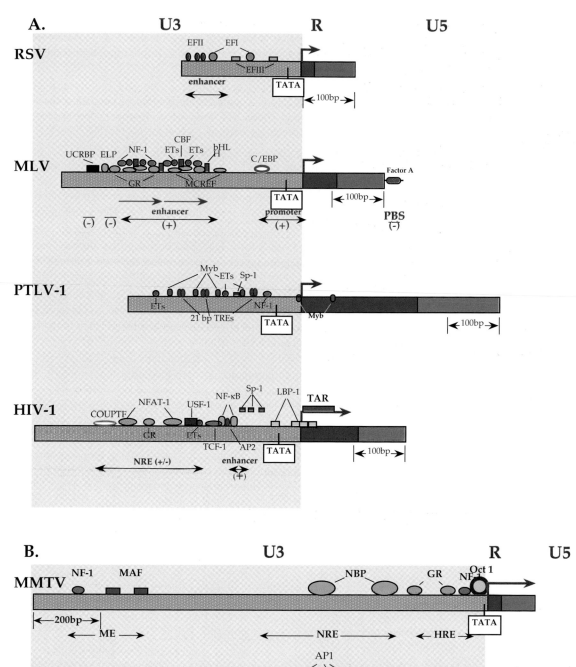

FIGURE 5.5 Transcription signals in retroviral LTRs. Where the same factor binds to a number of LTRs, the same symbol has been used. The TRE elements in PTLV-1 are the *Tax*-responsive *e*lements. The blue arrow marks the beginning of transcription. Note that the scale of B is twofold different from that in A. [This figure is a composite of Figs. 4, 5, 7, 8, 9, and 11 in Chapter 6 of Collins *et al.* (1997).]

Some fraction of the genomic RNA is exported to the cytoplasm without further processing. There it serves as mRNA for the synthesis of Gag, Gag–Pro, and/or Gag–Pro–Pol, as described below. Alternatively, it can be packaged into progeny viruses. Genomic RNA that serves as mRNA and genomic RNA that serves as the source of RNA for packaging are maintained in separate pools.

Some fraction of the genomic RNA is spliced before export to the cytoplasm. Only one spliced RNA is made in the simple retroviruses, which serves as mRNA for Env. In most retroviruses, the entire Gag–Pro–Pol region is spliced out and the initiation codon for Env is encoded in *env*. In the avian retroviruses, however, the upstream splice site is located within the Gag coding sequence so that Env begins with the first six codons of Gag.

The need for both spliced and unspliced versions of the viral RNA means that mechanisms must exist to ensure that both are produced and that the ratio of spliced to unspliced RNA is optimal for virus replication. In the simple retroviruses, the splice sites are suboptimal, so that not all RNA is spliced. Experiments have shown that the result is an optimal ratio of spliced to unspliced RNA. Mutations that make splicing more efficient are deleterious for virus growth and revertants quickly arise that restore the proper ratio. The second problem faced by these viruses is the need to export unspliced RNA to the cytoplasm. Eukaryotic cells have control mechanisms to ensure that RNA containing splice sites is not exported from the nucleus. It has been found that sequence elements in the unspliced RNA are required for its export, and it is assumed that these elements interact with cellular proteins that promote export. In the simian retrovirus Mason–Pfizer monkey virus, this element is called the constitutive transport element. It is 154 nucleotides long and is located in the 3′ nontranslated region of the RNA. In the avian retroviruses, an apparently unrelated element of about the same size, also present in the 3′ nontranslated region, provides the same function for unspliced avian retroviral RNA. Interestingly, the avian sequence does not work in mammalian cells. The monkey virus sequence works in both mammalian and avian cells, but works better in mammalian cells. These findings are consistent with the hypothesis that these transport elements interact with cellular proteins. The inability of unspliced RNA to be exported from the nucleus is one of the reasons that the avian retroviruses will not replicate in mammalian cells.

Translation of Viral Genomic RNA

Retroviral genomic RNA is translated into two or three polyproteins that are eventually processed by the viral PR. The order of genes along the genomic RNA is *gag–pro–pol*, encoding the proteins Gag, Pro, and Pol. Stop codons are present between Gag and Pro, or between Pro and Pol, or in both places, as illustrated in Fig. 5.6. Termination of the polypeptide chain occurs at these stop codons most of the time during translation. These stop codons are suppressed some of the time, however, either by read-through or by frameshifing (Chapter 1), so that the amount of Pol produced is usually about 5% that of Gag. In viruses with one stop codon, the frequency of suppression is about 5%, but in viruses with two stop codons, the frequency of suppression of each stop codon is higher so that significant amounts of Pol are produced even though suppression of two stop codons is required. This means that the frequency of suppression is variable and can be controlled by changes in the sequence of the viral RNA. Because reinitiation does not occur once the chain is terminated, the polyproteins produced are Gag and/or Gag–Pro and/or Gag–Pro–Pol, depending on the positions of the stop codons (Fig. 5.6).

Pro is produced in three different ways in different retroviruses, as illustrated in Fig. 5.6. In the avian viruses, such as ALV, there is no stop codon between Gag and Pro so that a Gag–Pro polyprotein is produced. Gag and Pro are thus produced in equal amounts. Frameshifting results in the production of a longer polyprotein, Gag–Pro–Pol. Most of the mammalian viruses also only have a single stop codon in the ORF, but it is positioned between Gag and Pro. Read-through of a UAG stop codon (murine leukemia viruses) or frameshifting (other mammalian viruses with a single stop codon) results in the longer polyprotein. The two polyproteins produced are Gag and Gag–Pro–Pol, and Pro and Pol are produced in the same low amounts. Finally, several mammalian retroviruses, for example, MMTV and PTLV-1, have two stop codons in the ORF, both of which can be suppressed by frameshifting. Thus, three polyproteins are produced, Gag, Gag–Pro, and Gag–Pro–Pol. In this case Pro is produced at intermediate levels.

Processing of these various polyproteins occurs during assembly of progeny virions. The viral Pro is an aspartate protease whose active site contains two aspartic acid residues (Chapter 1). The protease domain is functional in polypeptides containing the Pro sequence as well as after its release by proteolysis as a small protein of about 100 residues. The enzyme is active only as a homodimer, with each chain in the dimer supplying one of the aspartic acids in the active site. Because the monomer is not active, there is a delay in processing. The high concentration of viral polyproteins that occurs in viral particles or in previral particles is required to achieve efficient dimerization of the protease and its activation. Experiments have shown that premature activation of the protease, which can be achieved by using genetic tricks, is deleterious for virus assembly. Thus, it is important that processing be delayed until assembly begins or is completed.

During processing of Gag, several different proteins are produced, some of which are quite small, whereas others are larger. The proteins produced from Gag are illustrated schematically in Fig. 5.7 for a number of retroviruses. Gag

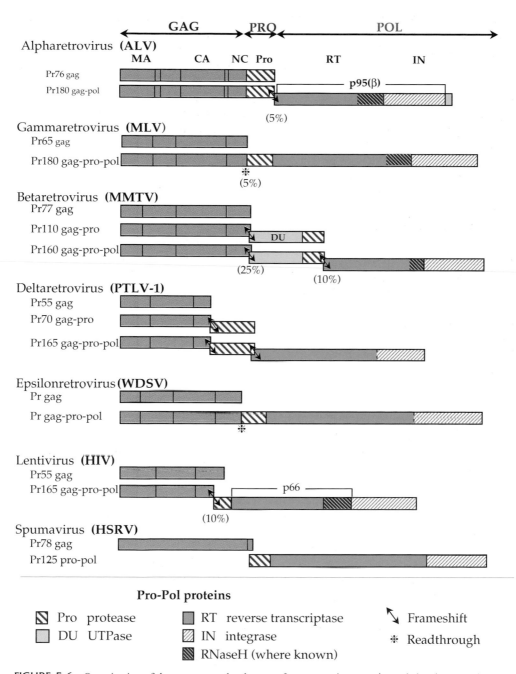

FIGURE 5.6 Organization of the *gag, pro,* and *pol* genes of representative retroviruses belonging to each genus. In some cases two frameshifts are required to generate a complete Gag–Pro–Pol precursor. The *gag* proteins are illustrated in more detail in Fig. 5.7. [This figure is a composite of Goff (1997, Fig. 3.16, p. 157) and Coffin *et al.* (1997, pp. 45, 269, 795, 799).]

is cleaved to produce at least three protein products in all retroviruses except the spumaviruses, called MA (membrane-associated or matrix protein), CA (capsid protein), and NC (nucleocapsid protein). These three peptides are always present in that order from the N terminus to the C terminus in the Gag polyprotein. Gag (and thus MA) is myristylated in most retroviruses and associates with membranes. Myristylation may serve to recruit Gag to membranes for assembly or budding. NC is a small basic protein that binds RNA and Zn^{2+}. It has a number of functions, including binding to the packaging signals in the viral RNA that lead to its incorporation into virions, the facilitation of

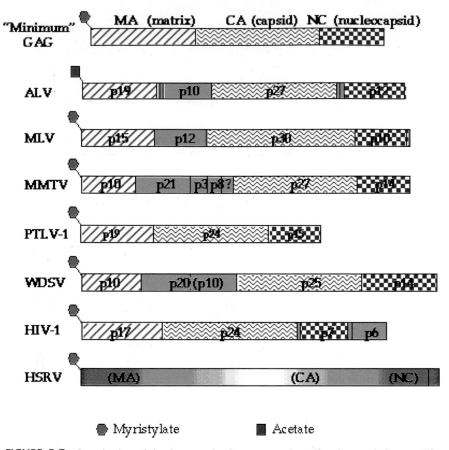

FIGURE 5.7 Organization of the Gag proteins in representatives of each retroviral genus. The viruses are shown in the same order as those in Fig. 5.6. Vertical solid lines mark sites of cleavage by the viral protease. Sequences representing the mature matrix, capsid, and nucleocapsid proteins are indicated with different shadings, and the approximate molecular weights of the processed proteins are shown. Note that the Gag polyprotein of HSRV is not processed. [From Coffin *et al.* (1997, pp. 44, 798).]

binding of the tRNA primer to the genomic RNA, the formation of the genomic RNA dimer, and strand transfer during reverse transcription. CA is a larger protein that is believed to form a shell around the viral RNA and its associated internal proteins (Fig. 2.17). In addition to these three proteins, in most retroviruses other proteins, whose functions are not well understood, are also produced from Gag (Fig. 5.7).

Processing of Pol varies in different viruses. Three patterns of cleavage can be distinguished. In some viruses, RT, consisting of the polymerase domain and the RNase H domain, is released from the upstream Pro and the downstream IN. The active reverse transcriptase is a monomer or a homodimer of RT. In other viruses, cleavage between RT and IN is incomplete, and the active enzyme is a heterodimer of

RT and RT-IN. A third pattern occurs in HIV, in which partial cleavage occurs between the polymerase domain and the RNase H domain. In this case, the products of Pol are a truncated RT lacking the RNase H domain (called p51), the full-length RT (called p66), and IN. The active reverse transcriptase is a heterodimer between p51 and p66.

Production and Processing of Env

Translation of the mRNA for Env produces the envelope glycoprotein precursor. Processing of Env and the location of key features are illustrated in Fig. 5.8. The precursor has an N-terminal signal sequence, which leads to its insertion into the endoplasmic reticulum, a membrane anchor sequence near the C terminus, and a C-terminal cytoplas-

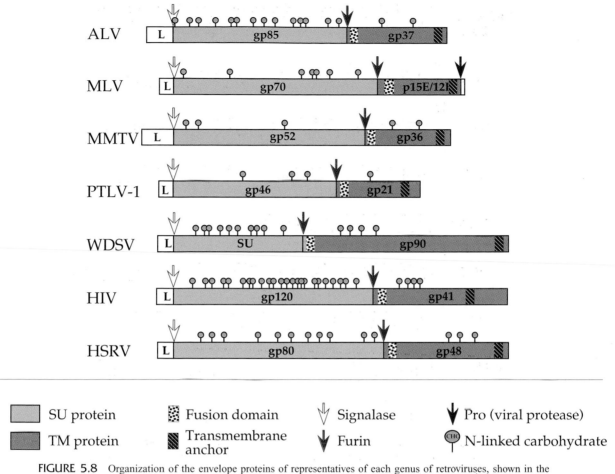

	SU protein		Fusion domain		Signalase		Pro (viral protease)
	TM protein		Transmembrane anchor		Furin		N-linked carbohydrate

FIGURE 5.8 Organization of the envelope proteins of representatives of each genus of retroviruses, shown in the same order as in Fig. 5.6 and 5.7. The domains corresponding to the mature SU and TM proteins are shown in different shades of color. Cleavages by signalase to remove the leader peptdie (L) are indicated with white arrows, and those due to furin by yellow arrows. The red arrow marks the site of cleavage in MLV by the viral protease. The fusion domains, the transmembrane domains, and the sites of predicted N-linked carbohydrate addition are shown. [Adapted from Coffin *et al.* (1997, Fig. 10, p. 56).]

mic domain (i.e., the protein is a type I integral membrane protein). The signal sequence is removed during translocation into the endoplasmic reticulum, and the protein is glycosylated and transported to the plasma membrane by conventional cellular pathways. As with many other viral glycoproteins, the precursor is cleaved by furin during transport to produce an N-terminal extracellular component called SU (for surface) and a C-terminal membrane-spanning component TM (for transmembrane). SU and TM remain associated. In some cases, the association is stabilized by disulfide bonds, but in other cases there is no covalent linkage. SU is always glycosylated, whereas TM may or may not be glycosylated. Cleavage to produce SU and TM is required for the activation of the fusion activity, which is located near the N terminus of TM. Thus, the production of the envelope glycoproteins of the retroviruses parallels that of the envelope glycoproteins of many enveloped RNA viruses.

Accessory Genes of Complex Retroviruses

In addition to the core genes *gag*, *pro*, *pol*, and *env* that are present in all retroviruses, members of the four genera of complex retroviruses possess additional genes that allow them to regulate the development of the infection cycle. Accessory genes are also present in the betaretroviruses, but these are not regulatory in nature. A listing of the accessory genes in different retroviruses is given in Table 5.3, and a complete listing of all of the proteins of HIV is shown in Table 5.4

The accessory genes of the complex retroviruses are located upstream or downstream of *env* and are translated

TABLE 5.3   Accessory Genes in Retroviruses

Gene	Functions
Betaretrovirus (MMTV)	
sag	Superantigen
dut	dUTPase
Deltaretrovirus (HTLV/BLV)	
tax	Transcription activator (like *tat*)
rex	Splicing/RNA transport regulator (like *rev*)
Lentivirus (HIV-1)	
tat	Transcription activator (like *tax*)
rev	Splicing/RNA transport regulator (like *rex*)
vif	
vpr/vpx	
nef	See Table 5.4
vpu	
dut	dUTPase (in nonprimate lentiviruses) Facilitates replication in certain cell types.
Spumavirus (HRSV)	
bel1	Activates transcription
bel2	?
bet	?
Epsilonretrovirus (WDSV)	
Orf A	?
Orf B	?
Orf C	?

Source: Adapted from Coffin *et al.* (1997, Table 1, p. 36).

from spliced mRNAs, with some of the genes requiring multiple splicing for expression. The genome organizations of the different genera are diagrammed in Fig. 5.9 to show the location of these accessory genes. The different accessory genes have been inserted into the retroviral genome at different times during the evolution of these viruses (Fig. 5.1), and insertion of such genes appears to be a dynamic process that is ongoing. As one example, the *vpx* gene of HIV-2 and the *vpu* gene of HIV-1 appear to have been inserted into their respective viruses after the separation of HIV-1 and HIV-2.

The presence of the accessory genes allows more vigorous replication of the retrovirus that possesses them, which can be fatal to the host cell. As described above, the simple retroviruses do not kill the host cell, but instead establish a persistent infection in which the cell survives and produces low levels of virus indefinitely. However, at least some of the complex retroviruses, such as HIV-1, can replicate to high titer in some cell types with the result that the cells die. It is interesting that many endogenous retroviruses are present in the germ line of different vertebrates, as described below. However, none of these are complex retroviruses. It is possible that the regulated lifestyle of the complex viruses would make complex endogenous viruses difficult to silence. The inability to silence such viruses

TABLE 5.4   HIV Proteins

Protein	mRNA	Size (kDa)	Post-translational modifications	Functions
Gag	Genomic RNA	p25 (CA)	None	Capsid structural protein
		p17 (MA)	Myristylated at Gly-2	Matrix protein
		p7 (NC)	?	RNA-binding protein
		p2	?	RNA binding protein
Pro	Genomic RNA, frameshifted	p10 (Pro)		Viral protease, processes Gag and Gag-pol polyproteins
Pol	Genomic RNA, frameshifted	p66/p51 RT	Heterodimer, p51 lacks RNase H domain present in p66	Reverse transcriptase
		p32 (IN)	Cleaved from Gag–Pro–Pol by Pro	Integrase
Vif	vif mRNA	p23		*viral infectivity factor*, essential for spread in macrophages
Vpr	vpr mRNA	p15	Associates with p7	Augments replication
Tat	tat mRNA	p14		Required for replication, transactivates RNA synthesis, binds to TAR RNA
Rev	rev mRNA	p19		Regulates splicing/RNA transport; binds RRE element and facilitates *env* translation
Vpu	vpu/env mRNA	p16	Phosphorylated on Ser	Helps in virion assembly and release, dissociates gp160/CD4 complex
Env	vpu/env mRNA	gp 120 (SU)	24 sites for N-linked glycosylation	Surface glycoprotein, mediates cellular attachment
		gp 41(TM)	7 sites for N-linked glycosylation	Transmembrane glycoprotein
Nef	nef mRNA	p27	Myristylated at Gly-2, phosphorylated at Tyr-15	Homodimer, causes pleiotropic effects, including downregulation of CD4

Source: Adapted from Levy (1994, p. 8).

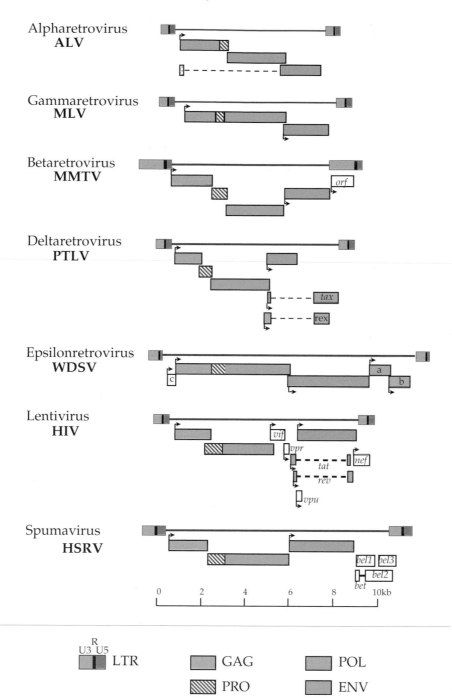

FIGURE 5.9 Coding regions of representatives of each retrovirus genus. Sites of translation initiation are shown with arrows. The locations of the four major genes are shown in different colors. Accessory genes are named. [Redrawn from Coffin *et al.* (1997, Fig. 5, p. 37) and from Fields *et al.* (1996, p. 1776).]

could lead to selection against organisms that contain them in the germ line.

Many of these proteins encoded by the accessory genes are multifunctional and their functions are only partially understood. The intense interest in PTLV and HIV has led to extensive study of their accessory genes, *tax* and *rex* in

the case of PTLV/BLV and *tat, rev, vif, vpr/vpx, nef,* and *vpu* in the case of lentiviruses. The accessory genes of the spumaviruses, *bel1, bel2,* and *bet,* as well as those of the epsilonretroviruses, have been less well studied. The functions of these genes can be conveniently grouped into four categories: (1) transport across the nuclear membrane of subviral particles that synthesize viral DNA after infection, allowing the virus to infect quiescent cells; (2) activation of the transcription of the provirus to greatly increase the rate of production of viral RNA; (3) export of unspliced viral RNA to the cytoplasm; and (4) promotion of virus assembly or increase in the infectivity of the virion, whether directly or indirectly.

Transport of Viral DNA into the Nucleus

HIV-1, and probably other lentiviruses as well, can infect quiescent cells because the viral DNA and associated proteins can cross the nuclear membrane. The mechanism by which this occurs is not known but the product of the *vpr* gene is thought to be required. MA and IN have also been implicated in the process.

Transactivation of the Transcription of Viral RNA

All complex retroviruses encode a protein that transactivates transcription of viral RNA from the provirus. In PTLV/BLV, the gene for this protein is called *tax*. The Tax protein activates transcription by means of sequence elements in the U3 region of the viral DNA called *Tax-responsive elements* (TREs), whose locations are shown in Fig. 5.5. Tax interacts with cellular transcription factors that bind to TREs, and this interaction results in increased activity of the transcription factors. One such transcription factor appears to be the cAMP response element/activating transcription factor (CREB/ATF). Tax also increases the transcription of certain cellular genes, in some cases through its interactions with CREB/ATF, and in other cases by stimulating the transcription of genes regulated by NF-κB. Many of these cellular genes are important in the regulation of T cells, and their stimulation may relate to the pathology of disease caused by the virus.

In the lentiviruses, the gene for the transcriptional activator is called *tat*. The Tat protein works in one of two different ways. Tat of visna virus appears to interact with cellular transcription factors in a manner similar to Tax, using a sequence element in U3, although the cellular factors are different. Tat of HIV and its close relatives, however, stimulates transcription by binding to a sequence element called TAR at the 5′ end of the viral RNA. TAR is composed of the first 60 nucleotides of HIV-1 RNA, which form a stem-loop structure that is essential for the function of TAR. TAR of both HIV-1 and HIV-2 are shown in Fig. 5.10. How the binding of Tat to TAR activates transcription is not yet clear. One model is that Tat interacts not only with the nascent viral RNA but also with cellular transcription factors, and in so doing stabilizes the transcription complex or changes its composition. In this model, the altered transcription

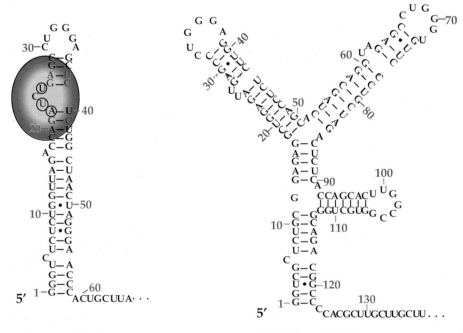

HIV-1 TAR   HIV-2 TAR

FIGURE 5.10 Structures of HIV-1 and HIV-2 TAR RNA elements. Nucleotides are numbered from the 5′ end of the RNA. For HIV-1, positions involved in binding to TAT protein (yellow oval) are circled, and the bases involved in tertiary structure alterations following Tat binding are shown in red. Less is known about the HIV-2 Tat blinding. [Adapted from Coffin *et al.* (1997, Fig. 12, p. 226).]

complex is more processive, allowing the production of complete RNA genomes rather than truncated transcripts. It may also initiate transcription more frequently.

Export of Unspliced Viral RNA to the Cytoplasm

Complex retroviruses encode proteins that promote the export of unspliced or partially spliced RNA from the nucleus. The proteins are Rex in the case of PTLV/BLV and Rev in the case of lentiviruses. Rex and Rev are translated from multiply spliced mRNAs, as are the transcriptional activators and Nef in the case of the lentiviruses. The multiple splicing events in HIV-1 are illustrated in Fig. 5.11. Early in infection, before Rex and Rev are present, only completely spliced mRNAs are exported from the nucleus to the cytoplasm. Thus, the proteins made early are the transcriptional activators, Nef, and the proteins that control the export of unspliced or partially spliced viral RNAs from the nucleus. The transcriptional activators accelerate the production of viral RNAs, and Rex and Rev allow the export of

mRNAs for the other viral proteins, which includes the genomic RNA. These processes are illustrated schematically in Fig. 5.12.

Studies with HIV-1 have shown that Rev binds to a sequence element in the viral RNA called the Rev response element or RRE. HIV-1 RRE is 234 nucleotides in size and has multiple stem-loop structures that are important for function. It is found in the *env* gene region and is therefore spliced out of the multiply spliced mRNAs (Fig. 5.11). In addition to binding to RRE, Rev also interacts with cellular proteins involved with the nuclear export pathway. Although the detailed mechanisms are not known, the end result is that Rev promotes the export of RNAs containing RRE (i.e., unspliced genomic RNA and singly spliced mRNAs) from the nucleus to the cytoplasm (Fig. 5.12). Rev appears to accompany the RNA to the cytoplasm and then cycle back to the nucleus. Thus, after the appearance of Rev, the infection cycle switches to a late phase with the appearance in the cytoplasm of the mRNAs for Gag–Pro–Pol, Env, Vpu, Vif, and Vpr. Genomic RNA

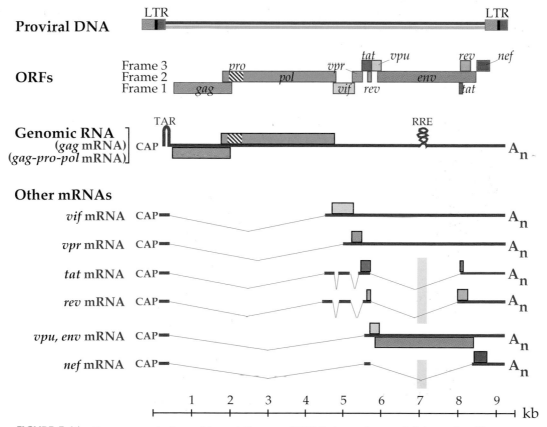

FIGURE 5.11 *Genome organization and transcription map of HIV, the human immunodeficiency virus. The genome is shown on the top line as the integrated provirus. The LTRs and all open reading frames (ORFs) are indicated. Below this, the unspliced genome RNA is shown, with TAR and RRE (the Rev response element in env) indicated. The various spliced mRNAs (and the ORFs translated from them) are diagrammed below the RNA genome. The pale blue shading indicates the location of the RRE, which is spliced out of tat, rev, and nef messages. [Redrawn from Coffin et al. (1997, p. 803).]*

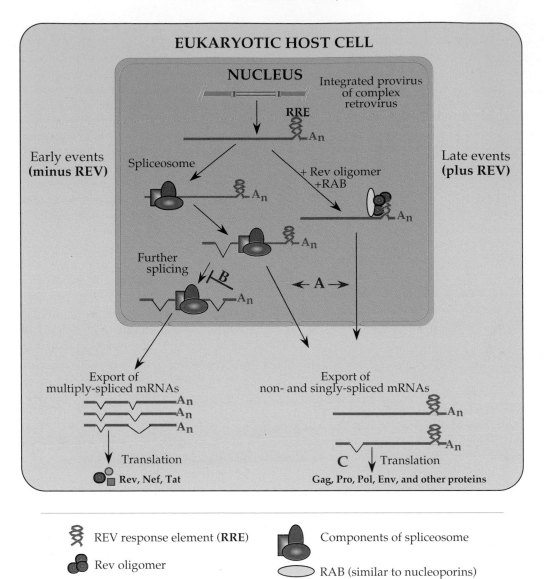

EUKARYOTIC HOST CELL

NUCLEUS

Integrated provirus
of complex
retrovirus

RRE

A_n

Early events
(minus REV)

Spliceosome

Late events
(plus REV)

+ Rev oligomer
+RAB

A_n

A_n

Further
splicing

B

A_n

← A →

A_n

Export of
multiply-spliced mRNAs

A_n
A_n
A_n

Export of
non- and singly-spliced mRNAs

A_n

A_n

Translation

Rev, Nef, Tat

C   Translation

Gag, Pro, Pol, Env, and other proteins

REV response element (**RRE**)   Components of spliceosome

Rev oligomer   RAB (similar to nucleoporins)

FIGURE 5.12   Model of Rev action. Rev is thought to (A) mediate export of nonspliced or singly spliced RRE-containing RNAs from the nucleus; (B) inhibit complete splicing of mRNA; (c) enhance translation of unspliced and singly spliced mRNAs. [Adapted from Collins *et al.* (1997, p. 244).]

exported to the cytoplasm can also be packaged into progeny virions.

In addition to its function in the export of viral RNAs to the cytoplasm, Rev also has other functions. These functions appear to include regulation of the splicing of HIV RNAs and increasing the efficiency of translation of HIV RNAs (Fig. 5.12).

The PTLV Rex appears to function similarly to Rev. It binds to an RNA element, called the RexRE, that is similar in size to the RRE, and binding of Rex promotes export of viral RNA from the nucleus. However, the RexRE is found in the U3–R region of the genome and is therefore present in all of the viral mRNAs. It is curious that Rex binds HIV-

1 RNA, although not in the same place as Rev, and will substitute for Rev in HIV-1. However, Rev does not bind PTLV RNA and will not substitute for Rex.

Assembly Functions

Other accessory proteins of the lentiviruses have multiple functions that are not well understood, but all seem to promote the assembly and release of infectious virus, either directly or indirectly. Vpr, found in most lentiviruses, including both HIV-1 and HIV-2, and Vpx, found in HIV-2 and its close relatives, but not in HIV-1, are structural proteins that interact with Gag. They are found in the virion in amounts similar to that of Gag. Vif is a membrane-associated protein that is also present in the virion, but in amounts similar to Pol rather than Gag. It has an effect on the processing of Gag and is required for the production of infectious virions. Nef, found in the primate lentiviruses, is a multifunctional protein that has a role in the assembly and release of infectious virus. Myristylated forms of Nef are associated with the virion, where they may be required for full infectivity of the virion. Another function of Nef is the removal of CD4, the receptor for HIV, from the surface of the infected cell. Removal of CD4 prevents the interaction of Env with CD4 and facilitates the assembly and release of viruses. It also prevents superinfection of the cell by released virus. Finally, Vpu, a small integral membrane protein found in HIV-1 but not in HIV-2, promotes virus release in two ways. It promotes the degradation of CD4 by a mechanism different from that used by Nef. It also facilitates the release of budding virus from the cell, a function fulfilled by Env in HIV-2.

The dUTPase Gene

The nonprimate lentiviruses and members of the betaretroviruses contain a gene encoding dUTPase, called *dut*. This gene was acquired independently by these two different lineages (Fig. 5.1). The enzyme dephosphorylates dUTP, thereby preventing its incorporation into DNA. A cellular gene usually provides this function for retroviruses (and for DNA viruses). However, possession of this gene allows these two genera of retroviruses to replicate in quiescent cells that do not express adequate levels of the enzyme, such as macrophages.

Assembly of Retroviruses

The retrovirus virion is approximately 100 nm in diameter and acquires its lipid envelope by budding through the plasma membrane of the infected cell. Two forms of budding occur, as illustrated in Fig. 2.17. In the betaretroviruses and the spumaviruses, the capsid is assembled in the cell cytoplasm and then buds through the plasma membrane. In the alpharetroviruses, the gammaretroviruses, and the lentiviruses, the capsid assembles during budding and is not visible as a distinct structure within the cell. Mason–Pfizer monkey virus, a betaretrovirus, changes from budding of preassembled capsids to assembly of capsids during budding as the result of a single amino acid change in MA. Thus, the distinction between the two types of budding does not represent a fundamental difference in budding pathway but simply a matter of the stability of the capsid in the absense of interactions with other components of the virion during budding.

The capsid is formed by the assembly of Gag, Gag–Pro, and/or Gag–Pro–Pol polyproteins into a structure having spherical symmetry. RNA appears to be required for assem-

bly but not the Env protein, even in viruses whose capsids assemble during budding. These polyproteins are incorporated into the capsid in approximately the same ratio as they are produced inside the cell. During or after budding, the viral protease cleaves the polyprotein precursors to the final products. These cleavages are required for the virus to be infectious and result in a change in the structure of the capsid. The final, fully cleaved virion may have the capsid centrally placed within it or it may be eccentrically placed, and the shape of the capsid may be spherical or it may have other shapes, depending on the virus. As described above, it is important that the cleavages be delayed until the virion is partially or completely assembled, or the assembly process will not work properly.

During assembly, the genomic RNA is recruited into the capsid and dimerized. There is a packaging signal in the RNA, usually referred to as ψ, that is often found in the 5′ region downstream of the LTR. The ψ signal is not present in the spliced RNAs of many retroviruses but even in those in which it is present, the spliced RNAs are not packaged. The packaging signal is not absolutely required for packaging, but increases the efficiency of incorporation into the virion by about 100–fold. Because Gag is the only protein required for assembly of capsids, the recognition of ψ must be a property of Gag. During maturation of the virion, the RNA dimer also matures from a less stable form to a more stable form. tRNA is recruited into the capsid by RT, but association with RT is not absolutely required for packaging of the primer tRNA.

The envelope glycoproteins, SU and its associated TM, are incorporated into the virion during budding. The mechanism by which SU/TM is recruited is uncertain. Evidence indicates that MA interacts with TM, but a model in which SU/TM is incorporated nonspecifically, perhaps because it is free to diffuse whereas cellular proteins are not, cannot be excluded. It is clear that SU/TM is not required for virus assembly. Glycoproteins from other viruses can substitute for SU/TM. More strikingly, capsids can bud to form bald particles free of glycoprotein and, thus, noninfectious. In a key experiment, it has been found that HIV-1 buds from the basolateral surface of polarized cells, where SU/TM are found. However, when SU/TM is absent, Gag-directed budding occurs from both basolateral and apical surfaces. This result provides evidence that specific interactions between the capsid and the glycoproteins occur during virus budding and are important, but these interactions are not essential for budding to occur.

Alpharetroviruses and Gammaretroviruses

Alpharetroviruses

The alpharetroviruses comprise a large collection of avian leukosis and sarcoma viruses. Dozens of ALVs are known. They are grouped into seven interference groups, A–F, in which members of a group use the same receptor. Infected cells cannot be superinfected by another member of the same interference group, because the receptors for it have been eliminated. Expression of the viral envelope protein is usually responsible for the downregulation of receptors, which is important for the budding and release of virus. It also prevents the cell from becoming infected by hundreds or thousands of progeny viruses. This may be important for the survival of the infected cell and may allow it to produce progeny viruses indefinitely.

Interference groups A, B, C, and D of the ALVs are exogenous viruses that are transmitted as infectious agents from chicken to chicken. Members of group E are endogenous viruses, resident in the germ line of the chicken. The endogenous viruses may be quiescent and not expressed, or parts of the provirus genome or even the entire genome may be expressed. In the latter case, progeny viruses are formed that can infect other cells, leading to widespread expression of the virus in the animal. The endogenous viruses of chickens, as well as those of mammals, tend not to be expressed. Expression often results in disease and a shortened life span, and the animals are selected to not express integrated proviruses. In chickens, the level of expression and the time in the animal's life when expression of an endogenous virus occurs is different for different strains of chickens.

ALVs are commonly present as infectious agents in chicken flocks around the world. The major illness caused by these viruses is a wasting disease characterized by anemia, immunosuppression, and poor growth. Perhaps of more interest to the molecular biologist, these viruses may also cause leukemia or sarcoma, as described below.

Interference groups F and G are viruses of pheasants. These viruses have not been as well studied as the viruses of chickens.

Gammaretroviruses

The gammaretroviruses consist of a large number of leukemia and sarcoma viruses of mice, cats, primates, and other mammals. Also included in the genus is reticuloendotheliosis virus of birds, which causes immunodeficiency. The murine leukemia viruses have been particularly well studied. Both exogenous viruses and endogenous viruses of mice are known. The mouse genome contains 500–1000 endogenous proviruses that are divided into four classes, but only two of these classes are known to encode infectious viruses. One class encodes betaretroviruses, described below, and the other class consists of 50 to 60 copies of gammaretroviruses. The host range of mouse endogenous gammaretroviruses depends on the *env* gene. These viruses may be ecotropic (able to infect only mice), xenotropic (unable to infect mice but able to infect other animals, such as rats), or polytropic (able to infect both mice and other

animals). The endogenous viruses are usually transcriptionally silent, but expression does occur in some mouse strains. In an extreme example, AKR mice usually become viremic at an early age and most of the animals eventually die of leukemia.

Feline leukemia virus (FeLV) is an important pathogen of cats. It is an exogenous virus that causes T- cell lymphomas and immunodeficiencies, as well as severe aplastic anemia, in cats. Sarcoma strains of FeLV are also known. Effective vaccines against FeLV have been developed for use in pet cats.

Gammaretroviruses of other mammals, including primates, are also known, as are sarcoma viruses derived from them. Interestingly, however, none of the known gammaretroviruses infect man, although the human genome does contain many retroviral-like elements.

Induction of Leukemia by Alpha- and Gammaretroviruses

Alpha- and gammaretroviruses are an important cause of cancer in chickens, mice, cats, and subhuman primates. The cancers are usually forms of leukemia or lymphoma, and arise only after a long latent period. The tumor cells are usually clonal, having developed from a single progenitor tumor cell. Not all infected animals develop cancer. Infection by these retroviruses does not directly produce tumors. Instead, rare insertional or recombinational events must occur that give rise to a tumor cell. Although these events are rare, the very large number of cells infected during the persistent infection established by the virus may render such events probable, and after a sufficiently long latent period the probability that a tumor will arise may be high.

It is only recently that the mechanisms by which unmodified alpha- and gammaretroviruses induce tumors have become at least partially understood. One mechanism involves the insertion of the provirus near to or within a cellular oncogene. This may bring the expression of the gene under the control of the strong viral promoters, and the resulting overexpression of the oncogene may result in a tumor cell. In the case of bursal lymphomas induced by infection of chickens by ALV, for example, it has been found that more than 80% of tumors have ALV provirus inserted near the c-*myc* gene and overexpress the c-*myc* product. In other cases, insertion of the provirus within the oncogene may result in the expression of an mRNA that lacks control sequences (such as sequences that cause the mRNA to be degraded rapidly) or that is translated into an altered protein product that has lost regulatory elements. Such a process is often seen in erythroblastosis induced by ALV in chickens, for example. Integration of the provirus between two exons of c-*erbB* separates the domain of the protein that binds epidermal growth factor and tumor growth factor α from the domain that leads to downstream signaling by means of protein tyrosine kinase activity. Unregulated signaling by the modified protein product results in deregulation of growth.

A different mechanism of tumor induction is often seen in mice infected by MLV. Most AKR mice spontaneously express an endogenous virus called Akv1 early in life, and die of thymic lymphomas in their first year of life. For the tumors to develop multiple recombination events between Akv1 and other endogenous viruses are required to produce an Env protein with altered host range. The altered SU may stimulate T-cell proliferation by binding the interleukin 2 (IL-2) receptor, and other altered interactions with T cells and their precursors may also be important in tumor induction. Alterations in Env are also found in tumors induced by FeLV.

It is obvious from this description that the properties of the virus (host range, the nature of the LTR promoters) as well as the properties of the host cell are important in determining whether a tumor will arise. Thus, different strains of the same virus may have different effects on infection of the same host, or one strain of virus may affect different strains of its hosts differently.

Alpha- and Gammaretroviruses That Express Oncogenes

Alpha- and gammaretroviruses undergo recombination with cellular oncogenes to produce viruses capable of causing tumors in their hosts. Recombination is thought to occur during reverse transcription of a hybrid genome in which a host mRNA replaces one copy of the viral RNA. Recombinant viruses that express a variety of oncogenes have been repeatedly isolated from spontaneous tumors in animals that are infected by leukosis/leukemia viruses. Such retroviruses will usually transform cells in culture and rapidly cause tumors when inoculated into a new host. Many of these viruses cause sarcomas and the oncogene-containing virus is then referred to as a sarcoma virus. Examples of the genomes of four oncogene-containing retroviruses are shown in Fig. 5.13. Two are derived from avian leukosis viruses and two from murine leukemia viruses. These examples have been chosen to illustrate a range of possibilities for the incorporation and expression of the oncogene in the transforming virus genome.

Most transforming retroviruses are defective, because the oncogene replaces part of the retrovirus genome. However, some isolates of Rous sarcoma virus (RSV) are nondefective. In this case, the v-*src* gene in present in the 3′ region of the genome and is translated from an independent spliced mRNA. Such a nondefective RSV is illustrated in the figure. Also illustrated is an example of avian myeloblastosis virus, which expresses the v-*myb* gene. This virus is defective because the v-*myb* gene replaces most of the *env* gene of the virus. It is translated as a v-myb-Δenv fusion protein from the spliced mRNA that would express the Env protein in ALV.

The Abelson murine leukemia virus expresses v-*abl*. This gene replaces the *pro* and *pol* genes and part of *gag*, in the example shown, and the virus is defective. In this case, Abl is produced as a fusion protein linked to the N-terminal domain of Gag. Molony murine sarcoma virus expresses v-*mos*. In the example shown, v-*mos* replaces *env* and is translated from a spliced mRNA normally used to express Env.

RSV, named for its discoverer Peyton Rous, was one of the earliest such retroviruses discovered. RSV causes sarcomas in chickens and transforms cells in culture. The transforming ability is due to the expression by the virus of the oncogene referred to as *src*. Many isolates of RSV have been made over the years, and all such viruses express *src* by definition. However, the v-*src* expressed by the different isolates differ because the recombination points are different, and the v-*src* genes usually have mutations that distinguish them from the cellular gene. Most RSVs are defective because the recombination event results in the replacement of parts of the viral genome by the cellular oncogene, although nondefective RSVs have also been isolated as shown in Fig. 5.13. In the case of defective RSVs, or of other defective oncogene-containing retroviruses, a replication competent ALV is required to supply the missing functions if the RSV is to produce progeny virions. However, a defective RSV virion, once formed, is capable of infecting a cell, making cDNA, and integrating the proviral DNA into the host genome without help. Subsequent expression of the oncogene under the control of the viral LTRs leads to transformation of the infected cell.

Since the discovery of RSV, many different oncogene-expressing retroviruses have been isolated, primarily from chickens, mice, and cats. The identification and study of oncogenes has resulted in the discovery of many proteins that are crucial for the regulation of cellular pathways. A sampling of oncogenes that have been found in retroviruses is given in Table 5.5 to illustrate the range of oncogenes that have been captured by retroviruses. Oncogenes encode protein kinases, receptors that respond to growth factors, transcription factors, and other proteins that function at critical points in the cell regulatory cycle. The regulation of cell growth is complex. In broad outline, cells express receptors that bind growth factors. On binding its ligand, the receptor signals the cell to synthesize DNA and divide. This signal is often transmitted by means of a phosphorylation cascade by which transcription factors are activated. The activated transcription factors cause mRNAs to be produced that result in the synthesis of proteins that, in turn, induce DNA synthesis and cell division. These activities are regulated in part by interactions with regulatory proteins called anti-oncogenes, some of which are described in Chapter 6. Thus, oncogenes all have in common that over-expression of the protein, or expression of an altered form

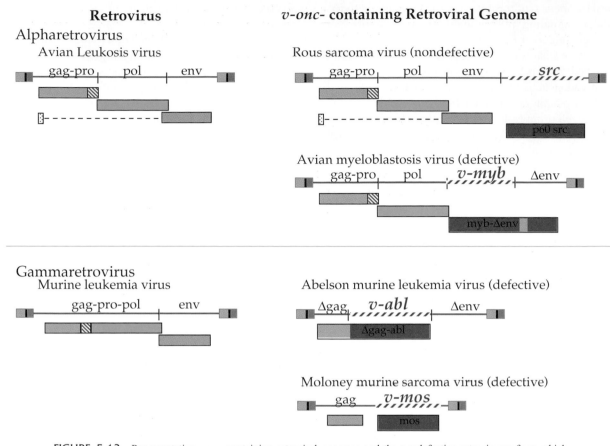

FIGURE 5.13   Representative *v-onc*-containing retroviral genomes and the nondefective retroviruses from which they were derived by capture of cellular oncogenes. In each case the inserted oncogene is shown as a patterned domain in the DNA and its product as a green box. For Rous sarcoma virus, the *src* gene is expressed from a new spliced message. For Abelson murine leukemia virus, the *abl* gene is present in a fusion protein with a deleted form of *gag*. *Myb* and *mos* are expressed from spliced Env messages, from which most or all of the *env* gene has been replaced by the oncogene. [Adapted from Coffin *et al.* (1997, pp. 5, 504) and from Fields *et al.* (1996, p. 1776).]

of the protein that no longer responds appropriately to regulatory signals, leads to unregulated growth. The regulatory signal could be, for example, the ligand to which a growth receptor normally responds, or domains that interact with inhibitory elements or that normally require phosphorylation for activity, or any one of many other mechanisms that regulate cell growth.

For some oncogenes, overexpression by the strong retrovirus promoter is sufficient for cell transformation. The large amount of oncoprotein overwhelms the cells' regulatory abilities, or the protein is expressed in tissues that normally do not express it and therefore lack the regulatory machinery to control it. For most oncogenes, however, overexpression is not sufficient and the viral oncogene, referred to as v-*onc*, differs from its cellular counterpart, referred to as c-*onc*. The changes in v-*onc* render the oncoprotein nonresponsive to regulatory controls so that the cell is always turned on. One mechanism for the loss of regulatory control is the loss of domains that respond to such control signals. A second mechanism is point mutations that arise during the replication of the recombinant retrovirus. Mutations occur at high frequency during the replication of retroviruses but at a very low frequency during the replication of cells.

The dozens of oncogenes found in retroviruses are too numerous to describe here, but a few examples are cited to illustrate specific points. More than one-third of erythroblastosis in chickens induced by oncogene-containing ALVs arises from viruses that have captured the *erbB* gene, and the viral *erbB* gene has changes in the C-terminal part of the protein (these viruses are called avian erythroblastosis viruses). It was noted above that ALV-induced erythroblastosis often results from the provirus integration into the *erbB* gene. Thus, this gene is very important for the induction of erythroblastosis. As a second example, Ras proteins are G proteins that are important in signal transduction.

TABLE 5.5   Selected Retroviral Oncogenes

Oncogene/ functional class	Retrovirus	Viral oncoprotein[a]
Growth factors		
sis	Simian sarcoma virus	p28$^{env\text{-}sis}$
Hormone receptor (thyroid hormone receptor)		
erbA	Avian erythroblastosis virus[b]	p75$^{gag\text{-}erbA}$
Tyrosine kinase growth factor receptors		
erbB	Avian erythroblastosis virus[b]	gp65^{erbB}
sea	S13 Avian erythroblastosis virus	gp160$^{env\text{-}sea}$
fms	McDonough feline sarcoma virus	gp180$^{gag\text{-}fms}$
kit	Harvey-Zuckerman-4 feline sarcoma virus	gp80$^{gag\text{-}kit}$
ros	UR2 avian sarcoma virus	p68$^{gag\text{-}ros}$
mpl	Mouse myeloproliferative leukemia virus	p68$^{gag\text{-}ros}$
eyk	Avian retrovirus RPL30	gp37^{eyk}
Nonreceptor tyrosine kinases/signal transduction factors		
src	Rous sarcoma virus	pp60^{src}
abl	Abelson murine leukemia virus	p460$^{gag\text{-}abl}$
fps[c]	Fujinami avian sarcoma virus	p130$^{gag\text{-}fps}$
fes[c]	Snyder–Theilen feline sarcoma virus	p85$^{gag\text{-}fes}$
fgr	Gardner–Rasheed feline sarcoma virus	p770$^{gag\text{-}actin\text{-}fgr}$
yes	Y73 avian sarcoma virus	p90$^{gag\text{-}yes}$
Serine-threonine kinases/signal transduction factors		
mos	Moloney murine sarcoma virus	p37$^{env\text{-}mos}$
raf[d]	3611 murine sarcoma virus	p75$^{gag\text{-}raf}$
mil[d]	MH2 avian myelocytoma virus	p100$^{gag\text{-}mil}$
G Proteins (GTPases)		
H-ras	Harvey murine sarcoma virus	p21^{ras}
K-ras	Kirsten murine sarcoma virus	p21^{ras}
Transcription factors		
jun	Avian sarcoma virus 17	p65$^{gag\text{-}jun}$
fos	Finkel–Biskis–Jenkins murine sarcoma virus	p55^{fos}
myc	OK10 avian leukemia virus	p200$^{gag\text{-}pol\text{-}myc}$
myb	Avian myeloblastosis virus	p45^{myb}
ets	Avian myeloblastosis virus[b]	p135$^{gag\text{-}myb\text{-}ets}$
rel	Avian reticuloendotheliosis virus	p64^{rel}
maf	Avian retrovirus AS42	p100$^{gag\text{-}maf}$

Source: Adapted from Fields *et al.* (1996, p. 309).
[a]In these names, p is for protein, gp is for glycoprotein, pp is for phosphoprotein, and the numbers are the approximate apparent molecular weight in kDa.
[b]This is a retrovirus with two oncogenes.
[c]*fps* and *fes* are the same oncogene derived from the avian and feline genomes, respectively.
[d]*raf* and *mil* are the same oncogene derived from the murine and avian genomes, respectively.

Three different *ras* genes have been found in retroviruses and all have changes in codon 12, illustrating the importance of this amino acid for the control pathways that regulate its activity. Mutations in this same codon have also been found in tumors that were not produced by retroviruses.

Two examples of transcription factor oncogenes will be cited. The c-*myc* gene product is tightly regulated and present in low amounts in normal cells. It is active on forming heterodimers with a second factor, present in larger amounts. Overexpression of v-*myc* seems to overwhelm the regulatory pathways and result in high-level expression of genes required for cell proliferation. Even in this case, however, mutations in v-*myc* upregulate its transforming ability. Another transcription factor, the product of *erbA*, is a receptor for thyroid hormone α. On binding the hormone, it becomes active as a transcription factor. v-*erbA* expresses a modified protein that no longer needs to bind the hormone in order to be active as a transcription factor, and thus is always on.

Often the expression of a single oncogene, whether modified or not, is not sufficient to achieve full tumorigenic potential. Some transforming retroviruses have been found to express two oncogenes, both of which are required for tumor induction. Furthermore, many oncogene-expressing retroviruses induce tumors only after a lag and in some cases the tumors are clonal. This suggests that for these viruses, the expression of v-*onc* is not in itself sufficient, and additional changes in the infected cell must occur before it becomes fully tumorigenic.

The production of an oncogene-expressing retrovirus that is capable of inducing a tumor is a rare event, but the presence of the virus is signaled by the tumor itself. Most such viruses have been isolated from chickens, mice, or cats because these animals are commonly infected by C-type retroviruses, but also because the occurrence of tumors is more likely to be detected in these animals than in many other animals. Chickens are processed for human food in very large numbers, and tumors, although rare, are often noticed during processing. Mice are used in large numbers in laboratory experiments and receive attentive care from investigators. Pet cats are regularly seen by veterinarians when they become ill.

The study of oncogenes that occur in transforming retroviruses has produced a wealth of information about cell regulatory pathways by which cell growth and the interactions of cells with one another and with various growth hormones are controlled. Many of these pathways were first discovered because of the presence of critical components in retroviruses. Furthermore, some of the oncogenes found in transforming retroviruses were later found to be overexpressed or to be expressed in a mutant form in human cancers, confirming the importance of these products in the production of human tumors. However, although of great importance for our understanding of the regulation of eukaryotic cells, and of importance for the animal in which the tumor produced by the oncogenic retrovirus arose, the transforming retroviruses have little significance for the persistence in nature of their retroviral parent. The production of a tumor shortens the life span of the host, and virtually all transforming retroviruses are defective and require a helper virus for multiplication. Thus, the probability of transmission of the virus to a new host is low. The appearance of a recombinant retrovirus expressing an oncogene is an accident that soon disappears from the population, along with the unfortunate host.

Betaretroviruses

The best studied member of the betaretroviruses is mouse mammary tumor virus (MMTV). The promoters in the LTR respond to estrogen stimulation, consistent with the growth of the virus in mammary epithelium. MMTV occurs as an endogenous virus, present in 0–4 copies in different strains of mice. It also occurs as an exogenous virus that infects mice by vertical transfer, usually from mother to daughter through the milk. After a latent period, the virus induces mammary tumors with a significant frequency.

MMTV encodes a *s*uper*a*ntigen (the *sag* gene in Table 5.3). As described in Chapter 8, T cells express a receptor that, when stimulated by binding to a peptide antigen recognized by it, causes the cell to become active and divide. Activated T-cells induce B cells to proliferate. Superantigens bind to T cell receptors and to molecules of the major histocompatibility complex. Binding is not to the domain of the receptor that interacts with peptide antigen, but to a region shared by all T-cell receptors of a given class. A peptide antigen, presented by major histocompatibility complex molecules, may lead to activation of 10^{-5} of all T cells. However, the Sag protein binds to receptors on as many as 10% of all T cells, leading to their activation. This, in turn, leads to proliferation of B cells, which are the first targets of the virus on infection. In this way, the pool of susceptible cells is increased in size and the infection is amplified. Infected B cells spread the infection to the mammary glands, where mammary epithelium is infected. The virus induces proliferative changes in mammary cells, which eventually leads to tumors.

Mason–Pfizer monkey virus and the so-called simian AIDS virus, or SAIDS virus, are also betaretroviruses. The SAIDS virus causes immunodeficiencies in monkeys but is distinct from simian immunodeficiency virus (SIV). SIV is closely related to HIV, described below.

PTLV Viruses

The two human T-cell leukemia (or lymphotropic) viruses are now called primate T-cell lymphotropic viruses (PTLV) (Table 5.1). PTLV-1 causes adult T-cell leukemia (ATL) in humans as well as a neurological disorder that has been termed HTLV-1–associated myelopathy (HAM) (from the former name for the virus). The distributions of PTLV-1 and PTLV-2 are shown in Fig. 5.14. PTLV-1 is prevalent in southern Japan, where transmission appears to be either through sexual intercourse or via breast milk. More than 1 million people in this region are infected by the virus. The virus is also found in other parts of Asia, in the Caribbean region, and in central Africa, as well as in the United States and western Europe, where it is increasing due to injecting drug use. Blood transfusions have also been implicated in the spread of the virus.

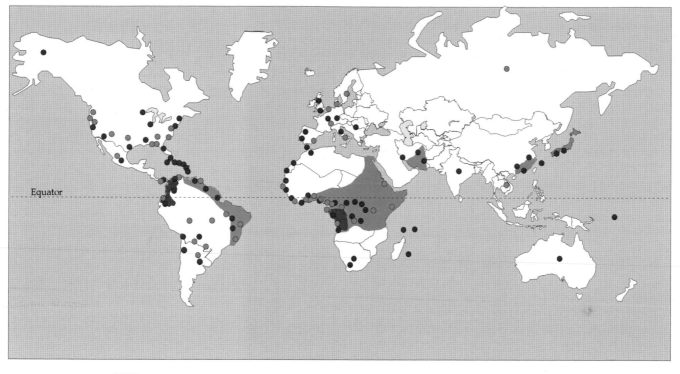

▨ Areas in which PTLV-1 is endemic	● Location of PTLV-1 isolates
▨ Areas in which PTLV-2 is endemic	● Location of PTLV-2 isolates

FIGURE 5.14 Global distribution of PTLV-1 and PTLV-2. Shading indicates the areas of endemism, while the dots are locations of isolates. For PTLV-1 most of the dots refer to subgroup A isolates, but the B subgroup is found in Zaire, the D subgroup is found in pygmies in Central Africa, and subgroup C is found in Australia and the islands of Melanesia. [Redrawn from Coffin *et al.* (1997, Fig. 21, p. 544).]

ATL has a long latent period, 20–30 years, and only about 1% of infections result in leukemia. Tax is thought to have a role in induction of leukemia, but the long latent period implies that other events, probably mutations in the affected cells, must occur before ATL appears. Patients with acute ATL live only about 6 months on average.

HAM is a serious neurological disease. It usually develops more rapidly than ATL, and about 1% of PTLV-infected humans develop the disease. Demyelination in the spinal cord occurs, as well as a vigorous inflammatory response involving T cells and other lymphocytes. It is not known whether the pathology is due to damage to virus-infected cells by CD8$^+$ T cells (see Chapter 8) or whether the release of cytokines and chemokines by the inflammatory cells is responsible for the damage.

The retroviruses are best known for their ability to cause tumors, including leukemia, or immunodeficiency, but it is important to note that many retroviruses are associated with neurological diseases. Some of these diseases are characterized by an inflammatory response, as is HAM, whereas in others there is no inflammation associated with the disease. As examples, many strains of MLV cause a spongiform encephalopathy in mice, in which inflammation is lacking, and many lentiviruses commonly cause neurological disease, as will be described below.

PTLV-2 is closely related to PTLV-1. It has been implicated as the causative agent of hairy-cell leukemia, but the number of cases is small and it is not certain that the virus is in fact responsible for the disease.

The PTLVs infect not only humans but also a number of nonhuman primates. In any area, the strains of viruses isolated from infected humans are more closely related to viruses isolated from monkeys in the region than they are to human strains isolated from another area. Thus, the viruses seem to be passed back and forth between monkeys and people.

A virus of cows called bovine leukemia virus (BLV) also belongs to this genus of complex retroviruses. This virus causes B-cell lymphomas in a small fraction of infected animals.

Lentiviruses

The lentiviruses are complex retroviruses whose core is cone shaped in the mature virion. The most important are the human immunodeficiency viruses, HIV-1 and -2. Other members of the genus include simian immunodeficiency virus (SIV), feline immunodeficiency virus (FIV), bovine immunodeficiency virus (BIV), visna virus (which causes disease in sheep), equine infectious anemia virus (EIAV), and caprine arthritis-encephalitis virus (CAEV) (Table 5.1). As described above, the regulation of lentivirus replication is complex because lentiviruses encode three to six accessory genes, depending on the virus, to regulate the replication cycle and aid in the assembly of progeny viruses.

All lentiviruses have a specific tropism for macrophages, which comprise the major reservoir of infected cells in an animal. Other cells can also be infected, and this can be important in the disease process (e.g., CD4$^+$ T cells in HIV infection). Lentiviruses establish a lifelong, chronic infection which elicits a vigorous immune response that is still unable to clear the infection. Most known lentiviruses cause serious disease in their native host after a long latent period. However, SIV produces an asymptomatic, lifelong infection of its natural host, African monkeys. Intriguingly, SIV causes AIDS when transmitted to Asian monkeys.

Human Immunodeficiency Virus

The two human lentiviruses, HIV-1 and HIV-2, cause the well-known disease called AIDS (*a*cquired *i*mmuno-*d*eficiency *s*yndrome). Although both HIV-1 and HIV-2 cause AIDS in people, HIV-2 is not as serious a pathogen. The latent period before disease develops is longer, the virus is not as easily transmissible, and it has not spread as extensively as has HIV-1. It is HIV-1 that is responsible for the vast majority of AIDS in people, and HIV-1 has correspondingly been much more exhaustively studied.

The primary cellular targets of HIV in infected humans are macrophages and CD4$^+$ T cells. These cells express CD4, the primary receptor for all primate lentiviruses, as well as chemokine coreceptors that are also required for entry of the virus (Chapter 1). The infection of macrophages and of CD4$^+$ T cells differs in several important respects. HIV can replicate in fully differentiated, nondividing macrophages, but infection of macrophages does not lead to extensive cytopathology. Efficient infection of CD4$^+$ T cells, in contrast, requires that the T cell be activated (Chapter 8) and results in the death of the cell. As described in Chapter 1, strains of HIV are known that infect macrophages (M-tropic virus) or T cells (T-tropic) more efficiently, which is a function of the *env* gene of the virus.

A representative time course of the infection of a human by HIV-1 is shown in Fig. 5.15. Primary infection may be asymptomatic (most infected people do not see a doctor on primary infection) or accompanied by nonspecific symptoms that appear 3–6 weeks after infection. Symptoms may include rash, fever, diarrhea, arthralgia, myalgia, nausea, sore throat, lethargy, headache, or stiff neck. The number of CD4$^+$ T cells declines during this phase. An immune response is mounted that includes both CD8$^+$ cytotoxic T cells (CTLs) and antibody production (Chapter 8). This reduces the amount of virus in the blood, the number of CD4$^+$ T cells recovers but does not reach preinfection levels, and symptoms of disease ameliorate. However, clearance of the virus is not complete, and a long period of clinical latency ensues. During this period, even though the infected individual is often asymptomatic, the virus continues to replicate actively, especially in lymph nodes, and the immune system continues to fight the infection. It has been estimated that during this clinically quiescent period 10^8 to 10^9 virus particles per day are released into the peripheral blood supply (and cleared). During this time the turnover of CD4$^+$ lymphocytes is also 10^8 to 10^9 cells per day.

In untreated individuals, the continuing replication of virus results in a steady decline in the number of CD4$^+$ T cells (Fig. 5.15) and the continuing appearance of mutations in the virus. In the Env protein, the rate of change in the amino acid sequence averages about 2.5% per year, with a large fraction of changes occurring in certain hypervariable regions. It seems clear that this accumulation of mutations is driven, at least in part, by immune selection.

AIDS, the Disease

With time the immune system is overwhelmed, with the result that the replication of HIV escalates, opportunistic fungal, protozoal, bacterial, and viral infections occur, characteristic cancers appear, and neurological symptoms become apparent. The time to progression to AIDS following infection by HIV is variable but averages about 10 years. Symptoms occur much sooner in infants infected at birth, perhaps because the immune system is not as well developed. In adults infected by the virus, the time to appearance of the symptoms of AIDS is correlated with the level of viral replication during the latent period, as illustrated in Fig. 5.16. In individuals in which viral replication is high, the CD4$^+$ T-cell count declines faster and the time to appearance of AIDS is shorter. The rate of virus replication is probably a function of the strength of the immune response against the virus, especially the CD8$^+$ CTL response. Individuals with a better immune response control virus replication better and longer. Once the symptoms of AIDS appear, the time to death is usually 1–2 years unless antiviral treatment is administered.

CD4$^+$ T cells, most of which function as T helper cells, are a critical part of the immune system. They are necessary for an immune response to an antigen, whether the response is to produce CD8$^+$ CTLs or to produce circulating antibodies (see Chapter 8). AIDS results when so few CD4$^+$ T cells are present that the body is unable to mount an effective immune response. A partial list of symptoms that appear and opportunistic infections that occur is shown in Table 5.6. The first symptoms usually occur when the CD4$^+$ T-cell concentration falls below 500/µl and may include reactivation of

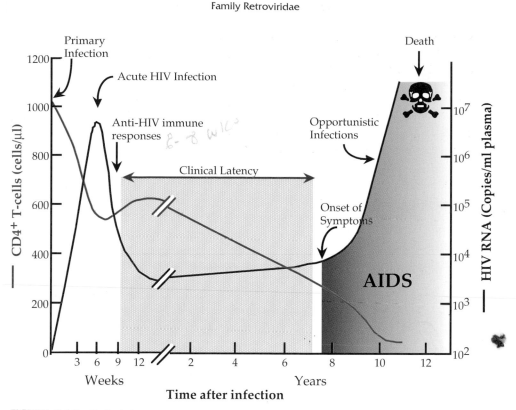

FIGURE 5.15 Typical time course of HIV infection and progression to disease. Patterns of CD4+ T-cell decline and virus load vary greatly among patients. See also Fig. 5.16 for mortality as a function of initial viral load. Some case definitions for AIDs specify a CD4+ count below 400 cells/μl (horizontal line). [From Coffin *et al.* (1997, Fig. 5, p. 600).]

viruses such as herpes zoster, reactivation of bacteria such as *Mycobacteria tuberculosis*, oral lesions caused by fungi, or lymph node pathology. These first symptoms are sometimes referred to as *A*IDS-*r*elated *c*omplex or ARC. Full-blown

AIDS is normally signaled by a drop in the CD4+ T-cell concentration below 200/μl. At this point the infected individual becomes susceptible to numerous opportunistic infections, including those of microorganisms such as *Pneumocystis*

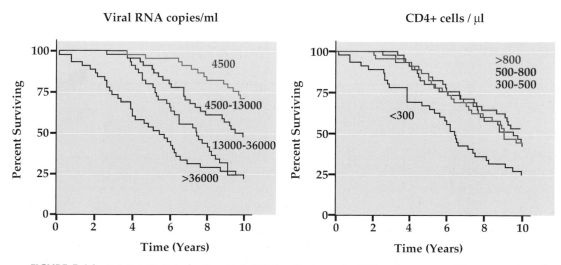

FIGURE 5.16 Relationship between the extent of HIV replication or the CD4+ cell count, two different measures of HIV/AIDS and clinical progression, and time to death. Kaplan–Meier curves are shown for survival versus viral load (left) and concentration of CD4+ T cells (right) at time of diagnosis, plotted by quartiles. [Redrawn from Mellors *et al.* (1996).]

TABLE 5.6 Pathology and Opportunistic Infections of Patients with HIV/AIDS

Stage of HIV/AIDS syndrome/symptoms	Type of Organism	Infectious agent
CD4$^+$ T-cell count 200–500 cells/μl		
Generalized lymphadenopathy		
Headache, fever, malaise		
Generalized weight loss		
Shingles	Viral	Reactivation of herpes zoster
Skin lesions	Viral	*Molluscum contagiosum*
Basal cell carcinomas of skin		
Oral lesions		
Thrush	Fungal	*Candida albicans*
Hairy leukoplakia	Viral	Epsein–Barr virus
Lung disease (tuberculosis)	Bacterial	Reactivation of *Mycobacterium tuberculosis*
CD4$^+$ T-cell count <200 cells/μl		
(Syndromes in addition to those mentioned above)		
Microbial infections		
Pneumonia	Taxonomy unclear	*Pneumocystis carinii*
Disseminated toxoplasma	Protozoan	*Toxoplasma gondii*
Severe diarrhea	Protozoan	*Isospora belli*
Chronic diarrhea	Protozoan	*Cryptosporidia*
Tuberculosis	Bacteria	*Mycobacterium tuberculosis*
Bacterial infections	Bacteria	*Salmonella, Streptococcus, Hemophilus*
CNS disease	Fungal	*Cryptococcus neoformans*
Viral infections and malignancies		
PML	Viral	JC polyomavirus
Disseminated disease of lungs, brain, etc.	Viral	Genital herpes, cytomegalovirus
B-cell lymphoma	Viral	Epstein–Barr virus, HHV-8
Kaposi's sarcoma	Viral	HHV-8
Anogenital carcinoma	Viral	Human papilloma virus
Other syndromes		
Wasting disease	??	??
Aseptic meningitis		
AIDS dementia complex	Viral	HIV encephalopathy

Source: Adapted from Coffin *et al.* (1997, Table 1, p. 597).
Abbreviations: CNS, central nervous system; PML, progressive multifocal leukoencephalopathy; HHV-8, human herpesvirus eight.

carinii, bacteria such as *Mycobacterium*, and fungi such as yeast. Viruses that are normally controlled by the immune system become a problem, such as cytomegalovirus and herpes simplex virus. Several virus-induced cancers become common, such as lymphoma caused by Epstein–Barr virus, Kaposi's sarcoma caused by human herpesvirus-8, or anogenital carcinoma caused by human papilloma virus. Many of these opportunistic diseases, such as *P. carinii* pneumonia or Kaposi's sarcoma, are rarely seen in non-HIV-infected people. Others are regularly seen in the general population but HIV-infected individuals have a much higher incidence of the disease, 100–fold higher in the case of lymphoma, for example. Other symptoms of AIDS include a wasting syndrome and neurological abnormalities, described below. The symptoms of AIDS become progressively worse as the CD4$^+$ count continues to drop until virtually no CD4$^+$ cells are present. The decline in numbers of CD4$^+$ cells is accelerated by the increased replication of HIV during this terminal phase, when the immune system can no longer control viral infection (Fig. 5.15).

The lymph nodes are organs that trap invading pathogens and present them to immune cells. Large numbers of macrophages and CD4+ T cells are present in lymph nodes, and the nodes are sites of active HIV replication. During the progression of disease following HIV infection, the lymph nodes deteriorate. During the final stages of disease, the architecture of the nodes is completely destroyed and this loss of lymph node function contributes to the loss of immune function.

A wasting syndrome is characteristic of late-stage HIV infection. Bowel involvement results in diarrhea and malabsorption. The wasting syndrome is thought to be a symptom of HIV infection itself, but opportunistic infections of the gut probably exacerbate the symptoms in many individuals.

Most HIV-infected individuals suffer, sooner or later, from neurological disease. Some disease is caused by opportunistic pathogens, such as progressive multifocal leukoencephalopathy caused by JC virus (Chapter 6). However, two-thirds of infected individuals develop an encephalopathy induced by HIV infection itself that produces symptoms including dementia, motor and behavioral abnormalities, and seizures. These symptoms are referred to as AIDS dementia complex. HIV is present in the brains of infected individuals in macrophages and microglia, but does not infect neurons. The cause of the neuronal loss induced by HIV infection is unknown.

Spread of AIDS

HIV is spread sexually, through contaminated blood, and from mother to child. The virus is present in semen, both as free virus and in infected cells, and in vaginal secretions of infected people, and can be transmitted by either homosexual or heterosexual intercourse. The probability of a woman becoming infected during unprotected vaginal intercourse with an infected male is estimated at less than 1/50. The risk of a man becoming infected during heterosexual intercourse is less. The risk is much higher if genital lesions resulting from sexually transmitted diseases are present. The risk of infection is also much higher during receptive anal intercourse. The use of condoms reduces the risk of transmission considerably.

The virus can also be spread by means of contaminated blood, whether through blood transfusions, the use of contaminated products by hemophiliacs, by needle stick in health care workers, the sharing of needles by injecting drug users, or via tattoo needles. Blood transfusion was an important source of infection before the development of tests for the virus and remains a problem in developing countries where blood tests may not be regularly available. Similarly, many hemophiliacs became infected before the development of blood tests. Use of blood tests and screening of donors for risk factors have greatly reduced the risk of transmission of HIV through blood products in developed countries. However, transmission among drug users remains an important source of infection.

Untreated, infected women transmit the virus to a newborn child about one-fourth of the time. Transmission can occur during delivery or during breast feeding. The use of anti-HIV drugs has reduced transmission to infants in countries where the drugs are available.

At the end of 1999, an estimated 43 million people were living with HIV/AIDS, with about 1 million of those in North America (Table 5.7). The cumulative death toll from AIDS since the beginning of the epidemic reached 20 million in the first half of 2000. More than 5 million new infections occurred worldwide in 1999, and about 5.4 million people died of AIDS that year. In the absence of treatment, virtually all infected people go on to develop AIDS and die within 2 years of the appearance of the symptoms of AIDS.

The focus of HIV infection is sub-Saharan Africa, where 25 million people, or more than 5% of the population, are thought to be infected and where medical care is still primitive. In the 15 years from 1984 to 1999, the number of HIV infections increased dramatically (Fig. 5.17; see also Table 5.7). The extent of the problem is illustrated by the fact that in some large cities in sub-Saharan Africa, an estimated 20–40% of adults may be infected. The spread of HIV has resulted in a marked decrease in life expectancy, which has reversed a long period of increasing life expectancy as health care standards have improved (Fig. 5.18).

In other areas of the world, HIV continues to spread but has not reached the extreme levels found in parts of Africa (Table 5.7). In Latin America and in southern and southeast Asia, about 0.6% of the population is infected. In North America, western Europe, and Australia, estimates are that 0.3% of the population is infected (see Fig. 5.19 for a breakdown by state in the United States). Table 5.7 documents the spread of the virus between 1996 and 1999.

In North America, HIV has been spread primarily through homosexual contacts and by sharing of needles by drug users who inject drugs, but the frequency of heterosexual transmission is increasing. In developing countries, heterosexual contact is the primary mode of transmission. The result is that in Africa, the male/female infection ratio is approximately 1 to 1, whereas in the United States the male/female ratio is almost 4 to 1 because of the importance of male homosexual contacts in the spread of the virus in this country (Table 5.7). In the United States, the number of newly diagnosed cases of AIDS, especially in white homosexual men, has dropped because awareness of the disease has led to the more frequent use of condoms and a lowered incidence of multiple sex partners, and because more effective drug therapies have been introduced (Fig. 5.20). This trend may be changing for the worse, however, as the development of more effective drugs to

TABLE 5.7   Characteristics of the Global HIV/AIDS Epidemic

Geographical region	Number of people with HIV/AIDS			Deaths from HIV/AIDS			% Women	% Adult prevalence rate, 1999	Primary mode of transmission[a]
	1996	1999	% Change 1996–1999	1996	1999	% Change 1996–1999			
North America	750,000	900,000	20	61,300	20,000	−67	20	0.58	1) MSM 2) IDU and hetero
South America	1.3 million	1.3 million	0	70,900	48,000	−32	25	0.49	1) MSM 2) IDU and hetero
Caribbean	270,000	360,000	33	14,500	30,000	107	35	2.11	Hetero
Sub-Saharan Africa	14 million	24.5 million	75	783,700	2.2 million	181	>50	8.57	Hetero
North Africa and Middle East	200,000	220,000	10	10,800	13,000	20	20	0.12	1) IDU 2) Hetero
W. Europe	510,000	520,000	2	21,000	6800	−68	25	0.23	1) MSM 2) IDU and hetero
C. and E. Europe and Central Asia	50,000	420,000	740	1000	8500	750	20	0.14	1) IDU 2) MSM
South and Southeast Asia	5.2 million	5.6 million	8	143,7000	460,000	220	35	0.54	Hetero
East Asia and Pacific	100,000	530,000	430	1200	18000	1500	13	0.06	1) IDU and hetero 2) MSM
Australia and New Zealand	13,000	15,000	15	1000	<200	−80	10	0.13	1) MSM 2) IDU and hetero

Note: These data are through the end of 1999, from the web site http://www.unaids.org/epidemic_update/report/ At the end of 1999, 43.3 million people were living with HIV/AIDS; during 1999 there were 5.4 million deaths, 47% of the infected were women, and the overall prevalence rate in adults was 1.07%

[a]*Abbreviations:* IDU, injecting drug use; hetero, heterosexual transmission; MSM, sexual transmission among men who have sex with men.

treat the disease appears to have resulted in an increase in unprotected sexual activity.

The capability of HIV to spread rapidly is illustrated by its recent spread in Thailand. In Thailand, every male is subjected to compulsory military service and is tested for HIV infection at the age of 18 when called up for military duty. Several years ago HIV was almost unknown in Thailand, but within a very few years the disease became prevalent, infecting some 3% of the male population. Thailand is a country that is very tolerant of sexual activity and many young males visit prostitutes. Most of the prostitutes became infected with HIV and began to transfer the infection to many of their male partners, resulting in the observed explosion of infections in the population. In some provinces in northern Thailand, as many as 20% of 18-year-old males were infected with HIV as of a few years ago. Recently, due in part to an aggressive education campaign by the government and to the distribution of condoms, the proportion of infected inductees has begun to drop.

Prevention and Treatment of AIDS

The most successful method for the prevention of most viral diseases has been the development of vaccines. However, the development of a vaccine against HIV has been a very slow, frustrating, and to date unsuccessful process. This is perhaps not surprising in light of the fact that the virus establishes a persistent lifelong infection in the face of a vigorous immune response on the part of the host that succeeds in controlling the virus for many years. Vaccine development continues to be a priority, but to the current time only public education to slow the spread of the disease and the development of antiviral drugs have had any success in control of the virus.

Much effort has been made to reduce the epidemic spread of HIV by educating the public and thereby altering patterns of behavior, especially sexual behavior. AIDS is a significant health problem in the United States, particularly in urban areas, and has become one of the major causes of death for persons in their most productive years (Fig. 5.21). However, safe sex practices, such as the use of condoms, have lowered the numbers of new infections, especially for HIV infections among gay white men. Education campaigns in Thailand and Uganda have succeeded in slowing the spread of the disease, and similar campaigns in other areas of the world have met with at least some success. However, although such campaigns slow the spread of the virus, they do not stop it and will never succeed in eliminating the virus from the population.

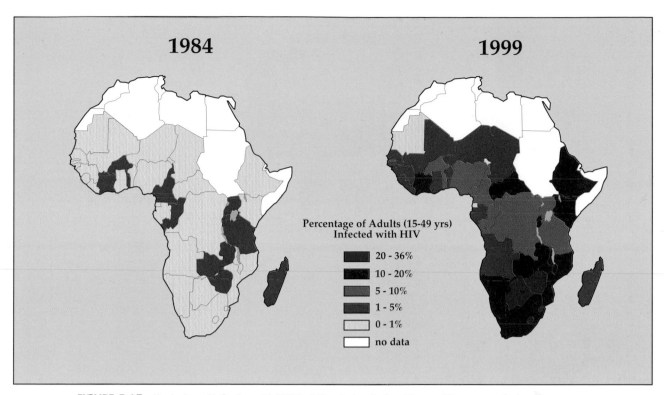

FIGURE 5.17 Explosion of infection with HIV in Africa during the last 15 years. The two panels show the percentage of adults (15–49 years old) infected with HIV in various countries of Africa in 1984 and in 1999. [Map is from Nations Programmes on HIV/AIDS Conference in Geneva, June 2000; reported in Schwartlander, B., Garnett, G., Walker, N., *et al.* (2000) *Science* **289**; 64–67.]

The development of drug therapies for viral diseases has met with partial success. Early drugs for treatment of HIV infection were disappointing at best. The drug 5′ azidothymidine (AZT) and related nucleoside analogues were first used in monotherapy, and were initially successful in decreasing virus load and allowing partial recovery of immune function. Incorporation of these analogues into DNA results in premature chain termination, and because the viral RT has a high affinity for these drugs whereas cellular enzymes have a low affinity for them, their primary effect is to disrupt replication of HIV. However, virus mutants that are resistant to the drugs rapidly appear, and control by this agent was found to be of limited duration. Second- generation inhibitors have now been developed that interfere with the activity of the viral protease, which is required for assembly of infectious virus. Variants resistant to these inhibitors appear much more slowly, probably because mutliple changes in the protease are required for resistance. However, resistant virus does eventually emerge when the inhibitor is used in monotherapy.

Third-generation drug therapy uses combinations of two different nucleoside analogues and a protease inhibitor. This treatment has been quite successful in more than half of HIV patients, in which the viral load declines to undetectable levels with full recovery of immune function. The introduction of combined therapy has led to a marked reduction of the death rate from AIDS in the United States (Figs. 5.20 and 5.21) and other developed countries. At first it was hoped that long-term treatment with this regime would result in the elimination of the virus and curing of the infection. However, results to date suggest that this may not be possible and that the therapy will have to be continued over the lifetime of the patient. To date, resistant variants have not arisen, presumably because of the very low

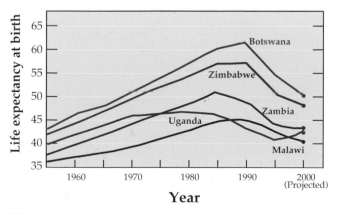

FIGURE 5.18 Changes in life expectancy in selected African countries with high HIV prevalence. [From UNAIDS publications, slide 12 of the "AIDS in Africa" series on the web site http://www.UNAIDs.org.]

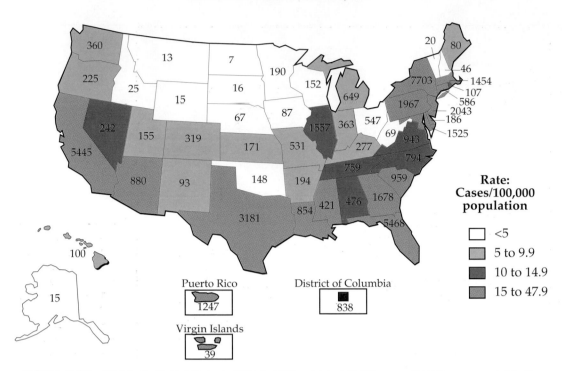

FIGURE 5.19 AIDS in the United States in 1998. In 1998, a total of 46,521 cases of AIDS were reported in the United States. In this map, the rate (cases per 100,000 population) is indicated by the color, and the total number of reported cases is given for each state. The rate for District of Columbia is so far off scale (189.1) that a special color is used. [Data from Summary of Notifiable Diseases in the United States for 1998. *MMWR* **47**; No. 53 (1999).]

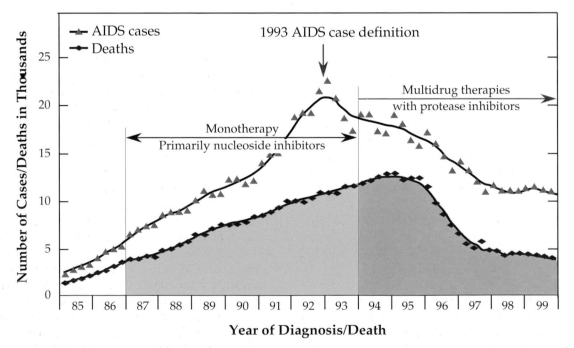

FIGURE 5.20 Estimated incidence of AIDS by quarter-year of diagnosis and number of deaths by quarter-year of death, among persons >13 years old in the United States from 1985 to 1998. [From *MMWR* **48**; Suppl. RR-13, p. 3 (1999).]

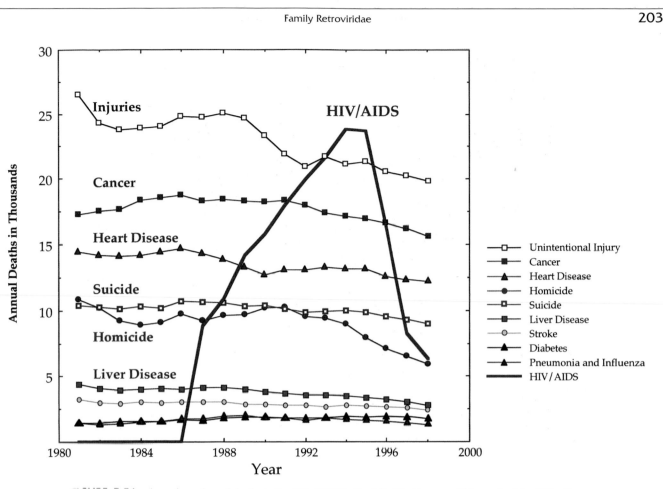

FIGURE 5.21 Annual number of deaths in the United States in adults 25–44 years old for each of the 10 leading causes of death, normalized for population growth. The data shown are the aggregate for all races and both sexes. [Data from CDC, National Center for Injury Prevention and Control.]

rate of replication of the virus under these conditions, but it remains possible that such variants will ultimately arise.

Although this therapy has been successful in many patients, it is clearly not a panacea. First, it is very expensive and suitable for use only in developed countries that can afford a very high level of health care. Second, it requires taking dozens of pills a day at precisely timed intervals, and compliance becomes a major issue when the therapy must be continued indefinitely. It may be for this reason that the therapy fails in a significant fraction of HIV patients. However, although this treatment is quite complex and of limited success, its very existence seemingly has led to a rise in unprotected sexual activity and to an upsurge in the rate of HIV infection among populations at risk.

Attempts to develop a vaccine continue and there is hope that a vaccine may ultimately be produced. It is clear that an immune response that involves both neutralizing antibodies and CD8+ T cells is important in the control of the virus during the clinically latent state. The T-cell response seems to be especially important. Stronger CTL responses result in lower levels of virus production during the latent state, which results in a slower rate of progression to AIDS. Neutralizing antibodies are also important, but it has been found that fresh isolates of virus from patients are difficult to neutralize, presumably because of the many carbohydrate chains attached to the HIV surface glycoprotein and because of the unusual folding properties of this protein. In addition, variants arise during prolonged infection that are resistant to the mix of antibodies that exist in the patient at the time that the variants arise.

In developing a vaccine, there is the significant question as to whether sterilizing immunity, which is very difficult to achieve is required, or whether protection from disease can be obtained with nonsterilizing immunity. For virtually all other viruses for which vaccines have been developed, the vaccine is nonsterilizing—subsequent infection with the virulent virus results in a subclinical infection that is quickly damped out. But with a virus that produces a lifelong persistent infection in the face of an immune response, the needs may be different. Very recent findings are encouraging that nonsterilizing immunity may suffice. Eight people were treated with anti-HIV drugs very early after infection and then taken off the drugs. Five of these patients may have cleared the virus—they have no detectable virus 8–11 months after stopping therapy. The other three are back on drug therapy. These results are very preliminary but suggest that it may be possible for the immune system to clear the virus if it gets a significant jump early in infection.

A second, related question is whether is it possible to boost the immune response in people infected with HIV and obtain better control or elimination of the virus. Because of the very long latent period before disease develops, this is an extremely difficult question to approach. However, a number of clinical trials are under way to test such possibilities, a topic that is covered in Chapter 9. Clinical trials have also been conducted that test the responses of uninfected people to immunization with the external glycoprotein of HIV. Immune responses have been generated, but it is unknown whether they are protective.

The possibility of a live virus vaccine is suggested by the finding that a small fraction of infected people exists who have not progressed to AIDS after 10–15 years of infection and whose CD4+ T-cell count has remained fairly stable. In at least some of these individuals, the infecting HIV strain has deletions in *nef*. Intriguingly, experiments with *nef* deletions in SIV in monkeys give comparable results—the monkeys do not develop AIDS and they are protected from infection by wild-type strains of SIV. However, any attempt to develop a live virus vaccine for a virus that establishes a persistent infection that is ultimately fatal, and for which the symptoms of disease are delayed for many years, faces formidable obstacles.

The Origins of HIV

A phylogenetic tree of the primate lentiviruses is shown in Fig. 5.22. There are two lineages of HIV-1, both of which are related to SIV isolated from chimpanzees. Of the two lineages of HIV-1, the M lineage, represented by YBF30 and U455(A) in the figure, is found worldwide and is responsible for the majority of human infections. The O lineage, represented by ANT70C, is found only in western Africa and in France. It is not known if these two lineages represent two different introductions of HIV into humans or whether they represent a single introduction that has subsequently diverged. Very few isolations of SIV from chimps have been made, and the prevalence of SIV in chimps is not known. In fact, it has not been demonstrated that SIV$_{cpz}$ is a natural infectious agent in wild chimp populations.

HIV-2 represents a distinct lineage that is closely related to SIV of sooty mangabey monkeys (smm) and of macaques (mac). SIV of African green monkeys (agm) and of mandrills (mnd) form other lineages that are more closely related

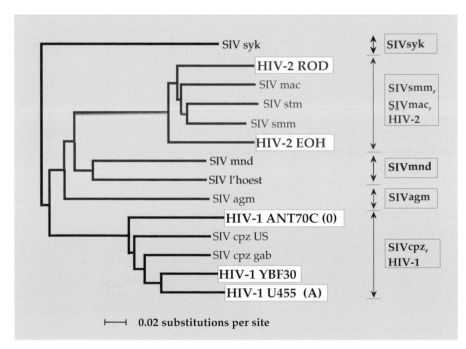

FIGURE 5.22 Phylogenetic tree of the primate lentiviruses, constructed using the neighbor-joining method on selected SIV and HIV *pol* sequences. Horizontal branch lengths are to the scale shown. HIV strain names are arbitrary. SIV names include the name of the host from which they were obtained: syk, Sykes monkey; smm, sooty mangabey; mac, rhesus macaque; mnd, mandrill, l'hoest, l'hoest monkey; agm, African green monkey; cpz, chimpanzee. The boxes at the right give the names of the five major lineages of primate lentiviruses identified to date. [Redrawn from Whetter *et al.* (1999, Fig. 1) and Coffin *et al.* (1997, p. 563).]

to HIV-2 than to HIV-1. It is clear that SIV$_{smm}$ and SIV$_{agm}$ are naturally occurring infectious agents that are widespread in Africa and have coevolved with their monkey hosts. Sequence comparisons have shown that different isolates of SIV$_{agm}$ group with their hosts rather than by geography, and they are therefore adapted to their hosts. They cause no disease in their natural host, but SIV$_{smm}$ does cause AIDS when transferred to Asian macaques in captivity. HIV-2 is found primarily in western central Africa, where its distribution is almost coincident with that of mangabey monkeys. It seems clear that HIV-2 represents a separate introduction of SIV from a different monkey host into humans.

Thus, SIV must be the source of HIV, with different viruses serving as the source of HIV-1 and HIV-2. When the viruses crossed the species barrier and became established in man as HIV is not clear. HIV-1 has been isolated from serum collected in 1959 in Zaire and antibodies to HIV have been found in serum collected in 1963 in Burkino-Faso, so HIV-1 has been in the human population at least that long. From sequencing studies of the glycoprotein gene and examination of the rate of divergence, one estimate is that the virus might have entered the human population about 70 years ago, although estimates of divergence rates are controversial. A likely scenario is that the viruses have been introduced into human populations many times in the past but failed to become epidemic. Infections in humans either died out because of the low transmissibility of the virus or smoldered in only a small fraction of the population. The different lineages of HIV-1 and, especially, of HIV-2 suggest, in fact, that they may result from independent introductions into humans, consistent with the hypothesis that multiple introductions of both viruses have occurred. Recent changes in human behavior, including more extensive travel by truck, bus, and plane, changes in sexual practices, and the use of injectable drugs, as well as the increase in the human population, could have allowed the virus to spread more extensively than in the past and to become epidemic worldwide. The spread of the virus could also have been aided by the appearance of mutants that were more easily transmissible from person to person. The large increase in population during the last century has certainly resulted in more opportunities for the introduction and spread of the virus in humans, and therefore for the selection of such transmissible mutants.

Nonhuman Lentiviruses

SIV, FIV, and BIV cause AIDS or an AIDS-like disease in animals. SIV causes AIDS in Asian monkeys but not in African monkeys, which are its natural hosts. The disease appears very similar to human AIDS, and SIV has been useful in laboratory studies as a surrogate for human AIDS. The primary receptor for SIV is CD4, as is the case for

HIV. FIV causes immunodeficiency in cats that appears to result by mechanisms similar to those that produce human AIDS, although the primary receptor for the virus is different. The number of CD4$^+$ T cells declines markedly in the late stages of the disease and the disease is characterized by opportunistic infections as well as by neurological symptoms. BIV infection results in persistent killing of lymphocytes and lesions in the central nervous system, symptoms that resemble those produced by HIV.

In contrast, equine infectious anemia virus (EIAV) in horses, caprine arthritis-encephalitis virus (CAEV) in goats, and visna in sheep produce quite different symptoms. EIAV causes recurrent anemia in horses and is an important veterinary problem. Primary infection by EIAV results in an acute disease that is soon controlled by an immune response. However, variants arise that are resistant to preexisting antibody and cause recurrences of fever and acute anemia. This continues for 6 months to 1 year, after which the infection becomes clinically inapparent. Infection is lifelong, however, as in all lentiviruses. It is thought that the virus binds to, but does not infect, red blood cells. These cells are then destroyed by a complement pathway or by engulfment by macrophages, resulting in the anemia. The virus also appears to suppress the differentiation of precursors to red blood cells. Although anemia is the characteristic disease produced by EIAV, in a small number of horses virus infection results in encephalomyelitis.

CAEV infection of goats produces an inflammatory response that often results in arthritis. The arthritis arises several years after infection and resembles rheumatoid arthritis in humans. Other symptoms produced by infection include encephalomyelitis in 20% of infected animals, which arises within 6 months of infection. Visna virus infection of sheep follows a similar course, but the characteristic disease produced is pneumonia, which arises several years after infection. Encephalomyelitis occurs in less than 5% of infected animals.

Spumaviruses

The spumaviruses comprise a genus of complex retroviruses. There are no known human spumaviruses. A virus previously isolated from human cells in culture, and referred to as human foamy virus or human spumaretrovirus, is now thought to be a virus of chimpanzees or other monkeys. The spumaviruses group with the retroviruses (Fig. 5.1), but differ in many properties from other retroviruses. The Gag polyprotein is not cleaved to produce MA, CA, and NC proteins. Spumaviruses appear to bud into the endoplasmic reticulum, rather than from the plasma membrane as do other retroviruses. In addition, Pol appears to be translated from its own mRNA, rather than from the genomic RNA as a fusion protein with Gag. Recent studies have also indicated that the

RNA of human spumavirus is converted to DNA during packaging of the virion. It is possible that virions may contain either DNA or RNA, since both are found in virus preparations. The packaging of DNA in the virion rather than RNA, the translation of Pol from its own mRNA, and budding into the endoplasmic reticulum, resemble the events that occur in the pararetroviruses such as the hepadnaviruses, described below. In other respects, the spumaviruses resemble retroviruses rather than pararetroviruses. They use a tRNA primer for reverse transcription, like the retroviruses, and their DNA integrates into the host genome as a provirus. Thus, they are retroviruses that are intermediate in their properties between the classical retroviruses and the pararetroviruses.

Retroelements in the Genomes of Living Organisms

Reverse transcriptase is an ancient enzyme. Its origins probably go back to the RNA world at the time of the invention of DNA, when such an enzyme would have been useful for converting informational RNA molecules into DNA, the form in which genetic information is now stored. It appears to resemble the RNA polymerases of RNA viruses more than DNA polymerases or DNA-dependent RNA polymerases, and may have evolved from a primordial RNA polymerase of the RNA world. The enzyme still exists in modern eukaryotes and plays a role in the replication of eukaryotic cells. Telomerase is a eukaryotic enzyme that repairs the ends of the linear chromosomes, which are progressively shortened during replication. Cells lacking

telomerase cease to divide after the ends of the chromosomes become too short to support replication. Telomerase is a ribonucleoprotein enzyme that synthesizes repeat sequence elements that are attached to the end of the chromosomes. These elements are specific for a given species. This process involves reverse transcription of an RNA template that is a component of the telomerase, and the ends of the chromosomes serve as primers. The composition of telomerase has not been completely elucidated, but it does have a reverse transcriptase activity that is related to the RT present in a variety of retroelements.

There has been a long period of time in which elements that utilize reverse transcription for insertion into eukaryotic and prokaryotic chromosomes have been able to evolve. The number of different kinds of retroelements is consequently very large and encompasses endogenous retroviruses, retrotransposons of various types, retrointrons, retroplasmids, and retrons, among others. A partial listing of these elements is given in Table 5.8, and the phylogenetic relationships among the RTs of those elements that possess RT are shown in the tree in Fig. 5.23.

As a class, retroelements are found in all organisms from prokaryotes to mammals. Many of these elements are capable of amplification and are mobile, able to move within the genome. Some encode RT, whereas others utilize RT of other elements for their propagation. These elements can be thought of as selfish DNA. They replicate to fill up the host genome to what is, in essence, its carrying capacity. These elements must be benign and have only limited effects on the replication potential of the organisms in

TABLE 5.8   Retroelements and Their Distribution in Nature

Class	Distribution
Eukaryotic retroelements	
Retroviruses	Vertebrates
Pararetroviruses	
Hepadnaviruses	Mammals, birds
Caulimoviruses	Plants
Retrotransposons	
LTR retrotransposons	Animals, plants, fungi, protozoa
Non-LTR retrotransposons	Animals, plants, fungi, protozoa
Mitochondrial elements	
Group II Introns (retrointrons)	Mitochondria of fungi and plants, plastids of algae
Mauriceville plasmid	Mitochondria of neurospora
RTL gene	Mitochondria of chlamydomonas
Prokayotic retroelements	
msDNA-associated RT[a]	*Myxococcus xanthus, E. coli*, other bacteria

Source: Adapted from Eickbush (1994), Table 1, p. 122.
[a]msDNA, multicopy single-stranded DNA.

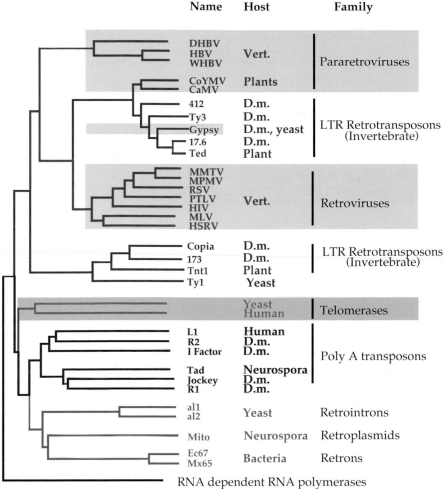

Name	Host	Family
DHBV HBV WHBV	Vert.	Pararetroviruses
CoYMV CaMV	Plants	
412 Ty3 Gypsy 17.6 Ted	D.m. D.m. D.m., yeast D.m. Plant	LTR Retrotransposons (Invertebrate)
MMTV MPMV RSV PTLV HIV MLV HSRV	Vert.	Retroviruses
Copia 173 Tnt1 Ty1	D.m. D.m. Plant Yeast	LTR Retrotransposons (Invertebrate)
	Yeast Human	Telomerases
L1 R2 I Factor Tad Jockey R1	Human D.m. D.m. Neurospora D.m. D.m.	Poly A transposons
al1 al2	Yeast	Retrointrons
Mito	Neurospora	Retroplasmids
Ec67 Mx65	Bacteria	Retrons

RNA dependent RNA polymerases

FIGURE 5.23 Phylogenetic tree of the retroelements, based on the sequences of the reverse transcriptases. The blue boxes enclose the retroviruses and pararetroviruses. (Gypsy and Ty3 are known to be infectious viruses and are provisionally classified in a new family, Metaviridae.) The hosts are shown; plant hosts are shown in green. The vertical distances are arbitrary; the length of the horizontal branches is proportional to the number of changes. The tree has been rooted with the RNA-dependent RNA polymerase sequences. Abbreviations of retrovirus names were given in Table 5.1; abbreviations of vertebrate pararetrovirus (Hepadnavirus) names are found in Table 5.9. CoYMV, commelina yellow mottle (Badnavirus); CAMV, cauliflower mosaic (Caulimovirus); d.m., *Drosophila melanogaster* (fruit fly). [Redrawn from Coffin *et al.* (1997, Fig. 17, p. 411).]

which they reside. Otherwise their host would be selected against during evolution.

The number of retroelements in eukaryotic genomes is often very large. In mammals, 5–10% of the genome appears to consist of elements that were introduced by reverse transcription. Of these, approximately one-tenth are provirus-like, containing LTRs and primer binding sites flanking regions with detectable relationships to *gag* and *pol*. Some of these are active endogenous viruses that have inserted into the genome recently, whereas others are inert. The time at which any element entered the germline can often be estimated from comparative studies of these elements in different animals, and these proviruses or other

retroelements constitute a kind of fossil record for these elements. As one example, chickens have one to four endogenous proviruses closely related to the ALVs, but there are no endogenous viruses in turkey or quail. Thus, insertion into the chicken genome occurred after chicken and turkey separated.

Below, four classes of eukaryotic elements, the endogenous retroviruses, the LTR-containing retrotransposons, the poly(A)-containing retrotransposons, and the group II retrointrons, are described in more detail. The relationships among these elements and their relationships to the infectious retroviruses described above, and to the pararetroviruses described below, are illustrated by the phylogenetic

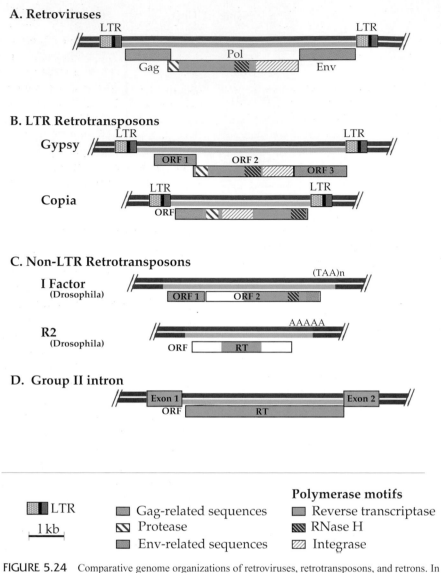

A. Retroviruses

B. LTR Retrotransposons

Gypsy

Copia

C. Non-LTR Retrotransposons

I Factor
(Drosophila)

R2
(Drosophila)

D. Group II intron

LTR

1 kb

Gag-related sequences
Protease
Env-related sequences

Polymerase motifs
Reverse transcriptase
RNase H
Integrase

FIGURE 5.24 Comparative genome organizations of retroviruses, retrotransposons, and retrons. In each case the integrated form of the element is shown, flanked with dark blue bars representing the host DNA. ORFs for each element are shown below the DNA, and boxes on different lines represent different reading frames. Note that Gypsy has been variously categorized as a retrotransposon and as a true endogenous *Errantivirus*, family Metaviridae. Various motifs related to retroviruses are indicated with different colors and shadings, as shown in the key. At the 3′ end of many non-LTR retrotransposons there are a variable number of TAA repeats (I factor) or adenylate residues (R2). [Adapted from Eickbush (1994).]

tree in Fig. 5.23. The genetic organizations of these elements are compared in Fig. 5.24.

Endogenous Retroviruses

Endogenous retroviruses are proviruses that have become established in the germ line of many different organisms at various times in the past. They align with the retroviruses in the tree in Fig. 5.23, and their genetic orga-

nization is identical to that of the simple retroviruses (see Fig. 5.24). However, many are defective, having deletions or mutations in them that prevent them from undergoing a full round of replication. Others are not defective and can, upon activation, give rise to progeny virus that can infect other cells. In most cases, endogenous retroviruses are silent—the genes that they encode are not expressed or are expressed only under restricted conditions. In some animals, however, one or more endogenous retroviruses are normally expressed during the lifetime of an animal.

The best studied endogenous retroviruses are those of chickens and of mice. Mice contain more than 1,000 endogenous viruses or elements that are thought to have originated from endogenous viruses. Most of these are defective and unable to undergo a complete cycle to give rise to infectious virus. However, up to 100 copies of non-defective endogenous retroviruses are present in some strains of mice. As described above, these viruses can be ecotropic, xenotropic, polytropic, or modified polytropic, depending on the receptor recognized by their envelope gene. In most strains of mice, these viruses are not expressed, but in many cases can be induced to replicate by treatment of the cell with mutagenic agents or agents that lead to demethylation of DNA. In some strains of mice, however, the endogenous viruses are transcriptionally active and expressed at some stage during the lifetime of the mouse. Early expression and vigorous replication in strains of mice such as AKR leads to the development of leukemia in most of these mice.

An interesting question deals with how the xenotropic viruses, which cannot infect mice, entered the mouse germline. The simplest explanation would be that these viruses entered the germ line of an ancestor of the modern mouse that did express receptors for the virus. In this model, during the evolution of mice these receptors were silenced or mutated to a form that was no longer usable by the xenotropic viruses. It is possible that modern mice were selected to not express a functional receptor for these viruses, in order to minimize the virus load from endogenous viruses.

LTR-Containing Retrotransposons

The LTR-containing retrotransposons are essentially endogenous retroviruses that lack the envelope gene (Fig. 5.24B). They contain LTRs and encode Gag-related proteins, protease, and RT with its associated activities. RNA transcribed from the elements is translated to produce Gag and Pol, and the RNA is assembled into intracellular capsid-like particles that reverse transcribe it back into DNA. Subsequent integration into the genome allows the elements to spread within the genome. Therefore, they are effectively intracellular viruses that replicate only within an individual cell.

The LTR-containing retrotransposons are obviously related to retroviruses. The LTRs that they possess are closely related to retroviral LTRs, including the conserved two terminal nucleotides of the provirus. They have tRNA primer binding sites and polypurine tracts, so that reverse transcription of the RNA appears to follow the same pathways as in retroviruses. These elements could be degenerate proviruses that have lost the envelope gene. Alternatively, they could represent an ancestral form of retroviruses in which an RT-containing element became mobile by associating with *gag*-like genes

that allowed it to spread within the cell. Later acquisition of an envelope gene would then give rise to a complete retrovirus.

Several lineages of LTR retrotransposons, or of elements that resemble them, are known. Two lineages have been found in fungi, plants, and invertebrates (Fig. 5.23), which represent, therefore, two independent evolutionary lines. One lineage contains Ty3 and Gypsy and is related to the pararetroviruses of plants. The second lineage contains Ty1 and Copia. The Ty3/Gypsy lineage is more closely related to retroviruses than it is to the Copia lineage (Fig. 5.23). There are also elements in mammals that resemble LTR retrotransposons, but these may belong to a distinct lineage that arose more recently.

Poly(A) Retrotransposons

Non-LTR-containing retrotransposons, often called poly(A)-containing retrotransposons because many have a poly(A) tract or an A-rich tract at the 3′ end, are a very large family of elements found in virtually all eukaryotes. Many of these elements encode RT, and of these many encode Gag-like proteins. The structures of two elements from *Drosophila melanogaster* are illustrated in Fig. 5.24C. Both encode RT. The I factor element encodes Gag-like proteins, whereas R2 encodes proteins that bind nucleic acid but are not obviously related to Gag. The elements that encode RT use it to move by reverse transcription, but because they lack LTRs, the mechanism of reverse transcription is different.

The human genome contains many non-LTR-containing retrotransposons. About 20% of the genome, in fact, consists of repeated elements called LINEs (*long interspersed nuclear elements*, often abbreviated L1). L1 belongs to the same lineage as the *Drosophila* elements R2 and I factor (Fig. 5.23). L1 elements encode Gag-like proteins and RT, but lack LTRs as well as an envelope gene. It is thought that reverse transcription is coincident with integration of L1 into a new location. In this model, the cell chromosomal DNA is nicked by an endonuclease encoded in the retrotransposon, and the nick site is used as a primer for reverse transcription of the L1 RNA.

In addition to the L1–like retrotransposons, a simpler class of poly(A) transposons exists that does not encode RT. This class includes elements in the human genome called SINEs, for *short interspersed nuclear elements*. SINEs include the human ALU sequences present in large numbers in the genome. It is thought that these elements probably borrow the transcription machinery of the L1–type elements during retrotransposition.

Group II Retrointrons

Group II introns are self-splicing introns that encode RT. The structure of such an intron is shown in Fig. 5.24D.

These introns are able to move and may have been the source of introns in nuclear genes. They are mostly found in prokaryotes and in organelles. The RT is translated as a fusion protein from unspliced RNA. Thus, the amount of RT produced, which determines the ability of the element to move, is regulated by the efficiency of splicing. The retrointrons group with the retroplasmids and retrons in the RT tree to form a distinct lineage (Fig. 5.23).

Effects of Retroelements on the Host

In many ways, retroelements are not that different from extracellular viruses that have learned to infect organisms and pass from organism to organism. These elements multiply to fill an ecological niche, but are limited to intracellular spread. However, these retroelements, as noted above, must be benign in order to avoid being selected against. The ability of retroelements to cause disease appears to be quite limited, and in many organisms they may cause no disease. In at least a few cases, they may even serve a useful function.

The ability of endogenous viruses to cause leukemia in AKR mice or mammary tumors in some mice seems to contradict this observation. However, these diseases are largely characteristic of inbred laboratory mice. Wild mice control their endogenous viruses much more successfully.

There is no evidence that endogenous viruses ever cause disease in humans. Humans appear to have no endogenous proviruses that are ever expressed, although humans do have defective proviruses that have been present in the germ line for a long time. The absence of endogenous viruses may be a function of the long life span of humans. Long-lived animals have a longer time span to express endogenous viruses, which would lead to selection against animals that contain such viruses.

FAMILY HEPADNAVIRIDAE

The hepadnaviruses (*hepa* from *hepa*totropic, dna from their *DNA* genome) share with retroviruses the property of encoding RT and replicating via an RNA-to-DNA step. They package DNA in the virion, however. The process of reverse transcription shares features with that described for the retroviruses but also differs in many important details, as described below. Because of the similarities in their mode of replication to that used by retroviruses, the hepadnaviruses and the plant viruses that replicate via RT are referred to as pararetroviruses. The hepadnaviruses form a distinct taxon in phylogenetic trees, however, and are not particularly closely related to the plant pararetroviruses or to the retroviruses (Fig. 5.23). In fact, the plant pararetroviruses, even though they package DNA in the virion, appear to be more retrovirus-like in their replication than are the hepadnaviruses, consistent with their position in the tree.

The hepadnaviruses include three viruses of mammals: hepatitis B virus of primates (HBV), woodchuck hepatitis virus (WHV), and ground squirrel hepatitis virus (GSHV); and two viruses of birds: duck hepatitis B virus (DHBV) and heron hepatitis B virus (HHBV) (Table 5.9). Other viruses are also known but have not been well studied. The mammalian viruses are closely related and share extensive nucleotide sequence identity (e.g., HBV and WHV are 60% identical in nucleotide sequence). The bird viruses form a distinct lineage (Fig. 5.23). They are closely related to one another but more distantly related to the mammalian viruses. As their names imply, all of the known hepadnaviruses are hepatotropic, infecting liver cells, and all can cause hepatitis in their native host. All have a very narrow host range that may be determined by the identity of the receptors used for entry.

TABLE 5.9   Hepadnaviridae[a]

Genus/members	Abbreviation	Natural host(s)	Transmission	Disease
Orthohepadnavirus				
Hepatitis B virus	HBV	Humans, chimpanzees, gibbons, woolly monkeys	Horizontal, IDU, sexual, infected blood, vertical	ACS, hepatitis, cirrhosis, HCC
Ground squirrel hepatitis B virus	GSHV	Ground squirrels, woodchucks, chipmunks	Horizontal, sexual, blood	ACS, hepatitis, HCC
Woodchuck hepatitis B virus	WHBV	Woodchucks	Horizontal, sexual, blood	ACS, hepatitis, HCC
Avihepadnavirus				
Duck hepatitis B virus	DHBV	Ducks Geese	Predominantly vertical	ACS, hepatitis
Heron hepatitis B virus	HHBV	Herons	Predominantly vertical	

Source: Adapted from Fields *et al.* (1996, Table 1, p. 2708).
Abbreviations: ACS, asymptomatic carrier state; HCC, hepatocellular carcinoma; IDU, injecting drug users.
[a]Hepatitis B virus has a worldwide distribution in humans, as show in Fig. 5.28A.

A. Genome organization

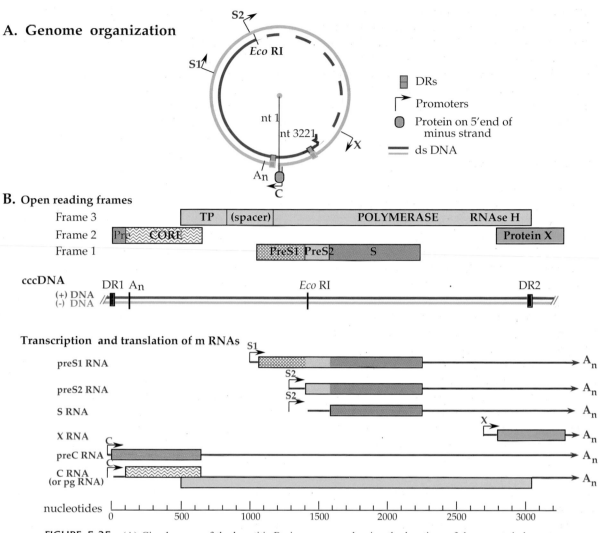

B. Open reading frames

Frame 3 — TP | (spacer) | POLYMERASE | RNAse H
Frame 2 — PreC | CORE | Protein X
Frame 1 — PreS1 PreS2 S

cccDNA
(+) DNA / (-) DNA — DR1 A$_n$ — Eco RI — DR2

Transcription and translation of m RNAs

preS1 RNA — S1 → A$_n$
preS2 RNA — S2 → A$_n$
S RNA — S2 → A$_n$
X RNA — X → A$_n$
preC RNA — C → A$_n$
C RNA (or pg RNA) — C → A$_n$

nucleotides — 0 | 500 | 1000 | 1500 | 2000 | 2500 | 3000

FIGURE 5.25 (A) Circular map of the hepatitis B virus genome showing the locations of the repeated elements (DR1 and DR2), the four known promoters (black arrows), and the gap of variable length in the plus strand of the DNA. This map is numbered from the beginning of the pgRNA. Some authors number the nucleotides from the unique EcoRI site (here at nt 1407). (B) Linearized map of cccDNA showing the open reading frames. Below this are shown the six mRNAs and the proteins (colored blocks) translated from them. Note that the pC RNA and C RNA (pgRNA) are of more than genome length, and that all mRNAs end at the polyadenylation site at nt 121 (but have been shown extended to the right for clarity). All transcripts are made from the (–) DNA (light blue) as template, in a clockwise direction on the circular map, or left to right in the linearized map. [Data for constructing this diagram come from Yen (1998), Hu and Seeger (1997), and Fields et al. (1996, p. 2706); details provided by Dr. James Ou.]

The hepadnaviral genome is circular and approximately 3.2 kb in size, as illustrated in Fig. 5.25A. It consists of DNA that is mostly, but not completely, double stranded. One DNA strand, the minus strand, is unit length and has a protein covalently attached to the 5′ end (the P protein). The other strand, the plus strand, is variable in length, but less than unit length, and has an RNA oligonucleotide at its 5′ end. Thus neither DNA strand is closed and circularity is maintained by cohesive ends.

Hepadnavirus virions are enveloped and about 42 nm in diameter. The nucleocapsid or core of the virion contains a major core protein called the HBV core antigen, abbreviated HBcAg. The external glycoproteins are called HBV surface antigens or HBsAg. Budding is through internal membranes.

In the description here, the focus is on HBV, but DHBV has been important for working out the mechanisms of replication. None of the established cell lines support the complete infection cycle of any hepadnavirus, making study of virus replication difficult. Much of what we know comes from studies of infected liver in experimental animals or studies in explanted primary hepatocytes. DHBV replicates well in primary duck hepatocytes, but the mammalian viruses replicate poorly in explanted hepatocytes. Interestingly, a number of hepatoma cell lines will support viral replication if transfected with viral DNA. However, attempts to infect them with virus do not result in replication, for reasons that are not clear.

Transcription of the Viral DNA

After infection of a cell, the viral genome is transported to the nucleus and converted to a covalently closed, circular, dsDNA molecule, called cccDNA. In this process, the P protein attached to the minus strand is removed, as is the RNA oligonucleotide at the 5′ end of the plus strand, gaps are filled in, and the ends of the DNA strands are closed. Host repair enzymes are assumed to carry out this process. The resulting cccDNA does not integrate into the host genome nor does it replicate as an episome; rather it is maintained as a single copy of circular DNA. Note that a primary site of replication of the virus, and the cells in which most of the studies of replication have been conducted, are terminally differentiated hepatocytes which divide only rarely and in which there is no ongoing DNA synthesis. Thus, the virus has evolved other means for amplification of its genome.

The cccDNA is transcribed by cellular RNA polymerase II to produce several mRNAs (Fig. 5.25B). Only one strand is transcribed. Four different promoters in the DNA of the mammalian viruses lead to the production of unspliced transcripts of lengths 3.5 kb (i.e., slightly greater than unit length), 2.4, 2.1, and 0.7 kb, all of which terminate at the same poly(A) addition site. More than one start point is used in the case of two of the promoters, and from these two promoters, RNAs with different 5′ ends are transcribed that serve different functions, as illustrated in Fig. 5.25B. The RNA transcripts are capped and polyadenylated. The polyadenylation signal in the mammalian hepadnaviruses is TATAAA rather than AATAAA, and the use of this suboptimal signal appears to require viral sequences upstream of this site.

Transcription of viral RNA is most efficient in hepatocytes. At least some of the promoters require transcription factors such as hepatocyte nuclear factor 1, present primarily in hepatocytes, for optimal activity. Furthermore, at least two enhancer sequences are known to be present in the DNA that function most efficiently in hepatocytes. In addition to the various cellular factors, the X gene product encoded in the mammalian viruses, also upregulates transcription of viral DNA.

The four classes of mRNAs are exported to the cytoplasm for translation and assembly of virions. Export is facilitated by a sequence element of about 500 nucleotides called the post-transcriptional regulatory element (PRE). This element is required because the major hepadnaviral RNAs are not spliced. Thus, PRE is functionally analogous to RRE of HIV or CTE of the simple retroviruses, but it is not known if the mechanisms by which these elements effect export are the same.

Synthesis of Viral Proteins

Synthesis of the viral proteins is complex. Four genes in mammalian viruses and three in avian viruses are usually recognized. The core gene gives rise to two products called precore (preC) and core (also referred to as HBcAg or simply C). The polymerase gene gives rise to RT (usually called the polymerase or P). The surface protein gene gives rise to three proteins called preS1, preS2, and S (also called HBsAg). The X gene and its encoded X protein are present only in the mammalian viruses. Four classes of mRNAs, produced by initiation at the four promoters, are produced, which form an overlapping set (Fig. 5.25B).

Both C and P are translated from the largest (3.5-kb) mRNAs, which are slightly longer than unit length. Two mRNAs are produced starting at this promoter. One mRNA is slightly longer and is translated to produce the protein called preC. The shorter form of the 3.5-kb mRNA, which is also called pgRNA, lacks the AUG used to initiate translation of preC. A downstream AUG in this mRNA, which is in the same reading frame as preC, is used to initiate translation of the protein called C, which is the major capsid protein of the virion. PreC has a different fate. It is inserted into the endoplasmic reticulum during synthesis and transported through secretory vesicles, undergoing cleavages to remove an N-terminal signal sequence and some C-terminal residues. It is secreted from the cell as a 17-kDa protein called HBeAg. HBeAg may be important for the establishment of a chronic infection in infants (see below). HBcAg, HBsAg, and HBeAg are all used as clinical markers of infection and virus replication. In general, HBeAg is associated with more aggressive clinical hepatitis.

The gene for P is downstream of C (Fig. 5.25B). It is in a different reading frame than C and partially overlaps C. The mechanism by which translation of P is initiated is not yet resolved. It is translated from the same mRNA as is C (i.e., pgRNA), but initiation is internal, using the start AUG of P, rather than being produced by some form of frame shifting and cleavage. Internal initiation does not appear to use an IRES but appears to be cap-dependent, and some form of ribosome scanning has been invoked in order to position the ribosome at this start site. This process is inefficient, and about 200 copies of C are produced for each copy of P.

Three forms of S are produced, a long version called L or preS1, a medium size version called M or preS2, and a short version S (Fig. 5.25B). These differ only at the N termini and are produced by using different in-frame AUG initiation codons. PreS1 is translated from the 2.4-kb mRNA, whereas preS2 and S are translated from two forms of the 2.1-kb mRNA in a manner similar to preC and C.

X is translated from the 0.7-kb mRNA. As described, it is absent in the avian hepadnaviruses.

The viral genome is very compact (Fig. 5.25). Over half of it is translated in two reading frames. The P gene requires about three-quarters of the coding capacity of the genome and this gene overlaps each of the other three genes. The S gene is completely contained within the P gene.

Replication of the Viral Genome

The 3.5-kb pgRNA serves not only as a messenger but also as an intermediate in viral genome replication. This process is illustrated in Fig. 5.26. Protein P (which has RT activity) uses this RNA as a template to make the (partially) dsDNA copy found in the virion. As is the case for retroviruses, transfer of initiated complexes from one end of the genome to the other occurs twice during reverse transcription. Unlike retroviruses, however, the primer for first-strand synthesis is not a tRNA but protein P itself, which remains covalently attached to the 5′ end of the first strand.

DNA synthesis takes place in capsids and the first step is therefore the encapsidation of the pgRNA. Only pgRNA is packaged. This RNA is the messenger for both C and P, and both C and P are required for encapsidation. P binds to a specific sequence called ε present in the 5′ region of the RNA (Fig. 5.26, step 1). The signal is found within a stem-loop structure present within direct repeats of about 200 nt at the two ends of the RNA, which is illustrated in Fig. 5.27. Interestingly, although ε is present at both the 5′ and 3′ ends of the RNA, only the 5′ signal functions for encapsidation. Once P binds to ε, C is recruited and the capsid assembles. In the absence of P, C assembles into capsids that package RNAs randomly. Thus, the specificity in packaging of viral RNA lies in the interaction of P with the RNA, unlike most viruses, including the retroviruses, in which it is the capsid protein that recognizes a packaging signal in the viral genome.

In the nucleocapsid, first-strand synthesis is initiated by using the –OH group of a specific tyrosine in P as the primer. This tyrosine is present in an N-terminal domain that is distinct from the domain that constitutes the RT. Four nucleotides are added, copied from ε (Fig. 5.26, step 1; see also Fig.5.27). P with its covalently attached chain is then transferred to the DR1 acceptor site at the 3′ end of the RNA (which has a sequence complementary to the four nucleotides used to start DNA synthesis) (indicated by the red arrow in step 1). DNA synthesis of first strand then continues until the 5′ end of the RNA is reached (Fig. 5.26, steps 2, 3 and 4).

The RNase H activity of P degrades the RNA strand during synthesis of first-strand DNA, but the extreme 5′ end of the RNA is not degraded. This 5′ piece, which is capped and about 18 nucleotides long, is transferred to the DR2 acceptor site near the 5′ end of the first-strand cDNA (step 4), where it serves as a primer for second strand DNA synthesis (step 5). After reaching the 5′ end of the first-strand cDNA, continued synthesis of second- strand cDNA requires translocation to the 3′ end of the first strand, as is the case for retroviruses (step 6). In the case of hepadnaviruses, cyclization of the DNA occurs during this translocation, promoted by terminal redundancies. Second-strand synthesis is usually not complete so that the genomic double-strand DNA has a single strand gap, of variable length, in it (step 7).

In the model for DNA synthesis shown in Fig. 5.26, the terminal protein domain is cleaved from the polymerase domain after initiation of DNA synthesis. However, reports conflict about whether cleavage does or does not occur. If no cleavage occurs, the entire P protein remains covalently attached to the 5′ end of the DNA, rather than just TP (terminal protein). If P remains attached to the 5′ end of the minus strand, it will keep the end of the first-strand DNA with it at all times. This could simplify the second jump and the cyclization of the DNA, since DR2 would be close by.

Some copies of the resulting viral DNA genome are transported back into the nucleus to amplify the replication cycle. This results in the accumulation in the nucleus of about 20 copies of viral DNA to serve as templates for mRNA synthesis. This is thought to occur primarily early in infection, before accumulation of large amounts of S protein.

Assembly of the Virion

Cores, with their partially dsDNA, bud through intracellular membranes to produce mature virions called Dane particles. The 42-nm virions contain both L and S in their envelopes, with at least one-fourth as much L as S, and both L and S are required for virion assembly. In the virion, both L and S have both their N and C termini outside. Thus, they must span the membrane at least twice, and some models propose that they span the membrane four times. L is myristylated at the N terminus. Myristylation is required for the infectivity of the virus, but not for assembly. Thus, myristylation may serve a function in entry. Virions also contain M in amounts equivalent to L. However, M does not appear to be required for assembly or for infectivity of virions.

In addition to the 42-nm Dane particles, 20-nm particles are also produced in abundance (10^4- to 10^6-fold excess over Dane particles). These particles contain S and M but little or no L, and lack the core. These particles form when S alone is expressed in cells, and thus S has the ability to produce a bud in the absence of other viral components. Why the virus produces vast quantities of such particles, which can result in concentrations as high as 10^{13}/ml in serum, is not clear. However, these particles do express HBsAg, whose concentration can reach more than 100 μg/ml in serum, and could be produced to tolerize the immune system.

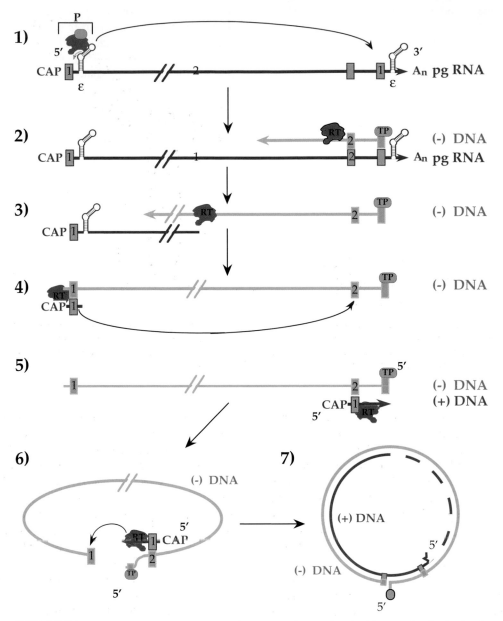

FIGURE 5.26 Mechanism of HBV DNA synthesis. Pregenomic RNA (pgRNA) is capped, polyadenylated, and greater than genome length (green line). It contains two copies each of DR1 and ε, the encapsidation sequence shown in Fig. 5.27. (Step 1) Priming of reverse transcription occurs when P protein (consisting of TP, RT, and RNase H domains) makes a tetranucleotide copy of the bulge in the ε structure. This tetranucleotide is covalently linked to P. The nascent DNA strand is then translocated to DR1 at the 3' end of pgRNA. (Step 2) TP (terminal protein) is cleaved from P (but see text for an alternative hypothesis) and remains attached to the 5' end of the minus strand, while the minus-strand DNA (light blue) is extended right to left by RT. (Step 3) During minus-strand DNA synthesis, RNase H activity degrades pgRNA until P (RT) reaches the 5' end of the template RNA. (Step 4) A short RNA oligomer is left annealed to a short terminal duplication. (Step 5) The RNA oligomer is translocated to DR2 where it primes plus-strand DNA synthesis (medium blue) left to right. (Step 6) During plus-strand elongation, a second template transfer circularizes the genome and plus strand is extended a variable length to give mature progeny viral DNA. [Modified from Locarnini *et al.* (1996) and Fields *et al.* (1996, p. 2719).]

HBV and Hepatitis in Humans

HBV has a pronounced tropism for hepatocytes, as do all hepadnaviruses, and causes hepatitis in humans. Infection of neonates or very young children is usually asymptomatic, but infection nevertheless has serious consequences. Infection of adults results in serious disease characterized by liver dysfunction accompanied by jaundice in

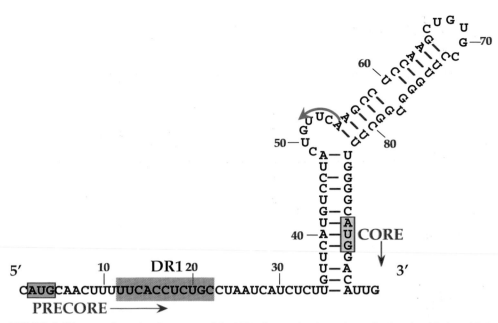

FIGURE 5.27   Two-dimensional structure of the RNA element that forms the packaging signal in hepatitis B RNA. Numbering is that for the sequence as presented in GenBank. The initiation codons for the precore and core proteins are marked, the *direct repeat* sequence (DR1) is shaded, and the blue arrow shows the initiation sequence for DNA replication (see Fig. 5.26). [Redrawn from Buckwold and Ou (1999, Fig. 2).]

about one-third of infections, although death from fulminant hepatitis is uncommon (the fatality rate is around 1% of acute cases). The incubation period is long, 45–120 days, and convalesence is usually extended (more than 2 months), but more than 90% of adults infected by HBV recover completely. In the United States, about 40,000 cases of hepatitis are reported annually, of which about one-third are due to HBV. It is reasonable to assume that the total number of new infections with HBV is perhaps 10-fold the number of reported cases of HBV-induced hepatitis.

In a small number of adult cases, less than 10%, the infection is not cleared and becomes chronic. Infection of neonates or young children results in high levels of chronicity, however. Up to 90% of neonates infected with HBV become chronically infected, and infection of 3-year-old children may result in 30% chronicity. It is thought that the immature state of the immune system in the very young is important in the development of a chronic infection. Chronic infection may remain asymptomatic and may even eventually clear in a small fraction of cases, especially if infection occurred as an adult. However, other patients develop chronic active hepatitis that may progress to cirrhosis and death.

HBV is spread primarily by contact with contaminated blood, by sexual intercourse, and from mother to child during delivery or breast feeding. Persistently infected individuals can have very high titers of virus in the blood, up to 10^{10}/ml, and the virus resists drying for up to 1 week. Thus, contact with infected blood need not be extensive to transmit the virus. It has been suggested that household contact leads to spread via sharing of razors, for example. Medical

personnel are at risk of contracting the virus from their patients, not only by needle stick, which is responsible for many cases in unvaccinated individuals, but through other contact with contaminated blood as well. The virus also spreads readily among institutionalized individuals. At one time, blood transfusion was a source of spread of virus, but with the development of sensitive assays for the presence of the virus, the risk of infection following blood transfusion in developed countries is now 1/200,000 per unit of blood.

Chronic infection acquired at birth is thought to be the major mechanism by which the virus persists in nature. Up to 90% of babies born to mothers who are acutely or chronically infected with HBV and positive for HBeAg will be infected by HBV, and most of these will become chronically infected. In the United States, there are an estimated 1.2 million carriers of HBV, and worldwide there are an estimated 350 million. The fraction of the population chronically infected with HBV varies from 0.1–0.5% in developed countries to 5–15% in Southeast Asia and sub-Saharan Africa. A map that illustrates the prevalence of HBV in different regions of the world is shown in Fig. 5.28.

HBV and Hepatocellular Carcinoma

Liver cancer causes more than 500,000 deaths a year worldwide, and about 90% of primary malignant tumors of the liver are hepatocellular carcinoma (HCC). HCC is more common in men than women, by 4 to 1, and is in the top 10 in frequency of cancers in humans. HCC is more common in

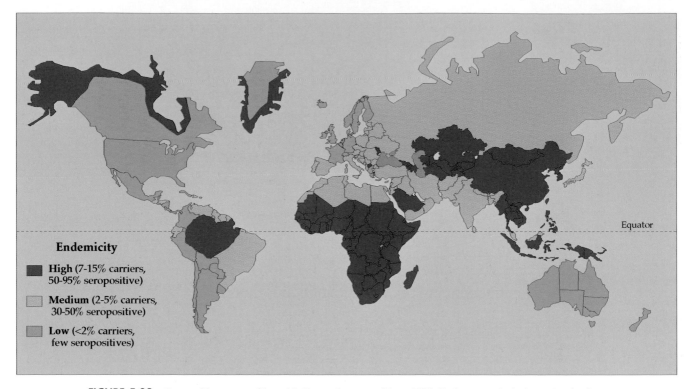

Endemicity

■ **High** (7-15% carriers,
    50-95% seropositive)

□ **Medium** (2-5% carriers,
    30-50% seropositive)

▨ **Low** (<2% carriers,
    few seropositives)

Equator

FIGURE 5.28 Geographic pattern of hepatitis B prevalence as of June 1998. Each country is designated as having high, medium, or low levels of endemicity, based on estimates of carrier frequency. The level of endemicity is correlated with the percentages of persons who show serological evidence of past infection. [From WHO Program on Diseases and Vaccines web page at http://www.who.int/vaccines-surveillance/graphics/htmls/hepb.htm.

regions that exhibit high chronicity for HBV (compare Fig. 5.29 with Fig. 5.28). The association of HBV with HCC is clearly shown by data such as the finding that in areas in which chronic infection occurs in 5–10% of the population, 50–80% of HCC patients are chronically infected with HBV. It seems reasonable to conclude that chronic HBV infection contributes to a large fraction of HCC cases.

The mechanisms by which chronic infection by HBV leads to HCC is not altogether clear, and may not be the same in all cases. One possibility is that long-term infection, characterized by continuing destruction of liver cells followed by regrowth, results eventually in the appearance and selection of tumor cells. Up to 90% of patients with HCC associated with HBV infection have cirrhosis, implying extensive liver damage. It also appears that HCC may result from other causes of liver disease such as alcohol-induced cirrhosis or chronic infection by HCV (see Chapter 3). Furthermore, HCC often appears only after 30–40 years of chronic infection by HBV. Thus, there is an association between HCC and continuing liver damage and regeneration over very long periods.

However, data for HBV, but not for HCV, suggest that HCC may result from a more direct effect of HBV infection. Chronic infection of woodchucks by WHV results in HCC, and in 40% of HCC cases in this system there is inte-

gration of the WHV DNA genome near N-*myc*. Furthermore, the HBV X gene can induce HCC in transgenic mice. This protein binds to p53, a known anti-oncogene (Chapter 6) that regulates signaling pathways the complex and modifies the activities of transcription factors. Thus, it is probable that the X gene product is responsible for induction of HCC in some fraction of cases.

The Immune System and HBV

HBV infection of itself does not lead to the death of infected hepatocytes. Whether *in vivo* or in cell culture, a persistent, noncytolytic infection is established by the virus. Liver damage during HBV infection results instead from the activities of cytotoxic T lymphocytes (CTLs) (see Chapter 8), which attempt to clear the infection by killing infected cells. It appears that the strength of the CTL response determines the course of infection. A vigorous response results in clearance and recovery, although often after frank hepatitis with jaundice. A weak response results in chronic infection with little symptomology. An intermediate response results in chronic infection characterized by chronic hepatitis.

Because of the potential seriousness of chronic infection by HBV, including the potential to infect others, continuing

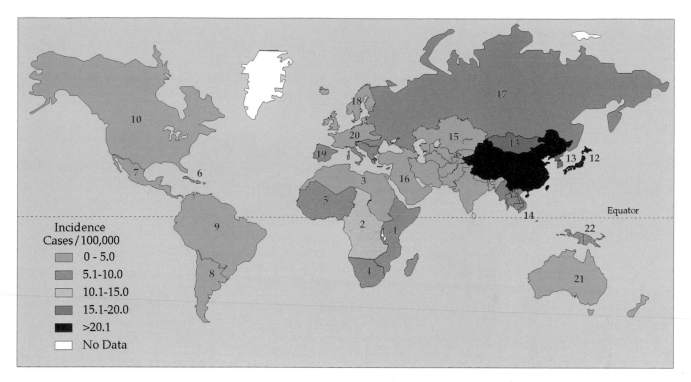

FIGURE 5.29 Average annual incidence (cases per 100,000) of liver cancer in various regions of the world in 1990. Of these cases, 67% are due to chronic hepatitis B and 26% are due to chronic hepatitis C infection. The names of the regions and the incidence for men/women are: (1) Eastern Africa, 9.4/3.1; (2) Middle Africa, 16.8/9.4; (3) Northern Africa, 2.6/1.5; (4) Southern Africa, 12.7/4.7; (5) Western Africa, 12.6/4.0; (6) Caribbean,6.7/4.0; (7) Central America, 3.1/2.5; (8) temperate South America, 2.7/1.8; (9) tropical South America, 2.1/1.5; (10) North America, 4.0/2.1; (11) China, 19.9/10.3; (12) Japan, 38.2/12.1; (13) other East Asia, 23.3/8.3; (14) Southeast Asia, 11.3/4.1; (15) South Central Asia, 1.8/1.0; (16) Western Asia, 3.0/2.1; (17) Eastern Europe, 5.7/4.7; (18) Northern Europe, 4.1/2.7; (19) Southern Europe, 12.6/6.1; (20) Western Europe, 6.9/3.1; (21) Australia/New Zealand, 3.3/1.4; (22) Melanesia, 11.3/6.3. [Data from Parkin *et al.* (1999).]

efforts are being made to develop methods of controlling or clearing the infection in chronically infected people. HBeAg was used as a marker of severity of infection until recently. This has now been replaced clinically by direct measurement of viral titers, which is used to assess the response to therapy.

The first treatment that showed at least partial success was use of high doses of interferon α for extended periods of time. This succeeds in clearing the viral infection in a small minority of patients. Interferon is a cytokine that boosts the immune response (see Chapter 8), and these doses of interferon appear to enable the immune system to eradicate the virus in those patients that respond. However, the drug is expensive and poorly tolerated, with significant side effects.

Trials are ongoing to test the effectiveness of nucleoside analogues in controlling HBV infection. Most analogues tested have not been effective with one or two exceptions. Lamivudine treatment resulted in improvement in more than 50% of patients in a large trial, and the apparent clearing of infection in 16% of patients. Resistant viruses appeared in more than 10% of cases, which limits the effectiveness of continuing treatment. It is hoped that combination therapy will increase the success rate.

Liver transplantation is offered to some patients with HBV infection. However, circulating virus invariably reinfects the graft.

Vaccination against HBV

Several vaccines have been developed to prevent infection by HBV. The first vaccine, which was licensed in 1981, was prepared from blood plasma from chronically infected individuals. It consisted of highly purified preparations of 20-nm particles that were treated to inactivate any residual virus infectivity (whether HBV or any other virus). This vaccine was effective and safe, but obvious difficulties accompany the production of large amounts of such a vaccine. Recombinant vaccines in which the S gene is expressed in yeast or in Chinese hamster ovary cells have now replaced this early vaccine. These vaccines are cheaper and can be produced in very large quantities.

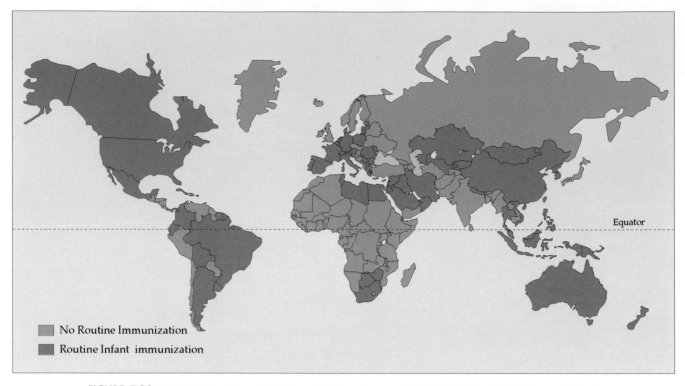

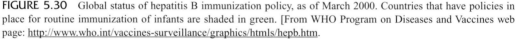

FIGURE 5.30 Global status of hepatitis B immunization policy, as of March 2000. Countries that have policies in place for routine immunization of infants are shaded in green. [From WHO Program on Diseases and Vaccines web page: http://www.who.int/vaccines-surveillance/graphics/htmls/hepb.htm.

The current vaccination schedule calls for three injections of S protein over a period of 6 months. Further boosting, assessed by anti-HBV titers in serum, may be necessary to achieve adequate immunity. These vaccines are quite effective if received prior to exposure or shortly after infection (the incubation period of HBV is quite long, as noted above). They are unable to eliminate the infection in a chronically infected individual, however. Adults in high-risk groups, such as medical personnel, laboratory personnel who handle potentially infected blood samples, sexually active individuals, drug users who inject drugs, or individuals in ethnic groups that have a high incidence of HBV such as Alaskan natives, are encouraged to become vaccinated. Vaccination of all infants at birth is also recommended by the American Academy of Pediatricians, in part to prevent the establishment of chronic infection in babies born to infected mothers, but also in order to eventually eliminate the virus from the population. Immunization of infants starting within 12 hr of birth and going through three or four shots has proven to be effective in preventing chronic HBV infection even in infants that would otherwise have become chronically infected (born to mothers positive for HBsAg). Many countries have now initiated programs to immunize infants, and the current status of vaccination against HBV on a worldwide scale is shown in Fig. 5.30.

FURTHER READING

Biology of Retroviruses

Banks, J. D., Beemon, K. L., and Linial, M. F. (1997). RNA regulatory elements in the genomes of simple retroviruses. *Semin. Virol.* **8**; 194–204.

Coffin, J. M., Hughes, S. H., and Varmus, H. E., eds. (1997). "Retroviruses." Cold Spring Harbor Laboratory Press, Cold Spring, Harbor, NY.

Goff, S. P. (1997). Retroviruses: Infectious genetic elements. In "Exploring Genetic Mechanisms" (M. Singer and P. Berg, eds.), pp. 133–202. University Science Books, Sausalito, CA.

Herniou, E., Martin, J., and Tristem, M. (1998). Retroviral diversity and distribution in vertebrates. *J. Virol.* **72**; 5955–5966.

Hindmarsh, P., and J. Leis (1999). Retroviral DNA integration. *Microbiol. and Molec. Biol. Reviews* **63**: 836–843.

Lentiviruses and HIV/AIDS

Berson, J. F., and Doms, R. W. (1998). Structure-function studies of the HIV-1 coreceptors. *Semin. Immun.* **10**; 237–248.

Cullen, B. R. (1998). Posttranscriptional regulation by the HIV-1 Rev protein. *Semin. Virol.* **8**; 327–334.

Freed, E. O. (1998). HIV-1 Gag proteins: Diverse functions in the virus life cycle—minireview. *Virology* **251**; 1–15.

Kohlstaedt, L. A., Wang, J., Friedman, J. M., *et al.* (1992). Crystal structure at 3.5Å aresolution of HIV-1 reverse transcriptase complexed with an inhibitor. *Science* **256**; 1783–1790.

Leitner, T., and Albert, J. (1999). The molecular clock of HIV-1 unveiled through analysis of a known transmission history. *Proc. Natl. Acad. Sci. U.S.A* **96**; 10752–10757.

Levy, J. A. (1998). "HIV and the Pathogenesis of AIDS," 2nd ed. ASM Press, Washington, DC.

Riddler, S. A., and Mellors, J. W. (1997). HIV-1 viral load and clinical outcome: Review of recent studies. *AIDS* **11**; S141–S148.

Special Report (1998). Defeating AIDS: What will it take? *Sci. Am.* **279**; 81–107, July.

Whetter, L. E., Ojukwu, I. C., Novembre, F. J., *et al.* (1999). Pathogenesis of simian immunodeficiency virus infection. *J. Gen. Virol.* **80**; 1557–1568.

Foamy viruses

Lecellier, C.-H., and Saïb, A. (2000). Foamy viruses: Between retroviruses and pararetroviruses. *Virology* **271**; 1–8.

Retroelements

Eickbush, T. H. (1994). Origin and evolutionary relationships of retroelements. *In* "The Evolutionary Biology of Viruses" (S. S. Morse, eds.), pp. 121–157, Raven Press, New York.

Hepadnaviruses

Buckwold, V. E., and Ou, J.-H. (1999). Hepatitis B C-gene expression and function: The lessons learned from viral mutants. *Curr. Top. Virol.* **1**; 71–81.

Feitelson, M. A. (1999). Hepatitis B in hepatocarcinogenesis. *J. Cell. Physiol.* **181**; 188–202.

Tavis, J. E. (1996). The replication strategy of the hepadnaviruses. *Viral Hepatitis Rev.* **2**; 205–218.

Yen, T. S. B. (1998). Posttranscriptional regulation of gene expression in hepadnaviruses. *Semin. Virol.* **8**; 319–326.

CHAPTER
6

DNA-Containing Viruses

INTRODUCTION

Twenty families of viruses that contain double-stranded DNA (dsDNA) as their genome are currently recognized by the International Committee on Taxonomy of Viruses (ICTV). These families are listed in Table 6.1 together with their hosts. The dsDNA viruses infect bacteria, mycoplasma, algae, fungi, invertebrates, and vertebrates, but interestingly there are no known dsDNA viruses of plants. Members of five of these families, the Poxviridae, the Herpesviridae, the Adenoviridae, the Polyomaviridae, and the Papillomaviridae, infect humans and cause disease, and only these five families are considered further. The genomes of these viruses vary in size from about 5 to 375 kb.

TABLE 6.1 Double-Stranded DNA Viruses

Family	Genera[a]	Genome size (kbp)	Type species	Host
Myoviridae	"T4-like phages"	~170	Enterobacteria phage T4	Bacteria
Siphoviridae	"λ-Like phages"	48.5	Enterobacteria phage λ	Bacteria
Podoviridae	"T7-like phages"	40	Enterobacteria phage T7	Bacteria
Tectiviridae	Tectivirus	147–157	Enterobacteria phage PRD1	Bacteria
Corticoviridae	Corticovirus	9	Alteromonas phage PM2	Bacteria
Plasmaviridae	Plasmavirus	12	Acholeplasma phage L2	Mycoplasma
Lipothrixviridae	Lipothrixvirus	15.9	Thermoproteus 1	Archaea
Rudiviridae	Rudivirus	?	Sulfolobus SIRV 1	Archaea
Fuselloviridae	Fusellovirus	15.5	Sulfolobus SSV 1	Archaea
Poxviridae		130–375		
Chordopoxvirinae	See Table 6.3			**Vertebrates**
Entomopoxvirinae	Entomopoxvirus A, B, C			Invertebrates[b]
Asfarviridae	**Asfivirus**	170	African swine fever	**Vertebrates**
Iridoviridae	Iridovirus	170–200	Invertebrate iridescent 6	Invertebrates
	Chloriridovirus		Invertebrate iridescent 3	Invertebrates
	Ranavirus		Frog 3	**Vertebrates**
	Lymphocystivirus		Lymphocystis disease 1	**Vertebrates**

(continues)

221

TABLE 6.1 (*continued*)

Family	Genera[a]	Genome size (kbp)	Type species	Host
Phycodnaviridae	Three genera of algal viruses	>300	Paramecium bursaria chlorella 1	Algae
Baculoviridae	Nucleopolyhedrovirus	90–160	Autographa californica nucleopolyhedrovirus	Invertebrates
	Granulovirus		Cydia pomonella granulovirus	Invertebrates
Herpesviridae		~125–235		
Alphaherpesvirinae	See Table 6.6			**Vertebrates**
Betaherpesvirinae	See Table 6.6			**Vertebrates**
Gammaherpesvirinae	See Table 6.6			**Vertebrates**
Adenoviridae	See Table 6.9	20–25		**Vertebrates**
Polyomaviridae	See Table 6.11	5		**Vertebrates**
Papillomaviridae	See Table 6.13	8		**Vertebrates**
Polydnaviridae	Ichnovirus	N.A.		Invertebrates
	Bracovirus			Invertebrates
Ascoviridae	Ascovirus	?	Spodoptera frugiperda ascovirus	Invertebrates

[a]Quotes are used to denote taxa without internationally approved names; kbp, kilobase pairs.
[b]Viruses in families in boldface type have vertebrate hosts, which include human hosts except for the genera *Asfivirus*, *Ranavirus*, and *Lymphocystivirus*.

There are also five families of DNA viruses that contain single-stranded DNA (ssDNA) as their genome, as listed in Table 6.2. These viruses infect bacteria, mycoplasma, spiroplasma, plants, invertebrates, and vertebrates. Members of only one family, the Parvoviridae, infect humans and cause disease, and only this family is considered further. The parvoviruses have a small genome (4–6 kb), whence they receive their name (Latin *parvus* = small). Note that in the classification used here, viruses with a DNA genome that replicate through an RNA intermediate, such as the hepadnaviruses and the caulimoviruses, are not referred to as conventional DNA viruses and were considered in Chapter 5.

During infection, most vertebrate DNA viruses stimulate host-cell DNA replication, or at least the early stages of DNA replication, in order to prepare a suitable environment for their own DNA replication. Such a favorable environment includes the presence of cellular factors required for DNA replication as well as an increase in the amount of substrates required for making DNA. Further, some viruses cause cells to proliferate, at least early in the infection cycle. For this reason, most DNA viruses are at least potentially transforming. If an incomplete replication cycle occurs and the early genes that stimulate DNA replication continue to be expressed, a cell may become transformed. Many DNA viruses are known to cause tumors in humans or in other animals.

In contrast, the parvoviruses do not encode proteins that stimulate cellular DNA replication. For this reason, they can only replicate in cells that are actively dividing and their target tissues in the host are organs that undergo continual renewal. Members of one genus of the Parvoviridae, the Dependoviruses, require a helper virus for replication, and replication will occur only in cells coinfected with the helper.

The larger DNA viruses interfere with the defenses of the vertebrate host against viruses. Some inhibit the interferon system, some inhibit immune effector functions, some interfere with the complement system, and most large DNA viruses interfere with more than one host defense pathway. In part because of this ability to interfere with host defense systems, infection with some DNA viruses results in a latent or a persistent infection that can endure for the life of the infected individual. Some aspects of these interference pathways are described in this chapter, where needed to understand the replication cycle and epidemiology of the virus, but a more detailed description of these defense mechanisms is presented in Chapter 8.

Some DNA viruses, such as the poxviruses and the adenoviruses, cause epidemics of symptomatic disease in vertebrates from which recovery is complete (if the infection is not fatal) and immunity is established. Other vertebrate DNA viruses, such as the herpesviruses and the polyomaviruses, establish long-term infections that persist despite a vigorous immune response. For such a strategy of long-term persistence to be successful, infection in the majority of hosts must be inapparent or cause only moderate symptoms that are not unduly deleterious. Spread may be epidemic and accompanied by symptoms during primary infection, but for some herpesviruses, vertical transmission to infant progeny occurs

TABLE 6.2   Single-Stranded DNA Viruses

Family	Genera	Genome size (kb)	Type species	Host[a]
Inoviridae	Inovirus	4.4–8.5	Enterobacteria phage M13	Bacteria
	Plectrovirus		Acholeplasma phage MV-L51	Mycoplasma
Microviridae	Microvirus	~4.4–6.0	Enterobacteria phase ϕX174	Bacteria
	Spiromicrovirus		Spiroplasma phage 4	Spiroplasma
	Bdellomicrovirus		Bdellovibrio phage MAC1	Bacteria
	Chlamydiamicrovirus		Chlamydia phage 1	Bacteria
Geminiviridae	Mastrevirus	2.5–3.0	Maize streak	Plants
	Curtovirus		Beet curly top	Plants
	Begovirus		Bean golden mosaic-Puerto Rico	Plants
Circoviridae	**Circovirus**	1.7–2.3	**Chicken anemia**	**Vertebrates**
Circinoviridae[b]	**Circinovirus**	3.8	**TT**	**Vertebrates**
Parvoviridae		4.0–6.0		
Parvovirinae	See Table 6.15			**Vertebrates**
Densovirinae	Densovirus		*Junonia coenia* densovirus	Invertebrates
	Iteravirus		*Bombyx mori* densovirus	Silkworms
	Brevidensovirus		*Aedes aegypti* densovirus	Mosquitoes

[a]Viruses in families in boldface type have vertebrate hosts; only members of the Parvovirinae and Circinoviridae infect humans; kb, kilobases.
[b]The taxonomic status of the Circinoviridae is provisional.

without producing symptoms and persists for the life of the animal. For such viruses, transmission needs to occur only once per generation for the virus to persist.

FAMILY POXVIRIDAE

The poxviruses are a very large family of dsDNA-containing viruses that infect mammals, birds, and insects. Eleven genera are recognized, eight of which infect vertebrates and three of which infect invertebrates. Viruses that infect vertebrates are classifed in the subfamily Chordopoxvirinae, and Table 6.3 lists the genera in this subfamily. Only two human poxviruses are known, variola or smallpox virus, a member of the Orthopox genus, and molluscum contagiosum virus, the only member of the genus Molluscipoxvirus.

The host range of any particular poxvirus is usually narrow. The two human poxviruses infect only humans in nature and other poxviruses are similarly limited in their natural host range. However, a number of mammalian poxviruses whose primary host is not man can cause natural, albeit usually limited, infections of humans, and still other poxviruses, including the avian poxviruses, can infect humans under experimental conditions. Such mammalian and avian poxviruses have been used as agents for vaccination against virulent human viruses (smallpox, described in detail below) or as vectors to express foreign antigens for the purposes of immunization (Chapter 9), because they normally cause only a limited or an abortive infection of humans and can be engineered to express foreign antigens.

The poxviruses are exceptional among DNA viruses because they replicate in the cytoplasm. The structure of the virion is also unusual and poxviruses may well have had an origin separate from that of other DNA viruses.

Structure of the Virion

Poxvirions are large and brick shaped. Vaccinia virus, which serves as a model for the family, is approximately 350 by 270 nm. Two forms of infectious vaccinia virions are known, an intracellular form and an extracellular form, which are illustrated in Fig. 2.20. The intracellular form has a lipid-containing surface membrane and can be released by disruption of the infected cell. The extracellular form has a second, lipid-containing envelope around it, which contains

TABLE 6.3   Poxviridae (Chordopoxvirinae)

Genus/members	Virus name abbreviation	Usual hosts(s)[a]	Transmission	Disease	World distribution
Orthopoxvirus					
Variola virus	VARV	**Humans**/none	Contact	Smallpox (now extinct)	Formerly worldwide
Monkeypox	MPXV	Squirrels/humans, monkeys	Contact	Smallpox-like	West and Central Africa
Vaccinia	VACV	Unknown/humans, bovines	Contact	Localized lesions	Worldwide
Cowpox	CPXV	Rodents/humans, cats, bovines, zoo animals	Contact	Localized lesions	Europe, W. Asia
Camelpox	CMLV	Camels/none	Contact, aerosols	Localized lesions	Africa, Asia
Ectromelia	ECTV	Unknown/laboratory mouse colonies, foxes, mink	Contact, aerosols	Lesions plus disseminated disease	Europe
Volepox virus	VPXV	Voles/none	Contact	?	Western US
Parapoxvirus					
Orf virus	ORFV	Sheep/humans, ruminants	Contact	Localized lesions	Worldwide
Bovine papular stomatitis	BPSV	Cattle/humans	Contact	Localized lesions	Worldwide
Pseudocowpox	PCPV	Cattle/humans	Contact	Localized lesions	Worldwide
Parapox of red deer	PVNZ	Red deer/none	Contact	?	New Zealand
Yatapoxvirus					
Yaba monkey tumor virus	YMTV	?Rodents/primates, humans	MTBA[b]	Localized lesions	East and Central Africa
Molluscipoxvirus					
Molluscum contagiosum	MOCV	**Humans**/none	Contact, including sexual transmission	Many nodular lesions	Worldwide
Capripoxvirus					
Sheeppox	SPPV	Sheep/none	Contact, fomites, MTBA		
Goatpox	GTPV	Goats/none			Asia, Africa
Suipoxvirus					
Swinepox	SWPV	Swine/none	MTBA, primarily lice	Generalized skin disease	Worldwide
Leporipoxvirus					
Myxoma	MYXV	*Sylvilagus* rabbits	MTBA, primarily mosquitoes	Benign tumors in natural hosts, severe disease in European rabbits	South America, Western United States, introduced into Australia
Rabbit fibroma	SFV	*Sylvilagus* rabbits	MTBA	Benign tumors in natural hosts	Eastern United States
Hare fibroma	FIBV	European hare			Europe
Squirrel fibroma	SQFV	*Sciurus* squirrels			Eastern and Western United States
Avipoxvirus					
Many species including fowlpox		Birds	Contact, MTBA	Lesions of skin and digestive tract	Worldwide

[a]Hosts are listed as "reservoir host/other naturally infected hosts"; viruses for which humans are the sole host are shown in bold face.
[b]MTBA, mechanical transmission by arthropods.

glycoproteins that are not present in the intracellular form. Within the virion is a core, which in mammalian viruses appears dumbbell–shaped, and one or two flanking lateral bodies. The core contains the viral DNA complexed with protein. The lateral bodies are proteinaceous. Altogether, the virion contains 30 or more structural proteins.

The cellular receptors used by poxviruses to enter cells have not been characterized. Evidence conflicts as to whether the virus enters by fusion at the cell plasma membrane or via the endosomal pathway, and it is possible that both pathways are used.

Replication of Poxviruses

An overview of the poxvirus replication cycle is shown in Fig. 6.1. Following infection, an RNA polymerase within the core is activated by the release of the core into the cytoplasm. This polymerase, together with accessory enzymes such as the capping enzyme and poly(A) polymerase, synthesizes mRNAs that are capped and polyadenylated. These are extruded from the core and translated by the host-cell machinery. Translation of early mRNAs leads to further uncoating of the virus and the development of a regulatory

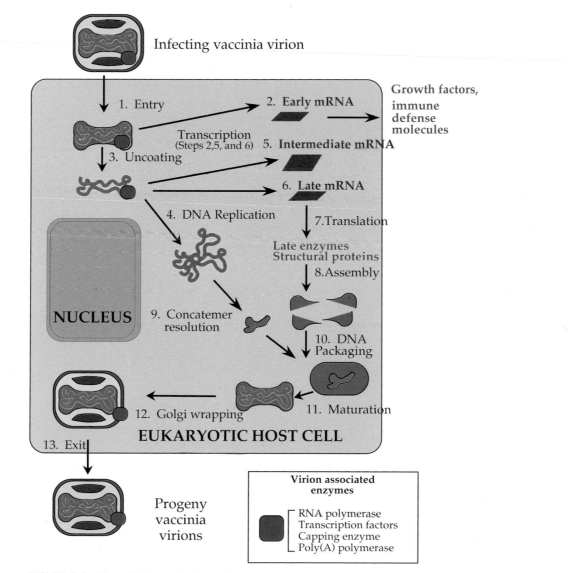

FIGURE 6.1 The replication cycle of vaccinia virus. Sequential steps are numbered in order. Note that despite the fact that vaccinia is a DNA virus, the entire replication cycle takes place in the cytoplasm. [Adapted from Fields *et al.* (1996, p. 2638).]

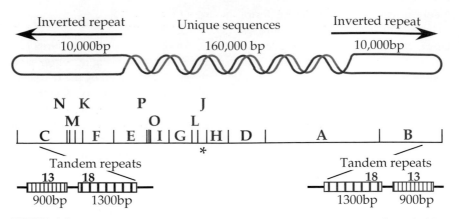

FIGURE 6.2 Schematic representation of the DNA of vaccinia virus. Upper part: linear double-stranded DNA with terminal hairpins and inverted repeats (not to scale). Center line is the *Hind*III restriction map (* indicates the TK gene). Color coding of the *Hind*III fragments is the same as that used in Tables 6.4 and 6.5. Lower line diagrams the internal structure of the terminal repeats. [From Fenner *et al.* (1988).]

pathway by which midcycle genes and finally late genes are expressed. Early and mid-cycle functions include interference with host defense mechanisms and the replication of the viral genome, whereas late genes are primarily involved with formation of progeny virions.

Poxviruses have large genomes, from 130 to 380 kb, and encode hundreds of proteins. A schematic of the vaccinia virus genome is shown in Fig. 6.2. It is double stranded and linear but the ends of the genome are covalently closed so that the genome consists, in essence, of a very large single-stranded circular molecule that is self-complementary. The ends of the genome possess inverted terminal repeats that are involved in the initiation of DNA replication.

Because virus replication, including DNA replication, occurs in the cytoplasm, poxviruses must encode all enzymes required for DNA replication and production of mRNAs. Thus the encoded proteins include DNA and RNA polymerases, a poly(A) polymerase to polyadenylate mRNAs, a capping enzyme, several enzymes with functions in nucleotide metabolism, protein kinases, DNA topoisomerases, as well as the proteins that form components of the virion and proteins that interfere with antiviral host defense mechanisms. Table 6.4 lists many enzymatic proteins encoded by vaccinia virus and Table 6.5 lists structural proteins and proteins without enzymatic activity. Poxvirus-encoded molecules that interfere with host defenses are discussed in Chapter 8.

A model for vaccinia DNA replication is shown in Fig. 6.3. The mechanisms by which DNA replication is initiated have not been completely worked out, but it is thought that a viral enzyme specifically nicks the DNA in or near the terminal repetitions, and the 3′ end of the nicked DNA forms a primer for DNA synthesis. Continued elonga-

tion of the DNA chain leads to production of concatenated progeny DNA. These concatamers must subsequently be resolved into genome-length segments whose ends are then covalently closed.

Replication of DNA and assembly of progeny virions occurs in what have been called factories in the cytoplasm. These are electron-dense areas that contain viral DNA and membranes. Progeny DNA is assembled into cores and assembly of virions occurs by condensation of membranes around the viral core. It is not clear whether the membranes that are used to assemble virions are derived from previously existing cell membranes or are newly synthesized. Intracellular forms of the virus, which are infectious, are surrounded by a lipid-containing membrane, but extracellular virions have an additional lipid-containing envelope. It is not known how the envelope is acquired. Poxvirions are illustrated schematically in Fig. 2.20.

Interactions with the Host

The poxvirus genome encodes many proteins that are not required for replication of the virus in cell culture. These include proteins that subvert the antiviral defenses of the host and that may extend the host range of the virus. Vaccinia virus, variola virus, and other poxviruses encode proteins that interfere with the complement system and the interferon system, that prevent an inflammatory response to the virus or the induction of apoptosis (programmed cell death), and that interfere with the activity of TNF (tumor necrosis factor), among others. The discovery of these proteins that interfere with host antiviral defenses is recent, and more study is necessary to understand the full extent of viral inhibition of host

TABLE 6.4   Vaccinia-Encoded Enzymes

Functional group name of enzyme	ORF[a]	kDa	Properties
DNA replication			
DNA polymerase	**E9L**	110	
Protein kinase	**B1R**	34	Phosphorylates H5R
Unknown	**D5R**	90	Replication fork, ATP/GTP binding motif A
Uracil DNA glycosylase	**D4R**	**25**	
Nicking-joining enzyme	??	50	Concatemer resolution
DNA toposiomerase	**H6R**	32	
ssDNA binding protein	**I3L**	**30**	
DNA ligase	**A50R**	63	Nonessential
DNA helicase	**A18R**	57	DNA-dependent ATPase
Early DNA-related metabolism			
Thymidine kinase	**J2R**	20	
Thymidylate kinase	**A48R**	23	
Ribonucleotide reductase			Provide dNTPs
M1	**I4L**	87	Large subunit
M2	**F41**	37	Small subunit
dUTPase	**F2L**	15	dUTP->dUMP, downregulate dUTP
DNA repair?	**D9R**	25	
	D10R	29	Hydrolyze 8-oxo-GTP
DNA dependent NTPase	**D11L**	72	NPH[b] I
RNA transcription			
RNA polymerase			Multisubunit enzyme
RPO147	**J6R**	147	
RPO132	**24R**	133	
RPO35	**A29L**	35	
RPO30	**E4L**	30	Transcription factor
RPO22	**J4R**	22	
RPO19	**A5R**	19	
RPO18	**D7R**	18	
RNA polymerase-associated protein	**H4L**	94	RAP94, early promoter-specificity factor
Early transcription factor			DNA-dependent ATPase
	A7L	82	ETR subunit I
	D6R	74	ETR subunit II
Poly(A) polymerase	**E1L**	55	Catalytic subunit
	J3R	39	Stimulatory subunit, methyl transferase
Capping enzyme			RNA triphosphatase, guanyl transferase
	D1R	97	Large subunit, catalytic activities
	D12L	33	Small subunit, stimulates transferase
RNA/DNA-dependent NTPase	**I8R**	77	NPH II[b]
Protein kinase 2	**F10L**	52	Phosphorylates, serines, and threonines
Glutaredoxin	**O2L**	12	Thioltransferase, dehydroascorbate reductase

Source: Adapted from Fields *et al.* (1996, Table 3, p. 2645) and data from Goebel *et al.* (1990).

[a]ORFs are named and color coded according to the restriction map shown in Fig. 6.2.

[b]NPH, nucleoside triphosphate phosphohydrolase.

TABLE 6.5   Vaccinia-Encoded Nonenzymatic Components

Location in virion	ORF[a]	kDa	Properties
Membrane of intracellular mature virus	I5L	8.7	Hydrophobic
	L1R	27.3	Myristylated, hydrophobic
	H3L	37.5	Hydrophobic
	H5R	22.3	Phosphorylated by B1R to give 34–36 kDa
	D8L	35.3	Cell-surface binding, virulence
	D13L	61.9	Rifampicin resistance
	A13L	7.7	Oligomeric
	A14L	10.0	Oligomeric
	A17L	23.0	Dimer, neutralizing epitope
	A27L	12.6	Fusion protein, neutralizing epitope, required for EEV, nonessential
Core of intracellular mature virus	F17R	11.3	Phosphoprotein, DNA binding
	I7L	49.0	Homology to topoisomerase II
	G7L	41.9	Processed
	L4R	28.5	Structural protein VP8
	D2L	16.9	
	D3R	28.0	
	A3L	72.6	Major core protein P4b
	A4L	30.8	
	A10L	102.3	Major core protein P4a
	A12L	20.5	Processed
Specific to enveloped extracellular virus (EEV)	F13L	41.8	Envelope antigen
	A34R	19.5	*N*-glycosylated, homologous to lectin, EEV release
	A36R	**25.1**	
	56R	34.8	*N*- and *O*-glycosylated, hemagglutinin, nonessential
	B5R	35.1	Complement control protein, required for EEV, homologue to C3L

Source: Adapted from Fields *et al.* (1996, Table 2, p. 2643) with additional information from Goebel *et al.* (1990).
[a]ORFs are named and color coded according to the restriction map show in Fig. 6.2.

defenses. It is clear that these various viral functions are required for successful viral infection of their hosts in nature, and the existence of these viral activities has been very useful for our understanding of host defenses against viral infection. We will return to this topic in Chapter 8.

Genus Orthopoxvirus

The best known poxviruses are the orthopoxviruses. Vaccinia virus has been widely studied in the laboratory as a model for the replication of poxviruses and has been used to immunize hundreds of millions of people against smallpox virus, also known as variola (from the Latin word for "spotted"). The extensive knowledge of vaccinia virus gained from laboratory and clinical studies has also led to its use as a vector to express foreign antigens in cultured cells or in animals (Chapter 9). Other members of the genus infect a variety of domestic and wild animals (Table 6.3). A similarity tree that illustrates the relationships among the orthopoxviruses is shown in Fig. 6.4. Notice that the human smallpox viruses, variola major and its close relative variola minor, are related to cowpox virus, but the lineage formed by these viruses is distinct from the lineage that contains vaccinia virus and monkeypox virus. Nonetheless, all of these viruses are cross-protective.

Smallpox Disease

Smallpox once caused vast epidemics in human populations. It was already endemic in India 2000 years ago and had spread to China, Japan, Europe, and northern Africa by 700 A.D. It was introduced into the New World by the Europeans during their explorations. Infection resulted in a fatality rate of 20–30% in most populations and at one time the virus infected virtually the entire population of Europe. Thus, the virus was responsible for a significant fraction of all human deaths on the Continent. When introduced into virgin populations in the New World, the mortality rate was much higher, probably due in part to a lack of previous selection for resistance to smallpox and in part due to the breakdown of the social system caused by the simultaneous infection of most of the population. The importance of smallpox, measles, and other Old World plagues in the conquest and settlement of the Americas by the Europeans was discussed in Chapter 4. A detailed description of the effect of smallpox virus on human civilization through the ages can be found in the book *Princes and Peasants: Smallpox in History* by Hopkins. The history of nations has often been changed by the early death of rulers from smallpox or from the appearance of smallpox in armies. As one example of interest to Americans, an invasion of Canada by American

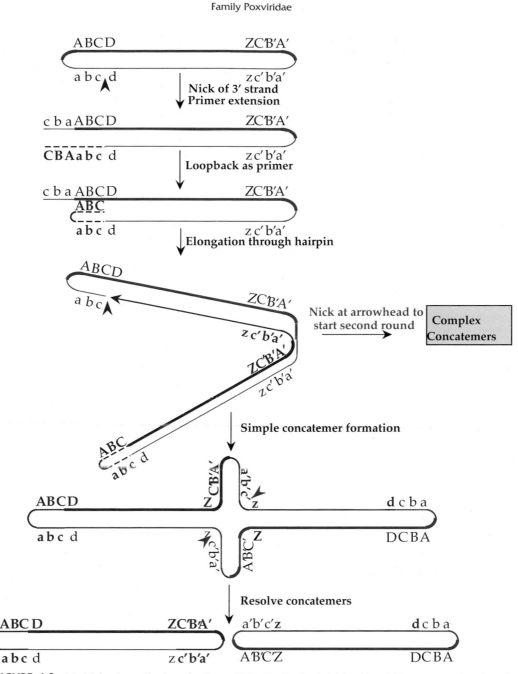

FIGURE 6.3 Model for the replication of orthopox DNA. Replication is initiated by nicking one strand at the red arrow near the left end of the DNA. This is followed by primer extension and loopback to form an internal primer. Extension then occurs through the hairpin at the right end of the molecule to form a concatemer. Concatemer resolution occurs by nicking at the blue arrows. Parental DNA is shown in blue, new strands in red. [Redrawn from data in Moyer and Graves (1981) and Traktman (1990).]

troops during the Revolutionary War, with the idea of adding Canada to the American colonies, failed when an epidemic of smallpox swept through the American troops.

Smallpox virus was exclusively a human virus, maintained by person-to-person contact, and recovery from the disease resulted in lifelong immunity to the virus. Thus, it was a disease of civilization. As was described for measles virus, it must have originated from a nonhuman poxvirus, possibly a poxvirus of a domesticated animal, at the time

that the development of human civilization led to the presence of large population centers and the domestication of animals.

Although infection by smallpox began as an upper respiratory disease, infection became systemic and many organs were infected. The virus replicated extensively in the skin, leading to pocks that could cover the entire body. These pocks left scars, particularly in the facial area, which marked the person for life, and at one time the fear of scar-

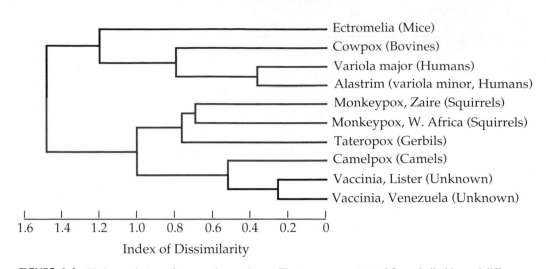

Ectromelia (Mice)
Cowpox (Bovines)
Variola major (Humans)
Alastrim (variola minor, Humans)
Monkeypox, Zaire (Squirrels)
Monkeypox, W. Africa (Squirrels)
Tateropox (Gerbils)
Camelpox (Camels)
Vaccinia, Lister (Unknown)
Vaccinia, Venezuela (Unknown)

| 1.6 | 1.4 | 1.2 | 1.0 | 0.8 | 0.6 | 0.4 | 0.2 | 0 |

Index of Dissimilarity

FIGURE 6.4 Phylogenetic tree of some orthopoxviruses. The tree was constructed from similarities and differences in *Hind*III restriction sites. Presence, absence, and impossibility of sites (because the fragments were too small) were analyzed. Index of dissimilarity has no absolute value, but illustrates that pairs of poxviruses isolated from the same reservoir hosts (for example, variola and alastrim) resemble one another more closely than poxviruses from divergent hosts. It also illustrates that the vaccinia virus used in recent times as a vaccine against variola, and whose natural reservoir host is unknown, is only distantly related to cowpox virus (see the discussion in the text). [Redrawn from Fenner *et al.* (1988, p. 94).]

ring on recovery from smallpox appears to have been at least as terrifying as the fear of death caused by the disease. A photograph of an infected child is shown in Fig. 6.5.

Immunization against Smallpox

So great was the fear caused by smallpox that a technique of immunization was developed by the 10th century, called variolation, in which people were deliberately infected by smallpox using dried skin from pox lesions. The infecting virus was delivered by blowing the inoculum up the nose or by inoculation into the skin. Infection via either method led to relatively low mortality, about 1–2% compared with 20–30% following natural infection, and the disease was milder with less scarring.

Jenner developed the modern concept of vaccination when he introduced and popularized the use of cowpox virus as a vaccine against smallpox. It was known that milkmaids had beautiful complexions because they never contracted smallpox, and there was some evidence that their resistance to smallpox resulted from infection with cowpox virus, which was occupationally acquired. In milkmaids, cowpox virus caused localized lesions, usually on the hands where they came in direct contact with the virus during milking, but the lesions did not spread. The virus is antigenically related to smallpox virus and will protect against infection with smallpox. Jenner's investigation in the 1790s included inoculation of a young boy with cowpox virus and then challenging him with virulent smallpox virus, to which the boy was found to be resistant, an experi-

ment that could not be performed today because of ethical considerations.

Jenner then introduced cowpox virus as a vaccine against smallpox in the general population. This concept

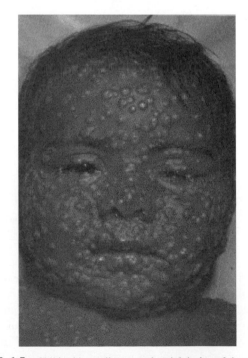

FIGURE 6.5 Child with smallpox, on the eighth day of the rash. The rash began to develop one day after the onset of fever. (From Fenner *et al.* (1988, p. 15).]

was controversial at first, but gradually won widespread acceptance. The virus now used for immunization is referred to as vaccinia, derived from the Latin name for cow. However, the relationship between modern vaccinia virus and cowpox virus is a mystery. At some time in the last century, the source of the virus was changed and no one knows where vaccinia virus came from. Modern vaccinia virus is more closely related to camelpox virus than to cowpox (Fig. 6.4) but exhibits a fairly wide experimental host range. It has been suggested that it was derived from a domestic animal other than the cow, perhaps a horse.

Jenner "vaccinated" people by placing a drop of vaccinia-containing solution on the skin and then scarifying the skin in some way, allowing the virus to replicate in this region. Vaccinia virus infection is localized to the area inoculated in the vast majority of people, leading to a localized lesion that results in a pock. This vaccination procedure gave solid immunity against smallpox but because of the limited replication of the virus in humans, the immunity was not considered to be lifelong and periodic reimmunization was practiced. Vaccinia virus has been used to immunize hundreds of millions of people over the years. It is generally safe, but infection of a small fraction of people results in a more serious disease. In people whose immune system is impaired, progressive vaccinia may develop because the individual cannot control the replication of the virus, which results in a fatal illness. Generalized cutaneous vaccinia can also result from inoculation or, very rarely, encephalitis.

Eradication of Smallpox

Following the introduction of vaccination, the incidence of smallpox declined and the virus was largely eliminated from developed countries. Interestingly, there also appeared a variant of smallpox that produced only a 1% mortality rate, called variola minor or alastrim (from the Portuguese word meaning something that "burns like tinder, scatters, and spreads from place to place"). Variola minor was endemic in Africa and the Americas and coexisted with variola major. Beginning in the 1960s, the World Health Organization (WHO) began an intensive campaign to eradicate smallpox virus from human populations. Such a campaign was possible because smallpox was exclusively a human virus, there was only a single serotype, vaccination with vaccinia virus was safe and effective, and inapparent infection was virtually unknown. Every case of smallpox was tracked down, infected patients were quarantined, and all contacts were immunized against smallpox. Over a period of about 10 years, smallpox epidemics became less frequent and more completely contained until finally the virus was, in fact, eliminated (Fig. 6.6). The last case of natural smallpox occurred in 1977, and vaccination against the virus is no longer practiced because of the possible side effects of the vaccine. Smallpox is the first virus to be

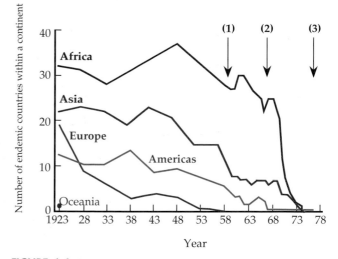

FIGURE 6.6   Number of countries reporting the occurrence of endemic smallpox between 1923 and 1978, grouped by continents. (1) marks the WHO resolution on global smallpox eradication in 1959; (2) shows the start of the intensified eradication program in 1967; (3) marks the last case of smallpox on October 22, 1977, in Somalia. [From Fenner (1983).]

deliberately exterminated, and the effort and dedication required to accomplish this were remarkable.

The remaining, known stocks of smallpox virus are now stored in two laboratories, the Centers for Disease Control and Prevention in Atlanta and the Vector Laboratories in Novosibirsk. These viruses are still being worked with and there is an effort being made to sequence many different strains of the virus and to obtain cDNA clones of them. The WHO has adopted the principle that all remaining stocks of smallpox should at some time be destroyed, but the earliest date for this has already passed as the deadline has been extended more than once. This issue has been controversial because many scientists believe that further useful information can be obtained from the study of smallpox virus. The genes that block human antiviral defenses have only recently been described, for example, and there are surely secrets of viral virulence that we do not understand. Furthermore, new forms of smallpox virus might arise. For example, monkeypox virus causes a disease in humans that is similar to smallpox but with greatly reduced transmissibility. Variants or recombinants able to spread more readily could well arise. (Could this have been the origin of smallpox in the first place?) Thirdly, there is no guarantee that if the known smallpox stocks in the United States or in Russia were destroyed, smallpox virus would cease to exist. There may be other stocks in laboratories around the world (the virus was very widespread at one time and extensively studied). If the virus surfaced from such a source at some time in the future, having ready stocks of virus for study would be useful. The other side of the argument is that smallpox virus represents a great threat to the human population, which is now largely nonvaccinated and within another generation will be completely susceptible to infection by smallpox. If the virus were to be released, whether accidentally or deliberately, it could again cause devastating epidemics.

Monkeypox Virus

Monkeypox virus was thought to be a virus of monkeys but is now known to be associated with squirrels. It infects humans under natural conditions but was thought to be only a rare zoonosis. The disease in humans caused by monkeypox virus is clinically very similar to smallpox, and it was not recognized as a distinct human pathogen until the eradication of smallpox in Africa. A recent epidemic of monkeypox in Zaire from February to August 1996, involving 89 cases of clinical disease with six deaths, challenges the prevailing view that

monkeypox is rare in humans and does not spread in humans. A follow-up investigation of monkeypox in the area in February 1997, which included a hut-by-hut search for active cases in 12 villages, gave evidence that up to 73% of monkeypox cases resulted from secondary human-to-human transmission (Fig. 6.7). Furthermore, three of the deaths were in children less than 3 years old and a large proportion of cases were in persons <15 years old. Thus, it is possible that the outbreak may be due to the cessation of vaccination against smallpox in the late 1970s, since smallpox vaccination is also protective against monkeypox. Monkeypox virus is therefore a human pathogen that can cause fatal illness, spread from person to person, and cause outbreaks of disease in susceptible populations. To date, spread has been limited, but the potential for wider spread exits. There is continuing need to monitor the incidence of monkeypox.

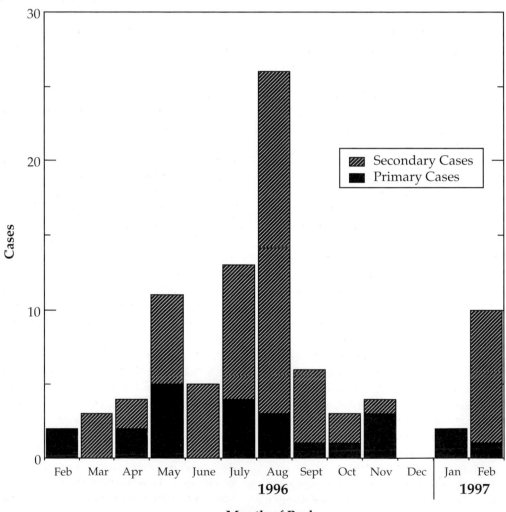

FIGURE 6.7   Number of monkeypox cases by date of rash onset in 12 villages in the Katako-Kombe health zone, Kasai Oriental, Zaire, February 1996–February 1997. Note that most of the cases are secondary cases resulting from human-to-human transmission. [From *MWWR* **46**; 306 (1997).]

Molluscum Contagiosum

Molluscum contagiosum virus is the sole representative of the genus Molluscipoxvirus. It is a widely distributed human virus that is spread by contact, including sexual contact. The virus causes a skin disease characterized by raised lesions. The disease is chronic but usually resolves within a few months, and the illness is considered to be trivial in immunocompetent individuals. In HIV-infected patients, however, the disease can be more troublesome. It has not been possible to grow the virus in cultured cells, and all virus for laboratory study has been derived from lesions of infected individuals, which usually contain large amounts of virus. Despite this limitation, the genome of the virus has been completely sequenced and molecular biological studies are under way.

Rabbit Myxoma Virus

Rabbit myxoma virus, a member of the genus *Leporipoxvirus*, has been widely used in Australia to control populations of the European rabbit, which was introduced with disastrous results into Australia after its settlement by Europeans. The history of the virus in Australia represents a facinating story of viral evolution in the field and is one of the best studied examples of coevolution of both a virus and its host. The story has important implications for our understanding of virus–host interactions.

Rabbit myxoma virus is native to the Americas, where it causes a localized skin fibroma in American rabbits. The virus is mechanically transmitted by mosquitoes. It does not replicate in the mosquito (it is not an arbovirus), but the mosquito can transmit the virus when mouth parts become contaminated by feeding on an infected rabbit with skin lesions. It was early discovered that, although the virus causes a trivial illness in American rabbits, it causes a systemic infection in European rabbits that is fatal more than 99% of the time, a disease called myxomatosis.

In Australia, a land of marsupials, the European rabbit was introduced as potential food for foxes, which in turn had been introduced for the pleasures of fox hunting. The rabbits multiplied as rabbits proverbially do and became a plague. The enormous populations of these introduced rabbits threatened agricultural crops and also threatened to overwhelm a number of Australian marsupials by competing with them for food supply. Campaigns in which people went out in large groups and killed as many rabbits as possible were undertaken in order to reduce the rabbit population, but it was a losing game. To make matters worse, the introduced foxes preferred the Australian fauna to the introduced rabbits anyway, and constituted a second plague on the native fauna, which have not evolved to deal with mammalian predators.

In an effort to eliminate or control the rabbit population, rabbit myxoma virus was introduced. At first this strategy was very successful and vast numbers of rabbits died. It was believed that the virus would have to be broadcast every year in order to continue the campaign against the rabbits, but there was hope that the rabbit plague would end. However, soon after its release, new strains of virus arose that were less virulent, as illustrated in Fig. 6.8. These strains prevailed because they persisted more successfully in the rabbit population—less virulent viruses remained within a host longer and were transmitted to new rabbits more successfully than virus that rapidly killed its host. Furthermore, rabbits that survived infection with the less virulent strains were now immune to the virulent virus, making it more difficult to control the rabbits by reintroduction of a virulent strain (although rabbits have a short life span and herd immunity is not very important in resistance to disease). With time the virus became progressively less virulent but, interestingly, strains of very low virulence never became dominant components of the virus population. Very low virulence strains did arise (Fig. 6.8, grade 5) but did not persist, presumably because they were transmitted less efficiently than were strains of moderate virulence. After a few years, the dominant strains were grades 3 and 4, which kill 50–95% of nonimmune, wild-type (i.e., not selected for resistance to myxomatosis) European rabbits.

At the same time that less virulent virus strains were being selected, rabbits that were more resistant to myxomatosis were also being selected. The enormously high death rate caused by viral infection coupled with the short generation time of rabbits rapidly led to the selection of rabbits that exhibited increased resistance to the disease caused by the virus, as illustrated in Fig. 6.9. Perhaps this is the reason that strains of low virulence first arose and then faded from the virus population. Notice that after seven epidemics of myxomatosis, virus strains of grade 3, the dominant strain in the virus population, now caused severe disease in only 60% of rabbits, compared with 95% in unselected rabbits. With continuing passage of time, the virus and rabbit populations evolved such that rabbit myxoma virus was approximately as serious for the rabbit population as smallpox was for man, that is, 20–30% mortality following viral infection. At this point, efforts were made to select more virulent strains of virus that would kill a larger percentage of the rabbits. This approach has been only modestly successful and although it continues to be pursued, control of the rabbit populations is only partial and this method of biological control is failing.

The case history of myxomatosis in Australia makes clear that a situation in which a virus rapidly kills the vast majority of its hosts is inherently unstable. There is selective pressure on the virus to attenuate its virulence and on the host to become resistant to the virus. Because both rabbits and rabbit myxoma virus multiply rapidly, the accommodation of the virus and host occurred rapidly.

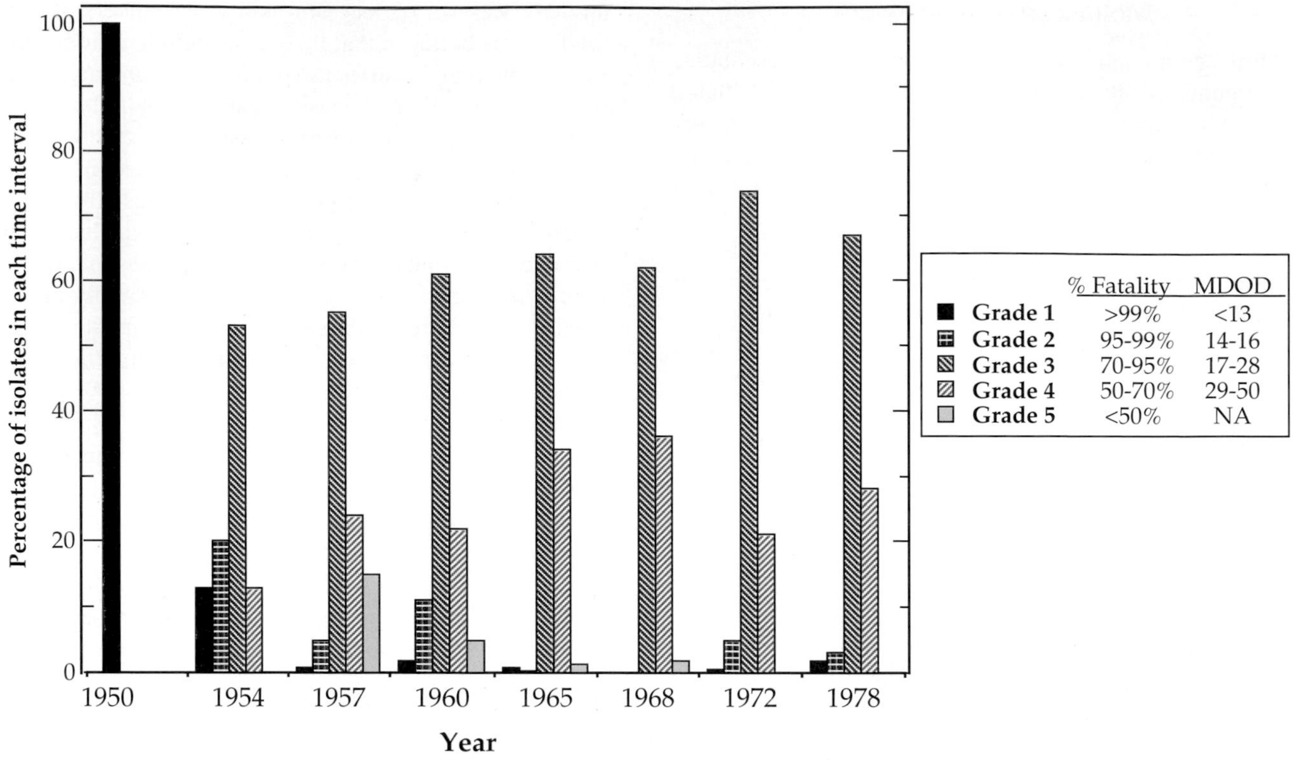

FIGURE 6.8 Virulence of field isolates of myxomatosis virus over a number of years in Australia after the introduction of a grade 1 virus for biological control in 1950. Field isolates were classed as belonging to one of the five grades of virulence on the basis of average survival times (MDOD = mean day of death) of six laboratory rabbits inoculated with each isolate. This measure closely correlates with case fatality rates (which defined the original grades of virulence). Isolates were collected over intervals of 3–5 years, and the collective data for each interval are plotted at a year in the midpoint of the interval. [Data from Fenner (1983).]

FAMILY HERPESVIRIDAE

The herpesviruses, of which more than 100 are known, infect vertebrates. A partial listing of these viruses is given in Table 6.6, together with their hosts and the diseases they cause. Most herpesviruses infect mammals or birds, but reptilian, amphibian, and fish herpesviruses exist as well. The viral genome is large, 120–230 kb, and the viruses encode many dozens of proteins, which allows them to finely regulate their life cycle. Virions are enveloped, 100–300 nm in size, with an icosahedral nucleocapsid (Figs. 2.1 and 2.16).

Herpesviruses are ancient viruses that have coevolved with their vertebrate hosts. Phylogenetic trees of a number of viruses belonging the subfamily Alphaherpesvirinae are shown in Fig. 6.10. Figure 6.10A shows a tree of 14 viruses representing four genera. Genera *Simplexvirus* and *Varicellovirus*, both of which contain important human viruses as well as viruses of other animals, form well-defined taxa. Figure 6.10B compares the tree for four simplexviruses with the tree of their host species. The trees are congruent. Similarly, Fig. 6.10C compares the tree of four varicelloviruses with that of their hosts, and again the trees

are congruent. Thus, it is clear that these herpesviruses have coevolved with their hosts.

The tree in Fig. 6.10A also contains another interesting feature. All of the simplexviruses are primate viruses, infecting man and various species of monkeys, except for bovine herpesviruses (BoHV) 2. BoHV-2 fits well in the virus tree, but cattle obviously do not fit in the primate tree of Fig. 6.10B. Thus, this bovine virus has not coevolved with cattle but appears to have been obtained from a primate. An obvious hypothesis is that this represents the spread of a virus from humans to their domestic animals, the other side of the coin from the transfer of viruses like measles to humans from domestic animals. Intriguingly, BoVH-2 causes lesions largely confined to the udders of dairy cattle, which could have come into contact with a human virus during milking.

The herpesviruses have a narrow host range and any particular herpesvirus is adapted to use only a single vertebrate host in nature. All herpesviruses are capable of establishing a latent infection in their natural host whereby they persist for the life of the animal. Latent infection is established in one specific set of cells that are nonpermissive or

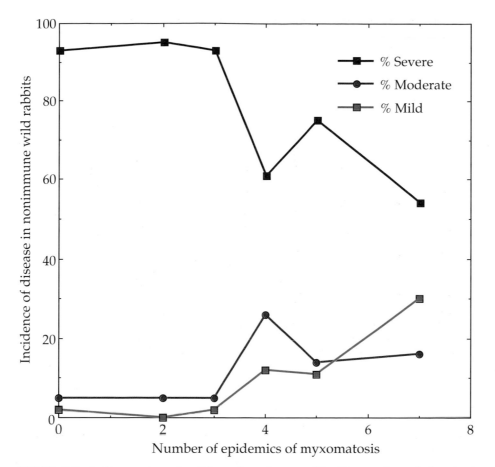

FIGURE 6.9 Incidence and severity of disease in nonimmune wild rabbits experimentally inoculated with a strain of myxomatosis of virulence grade 3 (which induces 70–95% mortality in laboratory rabbits) following a given number of epidemics of myxomatosis. [Data from Fenner (1983).]

TABLE 6.6   Herpesviridae[a]

Subfamily/genus/members[b]	Virus name abbreviation	Usual host(s)	Transmission	Disease
Alphaherpesvirinae				
Simplexvirus				
Herpes simplex 1	HHV-1 or HSV-1[b]	Humans	Contact	Cold sores on face and lips
Herpes simplex 2	HHV-2 or HSV-2	Humans	Contact	Genital ulcers
Herpes B[c]	CeHV-1 or HBV	Monkeys	Saliva	Cold-sore-like lesions in macaques, fatal infection in man
Ateline herpesvirus 1	AtHV-1	Spider monkeys		
Simian agent 8[c]	CeHV-2 or SA8	Vervet monkeys		
Bovine herpesvirus 2	BoHV-2	Cattle		
Varicellovirus				
Varicella-zoster	HHV-3 or VZV	Humans	Aerosols	Chicken pox, shingles
Pseudorabies virus[d]	SuHV-1 or PRV	Swine		
Equid herpesvirus 1 and 4	EHV-1,-4	Horses	Aerosols, contact	Respiratory disease, abortigenic disease
Bovine herpesvirus 1	BoHV-1	Cattle		

(continues)

TABLE 6.6 (continued)

Subfamily/genus/members[b]	Virus name abbreviation	Usual host(s)	Transmission	Disease
"Marek's disease-like viruses"[e]				
Marek's disease virus[f]	GAHV-2, -3	Chickens	Aerosols, contact	T-cell lymphoma
Herpes virus of turkeys[g]	MRHV-1	Turkeys		
"ILTV-like viruses"[e]				
ILTV[h]	GaHV-1 or ILTV	Chickens		
Numerous alphaherpeviruses unassigned to genera.				
Betaherpesvirinae				
Cytomegalovirus				
Cytomegalovirus	HHV-5 or CMV	Humans	Saliva, urogenital excretions	Disseminated disease in neonates or immunocompromised hosts leading to CNS involvement, hearing loss, and fatal pneumonitis
Muromegalovirus				
Mouse cytomegalovirus[i]	MuHV-1 (MCMV-1)	Mice	Saliva	
Roseolovirus				
Human herpesvirus 6	HHV-6	Humans	Contact, saliva	Exanthum subitum or sixth disease, may be associated with chronic fatigue syndrome and/or multiple sclerosis
Human herpesvirus 7	HHV-7	Humans	Saliva, urogenital excretions	Unknown
Gammaherpesvirinae				
Lymphocryptovirus				
Epstein–Barr	HHV-4 or EBV	Humans	Saliva, contact	Infectious mononucleosis, Hodgkin's lymphoma, Burkitt's lymphoma
Rhadinovirus				
Ateline herpesvirus 2	AtHV-1	Spider monkeys		
Human herpesvirus 8	HHV-8	Humans		Kaposi's sarcoma
Numerous gammaherpesviruses unassigned to genera, such as herpesvirus saimiri (SaHV-2) and equid herpesviruses 2 and 5 (EHV-2 and EHV-5)				

[a]Herpesviruses are generally worldwide in distribution.

[b]Although the newer nomenclature lists an adjective describing the species, followed by "herpesvirus" and a number, e.g. human herpesvirus 3 or HHV-3, many authors still use the former names, "varicella-zoster" or VZV, so both forms are shown for the commoner viruses.

[c]Herpes B is cercopithecine herpesvirus 1; simian agent 8 is cercopithecine herpesvirus 2.

[d]Suid herpesvirus 1.

[e]Quotes are used to denote taxa without internationally approved names.

[f]Gallid herpesviruses 2 and 3.

[g]Meleagrid herpesvirus 1.

[h]ILTV, infectious laryngotracheitis virus or gallid herpesvirus 1.

[i]Murid herpesvirus 1.

semipermissive for virus growth and which differ from virus to virus. A different set of cells is lytically infected so as to produce a progeny virus that is capable of spreading to new hosts. Lytic infection is invariably fatal for the infected cell. Reactivation of latent virus and lytic infection of permissive cells allow the virus to reemerge unchanged and infect new hosts. For some herpesviruses reactivation occurs only sporadically, sometimes only at very long intervals, whereas for other herpesviruses reactivation occurs more or less continuously and infectious virus is usually present. The ability of the viruses to latently infect their hosts for life presents a different set of constraints and opportunities for the spread of the virus in nature. The disease caused by the virus must be relatively innocuous, or at least not life threatening, in an immunocompetent native host if lifelong latent infection is to be a viable strategy for the virus. However, the ability of

A. Phylogenetic tree of alphaherpesviruses

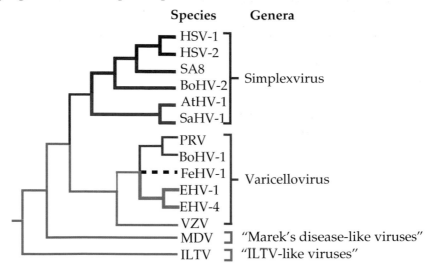

Species   Genera

HSV-1
HSV-2
SA8
BoHV-2   Simplexvirus
AtHV-1
SaHV-1

PRV
BoHV-1
FeHV-1   Varicellovirus
EHV-1
EHV-4
VZV

MDV   "Marek's disease-like viruses"

ILTV   "ILTV-like viruses"

B. Tree of some Simplexviruses and their hosts

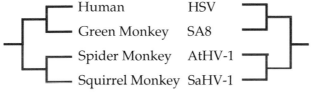

Human   HSV
Green Monkey   SA8
Spider Monkey AtHV-1
Squirrel Monkey SaHV-1

C. Tree of some Varicelloviruses and their hosts

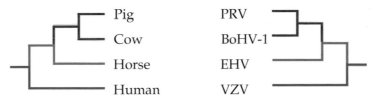

Pig   PRV
Cow   BoHV-1
Horse   EHV
Human   VZV

FIGURE 6.10 Evolutionary relationships among the alphaherpesviruses and their hosts. (A) Phylogenetic trees of the Alphaherpesvirinae, based on the analysis of deduced amino acid sequences from the glycoprotein B gene, using the neighbour-joining method. The most robust portions of the tree are shown in heavy lines. (B) Comparisons of the host and viral trees for several simplexviruses. (C) Comparison of host and viral trees for several members of the *Varicellovirus* genus. Virus abbreviations: HSV includes both herpes simplex type 1 (HSV-1) and herpes simplex type 2 (HSV-2); SA8 is simian agent 8 or cercopithecine herpesvirus 2; BoHV-1 and BoHV-2 are bovine herpesviruses; AtHV-1 is ateline herpesvirus 1; SaHV-1 is saimiriine herpesvirus 1; EHV includes both EHV-1 and EHV-4, equid herpesviruses; FeHV-1 is feline herpesvirus; PRV is pseudorabies virus or suid herpesvirus 1; VZV is varicella-zoster virus or human herpesvirus 3; MDV is Marek's disease virus; ILTV is infectious laryngotracheitis virus of chickens. [From McGoech and Cook (1994).]

the virus to remain latent and reemerge after long intervals means that a minimum sized population is not necessary to maintain the virus in nature. Thus, these viruses could have been present in human populations since humans arose, being passed on from their nonhuman ancestors, as suggested by Fig. 6.10. The interplay between the virus and the host required to establish lifelong infection in the face of a vigorous immune response as described below and in Chapter 8, is further evidence that the herpesviruses have coevolved with their hosts. Further support for this idea is the fact that most herpesviruses are worldwide in distribution. In the case of the human herpesviruses, most are present in almost all people on earth, including the most isolated and remote tribes of people that have been examined.

Classification of Herpesviruses

Herpesviruses have been divided into three subfamilies, alpha, beta and gamma (Table 6.6). Classification has been based on biological properties, as shown in Table 6.7, but as more sequence data become available these data are also used. Alphaherpesviruses have a broad host range in the laboratory and will infect a wide variety of cultured cells or experimental animals. They spread rapidly in cultured cells with a short reproductive cycle and efficiently destroy infected cells. In their natural host, latent infections are usually established in sensory neurons and lytic infection often occurs in epidermal cells. The human viruses belong to two genera, *Simplexvirus* and *Varicellovirus*, as shown in the table. Betaherpesviruses have a restricted host range and a long infection cycle in culture, and infected cells often become enlarged (cytomegaly). In the natural host, the virus is maintained in latent form in secretory glands, lymphoreticular cells, kidneys, and other tissues. The human viruses belong to two genera, *Cytomegalovirus* and *Roseolovirus*. Gammaherpesviruses have the narrowest host range and infect only the family or order to which the natural host belongs. They replicate in lymphoblastoid cells, and some can lytically infect epithelium and fibroblasts. They are specific for B or T cells and infection is frequently latent. HHV-4 belongs to genus Lymphocryptovirus and HHV-8 belongs to the genus *Rhadinovirus*.

Herpesviruses can also be grouped according to the structure of their genome. In some viruses, including herpes simplex, inverted repeats in the DNA lead to inversions of parts of the DNA sequences relative to one another during the process of replication. These different orientations are called isomers. In others, the linear sequence of the DNA is fixed (only one isomer exists). Herpesviruses have been grouped into six classes on the basis of the genome organization, in particular the location of reiterated domains, as illustrated in Fig. 6.11A. The repeated domains in the viruses in Class E of this figure give rise to four isomers; those in Class D give rise to two isomers. The genome organizations are important for the expression and replication of the genome, but the grouping according to genome organization does not coincide with the taxonomy of herpesviruses based on biological criteria (Table 6.6). Thus, the development of the constellation of repeated sequences may be a more recent occurrence, consistent with the multiple rearrangements in the genomes of herpesviruses documented by Fig. 6.11B. In this figure the genomes of three human herpesviruses, each of which belongs to a different subfamily, are compared. All share a substantial number of genes, but extensive genomic rearrangements have occurred during the evolution of these viruses.

The eight known herpesviruses of humans can be referred to as human herpesviruses (HHV) 1 through 8, but the older names for HHV-1 to 5 are still in common use and are used below. Although they generally cause inapparent or innocuous disease, serious illness can result, particularly in immunocompromised people. Some cancers are also associated with certain of these herpesviruses. In addition to the human herpesviruses, at least one herpesvirus of monkeys, Cercopithecine herpesvirus-1 or B virus, causes a serious, usually fatal illness of man that is of concern to animal handlers. These nine viruses and the diseases they cause are described next.

TABLE 6.7   Biological Characteristics of the Three Subfamilies of Herpesviruses

Characteristic	Alphaherpesvirinae	Betaherpesvirinae	Gammaherpesvirinae
Host range	Variable, often broad	Restricted	Limited to family of natural host
Reproductive cycle	Short	Long	Relatively long
Infection in cell culture	Spreads rapidly Infects many cell types	Progresses slowly	Infects primarily lymphoblastoid cells
Cytotoxicity	Much cell destruction	Enlarged cells form	Some lytic infections of epithelial and fibroblastic cells
Latency	Primarily in sensory ganglia	Maintained in many cells including secretory glands, lymphoretic ular cells, kidneys, and others	Specific for either B or T lymphocytes
Genera/human representatives	*Simplexvirus* HHV1 (HSV-1) HHV-2 (HSV-2)	*Cytomegalovirus* HHV-5 (CMV)	*Lymphocryptovirus* HHV-4 (EBV)
	Varicellovirus HHV-3 (VZV)	*Roseolovirus* HHV-6 HHV-7	*Rhadinovirus* HHV-8

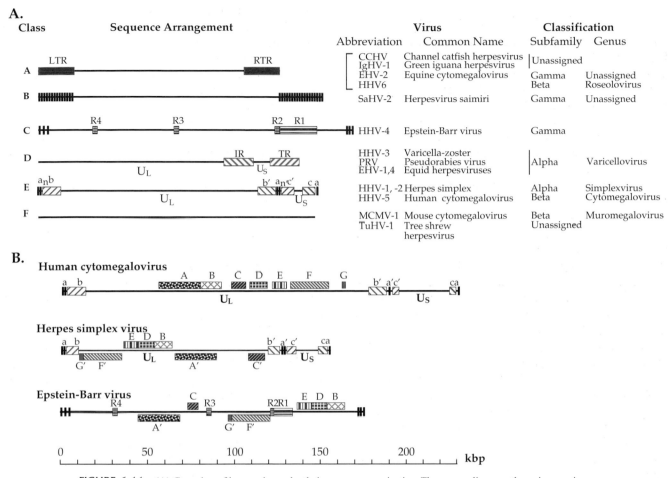

FIGURE 6.11 (A) Grouping of herpesviruses by their genome organization. The narrow lines are the unique regions of the genomes and the rectangles are repeated domains. These are designated the left and right terminal repeats (LTR and RTR) in Class A, internal repeats R1 to R4 in Class C, and the internal and terminal repleats (IR and TR) in Class D. In Class E, both the long and short unique regions are flanked by inverted terminal repeats (shown as **ab** and **b'a'**). Note that this grouping does not correspond exactly with the taxonomy of these viruses as shown in Table 6.6, which is based on a number of biological characters. [From Murphy *et al.* (1995, p. 116).] (B) A more detailed view of human cytomegalovirus, herpes simplex type 1, and Epstein–Barr virus showing conserved sequence blocks, which are distinguished by color and pattern. Blocks shown below the midline in HSV and EBV are those in which the orientation is reversed from that in CMV. [From Chee *et al.* (1990).]

Structure of the Virion

Herpesviruses are enveloped and approximately spherical, with a diameter of 100–300 nm (Fig. 2.1). They possess an icosahedral nucleocapsid 100 nm in diameter that is surrounded by or embedded within a structure known as the tegument (Figs. 2.5 and 2.16). The tegument is composed of virus-encoded proteins and its thickness can vary, even within a single virion. Outside the tegument is the envelope containing about a dozen virus glycoproteins. The mechanisms by which the extracellular virion obtains its envelope are still not well understood. The nucleocapsid and tegument assemble in the nucleus and bud through the nuclear membrane to obtain an envelope (Fig. 2.21A). However, it is not known whether this enveloped particle is then excreted by an unknown pathway, or the particle subsequently buds into the cytoplasm, losing the envelope, and then buds again at the plasma membrane to regain an envelope. Both theories have their proponents, and experimental data exist to support either point of view. Perhaps both occur.

Replication of Herpesviruses

The best studied herpesvirus is herpes simplex virus type 1, which has been used as a model for the entire family, and a more detailed description of its replication will be presented in the section that discusses this virus. Here some generalities of the replication cycle of herpesviruses are described.

Herpesvirus DNA is linear in the virus but circularizes on infection. As described earlier, a circular genome obviates the need for a special mechanism to repair the ends of the DNA during replication. Replication of the DNA occurs in the nucleus, and all herpesviruses encode a large number of enzymes that are involved in nucleic acid metabolism. Among these proteins are a DNA polymerase and a protein that binds to the viral origin of replication in order to initiate DNA replication.

Herpesviruses encode more than 70 proteins. The promoters used to transcribe mRNAs fall into different classes such that there is a temporal program for the expression of genes during the lytic cycle. The proteins encoded by the first genes to be expressed, the immediate-early genes, have regulatory functions. Expression of these genes permits the expression of the early genes, most of which are involved in DNA replication. Expression of the early genes results in DNA replication and the expression of the late genes, most of which encode structural proteins required for assembly of virions. The mRNAs are transcribed by host RNA polymerase II and the promoters are in general 40–120 nucleotides in length. Most promoters reside outside the open reading frames, but some are found within them. The genome organization is complex. Some genes lie within other genes, and a few genes are antisense to another gene.

The use of host RNA polymerases to transcribe mRNAs limits the host range of a herpesvirus, since cells that do not express the factors required for recognition of the viral promoters are not susceptible to a complete replication cycle. The mRNAs are polyadenylated following the cellular AATAAA consensus poly(A) signal and exported to the cytoplasm for translation. From most transcripts only one protein is translated. Most mRNAs are not spliced, but some transcripts are spliced or multiply spliced and partially overlapping transcripts may exist that form nested sets, using a common polyadenylation signal. Many of the proteins synthesized are dispensable in cell culture and function to extend the host range and tissue tropism of the virus or to subvert host antiviral defenses.

Herpes Simplex Viruses (HHV-1 and 2)

Herpes simplex virus (HSV) causes fever blisters or cold sores around the lips or in the genital area. The disease caused by the virus has been known for thousands of years. The characteristic vesicles were described by the ancient Greeks and have been noted in writings through the ages. The name of the virus comes from the Greek word *herpes*, to creep or crawl, referring to the lesions caused by the virus. HSV exists as two serotypes, HSV-1 and HSV-2, also known as HHV-1 and HHV-2. These alphaherpesviruses share about 50% sequence identity and produce a similar disease, but usually infect different parts of the body. HSV-1 infects the facial area, whereas HSV-2 infects the genital area. In humans, the viruses lytically infect epidermal and mucosal cells; they latently infect neurons. In the laboratory they lytically infect cells of many different origins and will infect many experimental animals. Because of the relative ease of experimental manipulation, HSV has been intensively studied as a model for the entire group of herpesviruses. The large size of the genome, 150 kb, and the large number of encoded genes have made a detailed understanding of the virus genome organization very complicated. However, due to the efforts of large numbers of workers, maps such as that shown in Fig. 6.12 can now be constructed. This map illustrates the different genes of HSV, their functions, and their classification into three temporal groups described below. The complexity of the genome and the existence of very many genes are clear from this diagram.

Lytic Infection by the Virus

A schematic representation of the lytic cycle of herpesvirus infection is shown in Fig. 6.13. Infection is normally initiated by fusion of the viral membrane with the cell plasma membrane and is mediated by specific viral glycoproteins on the surface of the virion. The nucleocapsid is then transported to nuclear pores where the viral DNA is released and enters the nucleus (stages 1–3). There the viral DNA is transcribed by RNA polymerase II to produce mRNAs. Of the 80 or so mRNAs produced, only 4 appear to be spliced. Transcription is regulated and can be divided into three phases, α, β, and γ. Transcription of α genes, also called immediate-early genes, occurs immediately on entry of the DNA into the nucleus. Transcription of these genes is transactivated by αTIF, a virus-encoded protein that is present in the tegument of the virion. αTIF interacts with a cellular transcription factor called Oct-1, which recognizes octomer sequences, and, to simplify somewhat, a complex of Oct-1, αTIF, and another cellular factor called C1 binds to the consensus sequence TAATGARAT in the HSV genome (R=A or G). This binding results in transcription of the α genes (stage 4). Five α genes are transcribed (stage 5), called ICP0 (ICP = intracellular protein, to distinguish them from virion proteins), ICP4, ICP22, ICP27, and ICP47. These proteins have regulatory roles in viral replication and are required to activate the β genes (stage 6). ICP4 transactivates HSV genes and functions together with ICP0. ICP27 activates expression of β genes and blocks splicing of cellular pre-mRNAs, thereby interfering with host protein synthesis. It shuttles between the nucleus and the cytoplasm at late times. ICP22 is not required in cultured cells and its function is unknown. ICP47 blocks presentation of antigens to cytotoxic T cells (CTLs) (see Chapter 8).

Most of the β genes encode proteins required for DNA replication, which include a DNA polymerase, a primase/helicase, a DNase, both double-strand and single-strand DNA binding proteins, thymidine kinase, ribonu-

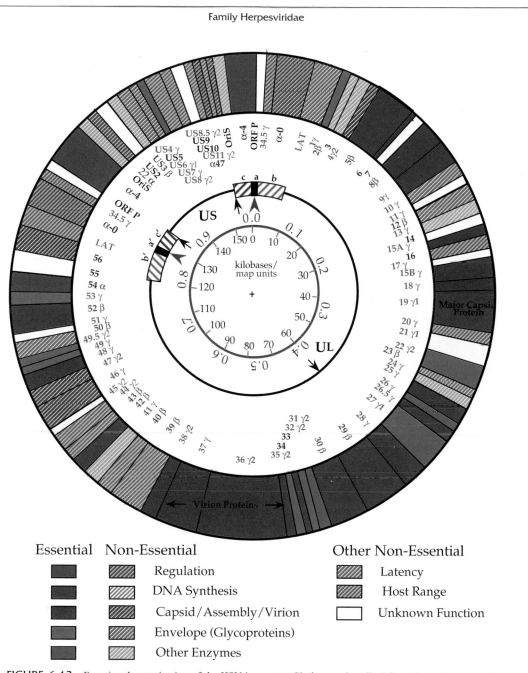

FIGURE 6.12 Functional organization of the HSV-1 genome. Circles are described from the center outward. Circle 1 shows kilobase pairs in black and map units in green. Circle 2 shows the general organization of the genome: U_L, the long unique region; U_S, the short unique region; and a,b,c, the inverted repeats. The black arrows are the three origins of DNA replication, and the green arrowheads show the sites of cleavage/linearization of concatameric or circular DNA. Circle 3 identifies the ORFs, color coded according to their kinetic class (α in red, β in ochre, γ in blue). Black numbers are ORFs not belonging to a particular class. The outer circle indicates the functions of the ORFs, where known, by color coding as indicated in the key below. Solid colors indicate ORFs required for replication in tissue culture cells, whereas the corresponding patterned ORFs can be deleted without affecting replication in culture. Only one of the two copies of α-4 (diagonal red stripes) is required. [Adapted from Fields *et al.* (1996, p. 2245).]

cleotide reductase, dUTPase, uracil DNA glycolase, and a protein kinase. Synthesis of the proteins encoded by these genes (stage 7) allows DNA replication (stage 8) to begin. There are three origins of replication, which are 800–1000

nucleotides in length and have a 100- to 200-nt critical core, but only one origin appears to be required. Replication takes place by a rolling circle mechanism (at least late in infection) so that continual initiation of DNA replication is

Panel A. Early events   **Panel B. Late Events**

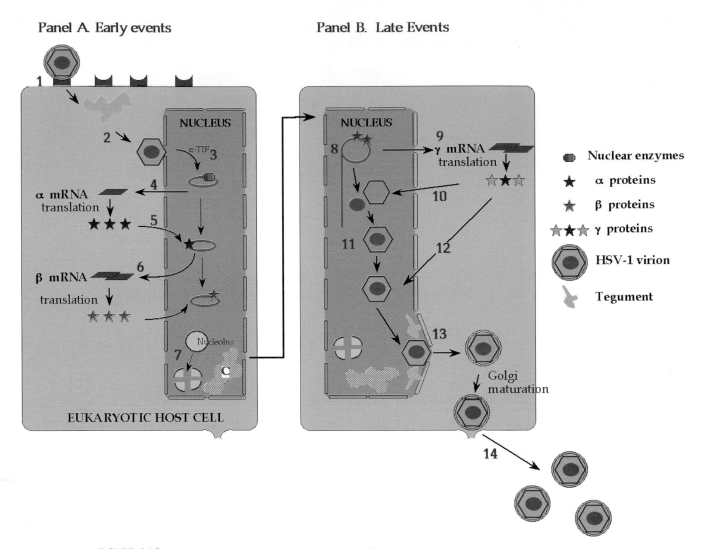

FIGURE 6.13 Replication of herpesviruses in permissive cells.

(A) Early events

    Stage 1) Attachment to cellular receptor and entry by pH-independent fusion with plasmalemma.

    Stage 2) Release of tegument proteins causing shutoff of host protein synthesis; nucleocapsid goes to a nuclear pore.

    Stage 3) α-TIF (VP16) is transported to the nucleus; viral DNA enters nucleus and circularizes.

    Stage 4) Transcription of early (α) genes by nuclear enzymes, and export of α mRNA.

    Stage 5) Immediate-early (α) proteins transported to nucleus.

    Stage 6) α proteins are involved in β mRNA synthesis.

    Stage 7) Chromatin (c) is degraded and nucleoli disaggregated.

(B) Late Events

    Stage 8) β proteins replicate DNA by rolling circle to give head-to-tail concatemers.

    Stage 9) β proteins lead to transcription of late (γ) mRNAs that are translated into structural proteins.

    Stage 10) Formation of empty capsids.

    Stage 11) Packaging of unit length DNA into capsids.

    Stage 12) Addition of further structural proteins.

    Stage 13) Particles receive envelope and tegument at nuclear membrane.

    Stage 14) Particles mature in Golgi and exit by exocytosis.

    [Redrawn from data in Reizman and Sears (1990).]

not required (see Fig. 1.7B). Unit length DNA is cut from the rolling circle by viral nucleases that recognize specific signals at the ends of the linear DNA. During replication, up to 50% of the whole-cell DNA becomes viral.

DNA replication is required for transcription of the γ genes, which mostly encode the proteins that form progeny virions, of which there are more than 30 (stages 9 and 10). The nucleocapsid and the tegument are assembled in the nucleus and bud through the nuclear membrane (stages 11, 12, and 13) (see also Fig. 2.21A). Cleavages in capsid proteins occur during assembly, associated with uptake of DNA, and may serve a maturation function.

More than half of the 80 or so genes in HSV are not required for replication of the virus in cultured cells. However, all, or almost all, appear to be required for infection of humans and maintenance of the virus in nature.

Host-cell macromolecular synthesis is shut off after lytic infection. A component of the virion called vhs (for *virion host shutoff*) inhibits host protein synthesis by mediating the degradation of cellular mRNA, so that inhibition begins very early. Later synthesis of new viral proteins leads to a more profound inhibition of host expression. Lytic infection is also accompanied by characteristic morphological changes in the nucleus, which include fragmentation of the nucleolus and degradation of the host cell chromosomes (stage 7). The profound inhibition of host macromolecular synthesis invariably results in the death of the host cell.

Latent Infection

After the infection of epithelial tissues, HSV infects sensory nerves that serve these tissues and establishes a latent infection. Virus, probably as nucleocapsids, is transported up axonal processes to sensory ganglia, where it is estimated that 5–10 copies of viral DNA, in the form of circular episomes, take up residence. Less than 1% of the neurons within a ganglion appear to be latently infected. Much of what we know about latent infection comes from studies of animal models. HSV will infect and establish a latent infection in mice, guinea pigs, and rabbits, but it is not known how faithfully these animal models reflect the situation in humans. Only one viral transcript is detected in latently infected neurons, called latency-associated transcript or LAT. This transcript does not appear to be required for establishment or maintenance of latency or for reactivation of the virus and its function is not understood, nor is it understood how latency is established and maintained. It is assumed that part of the answer is that neurons are nonpermissive for a lytic replication cycle.

Reactivation of virus occurs sporadically, in response to stressful stimuli such as fever, exposure to UV light, menstruation, or emotional stress. The frequency of recurrence varies in different people from monthly to less often than once per year. Stresses that are well known to induce reactivation include high fever resulting from infection with influenza virus and prolonged exposure to sunlight. On induction, limited replication of virus occurs in the neuron and it travels down the axon, where it infects epithelial cells served by that neuron. Thus, fever blisters erupt in the same tissues as were originally infected, and these lesions contain infectious virus. Classically, these lesions occur around the lips (HSV-1) or in the genital region (HSV-2). Virus replica-

tion in epithelial cells is quickly controlled by the immune system and the lesions heal within 2 weeks or so. Although not yet resolved, it appears that the limited replication of the virus in the neuron is probably fatal for that neuron.

Primary Infection and Maintenance of the Virus

Primary infection is established when a seronegative individual comes in contact with the virus, usually from a person who is secreting the virus at the time. Thus, a person with a reactivated infection can serve as the source of infection, but virus may be actively secreted even when no lesions are present. Acute infection is accompanied by the formation of vesicles that are sites of virus replication and contain infectious virus. In the case of HSV-1 these vesicles are normally found in the facial skin and in the oral mucosa, and latent infection is established in the trigeminal ganglion. HSV-2 is sexually transmitted and infects the genital area. Latent infection is established in neurons in the sciatic ganglion. At one time, HSV-2 was thought to be a causative agent of cervical carcinoma, but this disease is now known to be associated with papilloma virus infections, which are also sexually transmitted and produce warts in the genital region.

HSV-1 is extremely common in human populations. It has been found in every population examined, with 50–90+% of adults having been infected. Primary infection often occurs early in life, but may be delayed in some fraction of the population, especially in developed countries. Because of its mode of transmission, HSV-2 is less common and primary infection occurs later in life. Estimates of its prevalence in human populations center near 10%, but the only population found to be completely free of antibodies to HSV-2 was a group of Roman Catholic nuns. The acquisition of seropositivity to HSV-1 and HSV-2 in the United States in different populations is shown in Fig. 6.14.

Serious Disease Caused by HSV

Primary infection with HSV is usually inapparent or produces only minor illness. In one study of children, for example, 70% of infections were found to be asymptomatic. However, HSV infection of neonates is almost always symptomatic and frequently fatal. Infection *in utero*, during delivery, or shortly after birth usually leads to a disseminated infection often accompanied by encephalitis. The neurotropism of the virus also leads to serious illness on occasion in postneonatal individuals. The Centers for Disease Control and Prevention estimates that about 50 cases of herpes encephalitis occur yearly in the United States that are usually fatal if untreated, and this incidence may be underestimated. Of these cases, about half are due to primary infection and half to reactivated infection. HSV

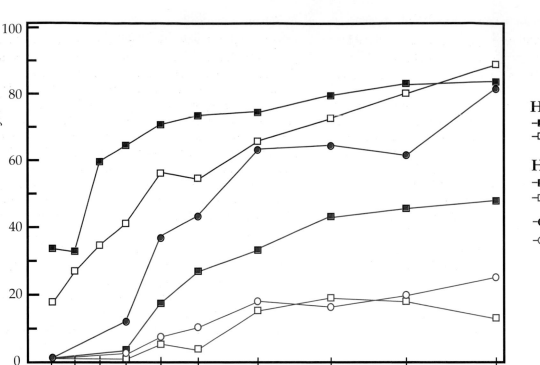

FIGURE 6.14   Seropositivity to herpes simplex virus types 1 and 2 as a function of age, sex, and race in the United States in 1978. [Data from Johnson *et al.* (1989) and Nahmias *et al.* (1990).]

keratoconjunctivitis also occurs and can lead to impairment of vision. Herpetic whitlow is an occupational hazard of dentists and other health care workers, characterized by painful herpetic infection on the fingers. And as is true of many viral infections, HSV infection or reactivation can be very serious in individuals whose immune function is compromised by suppressive therapy for organ transplant or by infection with HIV.

HSV infections can be treated with acyclovir. Acyclovir is a guanine analogue that is incorporated into DNA and results in chain termination. It is a specific inhibitor of HSV-1, HSV-2, and VZV that has little toxicity for host cells. It is less effective against other herpesviruses.

Interference with Host Antiviral Defenses

HSV interferes with several host defense mechanisms. These activities will be described in more detail in Chapter 8, in which the host defense mechanisms themselves are described, but a brief summary of HSV activities is presented here. The virus interferes with the interferon system, with the lysis of infected cells by cytotoxic T lymphocytes (CTLs), with complement-mediated lysis of infected cells, with antibody-dependent lysis of infected cells, and with the lysis of infected cells by a cell suicide pathway called apoptosis. The antiviral effect of interferon is blocked by a viral protein $\gamma 34.5$, which somehow causes a cellular phosphatase to remove the inactivating phosphate put on eukaryotic initiation factor (eIF)-2α (Chapter 8). CTL lysis of infected cells is blocked by $\alpha 47$, which prevents the presentation of peptide antigens to the T cells by major histocompatibility complex class I molecules. Complement-mediated lysis is blocked by HSV proteins that interact with complement. Antibody-dependent cellular toxicity is blocked by HSV-1 IgG Fc receptors, composed of gE and gI, that are expressed on the surface of the infected cells. Finally, HSV gene products suppress apoptosis by the infected cell. The net result is to allow the production of virus by the infected cell for an extended period before the host antiviral defenses can shut it down.

Establishment of latency in neurons is also an important part of the strategy evolved by HSV to avoid host defenses. Neurons are immunologically privileged, and elaborate mechanisms to protect the infected neuron for long periods of time are not necessary. Furthermore, because neurons are nonrenewing and long lived, the establishment of latency in these cells allows the virus to persist indefinitely even in the absence of reactivation.

Varicella-Zoster Virus (HHV-3)

Varicella-zoster virus (VZV, also known as HHV-3) is an alphaherpesvirus that is the prototype member of the genus *Varicellovirus*. Other members of this genus infect monkeys, horses, and pigs (see Table 6.6 and Fig. 6.10). The VZV genome is 125 kb in size and contains at least 69 different genes, of which all but 5 are homologous to genes in HSV. The homologous genes are almost all colinear with the corresponding genes in HSV, and thus the gene map of VZV is essentially the same as that for HSV. Molecular studies of VZV have been hampered by the inability to produce high titered virus stocks in cultured cells, because the virus remains cell associated. Where known, however, the VZV life cycle closely resembles that of HSV, as would be expected from their close relationship. This close relationship also exhibits itself biologically: VZV, like HSV, lytically infects a number of different cells but most characteristically epidermal cells resulting in skin lesions, and VZV, like HSV, establishes a lifelong latent infection in sensory ganglia. The vesicle fluid present in VZV skin lesions contains large amounts of free virus and can spread the disease to susceptible persons.

Chickenpox

VZV causes two different diseases known as chickenpox (varicella) and shingles (zoster). Chickenpox is a highly contagious childhood disease contracted by contact with other children with chickenpox or with an adult with shingles. The virus is transmitted by the respiratory route, and virus replication begins in the upper respiratory tract. It later disseminates through the bloodstream to other areas of the body. The characteristic feature of the disease is a rash of vesicular lesions in the skin that is often quite itchy. Up to 2000 occur in some patients, but fewer than 300 is the norm. Other symptoms include fever, malaise, and loss of appetite.

The disease, which has an incubation period of 10–21 days, is normally self-limited, lasting a week or less, but serious complications can occur. The most common complication in otherwise healthy children is bacterial infection of the skin lesions, which can become serious. Rare complications include viral pneumonia, central nervous system involvement leading to encephalitis or cerebellar ataxia (loss of muscle coordination during voluntary movements), involvement of the liver leading to hepatitis, or involvement of other organs. Primary varicella infection of adults is a more serious illness than primary infection of children. Viral pneumonia is not uncommon in adults, adult infection can lead to male sterility, acute liver failure can occur in adults, albeit rarely, and other complications are also more frequent in adults than in children.

Shingles

Following primary infection, which results in chickenpox, VZV sets up a latent infection in dorsal root ganglia in the spinal cord, where it may reactivate later in life. Reactivation is less common than for HSV-1 and is age related, occurring more frequently in older people, presumably as a result of waning immunity. Reactivation may occur without symptoms, but most commonly reactivation produces the disease known as shingles. Shingles is characterized by painful eruptions of vesicular lesions in skin, usually in the upper back, served by a single sensory ganglion (and therefore the lesions do not cross the midline). The disease normally resolves within a few weeks, but neuralgia (nerve pain, from *neuro* = nerve and *algia* = pain) can continue for up to a year or more and be quite debilitating. More extensive dissemination of the virus occurs in a significant percentage of patients, most of whom have some underlying immunological defect or are immunosuppressed. Dissemination can result in serious complications, as described for primary varicella infection. The vesicular lesions of shingles contain live virus that can infect children and give them chickenpox. Thus, the virus is able to remain latent for decades and then erupt in an essentially unchanged form to start a new epidemic of chickenpox.

Episodes of zoster are more frequent in older people but a second episode of zoster in a person is rare. It appears that the reactivation of the virus leads to a boost in immunity to the virus that prevents further episodes. Immunity to VZV is primarily a function of CTLs. Children that suffer from agammaglobulemia, who are unable to make antibodies, have a normal course of infection by VZV, but children that are deficient in CTL production often die. Furthermore, it has been found that an accelerated CTL response is associated with asymptomatic infection or a mild disease. Finally, reactivation of VZV to produce zoster is correlated with decreased CTL responsiveness against VZV, but not with decreased titer of IgG antibodies against the virus.

VZV in At-Risk Populations

Varicella or zoster is a serious illness in people with compromised immune systems, and zoster is a frequent complication in patients undergoing immune suppression or who have AIDS or leukemia. Before the introduction of antiviral drugs, in particular acyclovir, which is fairly effective for treatment of VZV infections, children who contracted varicella while undergoing immunosuppressive therapy for leukemia suffered a very high rate of visceral dissemination and pneumonia, with a fatality rate of about 10%.

Primary infection with VZV is also serious in pregnant women, leading to significant mortality in both the mother and the infant. Congenital varicella syndrome may occur when infection is in the first trimester of pregnancy, during active fetal organogenesis. Varicella infection of the neonate is serious as well, with a high mortality rate in the absence of treatment.

Epidemiology of VZV

The geographical pattern of infection by VZV is peculiar. The virus has a world-wide distribution but infection is much more common in temperate regions. In temperate regions, infection by VZV is almost universal and occurs mostly in early childhood, in association with epidemics that have a peak frequency in winter and spring. In tropical regions, however, only about half of the population contracts chickenpox. The difference in attack rate is not due to differences in susceptibility to the virus, because the attack rate is very high when uninfected adults move to temperate climates. Why this difference in attack rates occurs is a mystery.

It is interesting to consider the differences in the epidemiology of VZV and HSV-1 and the rationale for such differences. HSV reactivates fairly often and the virus is typically spread to young children by adults with reactivated HSV-1 when they fondle or kiss the child. There is no requirement that the virus spread from child to child in order for it to spread within a population. Because of this, virus perpetuation does not require large-scale production of virus during primary infection and primary infection is usually asymptomatic. In contrast, VZV reactivates very infrequently. Epidemics may begin with the exposure of a susceptible child to an adult with shingles, but the major mechanism for dispersal of the virus in a population is by epidemic spread among young children. The requirement for child-to-child spread for perpetuation requires that large-scale virus production occur during primary infection, and this results in the primary infection being symptomatic.

There is now a live attenuated virus vaccine for chickenpox, licensed in 1995, that was originally developed because of the severe complications of chickenpox in children undergoing chemotherapy for cancer. It remains to be seen how well received the vaccine will be by the general public, since the disease is not normally serious and the vaccine is not mandated by the authorities. Before 1995, there were 4 million cases of varicella annually in the United States, with approximately 100 deaths and 10,000 hospitalizations each year. Because chickenpox ceased to be a notifiable disease in 1997, the levels of case reporting are too low to assess the efficacy of the vaccine at present for controlling VZV in the United States. However, 20 states still reported cases to the Centers for Disease Control and Prevention in 1997, and the incidence of chickenpox by state is shown in Fig. 6.15.

Epstein–Barr Virus (HHV-4)

Epstein–Barr virus (EBV or HHV-4) is a gammaherpesvirus. It is named after Tony Epstein and Yvonne Barr, who first described the virus in tumor cells from patients with Burkitt's lymphoma. EBV is classified as a member of the genus *Lymphocryptovirus*. Several viruses infecting Old World primates also belong to this genus. Other members of the gammaherpesvirus subfamily belong to the genus *Rhadinovirus* and include herpesvirus saimiri (a virus of squirrel monkeys) and HHV-8, the virus responsible for Kaposi's sarcoma. Equine herpesviruses 2 and 5, originally classified as betaherpesviruses, are also now considered to be gammaherpesviruses. Gammaherpesviruses establish latent infection in B lymphocytes and cause these cells to proliferate. This ability to induce proliferation can result in cancers in their native host or in related species.

Primary Infection with EBV

Primary infection with EBV is thought to occur by transmission of virus present in oral secretions. In most human societies, the majority of the population is infected by age 3. In developed countries, however, infection is often delayed until the teens. Primary infection in infants is normally asymptomatic, but in young adults primary infection often results in the disease known as infectious mononucleosis or the "kissing disease." This disease is characterized by fatigue, fever, rash, and swelling of lymph nodes, the spleen, and, in a minority of patients, the liver, for extended periods of time.

Infection begins with replication of the virus in the oral mucosa. A latent infection of B lymphocytes quickly ensues that persists for life. As described in more detail below, infected B lymphocytes, a primary target of EBV, are stimulated to proliferate, especially cells infected early during primary infection by EBV. The symptoms of infectious mononucleosis may result from immunopathological activation of T lymphocytes that attempt to control the large number of B lymphocytes induced by viral infection. Lytic replication of virus in the oropharyngeal epithelium persists indefinitely, although the extent of shedding of virus into the saliva declines with time. There is evidence that persistently infected B cells continue to seed the oropharyngeal epithelium, and that this continued seeding is responsible for the maintenance of a chronic infection in this tissue throughout the life of the individual.

Replication of EBV

EBV will readily infect B cells in culture and establish a latent infection in which a limited set of viral genes is expressed but no production of progeny virus occurs. However, there is no fully permissive system to study the events that occur in lytic infection. Attempts to infect epithelial cells in culture have not been successful, and it is thought

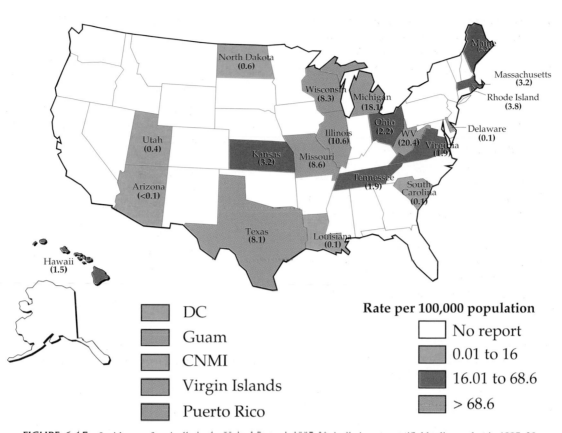

FIGURE 6.15 Incidence of varicella in the United States in1997. Varicella is not a notifiable disease but in 1997, 20 states, the District of Columbia, and four territories reported cases. Color shading indicates the incidence per 100,000 population and the numbers are the number of cases expressed as the percentage of the birth cohort. CNMI is the Commonwealth of Northern Mariana Islands. [Data from *Morbidity and Mortality Weekly Report (MMWR)* **48**; No. 3 (1999) and Summary of Notifiable Diseases in the United States for 1998. *MMWR* **47**; No. 53 (1999).]

that the differentiation status of the epithelial cell is important for the expression of the full lytic cycle. The nature of the receptor used to enter epithelial cells is also uncertain. The receptor used to enter B cells is a protein called CD21, which is a member of the Ig superfamily expressed at high levels in B cells. It has been suggested that this same protein may be expressed at low levels in oral epithelium, but it is possible that some other mechanism of entry into these cells is involved, such as fusion with infected B cells.

The latent infection of B cells involves complicated interactions of the virus with the host cell. Three different forms of latent infection have been distinguished in B cells. These were first described in cells isolated from tumors but can now be reproduced in cultured cells. These different forms of latency, referred to as latency (Lat) I, II, and III, differ in the extent to which the EBV genome is expressed, as illustrated in Fig. 6.16. Lat I cells express only one protein, EBNA1 (*E*pstein–*B*arr *n*uclear *a*ntigen). They also express two sets of RNAs that are not translated, called EBERs (*E*pstein-*B*arr *e*arly *R*NA) and *Bam*As (these last are transcribed from a region of the genome found in a BamHI restriction fragment

called the A fragment). Lat II cells express additional proteins called LMPs (*l*atent *m*embrane *p*rotein). Lat III cells express still more proteins called EBNAs 2, 3A, 3B, 3C, and LP. One model for the functions of these different types of latency is that Lat III is first established and serves to amplify the pool of infected B lymphocytes, since Lat III cells are stimulated to divide. Lat III cells are targets of CTLs, however, which kill these cells in an attempt to control the viral infection. In contrast, Lat I and II cells are resting cells. The more limited set of proteins produced does not stimulate B cells to proliferate nor make the cells targets of CTLs, and it is these cells that maintain the latent infection in the host. In this model, during the lifelong infection by the virus, Lat I and II cells may sporadically become permissive for lytic growth and produce virus, or may sporadically switch to the Lat III state, which results in cell division and the stimulation of CTLs that control these cells. Colonization of B cells is limited, and it is estimated that only about one in 10^5 B cells is infected by virus in asymptomatic carriers.

The RNAs and proteins expressed in latently infected cells are important for maintaining the latent state and

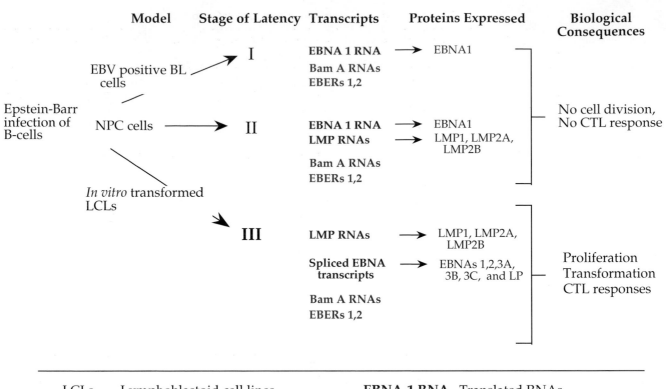

Model	Stage of Latency	Transcripts	Proteins Expressed	Biological Consequences

Epstein-Barr infection of B-cells

EBV positive BL cells → I

NPC cells → II

In vitro transformed LCLs → III

I
EBNA 1 RNA → EBNA1
Bam A RNAs
EBERs 1,2

II
EBNA 1 RNA → EBNA1
LMP RNAs → LMP1, LMP2A, LMP2B
Bam A RNAs
EBERs 1,2

No cell division, No CTL response

III
LMP RNAs → LMP1, LMP2A, LMP2B
Spliced EBNA transcripts → EBNAs 1,2,3A, 3B, 3C, and LP
Bam A RNAs
EBERs 1,2

Proliferation Transformation CTL responses

LCLs   Lymphoblastoid cell lines
BL   Burkitt's lymphoma cell lines
NPC   Nasophayrngeal carcinoma cell lines

EBNA-1 RNA   Translated RNAs
Bam A RNAs   Non-translated RNAs

FIGURE 6.16   Stages of latency after Epstein–Barr infection. Latency I was first described in BL cell lines, but can be reproduced by fusion of *in vitro* transformed LCLs with EBV-negative hemopoietic cell lines. Similarly, latency II was first described in NCP cells but can be reproduced by fusing LCLs with certain human epithelial, fibroblast or hemopoietic lines. [Data from Fields *et al.* (1996, pp. 2399–2402).]

inducing cellular proliferation. EBV DNA has three high-affinity binding sites for EBNA1 and this protein enables the circular EBV DNA to be maintained as an episome in the infected B cells. LMP1 is an integral membrane protein with six membrane-spanning domains. It has transforming activity when expressed in certain rodent cell lines and is presumably important for stimulation of B cell division. It also induces the production of cellular bcl-2, which protects the infected B cells from undergoing apoptosis. LMP2 is also an integral membrane protein. It appears to prevent complete activation of the B cell so that the infection remains latent. EBNA LP, 2, 3A, and 3C are required for continuing growth of the infected B cell.

The EBER RNAs are abundantly produced in infected cells. They are small, nonpolyadenylated RNAs that are located mainly in the nucleus. They are similar to the VA RNAs of adenoviruses. It has been suggested that they may counter the effects of interferon, but experimental evidence does not support such a role.

Latently infected B cells sporadically become permissive for virus replication. In the infected human, this results in production of virus and continued seeding of the oral mucosa. In the laboratory, treatments of latently infected B lymphocytes have been developed that result in the conversion of a substantial fraction of them to permissivity. This has allowed studies of lytic infection and production of progeny virus in cultured cells. Replication of EBV under these conditions appears to involve pathways similar to those of HSV, with which it shares many genes (Fig. 6.11).

The control of latently infected B lymphocytes by CTLs is obviously to the advantage of the virus as well as of the host. If the initial unlimited proliferation of infected B cells continued indefinitely, it would be lethal for the host. This in fact happens in people whose immune system is compromised, as described below.

Burkitt's Lymphoma

The ability of EBV to latently infect B cells and stimulate continuous cell division leads to an association with a number of human cancers (Table 6.8). The first to be described was Burkitt's lymphoma (BL), which was first

TABLE 6.8   Tumors Caused by Epstein–Barr Infection

Tumor	Subtype[a]	Latent period[b]	EBV positivity %	EB antigens expressed	World distribution
Burkitt's lymphoma	Endemic	3–8 years	100		Infection primarily of children <15 years of age in regions endemic for malaria (*P. falciparum*)
	Sporadic	3–8 years	15–85	EBNA1	
	AIDS-associated	3-8 years post HIV	30–40		
Nasopharyngeal carcinoma	—	>30 years	100	EBNA1, LMP1, LMP2	Mostly in SE Asia
Hodgkin's disease	Mixed cell	>30 years	80–90	EBNA1, LMP1, LMP2	Worldwide, but more common in Western Hemisphere
	Nodular sclerosing	>10 years	30		
T-cell lymphoma	Fatal IM	<6 months	?100	?	
	Nasal	>30 years	100	EBNA1, LMP1, LMP2	
	AILD pleomorphic	>30 years	?40	?	
Immunoblastic lymphoma	Fatal IM	<6 months	100	EBNA1, 2, 3A, 3B, 3C	
	Transplant-associated (early)	<6 months after transplant	100		
	Transplant-associated (late)	>1 year	100	LMP1, LMP2	
	AIDS-associated	5–10 years post-HIV	70–80		

Source: Adapted from Fields *et al.* (1996, Table 2, p. 2433).
[a]IM, infectious mononucleosis; AILD, angioimmunoblastic-lymphadenopathy-like.
[b]Post-EBV infection if not otherwise noted.

studied by Denis Burkitt in the 1960s in Africa. He attributed the disease to viral infection and it is now known to be associated with EBV. BL is a childhood malignancy that is worldwide but occurs predominantly in regions of Africa and New Guinea with a high incidence of malaria. The tumors arise in lymph nodes, frequently in submandibular nodes. The disease is fatal if not treated, but treatment with chemotherapeutic agents is effective and the majority of patients can be cured. The association with malaria was originally proposed to result from the suppression of CTLs induced by the microbe. Such suppression could result in an enlarged pool of EBV transformed cells, which are the progenitor cells that become malignanant. However, a more recent hypothesis proposes that it is the expansion of germinal centers that occurs in malaria that is responsible for the increased incidence of lymphomas. Germinal centers serve as sites in which somatic mutation occurs in the V genes of the Ig locus in order to increase the affinity of the antibody for antigen (see Chapter 8). BL cells are characterized by deregulated expression of the cellular oncogene c-*myc*, and the c-*myc* expressed consistently carries mutations. Chromosomal translocations have been found in all BL tumors that bring the c-*myc* oncogene upstream of the immunoglobulin (Ig) genes. Ig genes are expressed in B cells to high levels, and the translocation of c-*myc* to near the Ig locus leads to its deregulated expression in B cells. Perhaps its association with the Ig locus also leads to hyper-

mutation of c-*myc* and the development of a mutant form of the gene that results in the development of a tumor. In any event, it appears that overexpression of a mutated c-*myc* is essential for the development of BL.

Changes in genes other than c-*myc* also appear to be required for development of BL. The cellular tumor suppressor gene p53 is often altered in BL, and changes in other cellular oncogenes or in chromosome architecture also occur in some cases. These alterations may be important for the development of the full malignant phenotype. In addition, BL cells downregulate several functions that are required for recognition and lysis by CTLs. In these cells the expression of class I major histocompatibility complex proteins (MHC) and of the transporter proteins (TAPs) that are required to transfer antigenic peptides across the endoplasmic reticulum (see Chapter 8) are reduced, effectively downregulating the presentation of antigens to CTLs by class I MHC. BL cells also reduce or eliminate expression of cofactors required for efficient interaction with T lymphocytes. Thus, a BL cell resists lysis by CTLs, which are active in immune surveillance.

In summary, the transition from an infected B cell to a malignant cell that can form a fatal tumor in an immunologically competent person is a multistep process. Establishment of a latent infection in B lymphocytes by EBV is only the first step. Many other events must follow, most of which are rare.

Hodgkin's Disease

Hodgkin's disease is another form of malignant lymphoma associated with EBV. The disease often strikes young adults (hockey fans will remember that Mario Lemieux, a star of the Pittsburgh Penguins, underwent treatment for Hodgkin's disease). There is a second peak in incidence after age 45. The disease is worldwide, although more common in developed countries. It is estimated that perhaps 50% of all Hodgkin's disease is due to EBV. It is assumed that other events must occur in order for infected cells to become malignant, as is the case for BL.

T-Cell Lymphomas

The primary target of EBV in humans is B cells, as described above. However, EBV has also been associated with some T-cell lymphomas (Table 6.8). Nothing is known about the process by which the virus infects T cells and causes tumors.

Nasopharyngeal Carcinoma

EBV is also associated with nasopharyngeal carcinoma (NPC). This disease is worldwide but has a much higher incidence in Southeast Asia, in Eskimos, and among some populations in northern and eastern Africa. The available evidence suggests that both genetic and environmental factors are important for the higher incidence in these populations. Studies have found that a particular MHC haplotype (Chapter 8) is correlated with the relative risk of developing NPC. However, dietary factors are also important because immigrant Chinese in the United States have a lowered frequency of NPC than people in China, although the rate is still higher than that in Caucasians.

NPC is a carcinoma rather than a lymphoma, arising in epithelial cells of the nasopharynx. Little is known about how the carcinoma arises, but it presumably requires a non-lytic infection by EBV in which transforming genes are expressed. As for other tumors, several transforming events are probably required for the carcinoma to develop.

EBV Infection in People with Compromised Immune Systems

People who are immunodeficient because of infection with HIV or are pharmacologically immunosuppressed following organ transplant are at greatly increased risk for the development of lymphomas caused by EBV, as shown in Table 6.8. These lymphomas may develop after a very short latent period, less than 6 months. Of interest is the finding that AIDS patients develop two forms of lymphoma. The first arises early, while the immune system is relatively intact, and is a form of Burkitt's lymphoma, having the same c-*myc* chromosomal translocation as described above.

BL may arise at higher frequency in AIDS patients in comparison to people with normal immune systems because of expansion of the infected B-cell population, giving rise to an expanded pool of potential precursor cells. The second form arises late, when the immune system is highly compromised. The late form appears to result from failure of CTLs to control the EBV-infected B-cell population.

A fatal, infectious mononucleosis-like illness in young males has been described that is X linked (that is, the susceptibility gene is carried on the X chromosome, of which males have only one copy). The disease is apparently due to a defect in the immune system that allows EBV-infected B cells to proliferate out of control. The disease is fatal 75% of the time, and death usually results from uncontrolled immunoblastic lymphoma.

Cytomegalovirus (HHV-5)

The cytomegaloviruses (CMVs) are betaherpesviruses that, like all herpesviruses, are species specific in their natural host range. They replicate slowly in cultured cells and have a restricted host range in the laboratory. Human CMV (HCMV) will infect cultured human skin or lung fibroblasts as well as some peripheral blood monocytes. It will also infect chimpanzee cells. Lytic replication in cultured cells resembles that of HSV. There is regulated transcription of α, β, and γ genes, and many of the genes are shared with HSV (Fig. 6.11). However, the CMV replication cycle differs from HSV in one important aspect. CMV infection leads to the stimulation of host-cell DNA, RNA, and protein synthesis throughout infection, whereas infection with HSV results in the immediate shutoff of host-cell macromolecular synthesis.

Infection of Humans with CMV

Transmission of HCMV requires close contact between a susceptible person and a person shedding virus. Virus present in oropharyngeal secretions, breast milk, or other bodily secretions is probably responsible for transmission. It can also be transmitted by blood transfusion. The virus is ubiquitous, present in all human populations, and most humans become infected as infants. In different populations, 40–100% of persons become infected before the age of puberty. HCMV infections are usually asymptomatic, but primary infection of adults can result in infectious mononucleosis.

HCMV infects epithelial cells in many different tissues, in contrast to its restricted host range in cultured cells. Infection characteristically results in cell enlargement, from which the virus gets its name, and the presence of intranuclear inclusions. Shedding of infectious virus may persist for an extended period of time following primary infection, in fact, for years if infection is congenital or occurs very early in life. Following control of infection by CTLs, HCMV establishes latency in tissues or cells that have not been well defined, perhaps in leukocytes, although chronic infection rather than latent infection has not been ruled out. Infection is lifelong and, as for other herpesviruses, reactivation of viral infection can occur and result in renewed shedding of virus.

Infection in Populations at Risk for Disease

Congenital infection by HCMV can be very serious if the infant is not protected by maternal antibodies. About 1% of infants born in the United States are infected *in utero*, either as the result of reactivation of a latent infection in a seropositive mother or as the result of primary infection in a seronegative mother. In the case of mothers who are seropositive, maternal antibodies against HCMV, which are protective against disease, are transferred to the fetus. Congenital infection then occurs with a frequency of only 0.2–2% and symptomatic disease does not occur in the infected fetus. However, primary HCMV infection during pregnancy of women who were previously seronegative results in infection of the fetus up to 50% of the time, and about 10% of infections result in symptomatic infection in the newborn. Infection may be fatal or may result in long-term neurological sequelae, which may include defects in hearing or vision, seizures, microcephaly, or lethargy. Up to 80% of symptomatic infants suffer severe neurological problems, and neurological impairment may occur even in the absence of symptomatic infection. Hearing loss is the most common neurological sequela and congenital HCMV infection is the most common cause of hearing loss in the United States other than that caused by genetic factors.

Like many other herpesvirus infections, HCMV infection, whether primary or resulting from reactivation of latent infection, is extremely serious in patients with compromised immune systems. It is often the most common infection following organ transplant and can result in life-threatening systemic disease. It is also a major life-threatening disease in AIDS patients. Latent HCMV is present in most humans and systemic spread occurs when the CD4+ lymphocyte count falls to very low levels. Systemic disease affects virtually every organ in the body, but infection of the lungs, central nervous system, and the gastrointestinal tract are the most common and most serious. Infection of the lungs can lead to fatal pneumonitis. Infection of the central nervous system commonly results in retinitis, which develops in 20% of long-lived AIDS patients. Infection of almost any region of the gastrointestinal tract can occur and result in severe ulcerations that can lead to perforation of the gut.

The serious nature of disease in the immunocompromised shows that HCMV is an invasive virus that will infect many organs if not controlled by a vigorous immune response. Disease in transplant patients and in patients with AIDS is exacerbated by the expression of genes in HCMV that interfere with many aspects of the immune response. In immunocompetent people, these immunity-defeating mechanisms allow the virus to live in harmony with the host, establishing a lifelong infection that is associated with little or no disease. However, in the immunocompromised, the thwarting of an immune response that is at best weak leads to uncontrolled virus growth and serious illness. The mechanisms by which HCMV interferes with the immune response are described in Chapter 8. They include the synthesis of several proteins that block the presentation of antigens to CTLs by class I MHC and of a protein that interferes with chemokine responses.

Human Herpesviruses 6, 7, and 8

Three newly described human herpesviruses have come to light only recently and have simply been given the sequential numbers HHV-6, -7, and -8. They all appear to be typical herpesviruses that establish latent infections worldwide. These infections are normally accompanied by no disease or only mild disease symptoms. The silence of their infections caused them to be overlooked until recently, when the ability of HHV-6 and HHV-8 to cause disease in immunocompromised populations, especially in AIDS patients, led to their discovery.

HHV-6 and -7 have a tropism for lymphocytes, especially CD4+ T cells. On the basis of this tropism they were first considered to be gammaherpesviruses. However, they are genetically related to betaherpesviruses like CMV and are now classified as betaherpesviruses, in the genus *Roseolovirus*.

HHV-6 occurs as two major types, called A and B. The virus is probably transmitted by oral secretions. In one study, 90% of adults were reported to have infectious virus in their saliva, although other studies have given lower numbers. About half of children infected by HHV-6 suffer a disease called roseola infantum, exanthem subitum, or sixth disease, a mild disease of childhood that is characterized by fever and rash lasting 3–5 days. More severe symptoms or neurological complications occur only infrequently. Primary infection of adults is rare because most people are infected as children, but symptoms are more serious when it does occur.

As for other herpesviruses, primary infection or recurrence of infection in immunosuppressed people or people with AIDS can be life threatening. HHV-6 was first described in 1986 because of its association with lymphoproliferative disorders in AIDS patients. The virus can also cause serious problems in immunosuppressed populations, in particular, in patients undergoing bone marrow, kidney, or liver transplants.

HHV-7 was found during studies of HHV-6 in peripheral T cells. At present there is no clear evidence for the involvement of HHV-7 in human disease. Of possible clinical importance is the fact that HHV-7 may use the same receptor to infect CD4$^+$ T cells as does HIV, which may allow HHV-7 to be used as a vector to express anti-HIV genes in the specific target population infected by HIV.

HHV-8 is the most recently described human herpesvirus. It establishes a latent infection in B lymphocytes and is classified as a gammaherpesvirus, genus *Rhadinovirus*. It has a prevalence of 5% in the United States and is sexually transmitted. It was discovered through its association with Kaposi's sarcoma, the most common tumor found in patients with AIDS. Tumor cells are of endothelial origin and are multifocal. AIDS patients are much more likely to develop Kaposi's sarcoma than are immunosuppressed patients, and there must be a synergism between the infections of HIV and HHV-8. Intriguingly, Kaposi's sarcoma is 15–fold more common in homosexual male AIDS patients than in patients who acquired HIV by a nonsexual route. HHV-8 is also associated with body cavity-based lymphoma, a lymphoid tumor in some AIDS patients, and with multifocal Castleman's disease, a rare lymphoproliferative disorder. In common with a number of other herpesviruses, HHV-8 encodes factors that stimulate cells to divide, that interfere with the host immune response, and that block apoptosis.

Monkey B Virus

Many herpesviruses infect vertebrates other than humans, but only one nonhuman herpesvirus is known to be highly pathogenic for humans. Cercopithecine herpesvirus 1 or monkey virus B is indigenous to Old World monkeys in the genus Macaca. In its native host it causes a disease that is similar to that caused by HSV-1 in humans. A latent, lifelong infection is established in the monkey that seldom leads to serious illness. Sporadic reactivation of the virus occurs that results in the formation of vesicular lesions, particularly on the tongue and cheeks. However, infection of humans or of a number of monkeys other than macaques results in a very serious neurological disease that has a high fatality rate. As described earlier, HSV-1 is also neurotropic and occasionally causes fatal encephalitis.

B virus has usually been transmitted to humans through the bite of a monkey in which infectious virus was present in the saliva. However, transmission has also occurred by other means. In at least one case transmission resulted from contact of infectious material with the eye and two cases are thought to have resulted from exposure to aerosols. One case of transmission of virus from an animal worker to his wife has also been documented. The majority of human infections have resulted in fatal neurological disease, but some infections resulted in only mild disease and the establishment of a latent infection. Acyclovir has been used to treat persons infected by B virus with apparent success, but the number of cases is small and no controlled trials of efficacy have been conducted.

Recurrence of herpes infections is often associated with stress. Newly captured and shipped animals are subject to a great deal of stress, so that active infection in these animals is not uncommon. Animal handlers or researchers using these animals are at risk for the disease and must take proper precautions when handling macaques. Because of the dangers from B virus infection, most laboratories in the United States use only monkeys that lack antibodies to B virus, an indication that they are not infected. Although this greatly reduces the risk of handling the animals, it does not eliminate all risk because occasionally monkeys that are infected do not have detectable antibodies.

FAMILY ADENOVIRIDAE

Adenoviruses are widespread viruses of mammals and birds. The virions are a $T = 25$ icosahedron, 70–100 nm in diameter, with fibers projecting from the 12 fivefold axes of the icosahedron (Figs. 2.1 and 2.12). The virion contains 11 proteins, of which 4 are present in the core. The genome of adenoviruses is a linear 36-kb dsDNA. A terminal protein is covalently attached to the 5′ end of both strands that serves as a primer during DNA replication.

Adenoviruses are named after adenoids, a gland-like collection of lymphoid tissue in the nasopharynx. They establish a long-term infection in this tissue and were first isolated from human adenoids. Two genera are recognized. The genus *Mastadenovirus* contains viruses that infect mammals and the genus *Aviadenovirus* contains viruses that infect birds (Table 6.9). The viruses are species specific and in general will only undergo a complete replication cycle in cells isolated from their native host.

Fifty-one human adenoviruses have been distinguished on the basis of serological reactivity—an adenovirus is considered distinct if it resists neutralization by antisera against the other known adenoviruses. The 51 viruses are simply numbered in order of their isolation and are usually referred to as Ad1, Ad2, etc. The human viruses can be divided into a number of subgroups on the basis of several properties. The six groups, A through F, listed in Table 6.9 are based on serological cross reactions in a hemagglutination-inhibition assay. In this assay, the ability of an antiserum to bind to the virus and prevent it from agglutinating red blood cells is examined. An antiserum against one of the viruses of subgroup A, for example, inhibits hemagglutination by all members of that subgroup but not by members of other subgroups. The grouping in Table 6.9 correlates with a number of other properties of the viruses as well, such as their ability to form tumors in rodents, and is a convenient way of classifying relationships among these viruses.

TABLE 6.9   Adenoviridae

Genus/members			Virus name abbreviation	Usual host(s)	Disease in natural host[a]
Mastadenovirus					
Human adenoviruses (51 serotypes)				Humans	
Subgroup	A	types 12,18,31	HAdV-A		Enteritis
	B	types 3,7,11,14, 16, 21, 34, 35, 50	HAdV-B		Enteritis; military recruits' disease (3, 7, 14, 21); type 35 causes pneumonia in elderly and immunocompromised humans
	C	types 1,2,5,6	HAdV-C		Respiratory infection in children
	D	types 8,9,10,13,15, 17,19,20,22–30,32 33,36–39,42-49, 51	HAdV-D		Enteritis
	E	type 4	HAdV-E		Enteritis, pneumonia, and upper respiratory disease in military recruits
	F	types 40,41	HAdV-F		Infant diarrhea
Murine adenovirus A			MAdV-A	Mice	?
Bovine adenovirus A, B, C			BAdV-A, -B, -C	Cattle	
Ovine adenovirus A, B			OAdV-SA, -B	Sheep	Asymptomatic or mild respiratory disease
Porcine adenovirus A, B, C			PAdV-A, -B, -C	Swine	
Canine adenovirus, types 1 and 2			CAdV	Dogs	Hepatitis (type 1) Respiratory disease (type 2)
Adviadenovirus					
Fowl adenovirus A, B, C, D, E			FAdV-A, -B, -C, -D, -E	Chickens, ducks	Hepatitis, egg drop syndrome, duck hepatitis (rare)
Goose adenovirus			GoAdV	Geese	?
Pheasant adenovirus				Pheasants	?
Turkey adenovirus types 1–3				Turkeys	Bronchitis, enteritis

[a]In general adenoviruses are transmitted by both aerosols and fomites and by the oral/fecal route.

Because they replicate to high titer in cultured human cells, several human adenoviruses have been intensively studied, in particular Ad2, Ad5, and Ad12. Further interest has been generated by the fact that, although they will not undergo a complete replication cycle in rodent cells, they will infect and transform rodent cells in culture. Members of subgroup A will also cause tumors in rodents at a high rate and members of subgroup B at a moderate rate. Other human adenoviruses cause tumors in rodents at a low or undetectable rate. There is no evidence that adenoviruses are associated with tumors in man, however. Because of the extensive background of information about them and the fact that they usually cause only minor illness, attempts are being made to use them as expression vectors for gene therapy (Chapter 9).

Transcription of Adenovirus mRNAs

Adenovirus replication takes place in the nucleus. After the entry of the infecting genome into the nucleus, it is tran-scribed by host RNA polymerase II to produce a set of early RNAs. Later, a set of late RNAs is produced. A transcription map of Ad2 is shown in Fig. 6.17. Transcription of early RNAs occurs from five promoters, three on the so-called R strand and two from the L strand. The R strand is transcribed rightward on the chromosome as conventionally drawn, and the L strand is transcribed leftward. There are two other pro-moters for transcription of delayed early mRNAs. Multiple splicing of these transcripts leads to the production of about 30 mRNAs. The proteins translated from these early mRNAs are required for replication of the viral genome. The E1A and E1B gene products are oncogenes that stimulate the cell to enter S phase and thereby induce an ideal environment for the replication of the viral DNA. Their mode of action is described below. Proteins from the E2 region are directly involved in replicating adenovirus DNA and include a DNA polymerase, a ssDNA-binding protein, and a precursor to the terminal protein, which is involved in initiation of DNA replication as described below. E3 proteins modulate the host response to adenovirus infection, and this region is

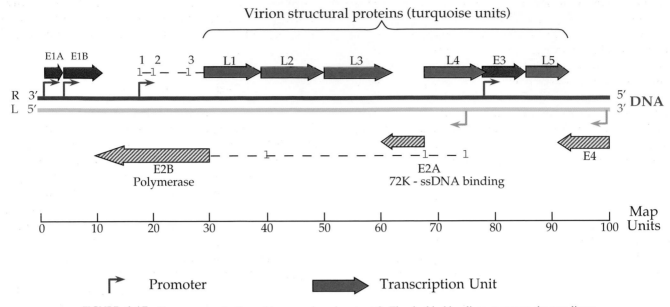

FIGURE 6.17 Genome organization of human adenovirus type 2. The double blue lines represent the two linear DNA strands that make up the genome of 36 kbp. The genes have been mapped by superimposing an arbitrary scale of 100 map units. Each arrow represents a transcription unit composed of a nested set of spliced messages, transcribed in the direction of the arrow. General functions of the various transcription units are shown. Proteins in the E3 cluster interact with the host immune system, and E4 genes are involved in DNA replication. The major late transcription unit includes the leaders "1," "2," and "3" and the L1, L2, L3, L4, and L5 families of genes. [Adapted from Wold and Golding (1991).]

nonessential in cultured cells. The functions of the E3 proteins are described in Chapter 8. Region E4 encodes proteins involved in transcription and transport of viral mRNAs and in DNA replication.

Late genes are all transcribed beginning from a single promoter on the R strand, and transcription leads to an RNA product that is about 80% the length of the adenovirus genome. Multiple splicing occurs to produce at least 18 different mRNAs that fall into five families, based on the use of five different polyadenylation sites. Each late mRNA has the same tripartite leader, formed by splicing. The late mRNAs are translated into the proteins required for the assembly of progeny virions.

The multiple splicing events that occur during processing of adenoviral mRNA, especially the late mRNA, which is made in abundance, led to the discovery of RNA splicing by Phillip Sharp. Upon examination of adenoviral mRNA–DNA hybrids with an electron microscope, he observed that regions of the genome were missing from the mRNA transcripts. For his codiscovery of RNA splicing, he was awarded a Nobel Prize, along with Richard Roberts, in 1993 (Table 1.1).

In addition to the many genes transcribed by RNA polymerase II, one or two adenovirus genes, called VA, are transcribed by host RNA polymerase III. Short VA RNA molecules are produced that are not translated. They func-

tion to inhibit the host interferon system and will be described in Chapter 8, after the description of the interferon system itself.

Replication of the Viral DNA

Inverted terminal repeats that contain the origins of replication are present at the ends of the adenovirus genome. DNA synthesis is initiated at one of the two ends and proceeds to the other end. There is no lagging strand synthesis and the partner of the strand being copied is displaced as a ssDNA (illustrated schematically in Fig. 1.7C). The precursor of the terminal protein serves as a primer during initiation. It forms a complex in solution with the adenovirus DNA polymerase, and it is assumed that these two proteins bind to the origin of replication as a complex during initiation of DNA replication. There are also binding sites within the terminal repeats for several cellular proteins that stimulate the initiation of DNA synthesis. The first step in initiation is the covalent linkage of dCMP, the first nucleotide in adenovirus DNA, to the preterminal protein. Subsequent chain elongation requires the activity of the adenovirus DNA polymerase and of the ssDNA binding protein, as well as of a cellular topoisomerase. The use of a protein primer eliminates the need for a primase to initiate DNA synthesis with an RNA primer, and thus solves the problem of how to maintain the ends of the linear adenoviral DNA molecule during replication.

The products of the first round of replication are a double-strand progeny genome and a single-strand copy of one of the two strands of the genome. Initiation of DNA synthesis can also occur on this ssDNA. It is proposed that a panhandle structure is formed by the terminal repeats so that an origin of replication is present that is identical to that in the dsDNA. Copying this ssDNA renders it double stranded and completes the production of two copies of the dsDNA genome from the parental genome.

Assembly and Release of Progeny Virions

Progeny viruses are assembled in the nucleus from preassembled hexons and pentons (see Fig. 2.12). Viral DNA is required for assembly. A packaging signal of about 260 nucleotides from the left end of the viral DNA leads to polarized encapsidation starting from this end. During assembly, the viral protease cleaves at least four viral products, and these cleavages are required to stabilize the particle and make it infectious. Release of virions from the cell is associated with the disruption of intermediate filaments. Vimentin is cleaved early after infection by an unknown protease, and cytokeratin K18 is cleaved

late by the viral protease. A schematic of the relative timing of the major events in the adenovirus life cycle is presented in Fig. 6.18. Virus infection is lytic and the cell eventually dies.

Adenovirus Oncogenes

Two early genes of adenovirus, E1A and E1B, encode proteins that induce the cell to enter S phase, in which cellular DNA is replicated. E1A targets a number of cell proteins that are involved in cell cycling, forming complexes with Rb, p107, p130, p300, and several other cellular proteins. These proteins are listed in Table 6.10 for two forms of E1A that result from differential splicing, called 12S E1A (because it is translated from an mRNA that sediments at 12S) and 13S E1A (because its mRNA sediments at 13S). Rb or retinoblastoma susceptibility protein is a tumor suppressor. It was first identified because it is absent in patients suffering from retinoblastoma. In its hypophosphorylated form, Rb binds a cellular transcription factor called E2F and causes the cell cycle to arrest in G1. Hyperphosphorylation of Rb causes it to dissociate from E2F. Free E2F activates the transcription of genes that cause the cell to enter S phase. The binding of Rb by E1A prevents it from complexing with E2F, and E2F is thus free

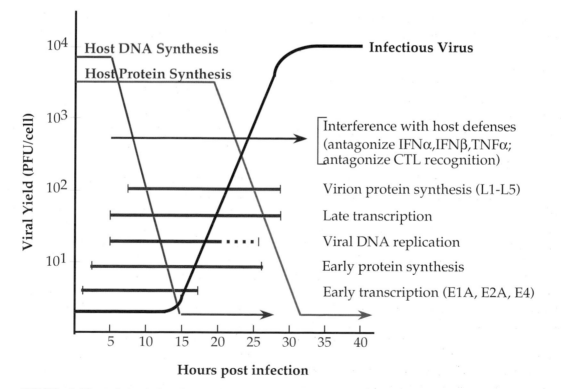

FIGURE 6.18   Relative timing of major events during the adenovirus life cycle. Example shown is for HeLa cells infected with adenovirus at a multiplicity of infection (MOI) of 10. L1–L5 are the major late transcription units mapped in Figure 6.17. [From Fields *et al.* (1996, p. 2119) and Ginzberg (1984, Fig. 1, p. 2).]

TABLE 6.10   Cellular Proteins That Bind Directly to Adenovirus E1A Proteins[a]

E1A binding protein	Binds to[b]		Other proteins in complex
	12S E1A	13S E1A	
pRb	+	+	None known
p107	+	+	Cyclin A, cdk2/cyclin E, cdk2
p130	+	+	Cyclin A, cdk2/cyclin E, cdk2
p300	+	+	None known
TATA binding proteins (TBP)	−	+	TBP-associated factors
ATF-2	−	+	None known
YY1	−	+	None known

Source: Adapted from Fields *et al.* (1996, p. 2122).

[a] E1A is the first viral transcription unit to be expressed. In the early phase of infection, differential splicing results in two mRNAs from this gene, which sediment at 12S and 13S, respectively. The translation products of these two mRNAs are referred to as 12S E1A and 13S E1A.
[b] Rb is the retinoblastoma susceptibility protein; cdk2 is cyclin-dependent kinase 2; ATF-2 is a member of the ATF family of transcription factors; YY1 is a human transcriptional repressor. These cellular proteins have been shown to bind directly to the E1A protein in a biologically relevant way. See also Fig. 6.19.

to induce the cell to enter S phase. p107 and p130 are other members of the Rb family that also interact with E2F, as well as with cyclins and cyclin-dependent kinases, whose activities are disrupted by binding to E1A. Binding of E1A to p300 appears to be an independent method of disrupting the cell cycle; p300 is thought to bind to DNA and activate transcription of factors involved in cell cycle progression.

The E1B 55kDa protein also targets a tumor suppressor protein called p53. p53 is another cellular protein that regulates cell cycle progression. It is the most commonly mutated gene associated with human tumors. It is both a transcriptional activator and repressor. It activates the transcription of genes whose function is to arrest cell cycle progression. E1B blocks the activity of p53, and when p53 is inactive or absent, cell cycle progression continues. Ad5 E1B binds p53 and sequesters it outside the nucleus, whereas Ad12 E1B does not appear to bind p53 but inhibits its activity in some indirect way.

The net result of the expression of E1A and E1B is the continued cycling of the cell. The expression of these two genes alone will transform cells in culture. Rat cells transformed by Ad12 (subgroup A) will produce tumors in syngeneic newborn rats but cells transformed by Ad2 or Ad5 (subgroup C) are not tumorigenic. The differences in the ability to induce tumors appear to be in the immune responses of the host. In particular, Ad12 appears to interfere more effectively with host CTL responses, as

described in Chapter 8. Although some adenoviruses cause tumors in rats, there is no evidence that they do so in humans. The regulation of cell cycling may be different in rodents and humans, or the adenovirus proteins may interact with rodent and human regulatory proteins in different ways.

The proteins targeted by E1A and E1B are key regulatory elements of the cell. The polyomaviruses and the papillomaviruses target many of these same proteins in order to induce cell cycling, emphasizing their importance in the control of the cell cycle. The oncogenes of these three families of viruses and their interactions with these key cellular proteins are illustrated schematically in Fig. 6.19.

A. Adenovirus

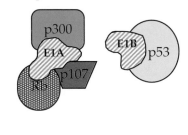

B. SV40

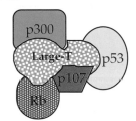

C. Papillomavirus

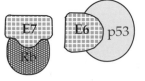

FIGURE 6.19   Known interactions between the oncogenic proteins (shaded with pink patterns) of an adenovirus, a polyomavirus, and a papillomavirus and cellular proteins that are regulators of cell cycle progression. (A) Protein E1A of adenoviruses binds to the Rb family, promoting entry into S phase. The 19-kDa form of E1B also binds to p53, blocking apoptosis. (B) The large T antigen of the polyomavirus, SV40, interacts with the Rb family of proteins as well as with p53 (see also Fig. 6.25). (C) The human papillomavirus proteins E6 and E7 bind Rb and p53, respectively, the latter promoting the destruction of p53. [Adapted from Berg and Singer (1997, p. 61).]

Interference with Host Defenses

Adenoviruses interfere with host antiviral defenses in multiple ways. This topic is covered in detail in Chapter 8, so only a summary is presented here. Adenoviruses have two independent mechanisms to suppress the interferon system. First, E1A inhibits the activation of interferon response genes by inhibiting the activity of ISGF3, a cellular transcription factor. Second, VA RNA prevents the activation of a protein kinase called PKR, which is one of the major effector products of the interferon pathway.

Adenoviruses also inhibit the lysis of infected cells by CTLs. Ad12 E1A protein blocks the transcription of the genes for class I MHC molecules, whereas the Ad2 or Ad5 E3 19-kDa protein prevents the export of class I MHC molecules to the cell surface. In either case, presentation of peptide antigens to CTLs is blocked.

Apoptosis is a cell suicide pathway by which cells die after infection by many viruses. Adenoviruses encode proteins that delay the advent of apoptosis in order to give the virus more time to replicate. First, E3 region proteins inter-fere with the action of TNF-α, blocking TNF-α-induced apoptosis. Second, E1B blocks, or at least delays, apoptosis otherwise induced by E1A.

Because adenoviruses block the antiviral defenses of the host, they have the ability to persist in the infected host for considerable periods of time. Virus may be present in tonsils and adenoids and may be shed in the stools for a year or more following primary infection (Fig. 6.20).

Adenoviruses and Human Disease

The human adenoviruses replicate primarily in the upper respiratory tract or in the gut. Some replicate well in both while others express a tropism for one or the other. Spread of the viruses is by a respiratory route or by an oral/fecal route. Many infections by adenoviruses appear to be asymptomatic, but about 5% of acute respiratory disease in children under 5 years old is due to adenovirus infection. Some serotypes can also cause gastroenteritis, but the overall

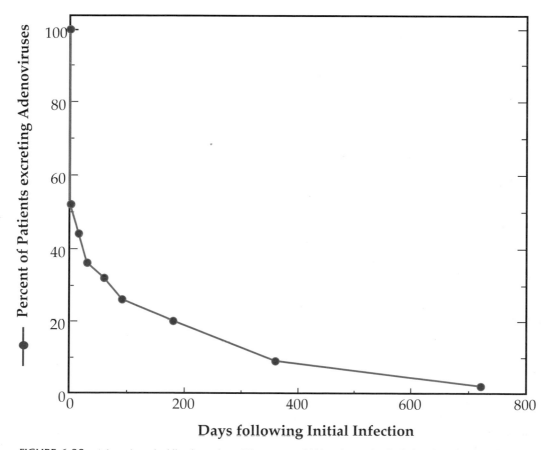

FIGURE 6.20 Adenovirus shedding by patients. The percent of 133 patients who shed virus in stools for at least the number of days indicated after an initial adenovirus infection is plotted. Note that the last point is almost 2 years. [Data from Strauss (1984, p. 459).]

importance of these viruses as causative agents of gastroenteritis is not resolved. Ad1, 2, and 5 are the most common viruses found in human populations, and antibodies to these viruses are present in about one-half of all children.

Virus can be shed for months following primary infection, especially in stools (Fig. 6.20). Children who are shedding virus are infectious and spread is particularly efficient between close family members. Figure 6.21 diagrams the spread of infection within four families in which older siblings were excreting adenovirus at the time of the birth of the youngest child. In all cases the new baby was infected within a year. Day care centers are also important in the spread of these viruses.

Adenoviruses also cause respiratory disease in adults and probably account for about 3% of such illnesses. The disease is usually mild, but Ad4 and Ad7 have caused epidemics of more serious respiratory illness in military recruits. Such epidemics of acute respiratory disease have resulted in the infection of 80% of the recruits in a unit and 20–40% of these have required hospitalization. The stress, crowding, and bringing together of young men from different backgrounds and from all over the country seems to potentiate the illness. Illness may be exacerbated if infection begins by deep inhalation of aerosolized virus, which results during the vigorous exercise that is required of recruits. Such epidemics result in protective immunity because seasoned troops do not suffer recurrent epidemics (Fig. 6.22). The military used a vaccine for many years to prevent the disease. This vaccine consisted of live nonattenuated virus that was encapsulated in coated capsules and taken orally. Infection of the enteric tract is asymptomatic for these viruses under these conditions and results in a good antibody response, presumably including mucosal antibodies to protect the respiratory tract. The vaccine is no longer available, and after the army stopped its vaccination program there have been several epidemics of adenovirus respiratory disease in recruits, in which a number of soldiers have died.

Adenoviral epidemics such as have occurred in military recruits are rare in the civilian sector but do occur. As one example, there was an epidemic of Ad11 in 1997 in a South Dakota job training facility. Like a military camp, this facil-

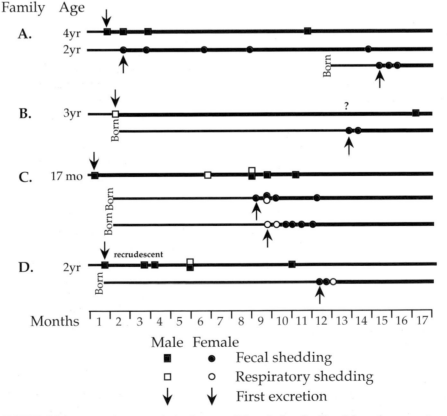

FIGURE 6.21 Adenovirus transmission between siblings in four families. Schematic representation of infection with adenovirus of newborn babies in families where one or more older siblings were already infected. The newborns became infected between 3 and 12 months after birth. It is believed that shedding of virus was continual (red lines) after the first virus-positive stool sample from a given child. [From Fox *et al.* (1969).]

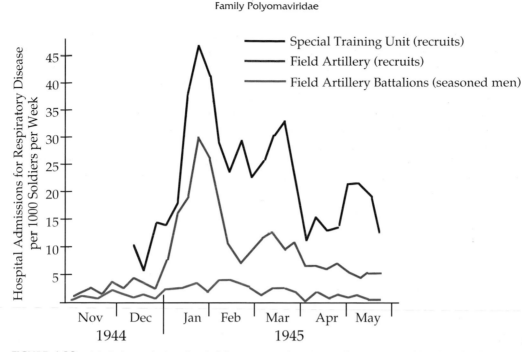

FIGURE 6.22 Admission to the base hospital for treatment of respiratory disease, presumably primarily adenovirus related, of two recruit groups and one group of seasoned troops. [From Dingle and Langmuir (1968).]

ity had young adults housed in crowded conditions with regular multiple introductions of new susceptible persons.

Adenoviruses are being developed as vectors to immunize people against other viruses and for gene therapy. Two approaches are being used. One is to use the military experience with the Ad4 and Ad7 oral vaccines to develop vaccines that could be delivered safely using this route. A second approach is to disable the virus by deleting some of the early genes. Such an attenuated virus has been used in clinical trials for treatment of cystic fibrosis. In these trials, virus that expresses the gene that is defective in such patients is delivered directly to the lungs. Trials that use adenoviruses in an attempt to treat other diseases are also ongoing. The use of adenovirus for gene therapy has not been very successful to date, and the recent death of a patient in a gene therapy trial from adenovirus has led to retrenchment in this area. This topic is discussed in more detail in Chapter 9.

FAMILY POLYOMAVIRIDAE

Until recently, the polyomaviruses and the papillomaviruses were considered to be two subfamilies within the family Papovaviridae. The name *papova* came from *pa*pilloma virus/*po*lyoma virus/simian *va*cuolating virus (= SV40), three characteristic members of the enlarged family. Recently, the ICTV has elevated the two subfamilies to the status of full families, and that is the treatment followed here.

A partial listing of the members of the family Polyomaviridae is given in Table 6.11. Polyomaviruses of humans, monkey, mouse, rat, hamster, cattle, rabbit, and parakeet are known. The two best studied viruses are mouse polyomavirus, so-called because it causes many (*poly*) different kinds of tumors (*omas*) in mice, and simian virus 40 (SV40), which infects monkeys. SV40 was first recognized as a contaminant in monkey kidney cultures used for the production of polio vaccine. Two, or possibly three, polyomaviruses infect humans. BK virus and JC virus are well-known human viruses, and SV40 or an SV40–like virus may also circulate in humans. The human viruses have received increasing attention as disease agents.

Polyomaviruses are icosahedral viruses 45 nm in diameter with pseudo-$T = 7$ symmetry (Figs. 2.1 and 2.5). The structure of two polyomaviruses, SV40 and mouse polyomavirus, have been solved to atomic resolution (Fig. 2.10). The polyomavirus genome is a circular dsDNA molecule that is 5 kb in size. In the virus it is complexed with cellular histones to form a supercoiled minichromosome.

The Early Genes

Transcription maps of SV40 and mouse polyoma virus are shown in Fig. 6.23, in which the genomes have been linearized for ease of presentation. The genomes are divided into two domains, an early domain that is transcribed from one DNA strand and a late domain transcribed from the other strand. An origin (Ori) that serves for both DNA

TABLE 6.11   Polyomaviridae[a]

Genus/members	Virus name abbreviation	Usual host(s)	Transmission	Disease
Polyomavirus				
Murine polyoma	MPyV	Mice	Reactivation of persistent infection, virus shedding in urine, contact, aerosols, sexual transmission	Tumors when inoculated into newborn mice, virus persists in the kidney
Simian virus 40	SV-40	Monkeys		PML-like disease[b] in rhesus monkeys and immunocompromised macaques; may cause childhood brain tumors in humans
Bovine polyomavirus	BPyV	Cattle		Common infection, persists in the kidney
BK polyomavirus	BKPyV	Humans	Transmission as above plus tissue transplantation in humans	Common infection of early childhood; virus persists; tumors in immunocompromised humans
JC polyomavirus	JCPyV	Humans		Common infection of late childhood; causes PML in immunocompromised hosts
Budgerigar fledgling polyomavirus	BFPyV	Parakeets		Fatal illness in fledgling parakeets
Plus other viruses infecting monkeys, hamsters, and rabbits				

[a]Most polyomaviruses are worldwide in distribution.
[b]PML, progressive multifocal leukoencephalopathy.

replication and for transcription by RNA polymerase II, and that also contains enhancer elements, is present at the junction of the two domains.

Following infection, the viral DNA is transported to the nucleus, whether as part of a virion or subviral particle or as free DNA is not clear, and early mRNA is transcribed by cellular RNA polymerase. The polyoma or SV40 promoters are strong promoters, active in many cells. Because of this, the SV40 promoter has been used to drive expression of foreign genes in many different expression systems. Early mRNA transcribed from the viral genome terminates about halfway around the genome and is differentially spliced to yield two mRNAs in SV40 or three mRNAs in polyoma (Fig. 6.23). The translation products of these mRNAs are called T antigens (tumor antigens) because their expression in the absence of productive viral infection leads to cell transformation and the formation of tumors in animals. The two SV40 proteins are called the small t antigen and the large T antigen, whereas the polyoma antigens are called small t, middle T, and large T antigens.

The large T antigens (size: about 700 amino acids) are multifunctional proteins that interact with viral promoters and several cellular proteins. A schematic representation of the large T antigen of SV40 that illustrates the locations of many of its activities within the protein sequence is shown in Fig. 6.24. These activities are differentially regulated by phosphorylation of serine and threonine residues, as shown. The large T antigen possesses a nuclear localization signal that causes it to be transported to the nucleus, a domain that binds to sequence elements within Ori, phosphorylation sites in two widely separated regions that serve to regulate its activities, domains that interact with DNA polymerase α, a domain that controls the host range of the virus, a domain that has DNA helicase activity, and domains that bind cellular proteins that play regulatory roles in cell cycling, among others.

Binding of large T antigen to the Ori regulates its own production. Binding is also required for DNA replication. Binding of SV40 large T antigen to Rb and p53, as well as to p107 and p300, is illustrated in Fig. 6.19. Binding to these tumor-suppressor proteins results in continued cell cycling and transformation of the cell. As described earlier, this induces a state suitable for viral DNA replication.

The activities in polyoma virus T antigens are distributed somewhat differently. Polyoma large T antigen binds Rb but does not bind p53. Binding to Rb induces cellular DNA synthesis. Cell transformation is a function of middle T antigen. Middle T antigen binds to *src*, a known proto-oncogene located at the cell surface, and activates its kinase activity (Fig. 6.25). The phosphorylated middle T antigen interacts with other cellular factors and induces cell transformation, and it has been shown that expression of middle T antigen alone is sufficient to transform a cell.

The oncoproteins of these small DNA viruses are excellent examples of how a viral protein may be multifunctional and mediate a variety of functions important for the outcome of infection.

A. SV40 (5243bp)

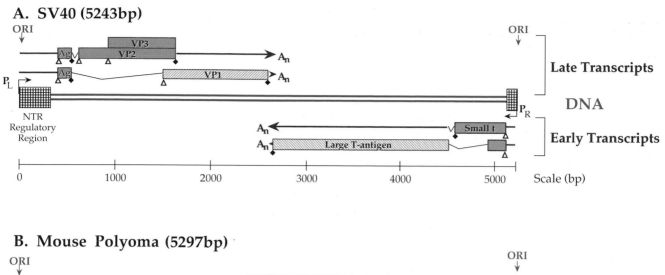

B. Mouse Polyoma (5297bp)

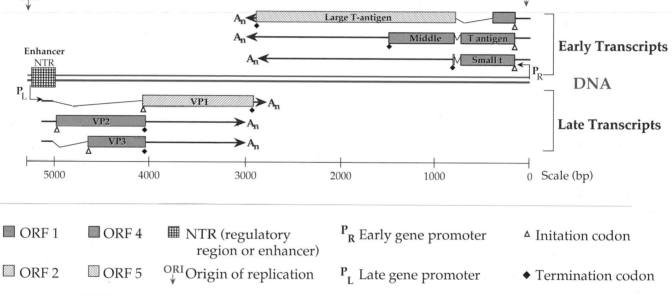

▦ ORF 1	▦ ORF 4	▦ NTR (regulatory region or enhancer)	P_R Early gene promoter	△ Initation codon
▨ ORF 2	▨ ORF 5	$\overset{ORI}{\downarrow}$ Origin of replication	P_L Late gene promoter	◆ Termination codon

FIGURE 6.23 Comparisons of the genome organizations of two polyomaviruses: SV40 and mouse polyoma. Both are circular genomes that have been linearized at the origin of replication (Ori) for ease of presentation. The genome of mouse polyoma virus is shown here in the opposite sense (right to left) to that of SV40. Ag is the agnoprotein. [Redrawn after Brady and Salzman (1986).]

DNA Replication

Large T antigen binds to specific sites within the Ori to promote replication of the viral genome. Binding first unwinds the DNA. Then T antigen associates with replication protein A followed by DNA polymerase α primase to form an initiation complex. Association with primase is species specific. As a result, SV40 productively infects only monkey cells and mouse polyomavirus only mouse cells. After initiation by primase, DNA polymerase takes over and replication proceeds. DNA synthesis is bidirectional and when the replication forks meet about halfway round the molecule, the daughter genomes separate, aided by topoisomerase II (Fig. 1.7A).

The Late Genes

Large T antigen also regulates the transcription of late mRNAs, which are transcribed from the opposite strand as the early mRNAs (Fig. 6.23). Differential splicing leads to

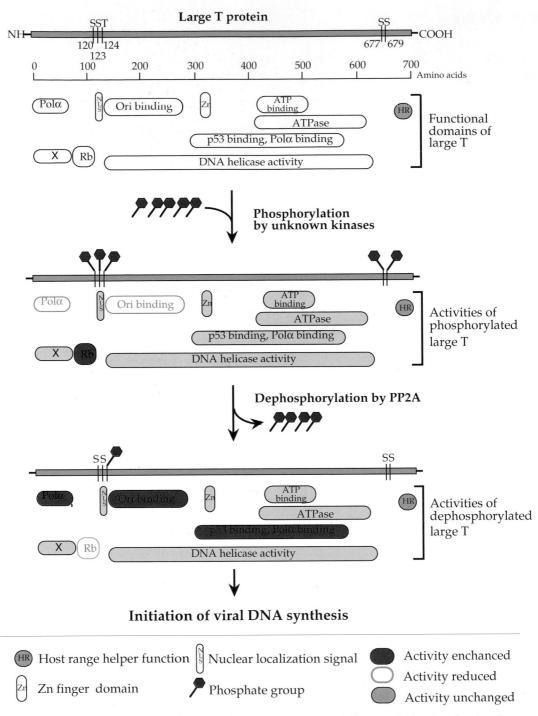

FIGURE 6.24 Functional domains of SV40 large T antigen. The top panel illustrates the location of various functional domains and of the serine and threonine residues that are phosphorylated. The second panel shows the functions of fully phosphorylated large T antigen. The bottom panel shows the activities of large T singly phosphorylated on threonine 124, which can bind to the origin of replication to initiate DNA synthesis. In the middle and lower panels, blue domains are unchanged in activity, gray domains are reduced in activity, and red functions are strongly increased. Other abbreviations: NLS, nuclear localization signal; Zn, zinc finger domain; HR, host range helper function. [From Berg and Singer (1997, Figs. 1.23 and 1.24) and Fields *et al.* (1996, Fig. 6, p. 2011).]

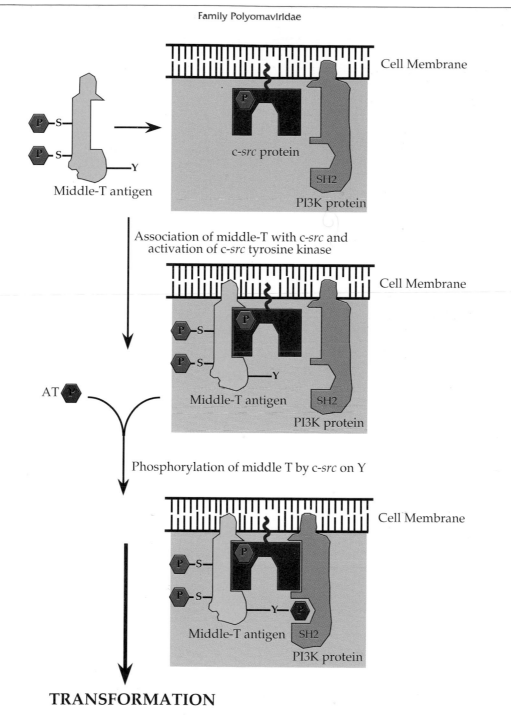

TRANSFORMATION

FIGURE 6.25   Interactions of polyoma middle T antigen with membrane proteins to initiate cellular transformation. Phosphatidylinositol-3 kinase (PI3K) and c-*src* are localized in the plasma membrane. The *src* kinase phosphorylates middle T on a tyrosine residue (Y). This generates binding sites for a variety of other cellular signaling proteins such as PI3K. The resultant complex closely resembles an activated growth factor receptor bound to signal transduction proteins. Serine residues (S) on middle T that are phosphorylated by other protein kinases are also indicated. [Drawn from data in DiMaio *et al.* (1998).]

two mRNAs in SV40 and three mRNAs in polyoma. One mRNA is translated into VP1, the major virion structural protein. In SV40 the second mRNA is translated into both VP2 and VP3, whereas in polyoma virus VP2 and VP3 are translated from different mRNAs. VP3 is a truncated form of VP2, consisting of the C-terminal 60% or so of VP2. Both are minor components of the virion. VP2 is myristylated and may serve an entry function. The three structural proteins are transported to the nucleus and assembly of virions takes place there.

VP1, when expressed alone, assembles into virus-like particles. If VP2 and VP3 are coexpressed with VP1, they assemble into virus-like particles whose composition is the same as that of the virion.

Release of virions is an active process. Membrane vesicles form and transport virions to the cell surface, where they are released. Infection by polyomaviruses is lytic. Expression of the late genes and assembly of progeny virions results in the death of the cell.

Human Polyomaviruses

The two well-known human polyomaviruses are BK virus and JC virus. These viruses were first isolated in 1971, JC from the brain of a patient with progressive multifocal leukoencephalopathy (PML) and BK from the urine of an immunosuppressed renal transplant patient. They were named after the initials of the patients from whom they were isolated. These two viruses share 75% nucleotide sequence identity. In addition to these two viruses, there are recent reports that an SV40–like virus may circulate in humans. A comparison of the proteins encoded by JC, BK, and SV40 is given in Table 6.12. Shown are the number of amino acids in each protein and the amino acid identity between any two of these viruses. This table makes obvious the close relationship among these viruses.

Most humans in the United States are infected with BK virus before the age of 10 (Fig. 6.26). Infection with JC virus usually occurs somewhat later, but by the age of 14 the majority of the population has been infected. Primary infection with BK virus has been associated with mild res-

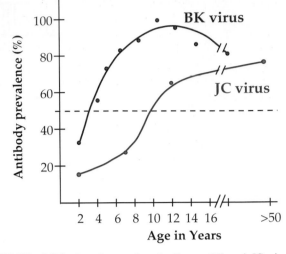

FIGURE 6.26 Prevalence of antibodies to BK and JC viruses in humans in the United States as a function of age. [From Fields *et al.* (1996), p. 2039).]

piratory disease or cystitis (bladder infection) in young children, but most infections with either BK or JC virus are not associated with illness. The viruses establish a latent infection that persists indefinitely, and the virus may reactivate after many years. The latent infection appears to be established in the kidneys, in B lymphocytes, and, for JC virus, perhaps in the brain. Reactivation may be brought about by immunosuppression or by factors such as pregnancy or diabetes, and results in the excretion of virus in the urine. Reactivation of JC virus resulting from immune suppression is serious because it can replicate in oligodendrocytes and result in PML. PML is a rare, subacute, demyelinating disease of the central nervous system that has a worldwide

TABLE 6.12   Virus-Coded Proteins of Primate Polyomaviruses

Protein	Number of amino acids			% Amino acid identity			Function
	JC	BK	SV40	JC/BK	JC/SV40	BK/SV40	
Late							
VP1	356	362	364	77.9	76.4	82.4	Major capsid protein, attaches to cellular receptors, hemagglutination, HI and NT epitopes
VP2	344	351	352	78.8	73.4	78.5	Minor capsid protein
VP3	225	232	234	74.5	67.2	73.6	Minor capsid protein
Agnoprotein	71	66	62	59.1	45.0	53.2	Facilitates capsid assembly
Early							
Large T	688	695	708	86.6	72.0	73.9	Initiation of replication; stimulates host DNA synthesis; modulates transcription, transformation
Small T	172	172	174	79.6	67.8	69.5	Necessary for efficient viral DNA replication

Source: Adapted from Fields *et al.* (1996, p. 2030) and additional data from Walker and Frisque (1986).

distribution. It is an infrequent complication of a wide variety of conditions, including Hodgkin's disease, chronic diseases such as tuberculosis, primary acquired immunodeficiency diseases such as AIDs, or immunosuppression following organ transplant. The frequency of PML has increased with the AIDS epidemic and PML is now recognized as one of the AIDS-defining illnesses—it occurs in 0.7% of all AIDS cases reported to the Centers for Disease Control and Prevention. PML may be more likely to occur when immunosuppression is due to infection by HIV because HIV-1 transactivates the JC late promoter. The disease progresses rapidly and can lead to mental deterioration and death within 3–6 months after onset.

Polyomaviruses cause tumors in laboratory animals and many attempts to associate BK or JC virus with human cancer have been made. Association of BK with certain human malignancies has been reported, but these associations are not consistent and no convincing evidence exists that the association is causative rather than adventitious.

Early lots of poliovirus vaccine were contaminated with SV40 virus, which shares 69% sequence identity with JC and BK viruses (Table 6.12), and many people were infected with this virus during the poliovirus vaccination campaigns. Extensive study of this cohort of people, both in the United States and in Europe, has not revealed any convincing evidence for tumors associated with SV40 infection. The fact that SV40 infects man is of interest because there are recent reports that an SV40–like virus of humans is circulating in the United States. There are suggestive data that this virus may be associated with certain brain tumors in children.

FAMILY PAPILLOMAVIRIDAE

Papillomaviruses resemble polyomaviruses in structure but are larger (Fig. 2.5). The virion is 55 nm in diameter, and the circular dsDNA genome is 8 kb in size. Replication and assembly of progeny virus occur in the nucleus. A partial listing of papillomaviruses is shown in Table 6.13. They are primarily mammalian viruses, but there are a few avian representatives in the family. The viruses are species specific. They infect epithelium and will undergo a complete replication cycle only in terminally differentiated cell layers.

On infection, papillomaviruses induce cellular proliferation that leads to the production of warts or papillomas. In most infections these eventually resolve, but in some cases tumors can result. Papillomaviruses of humans, cattle, sheep, and cottontail rabbit have been shown to be associated with cancers in their natural hosts. Our knowledge of the replication of the papillomaviruses is limited because none will undergo a full replication cycle in any simple tissue culture system. Bovine papillomavirus (BPV-1) has been the most extensively characterized because it proliferates in dermal cells as well as in terminal epithelial cells. It readily infects and transforms rodent cells, in which the early proteins are expressed and viral DNA replication occurs. In transformed cells, the BPV genome is maintained as a stable plasmid and this feature has permitted the use of BPV-1 as an expression vector. Much effort has also been put into the study of the human papillomaviruses (HPVs) because of their association with human cancer, but these studies have been hampered because human papillomaviruses will only grow in humans and human cells and will only undergo a full lytic cycle in terminally differentiated cells. Certain of the HPVs will immortalize human keratinocytes and will transform rodent cells in culture, albeit not as efficiently as BPV-1, which permits the study of the transforming genes. The recent development of tissue culture systems in which epidermal cells will differentiate now permits at least limited studies of a full growth cycle, as do studies using grafts of infected human tissue into immunocompromised mice.

TABLE 6.13   Papillomaviridae[a]

Genus/members	Virus name abbreviation	Usual host(s)	Transmission	Disease
Papillomavirus				
Cottontail rabbit papillomavirus	CRPV	Cottontail rabbit	Reactivation of persistent, infection, virus shedding in urine, contact, aerosols, sexual transmission	Benign papillomas; carcinomas
Human papillomavirus 1–82	HPV	Humans		Genital warts; dysplasias; cervical carcinoma (types 16, 18)
Bovine papillomavirus 1–4	BPV	Cattle		Fibropapillomas; warts; carcinomas
Plus papillomaviruses of numerous other species				

[a]Most papillomaviruses are worldwide in distribution.

Transcription of mRNAs

The genome organizations of two papillomaviruses, BPV-1 and HPV-11, are shown in Fig. 6.27. Also shown are transcription maps, with the HPV-11 map being presented in some detail. In the absence of cell culture systems for the virus, most studies of RNA transcription have used RNA extracted from papillomas or from carcinoma cell lines. Because of this, the maps are probably not complete.

The transcription of papillomavirus RNAs is complex. There are multiple promoters and splice sites and differential use of these in different cells. Furthermore, there is extensive overlap of genes in the genome. Different reading frames encoding different peptide sequences may be linked in different ways by alternative splicing. The complexity of the mRNAs produced from the HPV-11 genome is illustrated in Fig. 6.27C. In contrast to polyomaviruses, all papillomavirus mRNAs are transcribed in the same direction from only one of the two strands.

In BPV-1 there are at least seven promoters for transcription of RNA and more than 20 different mRNAs have been identified. Six promoters are used for the transcription of early mRNAs, all of which terminate at a poly(A) addition site at position 4180 (A_E). The seventh promoter is used for the transcription of the late mRNAs, which terminate at a poly(A) addition site at 7156 (A_L). Late genes are expressed only in terminally differentiated epithelial cells.

The early genes of BPV-1 include E1 and E2, required for DNA replication, and E5, E6, and E7, required for cell transformation. The late genes include L1 and L2, which encode proteins present in the virion. L1 is the major capsid protein and when expressed alone assembles into virus-like particles that appear identical to virions. If L2 is coexpressed with L1, it is also incorporated into the virus-like particles. Another late gene is E4, which although located in the early region is expressed from the late promoter.

The pattern of transcription in HPV-11 is slightly different. Only three promoters are known, two for the early genes and one for the late genes (Fig. 6.27C). There are poly(A) addition sites for early and late transcripts. Proteins corresponding to those of BPV-1 are produced during HPV-11 infection, but the complexity of the pattern of proteins produced and the difficulties in studying HPV replication make exact comparisons difficult.

DNA Replication

DNA replication requires the activities of E1 and E2. The E1 protein binds to the origin of replication, has helicase activity, binds α-primase, and is presumed to promote the initiation of DNA replication. Thus, it has many of the functions of the polyomavirus T antigens. E2 is a regulatory protein that is produced in multiple forms (Fig. 6.28). It can either transactivate or repress genes, depending on the location of binding sites for it within a gene. It plays an important role in the regulation of transcription, but its role in DNA replication is not well defined. It appears to be required only for efficient replication of DNA. It may interact with E1 to recruit host replication factors to the origin of replication or to promote the assembly of the preinitiation complex.

DNA replication occurs in two phases. In cells that the virus infects nonproductively (cells transformed by BPV-1 or cells in the dermal layer of warts), the DNA is maintained as a multiple copy plasmid (50–400 copies per cell). After replicating sufficiently to reach this number of copies, further DNA replication is limited to that required to maintain the copy number as cells divide. A complete replication cycle occurs in terminally differentiated cells, and large numbers of DNA genomes are produced for incorporation into progeny virions.

The Transforming Genes

Three genes of papillomaviruses, E5, E6, and E7, have been shown to be involved in transforming cells. E5 of BPV-1 is a small polypeptide (44 amino acids) that is believed to have a short N-terminal cytoplasmic domain, a transmembrane region, and a more extensive C-terminal extracellular domain. It activates the β receptor for platelet-derived growth factor, perhaps by binding to it and causing it to dimerize. Dimerization of many growth factor receptors present at the surface of cells results in activation of a protein kinase and phosphorylation of tyrosines, which leads to the activation of transcription factors whose activities stimulate cell proliferation. BPV-1 E5 also binds other cellular proteins that may be involved in transformation. Transformation of established rodent cells as defined by a number of criteria can be obtained by expression of E5 alone, but the expression of both E6 and E7 is required for the fully transformed phenotype.

The E5 encoded by HPVs has also been shown to induce some transforming alterations, but whether this protein plays an important role in transformation is uncertain. More is known about HPV E7. E7 from high-risk strains of HPV (strains that are often associated with human cancer) is a small zinc-binding protein that is phosphorylated. It has been shown to bind the cellular tumor suppressor protein Rb as well as p107 and p130 (Fig. 6.19). Rb undergoes changes in phosphorylation induced by cyclin-dependent kinases at the G/S1 boundary. In its hypophosphorylated form it inhibits cell cycle progression. The viral oncogene preferentially binds the hypophosphorylated form, thus preventing its inhibitory activity and inducing cycling of the cell (and therefore DNA synthesis). Genetic studies have shown that binding to Rb is required in order for E7 to transform cells. It is of considerable interest that E7 from low-risk HPVs binds Rb only one-tenth as efficiently as E7

A. Bovine Papillomavirus 1 (7945bp)

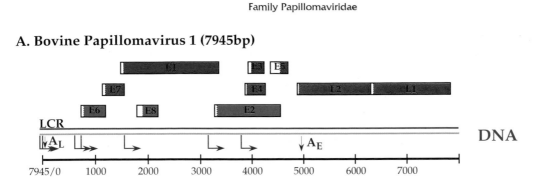

B. Human Papillomavirus 11 (7933 bp)

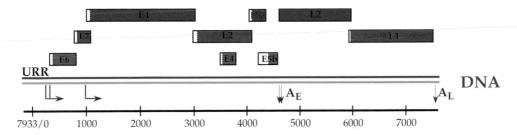

C. Transcription map of HPV 11

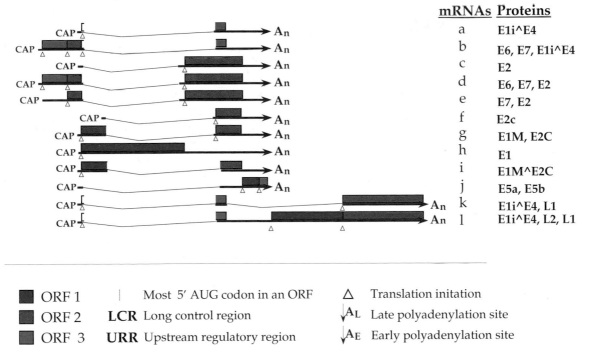

mRNAs	Proteins
a	E1i^E4
b	E6, E7, E1i^E4
c	E2
d	E6, E7, E2
e	E7, E2
f	E2c
g	E1M, E2C
h	E1
i	E1M^E2C
j	E5a, E5b
k	E1i^E4, L1
l	E1i^E4, L2, L1

■ ORF 1   ⋮ Most 5′ AUG codon in an ORF   △ Translation initation

■ ORF 2   **LCR** Long control region A_L Late polyadenylation site

■ ORF 3   **URR** Upstream regulatory region A_E Early polyadenylation site

                                            ⌐→ Transcriptional promoter

FIGURE 6.27 Genome organization and transcription map of papillomaviruses. (A) Genome organization of bovine papillomavirus 1. (B) Genome organization of human papillomavirus 11. (C) Transcription map of human papillomavirus 11. Genomes have been linearized at the upstream regulatory region (LCR or URR) for ease of presentation. All ORFs are transcribed from left to right from one DNA strand. Promoters are shown as turquoise arrows; sites of poly(A) addition with orange arrows labeled late (A_L) or early (A_E; initiation codons are shown as open red triangles. The symbol ^ joins two ORFs that are translated together as a fusion protein from a spliced mRNA; for example, E1i^E4 is a fusion protein translated from an mRNA in which the first 5 codons of protein E1 (ORF1) are spliced to the 85 C terminal codons of E4 (ORF2). [Adapted from Nathanson et al. (1996, p. 27) and from Fields *et al.* (1996, pp. 2051, 2052).]

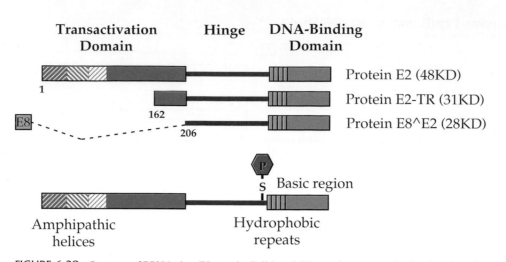

FIGURE 6.28 Structure of BPV-1 virus E2 protein. Full-length E2 contains a transactivation domain at the N terminus, linked by a hinge to a DNA-binding domain at the C terminus. There are three forms of E2, all of which contain the 85-amino-acid-long DNA-binding domain. The bottom diagram identifies other functional features of E2 such as the amphipathic helices, hydrophobic repeats, a basic region, and a phosphorylation site. [Data from McBride *et al.* (1989).]

from high-risk strains, and that the E7 from low-risk strains is inefficient in transformation assays.

E6 from high-risk HPVs, but not from low-risk HPVs, can complex with the tumor suppressor protein p53 (Fig. 6.19). SV40 and adenovirus oncoproteins also bind p53, but simply sequester it. In contrast, binding of HPV E6 leads to the degradation of p53 by the ubiquitin-mediated degradation pathway. Removal of p53 has the effect of inducing DNA synthesis, as described above. The importance of Rb and p53 in regulating cell cycling is made clear by the fact that three different viruses—high-risk HPVs, adenoviruses, and SV40—all target these proteins in order to provide an atmosphere conducive for DNA replication by the virus.

Human Papillomaviruses

More than 80 different HPVs have now been identified by analyses of the sequences of viral DNAs isolated from individual lesions. A listing of these viruses that shows the tissues infected and the probability that infection leads to cancer is shown in Table 6.14. HPVs cause warts in the skin, genital tract, mouth, or respiratory tract. The warts are normally self-limited proliferative lesions that regress after some time because of an immune response. CTLs are thought to play an important role in regression, and warts are often more numerous under conditions where the immune system is suppressed. HPVs are spread by direct contact and infection begins at the site of an abrasion in which the virus can contact the deeper epithelial layers.

HPVs are not only specific for humans but also for the tissues infected. HPVs cause either cutaneous lesions or mucosal lesions. More than 15 HPVs are known that cause skin warts. Only a few of these are responsible for most skin warts, which are common in school children. The remainder have been isolated only from patients who suffer from a rare disease called epidermodysplasia verruciformis. These patients are unable to resolve their warts, probably because of an inherited immunologic defect, and wart-like lesions appear all over the body. These warts often become malignant after many years, especially on areas of the skin exposed to sunlight, but these tumors are generally slow growing and do not metastasize.

More than 25 HPVs cause genital warts, which are among the most common sexually transmitted diseases. One study found that 46% of college women examined were positive for HPV DNA in the genital tract. Older women have a lower incidence of HPV, either because they have fewer recent partners or because they have acquired some immunity. In many cases the infection is cleared completely after some months, but the virus may remain in a latent or persistent form in apparently normal tissue adjacent to the wart and the lesions may recur. Immunosuppression results in an increased incidence of warts.

Genital HPVs are clearly associated with cervical cancer. Cancer is a rare complication of HPV infection that may take decades to develop and it requires additional genetic mutations. Because of the prevalence of HPVs, however, there are about 500,000 new cases of cervical carcinoma diagnosed annually worldwide and most, perhaps all, are associated with HPV. In developed countries with a

TABLE 6.14   Common Clinical Lesions Associated with Human Papillomaviruses

Type/isolates	Anatomical site	Disease		Risk	Cancers
		Common name	Medical term		
Cutaneous HPVs					
HPV 1,4,60,63	Sole, palm	Plantar warts	Verruca plantaris	None	None
HPV 2,4,26,28,57	Cutaneous	Common warts	Verruca vulgaris	None	None
HPV 3,10,20,21,27	Cutaneous	Flat warts	Verruca plana	None	None
HPV 5,8,12,50	Face, trunk, esophagus	Benign warts	EV	High	Skin carcinomas[a]
HPV 14,15,17	Cutaneous	Flat warts	Verruca plana, EV	Some	Some SCC
HPV 19,22–25,46,47	Cutaneous	Macular lesions	EV (benign)	None	
HPV 36	Cutaneous	Actinic keratoses	EV	?	
HPV 38	Cutaneous		Malignant melanoma	High	Skin carcinomas[a]
HPV 41,48	Cutaneous		SCC	Some	In ISP
HPV 49	Cutaneous	Plantar warts	Verruca plantaris	?	In ISP
Mucosal-Associated HPVs					
HPV 6,11,70	Anogenital, larynx	Genital warts	CD	Low	Rare[b]
HPV 13,34,40,42	Anogenital	Anogenital warts	IN	Low	Rare[b]
HPV 35,39,43–45, 61,62,64,67,68,69	Genital	Genital warts	CIN	Low	
HPV 72,73	Larynx	Oral papillomas		Low	In ISP
HPV 51,52,56,58	Genital mucosa	Genital warts	CIN	Some	Some malignant progression
HPV 16,18,31,33,66	Genital mucosa	Genital warts	CIN	High	1–3% progress to cervical carcinomas, cofactors unknown

Source: Adapted from Fields *et al.* (1996, pp. 2048, 2049, 2085) and data from Alani and Münger (1998).
[a]30–40% undergo neoplastic conversion in sun-exposed areas.
[b]On rare occasions these HPV types have also been found associated with carcinomas.
Abbreviations: SCC, squamous cell carcinoma; EV, epidermodysplasia verruciformis; CD, condyloma acuminatum; CIN, cervical intraepithelial neoplasia; IN, intraepithelial neoplasia; ISP, immunosuppressed patients.

high standard of health care, cervical cancer accounts for about 7% of cancer in women, but in developing countries cervical cancer accounts for 24% of all cancer in women. In the United States about 5000 deaths occur annually from cervical cancer.

Genital HPVs can be divided into low risk (rarely or never associated with cancer), intermediate risk, and high risk (often associated with cancer). HPV-16 and HPV-18 account for more than half of the cancers. As described above, papillomaviruses encode oncoproteins that interfere with the functions of tumor suppressor proteins, and the high-risk HPV-16 and HPV-18 interfere most strongly. This interference is almost certainly the basis of papillomavirus-induced cancer. In most cell lines derived from cervical cancer, the HPV genome has been integrated into the host chromosome, and this integration may result in higher expression of the viral transforming genes.

Genital HPVs 6 and 11 may also infect the mouth, nasal cavity, larynx, or lungs. Infection of the larynx may be prob-

lematic because of the resulting obstruction of the airways or because of hoarseness caused by infection of the vocal cords. Surgical removal of the papillomas may be required. These papillomas tend to recur, requiring further operations. There are in addition two HPVs that infect only the oral cavity.

Because of the association with cervical cancer, efforts have been made to develop vaccines. However, there is no animal model for the disease caused by these human viruses and, furthermore, it is not known whether natural infection leads to immunity, making development of vaccines problematic.

A phylogenetic tree of human papilloma viruses constructed using sequences in the E2 gene is shown in Fig. 6.29. There is no simple relationship between the relatedness of the different viruses as illustrated by this tree and the target tissue infected by the viruses or the risk of neoplastic transformation following infection by them. Note, for example, that HPV-16 and HPV-18, which cause the majority of cervical carcinoma, are widely separated in the tree.

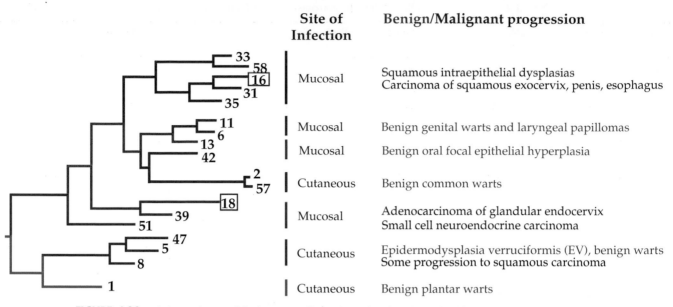

	Site of Infection	Benign/Malignant progression

FIGURE 6.29 Phylogenetic tree of the human papillomaviruses based on the nucleotide sequence of the amino-terminal half of the E2 gene. The sites of infection and the potential for neoplastic progression are shown. The numbers of the strains are shown (compare Table 6.14) and the two high-risk viruses that cause most cases of cervical carcinoma are boxed. Note that there is no simple relationship between position on the phylogenetic tree and either the site of infection or probability of progression to malignancy. [Adapted from Nathanson *et al.* (1996, p. 273).]

FAMILY PARVOVIRIDAE

The Parvoviruses are small icosahedral viruses that are 18–26 nm in diameter (Figs. 2.1 and 2.5). They contain ssDNA of about 5 kb as their genome. Different viruses variously package the minus strand (the strand complementary to the messenger sense) or a mixture of plus and minus strands. Two subfamilies are recognized, each containing three genera. The Parvovirinae are viruses of birds and mammals, whereas the Densovirinae are viruses of insects. A partial listing of the members of the Parvovirinae is shown in Table 6.15. These viruses are species specific and also specific for the spectrum of tissues that can be infected. Unlike other DNA viruses, the parvoviruses do not encode genes that induce the cell to enter S phase, and they can only replicate in cells that are actively replicating. The members of the *Dependovirus* genus of the Parvovirinae are further limited in their replication in that normally they can only replicate in cells that are infected by an adenovirus or a herpesvirus.

Transcription of the Viral Genome

The genome organizations and transcription maps for two human parvoviruses belonging to different genera are shown in Fig. 6.30. The parvovirus genome contains two genes, each of which is transcribed into multiple mRNAs. The nonstructural or replication gene is located at one end of the genome and the gene for the capsid proteins is located at the other end. The two genes are present in the same orientation so that only one strand is transcribed. B19 has only one promoter for transcription but two poly(A) addition sites, whereas AAV has three promoters for transcription but only one poly(A) addition site. The use of multiple promoters or poly(A) sites is combined with alternative splicing events to give rise to many mRNAs. The best understood translation products are a nonstructural protein of about 80 kDa and the two or three capsid proteins.

Replication of the Viral DNA

Replication of the viral DNA occurs in the nucleus. The DNA is linear and possesses palindromic sequences at the two ends. These palindromic sequences are 100–300 nucleotides long, depending on the virus, and can fold back to form a very stable hairpin structure, as illustrated in Fig. 6.31. The hairpin primes DNA replication, with the 3' end serving as a primer that is elongated to form a double-stranded intermediate, as illustrated in Fig. 6.32. How this intermediate is used to continue DNA replication and how it is resolved to give plus and minus DNA genomes is not

TABLE 6.15   Parvovirinae

Genus/members	Virus name abbreviation	Host(s)	Transmission	Disease
Parvovirus				
Mice minute virus	MMV	Mice	Contact, fomites	?
Aleutian mink disease	AMDV	Mink	Contact, fomites	Chronic immune complex disease
Bovine parvovirus	BPV	Cattle		Enteritis
Canine parvovirus	CPV	Dogs, cats	Contact, fomites	Enteritis in adults, myocarditis in pups
Kilham rat parvovirus	KRV	Rats		
Porcine parvovirus	PPV	Swine		Stillbirth, abortion, fetal death, mummification
Goose parvovirus	GPV	Geese	Vertical transmission	Hepatitis
Numerous viruses of other species				
Erythrovirus				
B19	B19V	Humans		Fifth disease, aplastic anemia, hydrops fetalis, arthritis, immunodeficiency
Dependovirus				
Adeno-associated virus 1–5	AAV	Humans	Transplacental (AAV-1), vertical (AAV-2)	None
Adeno-associated viruses of other species	BAAV CAAV	Cattle Dogs		None

clear, although it is known that the viral nonstructural protein is involved. Models have been proposed that involve either continued rolling of the hairpin or the formation of cruciform structures that might be resolved by cellular recombination enzymes. In some cases the palindromic sequence is present in two orientations. The flipping back and forth of this sequence makes it possible that a rolling hairpin is resolved by an endonuclease, a function that the nonstructural protein might perform. This mechanism is illustrated in Fig. 6.32. In other cases, however, the palindrome is found in only one orientation and models to suggest how intermediates might be resolved without flipping the sequence in the hairpin must be more complicated.

Structure of the Virion

The virion is a $T = 1$ icosahedron that is constructed of 60 molecules of capsid protein (see Fig. 2.5). The major capsid protein is called VP2 and is about 60 kDa in size. Smaller amounts of a larger capsid protein called VP1 (about 80 kDa) are present in all viruses, and some contain in addition a third capsid protein called VP3. The two or three structural proteins share significant sequence overlap (Fig. 6.30). The structure of the major capsid protein has been solved to atomic resolution and it possesses the same eight-stranded antiparallel β sandwich present in many RNA and DNA viruses (Chapter 2), which suggests that these capsid proteins share a common ancestry.

Genus Erythrovirus

B19 virus, the only member of genus *Erythrovirus*, is the only known human pathogen among the parvoviruses. Infection of humans with B19 is accompanied by nonspecific flu-like symptoms followed by symptoms of erythema infectiosum (fifth disease), which presents as a generalized erythematous rash with a "slapped cheek" appearance and inflammation of joints. Children infected by the virus are usually not very ill. However, illness in adults can be more serious because the joint inflammation may mimic rheumatoid arthritis and can persist for months or years.

B19 has a tropism for human erythroid progenitor cells, which are rapidly dividing cells capable of supporting virus replication. The virus is cytolytic and the infected cell dies. Thus, infection of erythroid progenitor cells results in a suppression of erythropoiesis for 5–7 days following infection by the virus. In healthy humans, whose red blood cells last for 160 days, this is a not a serious event. However, in people suffering from chronic anemias the inability to synthesize red blood cells for a week may be serious and occasionally fatal. In particular, patients with hemolytic anemia have a low hemoglobin concentration in the blood because their red blood cells have a short life span, only 15–20 days, so that arrest of erythropoiesis in the bone marrow leads to a sharp fall in hemoglobin concentration and worsening symptoms of anemia. Other populations at increased

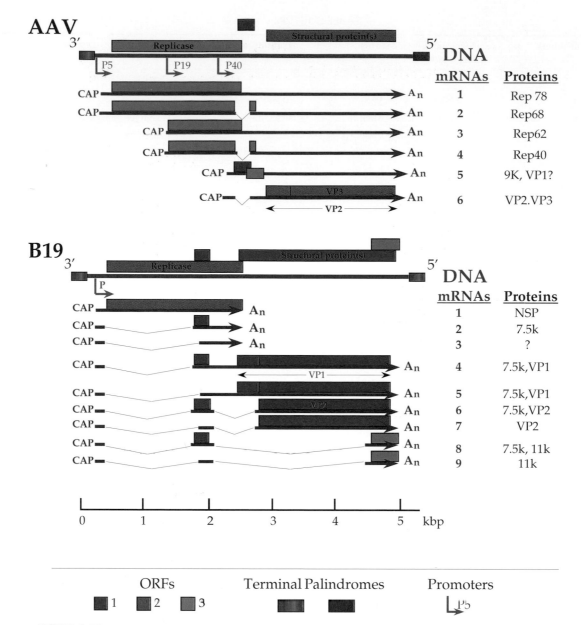

FIGURE 6.30 Genome organizations and transcription/translation schemes for two human parvoviruses belonging to different genera. Adeno-associated virus (AAV) is a dependovirus and B19 virus is an erythrovirus. These viruses use three and one promoters, respectively, to make a set of spliced and unspliced messages, all transcribed from one DNA strand, from which the various virus proteins are translated. Terminal palindromes are shown as shaded boxes. [Adapted from Cotmore (1990), Rhode and Iversen (1990), and Carter et al. (1990).]

risk following B19 infection include patients with compromised immune systems, in which infection by B19 can result in persistent anemia, and pregnant women. Congenital infection with B19 can be serious and can lead to fetal abnormalities or death, due to arrest of red blood cell formation and consequent anemia at critical times during development.

The receptor for the B19 virus is erythrocyte P antigen. This antigen is expressed on cells of the erythroid lineage, but only precursor cells can be productively infected. Mature erythrocytes are terminally differentiated, lack a nucleus, do not divide, and cannot support virus replication. In addition to possessing a receptor that allows the virus to enter, other factors required for replication are also furnished by erythroid precursor cells. Transfection of DNA into other cells does not lead to a complete replication cycle and the pattern of RNA transcription differs from that in permissive cells shown in Fig. 6.30.

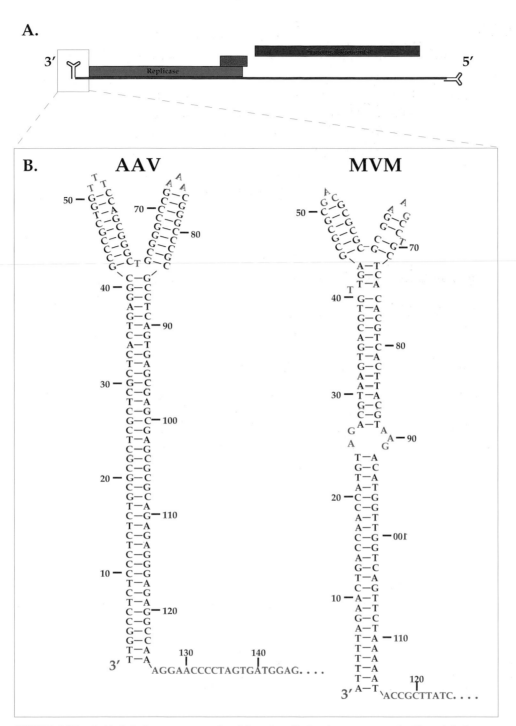

FIGURE 6.31 Stable hairpin structures predicted from the palindromic sequences at the 3′ termini of parvovirus virion DNAs. (A) Diagram of the genome, showing the location of the hairpin. (B) The most stable secondary structure predicted by the 3′ terminal nucleotide sequences of MVM DNA and AAV DNA minute virus of mice (MVM) is now called mice minute virus (MMV) as shown in Table 6.15. [Adapted from Fields *et al.* (1996, p. 2175).]

B19 is a common virus. About 50% of adults have antibodies against the virus, which show that they have been previously infected, and in the elderly this rises to more than 90%. Infection of healthy people with normal immune systems leads to a solid immunological response and subsequent immunity to the virus.

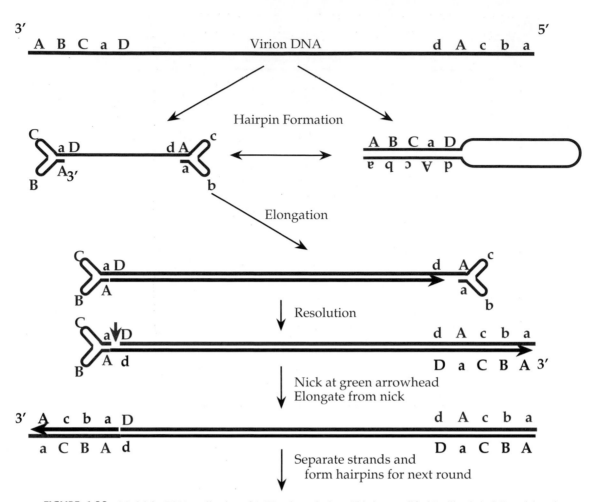

FIGURE 6.32   Model for DNA replication of AAV, a dependovirus. This is a modified "rolling hairpin" model, and results in inversion of both the repeated sequences at the termini. Models for replication of the autonomous parvoviruses like MMV are more complex, since in that case the 5′ terminal sequence of the virion DNA is inverted during replication, while the 3′ terminal sequence is not. [Adapted from Brister and Muzyczka (2000).]

Genus Dependovirus

The genus *Dependovirus* comprises viruses that require a helper virus in order to replicate. The helper may be an adenovirus or a herpesvirus, and because the dependoviruses were first found associated with adenoviruses, they are called adeno-associated viruses or AAVs. AAVs of man and of numerous other vertebrates are known. More than 90% of human adults have antibodies to AAV, which shows that the virus is widely distributed and common.

On infection of a nonpermissive cell (one that has not been infected by a helper), AAV establishes a latent infection in which its DNA integrates into the host chromosome. In human cells integration is specific—it occurs on chromosome 19 within a defined region. There appears to be a binding site for the nonstructural protein at this location that directs integration to a site within a nearby region of several hundred nucleotides. On infection of the cell by an adenovirus, the AAV DNA is excised and replicates. Latent infection in humans appears to be common.

Although dependoviruses normally require coinfection by an adenovirus or by a herpesvirus in order to undergo a complete replication cycle, it is possible to obtain independent replication of these viruses in cultured cells by treatment of the cells with toxic agents such as UV or one of several chemical carcinogens. This treatment must induce the production of cellular factors that are required for a full virus replication cycle, but the identities of these factors are not known. The helper function supplied by the helper virus may then be to independently supply these factors or to induce the production of these cellular factors. In light of the ability of the virus to replicate in cells treated with carcinogens, it is of considerable interest that retrospective studies have found that patients with cervical carcinoma were markedly deficient in antibodies to AAV as compared with normal controls. This suggests that, at least in some

A. Phylogenetic tree of the canine and feline parvoviruses

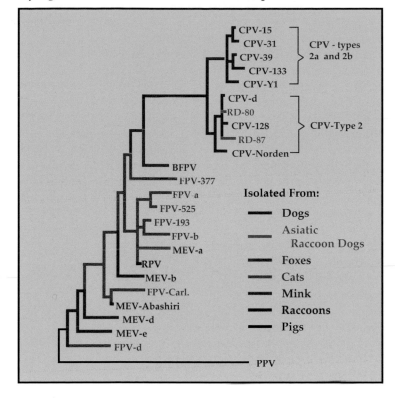

B. Evolution of the canine parvoviruses

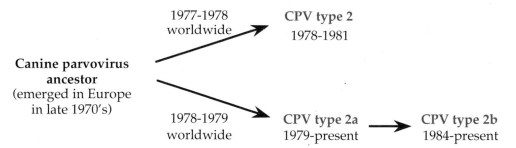

FIGURE 6.33 Phylogeny and evolution of feline and canine parvoviruses. (A) Phylogenetic tree of parvoviruses from dogs (CPV), Asiatic raccoon dogs (RD), cats (FPV), mink (MEV), raccoons (RPV), and foxes (BFPV), using porcine parvovirus (PPV) as an outgroup. The tree was constructed using the sequences of VP1 and VP2 and illustrates that all of the canine isolates form a distinct clade. (B) Diagram of the evolution of the canine parvoviruses into the two types circulating today. [From Parrish (1997).]

cancers, the cancerous cell might produce factors required for AAV replication, which results in the replication of AAV and the death of the tumor cell. AAV infection may therefore protect against the formation of tumors and be a virus that actually has a beneficial function in humans.

Genus Parvovirus

The genus *Parvovirus* contains viruses that infect a number of different mammals and birds. Several of these cause important diseases in domestic animals or in wildlife, including feline panleukopenia virus (FPV), porcine parvovirus, and canine parvovirus (CPV).

CPV is an example of an emerging nonhuman pathogen. It appeared suddenly in 1978 and spread around the globe in less than 6 months. The virus causes gastroenteritis in adult dogs and myocarditis in pups, which often leads to death from acute heart failure. It was called CPV-2 to distinguish it from a previously known canine parvovirus called minute virus of canines, which is sometimes referred to as CPV-1. CPV-2 is closely related to FPV and to parvoviruses that circulate in foxes and other wild animals. In fact, these various parvoviruses share 98% sequence identity. Molecular genetic studies have suggested that FPV, which does not replicate in dog cells, could have jumped the species barrier if as few as two amino acid changes occurred in the major coat protein VP2. Once the ancestor to CPV-2 arose from FPV or a related virus, further selection for growth in dogs could have produced the virus CPV-2, which is much better adapted to dogs as a natural host. Consistent with this hypothesis is the finding that antibodies to a CPV-2-like virus were first detected in serum from European dogs taken in the early to mid-1970s, but only in 1978 did explosive spread of the virus occur. Also of interest for this hypothesis is the fact that a variant of CPV-2 called CPV-2a replaced CPV-2 between 1979 and 1981, and a newer variant 2b appeared in 1984. Since this time CPV-2a and -2b have changed little and appear to be in worldwide equilibrium.

A dendrogram of these various parvoviruses is shown in Fig. 6.33. Although CPV-2 is closely related to FPV and its relatives, it clearly belongs to its own clade, which shows that the lineage of the canine virus is distinct. This is different from the pattern shown by the lineages of FPV and the viruses of foxes, mink, and other wildlife. Here the clades do not assort with species, which suggests that the various strains are freely transmissible to these various animals and circulate within this expanded group of animals. Thus it is of interest that CPV-2a and CPV-2b replicate well in cats, although CPV-2 does not. Also shown in the figure is a timeline for the emergence of CPV and its divergence into the different lines.

TT VIRUS: A NEWLY DESCRIBED HUMAN VIRUS

A new virus was isolated recently from the blood of a Japanese patient and called TT virus, after the initials of the patient. The viral genome was circular ssDNA, of size 3739 nucleotides. Studies of sera from other people around the world demonstrated that the virus was widespread and common. Recent estimates are that 12% of Japanese, 1% of Americans, and 2% of English are infected, but that a much larger fraction of the population in tropical countries such as Papua New Guinea (74% of people) and Gambia (83%) are infected. The nucleotide

sequences of different isolates from around the world exhibit up to 50% sequence diversity. It has been proposed that this virus represents a new family, tentatively named Circinoviridae.

TT virus was isolated from a patient with hepatitis. Thus, a major interest in the virus is a possible association with hepatitis in humans. However, there is no evidence at present that it does cause hepatitis in humans.

The isolation of a new virus that is widespread and common illustrates that there may yet be other viruses that infect humans but of which we are unaware. Whether TT virus or other viruses that are as yet unknown are associated with human disease remains to be determined.

FURTHER READING

Poxviridae

Beaud, G. (1995). Vaccinia virus DNA replication: A short review. *Biochimie* **77**; 774–779.

Goebel, S. J., Johnson, G. P., Perkus, M. E., *et al*. (1990). The complete DNA sequence of vaccinia virus. *Virology* **179**; 247–266.

Hopkins, D. R. (1983). "Princes and Peasants: Smallpox in History." University of Chicago Press, Chicago, Illinois.

Traktman, P. (1990). The enzymology of poxvirus DNA replication. *Curr. Top. Microbiol. Immunol.* **163**; 93–123.

Adenoviridae

Gray, G. C., Callahan, J. D., Hawksworth, A. W., *et al*. (1999). Respiratory diseases among U. S. military personnel: countering emerging threats. *Emerg. Infect. Dis.* **5**; 379–387.

Hay, R. T., Freeman, A., Leith, I., *et al*. (1995). Molecular interactions during adenovirus DNA replication. *Curr. Top. Microbiol. Immunol.* **199**; 31–48.

Leppard, K. N. (1998). Regulated RNA processing and RNA transport during adenovirus infection. *Semin. Virol.* **8**; 301–308.

Shenk, T. (1996). Adenoviridae: The viruses and their replication. *In* "Fields Virology" (D. N. Fields, D. M. Knipe, and P. M. Howley, eds.) pp. 2111–2148. Lippincott-Raven, Vol. 2, Philadelphia.

Herpesviridae

Boehmer, P. E., and Lehman, I. R. (1997). Herpes simplex virus DNA replication. *Annu. Rev. Biochem.* **66**; 347–384.

Campadelli-Fiume, G., Mirandola, P., and Menotti, L. (1999). Human herpesvirus 6: An emerging pathogen. *Emerg. Infect. Dis.* **5**; 353–366.

Fleming, D. T., McQuillan, G. M., Johnson, R. E., *et al*. (1997). Herpes simplex virus type 2 in the United States, 1976–1994. *N. Engl. J. Med.* **337**; 1105–1111.

Gallo, R. C. (1998). The enigmas of Kaposi's sarcoma. *Science* **282**; 1837–1839.

McGeoch, D. J., and Cook S. (1994). Molecular phylogeny of the alphaherpesvirinae subfamily and a proposed evolutionary timescale. *J. Mol. Biol.* **238**; 6–22.

Phelan, A., and Clements, J. B. (1998). Posttranscriptional regulation in herpes simplex virus. *Semin. Virol.* **8**; 309–318.

Rixon, F. J. (1993). Structure and assembly of herpesviruses. *Semin. Virol.* **4**; 135–144.

Schulz, T. (1998). Kaposi's sarcoma-associated herpesvirus (human herpesvirus-8) (review article). *J. Gen. Virol.* **79**; 1573–1591.

Polyomaviridae

Cole, C. N. (1996). Polyomavirinae. *In* "Fields Virology" (B. N. Fields, D. M. Knipe, and P. M. Howley, eds.) pp. 1997–2026. Vol.2, Lippincott-Raven, Philadelphia.

Agostini, H. T., Ryschkewitsch, C. F., and Stoner, G. L. (1998). JC virus type 1 has multiple subtypes: Three new complete genomes. *J. Gen.Virol.* **79**; 801–805.

Papillomaviridae

Chow, L. T., and Broker, T. R. (1994). Papillomavirus DNA replication. *Intervirology* **37**; 150–158.

Alani, R. M., and Münger, K. (1998). Human papillomaviruses and associated malignancies. *J. Clin. Oncol.* **16**; 330–337.

Oncogenesis

Nevins, J. R., and Vogt, P. K. (1996). Cell transformation by viruses. *In* "Fields Virology" (B. N. Fields, D. M. Knipe, and P. M. Howley, eds.) pp. 301–343. Lippincott-Raven, Philadelphia.

Levine, A. J. (1997). p53, the cellular gatekeeper for growth and division. *Cell (Cambridge, Mass.)* **88**; 323–331.

Friedlander, A., and Patarca, R. (1999). DNA viruses and oncogenesis. *Crit. Rev. Oncog.* **10**; 161–238.

Parvoviridae

"Handbook of Parvoviruses." (1990). P. Tijssen, (ed.). CRC Press, Boca Raton, FL, (esp. Chapters 3, 9, 11, and 12).

Parrish, C. R. (1997). How canine parvovrus suddenly shifted host range. *ASM News* **63**; 307–311.

CHAPTER

7

Subviral Agents

INTRODUCTION

This chapter considers a number of infectious agents that are subcellular, but that are not viruses in the strict sense of the term. Some of these are not capable of independent replication but require a helper virus, in which case the agent is effectively a parasite of a parasite. Others replicate independently but use unconventional means to achieve their replication and spread. Many of the agents discussed here cause important diseases in plants or animals, including humans.

The agents to be considered include defective interfering (DI) viruses that arise by deletions and rearrangements in the genome of a virus. DIs require coinfection by a helper virus to replicate. They may play an important role in modulation of viral disease or they may simply be artifacts that arise in laboratory studies. Related to DIs, at least conceptually, are satellites of viruses, which can replicate only in the presence of helper. Satellites are known for many plant viruses and are known to influence the virulence of the helper viruses. Completely different are agents called viroids. Viroids consist of small, naked RNA molecules that are capable of directing their own replication. They do not encode protein, but instead contain promoter elements that cause cellular enzymes to replicate them. Many are important plant pathogens. Related to viroids are virusoids, which are satellites of viruses that resemble packaged viroids. There are also agents that combine the attributes of both satellites and viroids, such as hepatitis delta virus, which is an important human pathogen. Finally, prion diseases, caused by infectious agents whose identity is controversial but which may consist only of protein, are discussed.

DEFECTIVE INTERFERING VIRUSES

Defective interfering viruses are a special class of defective viruses that arise by recombination and rearrangement of viral genomes during replication. DIs are *defective* because they have lost essential functions required for replication. Thus, they require the simultaneous infection of a cell by a helper virus, which is normally the parental wild-type virus from which the DI arose. They *interfere* with the replication of the parental virus by competition for resources within the cell. These resources include the machinery that replicates the viral nucleic acid, which is in part encoded by the helper virus, and the proteins that encapsidate the viral genome to form virions.

DIs of many RNA viruses have been the best studied. Because DI RNAs must retain all *cis*-acting sequences required for the replication of the RNA and its encapsidation into progeny particles, sequencing of such DI RNAs can provide clues as to the identity of these sequences. Identification of *cis*-acting sequences is important for the construction of virus vectors used to express a particular gene of interest, whether in a laboratory experiment or for gene therapy.

The most highly evolved DI RNAs are often not translated and consist of deleted and rearranged versions of the parental genome. In the case of alphaviruses, whose genome is about 12 kb (Chapter 3), DI RNAs have been described that are about 2 kb in length. However, they have a sequence complexity of only 600 nucleotides, because sequences are repeated one or more times. The sequences of two such DI RNAs of Semliki Forest virus are illustrated schematically in Fig. 7.1. From the sequences of these DIs as well as DIs of other alphaviruses, specific functions for

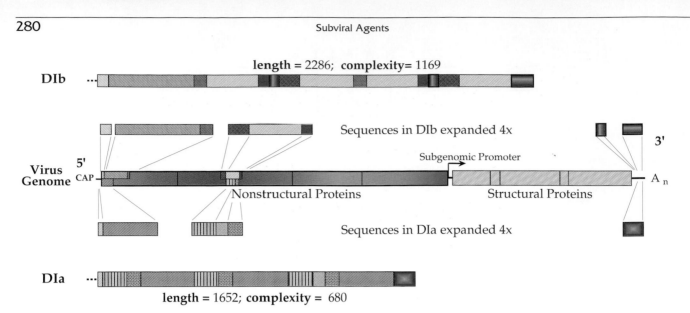

FIGURE 7.1   Schematic representation of DI RNAs found after high multiplicity infection by Semliki Forest virus. The central block shows the genome of the nondefective virus, with vertical lines denoting the four nonstructural and five structural polypeptides. The blocks of sequence found in two different DIs are expanded fourfold below and above. Their location in the DI genome is illustrated with blocks of identical shading. Note that some blocks of unique sequence are repeated three times in DIb and one block is repeated four times in DIa. [From Strauss and Strauss (1997, Fig. 1).]

the elements found in these DIs have been proposed, which have been supported by other approaches. The 3′ end of the parental RNA, which is retained in all alphavirus DI RNAs, forms a promoter for the initiation of minus-strand RNA synthesis from the plus-strand genome. The 5′ end of the RNA is also preserved in many DI RNAs, such as those illustrated in Fig. 7.1. Surprisingly, however, it has been replaced by a cellular tRNA in some DI RNAs. The complement of this sequence is present at the 3′ end of the minus strand, where it forms a promoter for initiation of genomic RNA synthesis. The finding that the DI RNAs with the tRNA as the 5′ terminus have a selective advantage over the parental genome during RNA replication suggests that this promoter is a structural element recognized by the viral replicase. It also suggests that the element present in the genomic RNA is suboptimal, perhaps because the genomic RNA must be translated as well as replicated. Finally, repeated sequences from two regions of the genome are present in all alphavirus DI RNAs. It is thought that one sequence (shown as red patterned blocks in Fig. 7.1) is an enhancer element for RNA replication and the second (shown as yellow and green patterned blocks) is a packaging signal. Repetition of these elements may increase the efficiency of replication and packaging of the DI RNA.

Vesicular stomatitis virus (VSV) (Chapter 4) DI RNAs vary in size from a third to half the length of the virion RNA. Some DI RNAs are simply deleted RNA genomes, but others have rearrangements at the ends of the RNAs. Representative examples are illustrated in Fig. 7.2A. During replication of the RNA, the sequences at the ends

must contain promoter elements for initiation of RNA synthesis. More genomic RNA (minus strand that is packaged in virions) is made than antigenomic RNA (which functions only as a template for genomic RNA synthesis) and therefore the promoter at the 3′ end of the antigenomic RNA is stronger than the promoter at the 3′ end of the genomic RNA. Thus, it is not surprising that some DI RNAs have the stronger promoter at the 3′ ends of both (+) and (−)RNA (as in Class II DIs), ensuring more rapid replication of the DI RNAs. The DI RNAs may have the luxury of doing this because they are not translated nor do they serve as templates for the synthesis of mRNAs.

The well-studied alphavirus DI RNAs and the VSV DI RNAs are not translated. For many DI RNAs, however, translation is required for efficient DI RNA replication. The best studied examples of this are DIs of poliovirus and of coronaviruses (these viruses are described in Chapter 3). DI RNAs of poliovirus are uncommon and contain deletions in the structural protein region. It has been suggested that in this case it is the translation product that is required for efficient replication of the RNA (the replicase translated from the RNA may preferentially use the RNA from which it was translated as a template). In contrast, for at least one well-studied DI of a coronavirus, translation of the RNA is required for efficient replication, but the translation product is not important. In this case, translation may stabilize the DI RNA, since there appears to be a cellular pathway to rid the cell of mRNAs that are not translatable. If so, how DI RNAs that are not translated avoid this pathway is uncertain. Some representative naturally occurring DIs of mouse

A. Vesicular stomatitis virus (VSV) and DIs derived from it

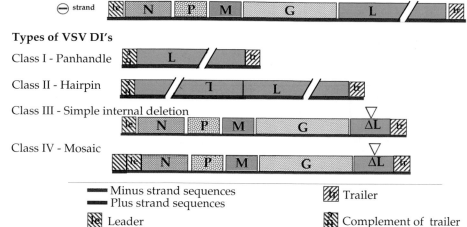

Types of VSV DI's

Class I - Panhandle

Class II - Hairpin

Class III - Simple internal deletion

Class IV - Mosaic

Minus strand sequences
Plus strand sequences
Leader
Trailer
Complement of trailer

B. Murine hepatitis virus (MHV) and DIs derived from it

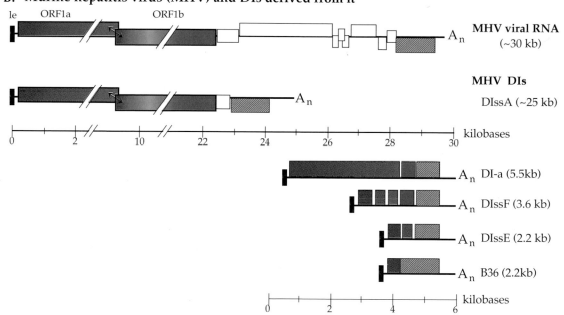

MHV viral RNA (~30 kb)

MHV DIs

DIssA (~25 kb)

kilobases

DI-a (5.5kb)

DIssF (3.6 kb)

DIssE (2.2 kb)

B36 (2.2kb)

kilobases

FIGURE 7.2 Types of DIs generated from a rhabdovirus and a coronavirus. (A) Diagrammatic representation of the VSV genome and RNAs of the four classes of DI particles. The leader and trailer are shown as patterned blocks. The genome is shown 3' to 5' for the minus strand (ochre). The parts of the DI RNAs corresponding to the complement of the minus strand are in green. A red triangle marks the internal deletion in the L gene, which is found in Class III and Class IV DIs. [Adapted from Whelan and Wertz (1997).] (B) Structures of naturally occurring DI RNAs of MHV (a murine coronavirus). DissA, DI-a, etc., were isolated from MHV-infected cells. The bottom line shows a synthetic DI replicon called B36. Sequences in the DIs are color coded by their region of origin in the parental virus genome. [Adapted from Brian and Spaan (1997, Fig. 1).]

hepatitis virus, a murine coronavirus, are illustrated in Fig. 7.2B.

Because DI RNAs are replicated by the helper virus machinery and encapsidated by the capsid proteins of the helper virus, they interfere with the parental virus by diverting these resources to the production of DI particles rather than to the production of infectious virus particles. It was first noted by von Magnus in the early 1950s that influenza virus passed at high multiplicity for many passages produced yields that cycled between high and low. This effect is illustrated schematically in Fig. 7.3A. We now know that this is due to the presence of DI particles. In early passages

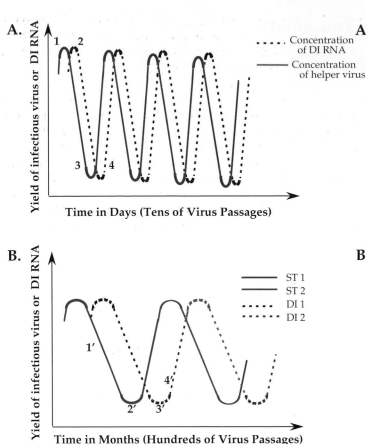

A.
1) Early in an undiluted passage series, standard virus replicates well, but DI's are beginning to accumulate.

2) DIs replicate at high efficiency, and interfere with standard virus.

3) So little standard virus is produced that there is little helper function, and DI replication drops.

4) With little DI replication, interference is reduced and standard virus titers rise again.

B.
1') DI 1 interferes strongly with replication of standard virus (ST 1).

2') A new variant of standard virus (ST 2) emerges that is resistant to interference by DI 1 and does not serve as a helper for DI 1.

3') Without helper assistance, DI 1 disappears, and ST 2 replicates vigorously.

4') New DIs of ST 2 (DI 2) appear and begin to depress ST 2 replication.

FIGURE 7.3 Stylized illustration of the influence of defective interfering particles on viral evolution. (A) Short-term generation of DI particles during undiluted passage and the cyclical fluctuations in infectious virus yield and concentration of DIs. This effect was first described by von Magnus in1952 during repeated passage of influenza virus at high multiplicity. (B) Role of DI particles in driving long-term evolution of viruses. The net result of hundreds of passages is that variant 2 and its DIs completely replace the original wild type or standard virus 1. [Adapted from Granoff and Webster (1999, Fig. 2, p. 373).]

virus yields are high. When DIs arise, they depress the yield of virus. Because high multiplicities of infection are required to maintain DI replication, so that cells are infected with both the helper and the DI, low yields of virus lead to a reduction in DI replication in the next passage or two. Reduced DI replication leads to higher yields of virus. Thus, the yield of infectious virus continues to fluctuate.

In a laboratory setting, at least, DIs can drive the evolution of the wild-type virus. This is shown schematically in Fig. 7.3B. When virus is passed at high multiplicity for very many generations, mutants often arise that have altered promoters that are recognized by mutant replication proteins. Such mutants are resistant to the DIs that are in the population at the time, because the mutant replication proteins do not recognize the promoters in the DIs. The mutant virus rapidly takes over the population of virus because of its selective advantage. However, new DIs then arise that will interfere with the mutant virus, and the cycle repeats.

It is unclear whether DI particles serve a biological role in nature or whether they are artifacts of abortive recombination that appear in the laboratory because of the high multiplicities of infection that are often used. It has been argued that DIs may arise late in the infection of an animal by a virulent virus and lead to attenuation of symptoms by reducing the yield of infectious virulent virus. As described in earlier chapters, such attenuation could be important for the persistence of a virus in nature. Hepatitis delta virus, described below, is a satellite that replicates only in a cell infected by HBV. Thus, it is clear that it is possible to achieve the multiplicities required to maintain a defective virus in at least some circumstances. Reconstruction experiments in mice have shown that it is possible to attenuate the virulence of lymphocytic choriomeningitis virus by injecting DIs along with the virus. However, it is not clear that DIs will arise in an acute infection in time to ameliorate symptoms. Thus, it has not been possible to provide firm evidence that DIs actually modulate the virulence of their parents in nature, and the question remains open.

SATELLITES AND SATELLITE VIRUSES

A number of satellites and satellite viruses are listed in Table 7.1. The dependoviruses, a genus in the family Parvoviridae, were considered in Chapter 6. Although the dependoviruses require a helper, the helper is only needed in order to stimulate the cell to enter a stage in which the dependovirus can replicate. The helper does not provide any function directly related to dependovirus replication or packaging. Satellites and satellite viruses, however, are usually more intimately dependent on the presence of a helper virus to furnish functions directly required for the replication of the satellite genome or for its encapsidation into progeny particles.

Many plant viruses have satellites associated with them. When a satellite encodes its own coat protein, it is sometimes referred to as a satellite virus. Otherwise it is simply called a satellite. One of the best-studied satellite systems is tobacco necrosis virus (TNV) and its satellite, tobacco necrosis virus satellite (TNVS). TNV has a plus-strand RNA genome of about 3.8 kb. The TNV virion is icosahedral with $T = 3$, and contains 180 copies of a single coat protein species of about 30 kDa. Associated with many isolates of TNV in nature is TNVS. TNVS has an RNA

TABLE 7.1   Some Subviral Agents

Group	Genome size	Helper virus	Host(s)	Comments
dsDNA satellite				
Bacteriophage P4	11.5 kb (10–15 genes)	Bacteriophage P2	Bacteria	All structural proteins from P2
ssDNA satellite viruses[a]				
Dependovirus (AAV)	4.7 kb	Adenovirus Herpesvirus	Vertebrates	See Table 6.15
dsRNA satellites				
M satellites of yeast	1–1.8 kb	Totiviridae	Yeast	Encode "killer" proteins; encapsidated in helper coat protein
ssRNA satellite viruses				
Chronic bee-paralysis virus-associated satellite	3 RNAs, each 1 kb	Chronic bee-paralysis virus	Bees	
Tobacco necrosis virus satellite	1239 nt	Tobacco necrosis virus	Plants	
ssRNA satellites				
Hepatitis δs virus	1.7 kb	Hepatitis B virus	**Humans**	Encodes two forms of δ antigen, encapsidated by helper proteins
B-type mRNA satellites	0.8–1.5 kb	Various plant viruses	Plants	Encode nonstructural proteins, rarely modify disease syndrome
C-type linear RNA satellites	<0.7 kb	Various plant viruses	Plants	Commonly modify disease caused by helper
D-type circular RNA satellites, "virusoids"	–350 nt	Various plant viruses	Plants	Self-cleaving molecules
Viroids				
Group A: ASBVd group	246–339 nt	—	Plants	Replicate by a symmetric strategy in chloroplasts of infected plants; can form self-cleaving hammerhead ribozymes in both plus and minus strands
Group B: PSTVd, HSVd, CCCd, ASSVd, and CbVd-1 groups	246–375 nt	—	Plants	Noncleaving, replicate by an asymmetric strategy in nucleus of infected cells

[a]When a satellite encodes its own coat protein, it is known as a satellite virus.
Abbreviations: Vd, viroid; ASBVd, avocado sunblotch viroid; PSTVd, potato spindle tuber viroid; HSVd, hop stunt viroid; CCCVd, coconut cadang-cadang viroid; ASSVd, apple scar skin viroid; CbVd-1, *Coleus blumei* viroid-1.

genome of 1239 nucleotides. The TNVS virion is a $T = 1$ icosahedral structure formed by 60 molecules of a single species of capsid protein encoded by the satellite RNA. The satellite RNA encodes only this single protein that encapsidates its own RNA. All of the functions required to replicate the RNA are provided by the helper TNV.

Some satellites of RNA plant viruses encode only a nonstructural protein required for RNA replication, and the RNA is encapsidated by the capsid protein of the helper virus. In other cases, the satellite is not translated into protein and depends on the helper for all of its functions, in which case it is functionally analogous to DI RNAs. A distinct class of satellite RNAs, called virusoids, consist of viroid-like RNAs that are encapsidated in the capsid protein of the helper virus. These are discussed in the next section.

Although satellites are quite common among plant viruses, they are almost unknown among animal viruses. The difference seems to lie in the means by which plant viruses are transmitted. It is very common among plant viruses to have the genome divided among two or more segments that are separately encapsidated into different particles, a situation that does not occur among animal viruses. Evidently, the mechanisms by which plant viruses are transmitted allow the infection of a plant, and of individual cells within a plant, by multiple particles that together constitute a virus or that constitute a virus and its satellites. Transmission of animal viruses between hosts or among the cells of a host does not appear to allow multiple infections with sufficient frequency to maintain virus systems that are constituted by multiple particles, with the exceptions of hepatitis δ virus, described below, and dependoviruses, which have evolved ways to persist within a cell until the helper virus comes along (Chapter 6). The defense mechanisms of the animal host may play a role.

VIROIDS AND VIRUSOIDS

Viroids are small, circular RNA molecules that do not encode any protein and that are infectious as naked RNA molecules. Sequenced viroids range from 246 to 375 nt and possess extensive internal base pairing that results in the RNA being rod-like and about 15 nm long. A partial listing of viroids is given in Table 7.1. All known viroids infect plants. However, hepatitis δ, which infects humans, has many viroid-like properties and may be related to viroids. Many viroids are important agricultural pathogens, whereas others replicate without causing symptoms. Viroids are often transmitted through vegetative propagation of plants, but can also be transmitted during agricultural or horticultural practices in which contaminated instruments are used. Some viroids can be transmitted through seeds and at least one viroid is transmitted by an aphid.

On infection of a plant cell, viroid RNA is transported to the nucleus. The circular RNA appears to be copied by host-cell RNA polymerase II, using a rolling circle mechanism in which multimeric antigenome sense RNA molecules are produced. The multimeric antigenome sense RNA can then be used as a template to make multimeric genome sense RNA. This synthesis may also be performed by RNA polymerase II. The concatenated RNAs are cleaved and cyclized to produce the progeny viroid RNA molecules. In an infected cell, as many as 10^4 viroid RNAs can accumulate, most of them in the nucleus.

Some viroids are capable of self-cleavage by the concatenated RNAs to produce genome-length RNAs, followed by self-ligation to cyclize the unit-length molecule. Other viroids are not. There are five groups of non-self-cleaving viroids, classified by the sequences in the central conserved region. The structures of these five groups are shown in Fig. 7.4. The conserved domains highlighted in the figure are thought to be important for the replication of the viroid (i.e., to form promoters recognized by RNA polymerase II) and for its cleavage to produce unit-length molecules. A pathogenesis domain is also highlighted. Changes in this domain affect the virulence of the viroid on infection of its plant host. For non-self-cleaving viroids, it is assumed that the concatenated RNAs are cleaved and ligated by host-cell enzymes.

The self-cleaving viroids possess a hammerhead ribozyme structure, illustrated in Fig. 7.5. The ribozyme activity cleaves the concatenated RNA at the points indicated by the arrows and ligates the ends to form circular molecules. The viroid RNA is very compact in its structure, with extensive secondary structure, including pseudo-knots.

There also exist a large number of satellites called virusoids. Virusoid RNAs are about 350 nt in length, and the RNA is a single-stranded, covalently closed circle. The mechanisms by which virusoid RNA is replicated have not been worked out, but they appear to be viroid-like and may replicate by the same mechanisms as viroids. At least some virusoid RNAs are capable of self-cleavage. Virusoid RNAs are encapsidated by the capsid protein of the helper virus of which the virus is a satellite. Thus, transmission occurs by conventional virus-like means, and virusoids may have arisen from viroids that evolved a mechanism for packaging using a helper virus.

HEPATITIS δ

The hepatitis delta (δ) agent (HDV) is a satellite of hepatitis B virus. It has a worldwide but nonuniform distribution. Regions of particularly high prevalence include the Mediterranean basin, the Middle East, Central Asia, West Africa, the Amazon basin, and certain islands in the South Pacific. The agent will only replicate in cells that are simul-

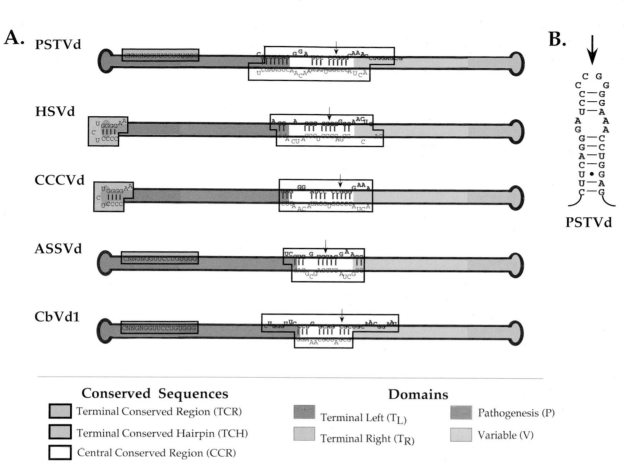

FIGURE 7.4 Models for the structures of different subgroups of non-self-cleaving viroids. (A) The RNA strand is shown as a green closed loop. Four functional domains (T_L, T_R, P, and V) are indicated with different colors of shading. Three conserved sequences are boxed. The central conserved region (CCR) is a white box, the terminal conserved region (TCR) is a lavender box, and the terminal conserved hairpin (TCH) is a green box. The nucleotides in the upper strand of the CCR (dark green) can in each case form a stable stem and loop structure with the top of the loop at the black arrow, as shown for PSTVd in (B). In this alternative configuration, the nucleotides that are invariant within all five groups are shown in red. The type member of each group is shown; the full names are listed in Table 7.1 at the bottom. [Adapted from Flores *et al.* (1997).]

taneously infected with HBV. The distribution of HDV is thus dependent on the distribution of HBV, which is shown in Fig. 5.28. However, as shown in Fig. 7.6, HDV is not uniformly distributed throughout the range of HBV. The percentage of hepatitis B patients that are also infected by HDV ranges from 5 to >60% in different geographic areas.

Infection of humans by HDV can either occur by simultaneous infection with both HBV and HDV, or by superinfection with HDV of a person who is chronically infected with HBV. In the case of coinfection, HDV establishes a chronic infection, which requires that HBV also establish a chronic infection, only 1–3% of the time. Most often the infection is completely resolved and recovery occurs. In contrast, superinfection of chronically infected HBV patients with HDV leads to chronic infection by HDV in 70–80% of patients. The different outcomes following

infection with HDV are illustrated schematically in Fig. 7.7, in which the symptomology at different times after infection is indicated.

The illness caused by HDV is usually more serious than that caused by HBV alone. The mortality rate from HDV infection is 2–20%, 10-fold higher than the rate for HBV infection alone, which is the next most severe form of viral hepatitis. Most cases of HDV infection are probably clinically important. More than 350 million people in the world are chronically infected with HBV. All are at risk for contracting HDV and suffering a more severe form of hepatitis.

The mechanisms by which HDV is transmitted are not understood. It is conjectured that poor hygiene together with intimate contact among people who are infected with the virus may be an important source of transmission in many parts of the world. In developed countries, contaminated

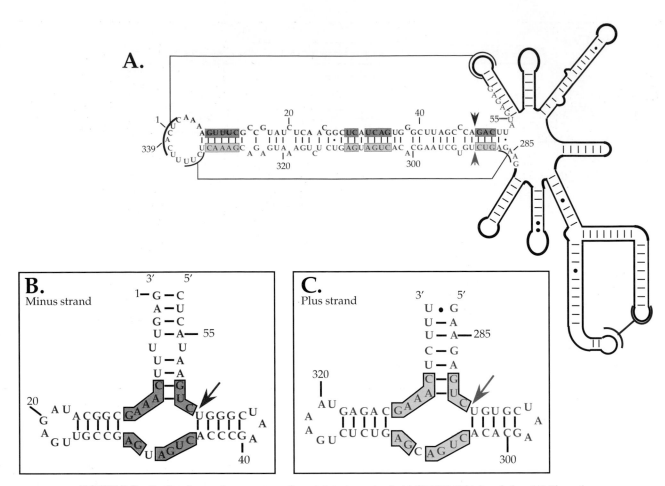

FIGURE 7.5 Predicted secondary structure of peach latent mosaic viroid (PLMV) RNA in solution. (A) The entire viroid. For most of the molecule, only the backbone is shown, with hydrogen bonds indicated by bars and G–U pairs indicated with dots. The numbering of the nucleotides is arbitrary. The structure is predicted to be even more compact due to pseudoknots formed by the regions joined by purple lines. The nucleotides making up the minus strand and plus strand hammerhead ribozymes are shown in green and blue letters respectively. The nucleotides conserved in all hammerheads in Group A viroids are boxed and shaded, and the sites of cleavage are marked by arrows. (B) and (C) The structures of the two hammerhead ribozymes using the same color conventions. The minus strand hammerhead is made up of the complement of the sequence shown in green in (A). [Adapted from Flores *et al.* (1997) and Pelchat *et al.* (2000).]

blood products and sharing of needles by drug users are important in the spread of the virus, but these are not important modes of transmission worldwide. Sexual transmission may occur, but again this does not appear to be an important component of the transmission of the virus on a worldwide basis. Since HDV depends on HBV for its propagation, control of HBV by vaccination will lead to control of HDV.

Replication of the HDV Genome and Synthesis of mRNA

The HDV genome is a single-stranded, covalently closed circular RNA molecule of 1.7 kb. The HDV genome can be thought of as a viroid into which has been inserted a gene

encoding a single polypeptide, the hepatitis δ antigen (HDAg). As is the case for viroid RNA, HDV RNA has a high degree of secondary structure, with about 70% of the molecule being base paired internally so that it forms a rod-like structure. The HDV genome has minus-sense polarity, that is, it is complementary to the sequence that is translated into the HDAg. The structures of the genomic RNA, the mRNA for the HDAg, and of the antigenomic RNA are illustrated in Fig.7.8A.

Following infection by the agent, the RNA is transferred to the nucleus. The HDAg is required for this. In the nucleus, the RNA is replicated by a mechanism similar to that used by viroid RNA. However, there is the added complication that an mRNA for HDAg must also be produced. Replication of the genome and synthesis of the mRNA are

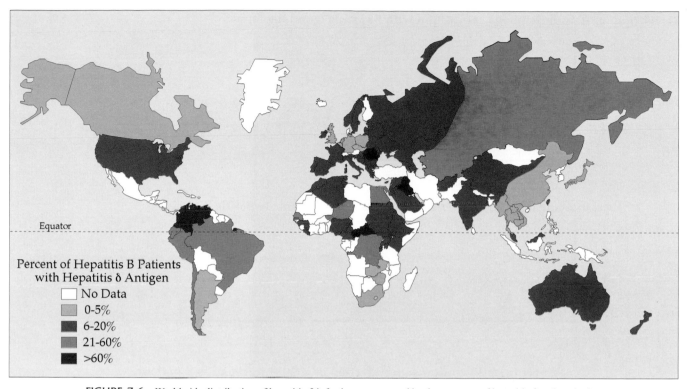

FIGURE 7.6   Worldwide distribution of hepatitis δ infection as measured by the presence of hepatitis δ antigen in the serum of hepatitis B-infected patients with hepatitis. [From Fields *et al.* (1996, p. 2826).]

**Percent of Hepatitis B Patients
with Hepatitis δ Antigen**
- No Data
- 0–5%
- 6–20%
- 21–60%
- >60%

Equator

carried out by RNA polymerase II. Other, currently unknown host factors also participate, and the HDAg is required. The genomic RNA is used as a template to produce two RNAs, an mRNA that is translated into the HDAg, and an antigenomic RNA that is a full-length complementary copy of the genomic RNA. The antigenomic RNA is also a covalently closed, single-stranded RNA molecule. The mechanisms by which these two RNAs are produced are not fully understood, but a possible model for synthesis of the mRNA and replication of the genome is shown in Fig. 7.9. The model postulates the presence of three key features in the genome: an origin for the start of RNA synthesis; a polyadenylation site following the open reading frame (ORF) for HDAg; and a self-cleavage site that is capable of self-ligation.

RNA synthesis is proposed to begin at the origin, nucleotide 1631 in Fig. 7.9, which is located upstream of the ORF for HDAg. Synthesis proceeds counterclockwise in the figure through the HDAg ORF, through the polyadenylation signal following this ORF, and through the self-cleavage site at position 903/904. Cleavage occurs and the released RNA fragment is processed by cellular enzymes that recognize the polyadenylation signal, cleave the RNA at the appropriate site, and polyadenylate it, to give the 0.8-kb mRNA for HDAg.

The polymerase continues to synthesize RNA, whose 5′ terminus is now nucleotide 903. Synthesis continues around

the circular genome and through the gene for HDAg a second time. It is proposed that the binding of the HDAg to a secondary structure present in the RNA prevents the recognition of the polyadenylation signal in the full-length RNA molecule. As synthesis proceeds through the self-cleavage site for the second time, cleavage releases a unit-length antigenomic molecule that cyclizes by a self-ligation activity.

Evidence that supports this model includes the fact that, like many viroids, hepatitis δ RNA, both genomic and antigenomic, is capable of self-cleavage. The linear, unit-length RNA is capable of self-ligation that results in cyclization. Ligation is inefficient, however, and cellular factors, perhaps even a cellular ligase, may also be required for efficient ligation. Thus, the broad aspects of the model are probably correct, but details may require modification as more becomes known about the replication of HDV.

The antigenomic RNA, once produced, can be used as a template to produce a full-length genomic RNA. The model proposed for this is the same as that above, but in this case there is no complication of a polyadenylation signal that results in the release of an mRNA, and the self-cleavage site is positioned at 688/689.

HDV Delta Antigen

The mRNA for HDAg is exported to the cytoplasm and translated into a polypeptide of 195 amino acids, referred to

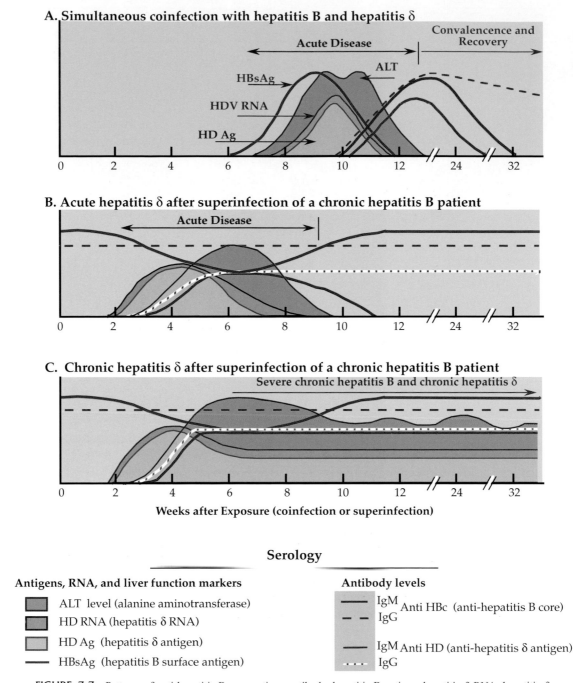

A. Simultaneous coinfection with hepatitis B and hepatitis δ

Acute Disease

Convalencence and Recovery

HBsAg

ALT

HDV RNA

HD Ag

0   2   4   6   8   10   12   24   32

B. Acute hepatitis δ after superinfection of a chronic hepatitis B patient

Acute Disease

0   2   4   6   8   10   12   24   32

C. Chronic hepatitis δ after superinfection of a chronic hepatitis B patient

Severe chronic hepatitis B and chronic hepatitis δ

0   2   4   6   8   10   12   24   32

Weeks after Exposure (coinfection or superinfection)

Serology

Antigens, RNA, and liver function markers	Antibody levels
ALT level (alanine aminotransferase)	IgM Anti HBc (anti-hepatitis B core)
HD RNA (hepatitis δ RNA)	IgG
HD Ag (hepatitis δ antigen)	IgM Anti HD (anti-hepatitis δ antigen)
HBsAg (hepatitis B surface antigen)	IgG

FIGURE 7.7 Patterns of anti-hepatitis B core antigen antibody, hepatitis B antigen, hepatitis δ RNA, hepatitis δ antigen, and ALT in patient serum during different types of coinfection with hepatitis δ and B. (A) Simultaneous infection by both types. (B) and (C) Superinfection by hepatitis δ of a patient with chronic hepatitis B infection. Many infections start with acute hepatitis δ as in (B). Some proportion of superinfections progresses to chronic hepatitis with elevated liver enzymes and sustained production of hepatitis δ RNA and protein as in (C). [Adapted from Fields *et al.* (1996, p. 2825).]

as the small δ antigen or S-HDAg. This protein is known to be required for RNA replication. Thus, for example, *in vitro* systems to study the replication of HDV RNA must be supplemented with S-HDAg for replication to occur. S-HDAg is a component of the infecting particle and is present in the infecting cell when RNA replication first begins.

Production of new protein after infection enables RNA replication to accelerate.

A second form of δ antigen is also produced during infection. An RNA editing event occurs in about one-third of the antigenomic templates, in which the termination codon, UAG, at position 196 of the ORF for the δ antigen is

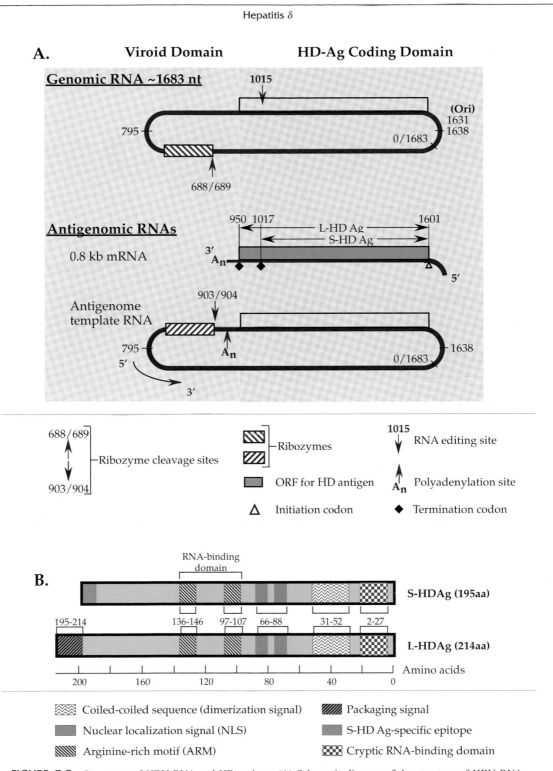

A.

Viroid Domain   **HD-Ag Coding Domain**

Genomic RNA ~1683 nt

Antigenomic RNAs

0.8 kb mRNA

Antigenome template RNA

688/689 ⟷ 903/904 — Ribozyme cleavage sites

Ribozymes

ORF for HD antigen

△ Initiation codon

1015 ↓ RNA editing site

A_n Polyadenylation site

◆ Termination codon

B.

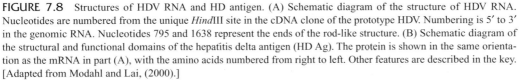

RNA-binding domain

S-HDAg (195aa)

L-HDAg (214aa)

195-214   136-146   97-107   66-88   31-52   2-27

Amino acids
200   160   120   80   40   0

Coiled-coiled sequence (dimerization signal)   Packaging signal

Nuclear localization signal (NLS)   S-HD Ag-specific epitope

Arginine-rich motif (ARM)   Cryptic RNA-binding domain

FIGURE 7.8 Structures of HDV RNA and HD antigen. (A) Schematic diagram of the structure of HDV RNA. Nucleotides are numbered from the unique *Hind*III site in the cDNA clone of the prototype HDV. Numbering is 5′ to 3′ in the genomic RNA. Nucleotides 795 and 1638 represent the ends of the rod-like structure. (B) Schematic diagram of the structural and functional domains of the hepatitis delta antigen (HD Ag). The protein is shown in the same orientation as the mRNA in part (A), with the amino acids numbered from right to left. Other features are described in the key. [Adapted from Modahl and Lai, (2000).]

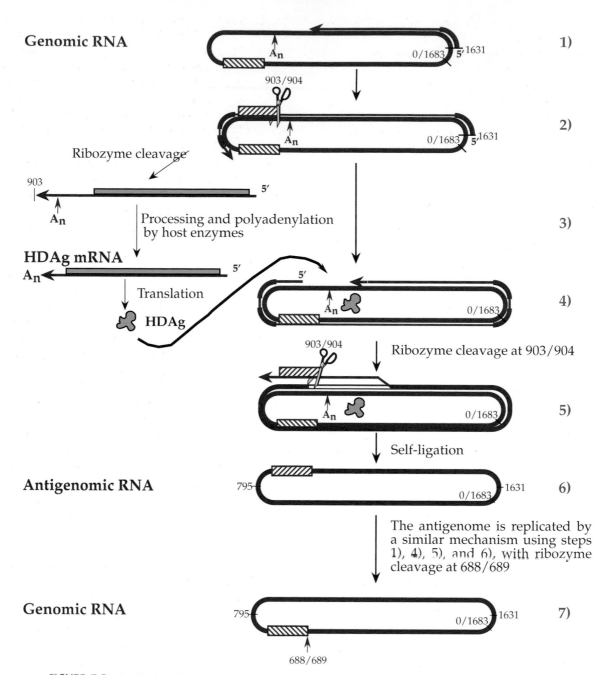

Genomic RNA 1)

Genomic RNA 2)

Ribozyme cleavage

Processing and polyadenylation by host enzymes 3)

HDAg mRNA

Translation

HDAg

Ribozyme cleavage at 903/904 4)

Self-ligation 5)

Antigenomic RNA 6)

The antigenome is replicated by a similar mechanism using steps 1), 4), 5), and 6), with ribozyme cleavage at 688/689

Genomic RNA 7)

FIGURE 7.9 Replication of hepatitis δ RNA. (1) RNA synthesis begins about 30 nt upstream of the HDAg ORF at nucleotide 1631. (2) Plus-strand synthesis proceeds through the ORF for HDAg. (3) The nascent strand is cleaved by the ribozyme at nucleotides 903/904 and the mRNA is subsequently processed and polyadenylated. (4) After translation, the HDAg binds to the RNA genome and promotes replication. (5) RNA strands of greater than unit length are produced, cleaved, and (6) self-ligated to form the plus-strand circular antigenomic RNA. (7) The plus strand is used in a similar manner as a template to make genomic RNA, which is cleaved by the ribozyme at nucleotides 688/689. [Adapted from Modahl and Lai (2000).]

changed to UGG, encoding tryptophan (see Fig. 7.8). This change is thought to be effected by deamination, in the antigenome, of the adenosine in the UAG codon to produce inosine. A cellular adenosine deaminase has been described that probably performs this function. Inosine pairs as guanosine, and continued replication of the RNA will lead

to the substitution of G for A. This RNA editing site is specific and requires specific sequences within the antigenomic RNA for it to occur.

Change of the termination codon to a tryptophan codon leads to the production of a polypeptide that is 19 residues longer, for a total length of 214 amino acids, referred to as the large δ antigen or L-HDAg. Because editing is required, it is only produced later in the infection cycle. L-HDAg suppresses RNA replication and leads to a shift from replication of RNA to encapsidation of RNA into progeny virions. A map of functional domains of the L- and S-HDAgs is shown in Fig. 7.8B.

The extent of editing is controlled, perhaps by S-HDAg. Obviously, only genomes that are not edited can give rise to infectious virions. S-HDAg is required for replication, and only nonedited genomes encode it.

Assembly of Virus

Assembly of HDV virions begins with the formation of a nucleocapsid or core that contains the HDV genome and both the L and S forms of the δ antigen. The core is 19 nm in diameter and matures by budding, using the HBV surface antigens. Budding appears to be the same as for HBV (Chapter 5), and the three surface antigens of HBV form the protein component of the outer envelope surrounding the HDV capsid. Thus, although the RNA of HDV can replicate independently of HBV, assembly of progeny virions requires the simultaneous infection of the cell by HBV to supply the surface glycoproteins needed to produce infectious particles.

Host Range of HDV

The only known natural hosts for HDV are humans, but HDV can be experimentally transmitted to chimpanzees and to woodchucks. Infection of chimps requires coinfection with HBV, and this provides a useful primate model system for the study of the agent. A second model system is furnished by woodchucks. Woodchucks can be chronically infected with woodchuck hepatitis virus (WHV, Chapter 5), a relative of HBV, and WHV can provide helper activity for HDV. Chronically infected animals can be infected with HDV, and in this case the surface properties of the HDV virion are determined by the helper WHV rather than by HBV.

Although HDV has a worldwide distribution, strains isolated from different regions of the world differ by up to 40% in their nucleotide sequence. Since the virus is a satellite, control of HDV is dependent on the control of HBV. Current HBV vaccines are highly effective at preventing HBV infection and this, together with increased screening for the presence of HBV in blood products, may be effective in controlling HDV in the future. Of special note is the fact that the agent is extremely prolific. The serum of an infected individual can contain up to 10^{12} RNA-containing HDV particles per milliliter.

PRIONS AND PRION DISEASES

Transmissible spongiform encephalopathies (TSEs), now often referred to as prion diseases, are progressive,

fatal diseases of humans and of other animals. A listing of TSEs is given in Table 7.2. TSEs of humans include kuru, Creutzfeldt-Jakob disease (CJD), Gerstmann-Sträussler-Scheinker syndrome (GSS), and fatal familial insomnia (FFI). TSEs are characterized by neuronal loss that appears as a spongiform degeneration in sections of brain tissue, often accompanied by amyloid plaques or fibrils. The most prominent symptoms of disease are usually dementia (loss of intellectual abilities) or ataxia (loss of muscle control during voluntary movement) that results from the progressive loss of brain function. The disease always has a fatal outcome. In humans, death usually occurs within 6 months to 1 year of the first appearance of symptoms.

TSEs can be contracted by inoculation with or ingestion of brain tissue or other tissues containing the infectious agent, and thus they can be transmitted as an infectious disease. Kuru first came to light as an infectious disease and many cases of CJD in man have been acquired by infection. However, TSEs can also occur as sporadic diseases for which there is no evidence of infection by an outside agent. In humans, CJD occurs sporadically with a frequency of about 10^{-6}. Finally, TSEs can appear as inherited diseases. GSS, most FFI, and some cases of CJD occur as dominant inherited diseases, associated with mutations in the gene for the prion protein. Inheritance of the mutant gene dramatically increases the probability of developing TSE, such that the probability of acquiring the disease over a lifetime may approach 100%. In most cases of sporadic or inherited TSE, the disease is transmissible as an infectious disease once it occurs.

There is now considerable evidence that all TSEs are related and result from defects in the metabolism of the prion protein. The pattern of symptoms associated with a particular TSE may vary, however, depending in part on how the disease was contracted; on the source of the infecting agent; and on the nature of mutations in the prion protein. Thus, although the prion protein is central to disease in every case, symptomology can differ, in part because the particular area of the brain most affected can vary.

There is no immune response associated with any TSE. No antibodies are formed and no inflammation marked by the infiltration of mononuclear cells is present. As stated, it is relentlessly progressive and always results in death.

Kuru

Kuru was a disease of epidemic proportions among the Fore people of New Guinea that reached a prevalence of about 1% of the population. The disease was characterized by progressive ataxia that led to total incapacitation and death, normally in 12–18 months of the appearance of symptoms in adults or 3–12 months in children. The demonstration that kuru was transmissible to primates by inoculation of brain tissue from people dying of the disease was the first demonstration of the transmissibility of a TSE

TABLE 7.2   Prion Diseases

Disease (abbreviation)	Natural host	Experimental hosts	Cause of disease
Scrapie	Sheep and goats	Mice, hamsters, rats	Infection in genetically susceptible sheep
Transmissible mink encephalopathy (TME)	Mink	Hamsters, ferrets	Infection with prions from sheep or cattle
Chronic wasting disease	Mule deer, white tail deer, elk	Ferrets, mice	Unknown
Bovine spongiform encephalopathy (BSE)	Cattle	Mice	Infection with prion-contaminated meat and bone meal
Feline spongiform encephalopathy (FSE)	Cats	Mice	Infection with prion-contaminated beef
Exotic ungulate encephalopathy (EUE)	Nyala, oryx, greater kudu	Mice	Infection with prion-contaminated meat and bone meal
Kuru	**Humans**	Primates, mice	Infection through ritual cannibalism
Creutzfeldt-Jakob disease		Primates, mice	
SCJD (sporadic)	**Humans**		Somatic mutation or spontaneous conversion of PrP^c to PrP^{Sc}
iCJD (iatrogenic)	**Humans**		Infection from prion-contaminated human growth hormone, dura mater grafts, etc.
nvCJD (new variant)	**Humans**		Ingestion of bovine prions?
fCJD (familial)	**Humans**		Germ-line mutation in PrP gene
Gerstmann-Sträussler-Scheinker syndrome (GSS)	**Humans**		Germ-line mutation in PrP gene
Fatal familia insomnia (FFI)	**Humans**	Primates, mice	Germ-line mutation in PrP gene (D178N, M129)
Fatal sporadic insomnia (FSI)	**Humans**		Somatic mutation or spontaneous conversion of PrP^c to PrP^{Sc}

Source: Adapted from Granoff and Webster (1999, p. 1389).

in humans. These transmission studies and other studies of kuru resulted in a Nobel prize for Carleton Gajdusek in 1976, the first of two prizes for work with TSEs (Chapter 1).

Kuru is believed to have been spread among the Fore people by cannibalism in which the bodies of relatives who had died were eaten in a ritualistic feast. Women and children were more often affected than men, and it is thought this was because they prepared the body for the feast and they ate the brains of deceased relatives. Men were less often affected, it is conjectured, because they ate primarily other body parts. It has been postulated that the epidemic began when a member of the tribe died of a sporadic case of CJD, and the disease was then spread to others through cannibalism. Through the efforts of missionaries, cannibalism ceased many years ago and the disease has become progressively rarer. Now only older people who contracted the infectious agent during the time of cannibalism continue to develop the illness. From studies of the continuing development of kuru in older Fore people, it is known that the disease can appear as long as 40 years after the event that resulted in infection with the agent.

Sporadic and Iatrogenic CJD

CJD in man is usually a sporadic illness that occurs with a frequency of about 10^{-6} that is uniform around the world.

However, once the disease has arisen it is transmissible by inoculation of infected material into experimental animals such as primates and transgenic mice. CJD has also been transmitted iatrogenically to humans. Iatrogenic cases have occurred in recipients of pituitary-derived human growth hormone obtained from cadavers, some of whom died of CJD; in recipients of homographs of *dura mater* derived from cadavers; through implantation into epilepsy patients of contaminated silver electrodes that had been incompletely sterilized; and through corneal transplants. The infectious agents of TSEs are extremely difficult to inactivate and require extraordinary sterilization techniques in order to destroy their infectivity. Better methods of sterilization have been introduced, and human growth hormone is now produced in bacteria from recombinant DNA plasmids, so that the iatrogenic spread of CJD has been greatly reduced.

Sporadic FFI has also been described. No case of sporadic GSS is known, however.

Inherited Forms of Human TSE

About 5% of CJD cases arise in a familial, autosomal dominant fashion and are associated with mutations in the gene for the prion protein. GSS and most cases of FFI are also inherited forms of TSE associated with mutations in the prion protein. Many of the responsible mutations are illustrated in Fig. 7.10, in which a schematic diagram of the human prion protein is presented.

A dozen single amino acid substitutions in the prion protein have been found to be associated with inherited CJD, GSS, or FFI. Additionally, an element normally containing five repeats of a 24-nucleotide squence (encoding an 8-amino-acid repeat, P-Q/H-G-G-G-W-C-Q) has been found to contain one to nine extra repeats, probably originating from unequal crossing over, in some cases of inherited CJD or GSS. These three diseases are distinguished on the basis of symptomology, which is overlapping. CJD is characterized by ataxia, dementia, and behavioral disturbances. GSS is usually characterized by cerebellar disorders accompanied by a decline in cognitive ability. FFI, as its name suggests, is characterized by abnormal sleep patterns, including intractable insomnia.

The penetrance of the different mutations varies but is usually very high. For example, CJD caused by the change from glutamic acid-200 to lysine (E200K), when residue 129 is homozygous for methionine, has been estimated to have a penetrance of 0.45 by age 60 and a penetrance of more than 0.96 above age 80. Thus, a person with this mutation is almost certain to develop CJD if he or she lives long enough.

Attempts have been made in many cases of inherited TSEs to transmit the disease to subhuman primates or to mice. Transmission has been achieved in most cases tested.

Thus once the disease arises, it is transmissible to animals that do not contain the mutation.

In addition to mutations associated with inherited TSEs, several polymorphisms in the prion protein are known that are not associated with disease (Fig. 7.10). The polymorphism at residue 129 is of particular importance. Homozygosity at this position affects the probability of contracting TSE.

TSEs in Other Animals

Naturally occurring TSEs of a number of other mammals are known. The oldest known TSE, in fact, is that of sheep, and is called scrapie. Scrapie has been known for more than 200 years and is widely distributed in Europe, Asia, and America. The name comes from the tendency of animals to rub themselves against upright posts, apparently because of intense itching that arises from this neurological disease. Scrapie appears to be transmitted horizontally in sheep flocks, but the mechanism by which it is transmitted is not understood. The infectious agent is very resistant to inactivation and may persist in pastures for a long time. It may be ingested, but other mechanisms for persistence have also been proposed.

Scrapie appears to have been transmitted to a number of other mammals. In some cases the spread has been to

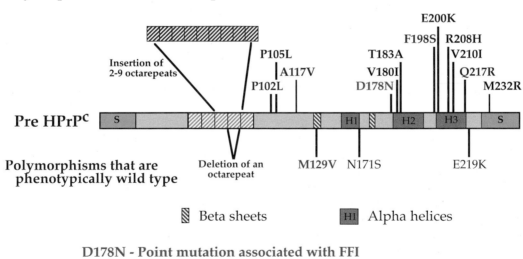

D178N - Point mutation associated with FFI

P102L - Point mutations associated with GSS

E200K- Point mutations and insertions associated with familial CJD

M129V - homozygosity at this locus increases susceptibility to sporadic CJD

FIGURE 7.10   Mutations found in the human prion protein gene. Polymorphisms that are phenotypically wild type are shown below the schematic of the gene; mutations that segregate with inherited prion diseases are shown above the gene. GSS, FFI, and CJD are defined in Table 7.2. [Adapted from Prusiner (1998) and Riek *et al.* (1996).]

animals that share pasturage with infected sheep, such as white-tailed deer, mule deer, and elk (where the disease is called chronic wasting disease). In these cases, it is thought that infection occurs by the same mechanisms that maintain scrapie in sheep flocks. In other cases, spread has occurred via food derived from infected sheep that was fed to mink (transmissible mink encephalopathy), domestic cats or exotic cats in zoos (feline spongiform encephalopathy), ungulates in zoos (exotic ungulate encephalopathy), and perhaps to cattle. However, there is no evidence that scrapie has ever spread to humans, despite the long history of human consumption of scrapie-infected sheep.

Bovine spongiform encephalopathy (BSE), also called mad cow disease, is a TSE of cattle that was recently epi-

demic in Britain. The epidemic was maintained by feeding cattle processed offal from cattle and other animals, that is, by a form of animal cannibalism as happened with kuru in man. Although this practice was of long standing, it did not cause trouble until recently, when a change in the rendering process was introduced. It is believed that this change allowed the BSE agent to survive the processing steps, whereas formerly it had been killed during rendering. The result was an epidemic of BSE that spread across all of Britain (Fig. 7.11). At the height of the epidemic, there were more than 35,000 cases of BSE per year in Britain (Fig. 7.12).

The original source of the BSE that led to the epidemic is uncertain. It may have arisen from a spontaneous case of BSE, similar to spontaneous CJD in man, although spontaneous BSE in cattle has never been described. A second

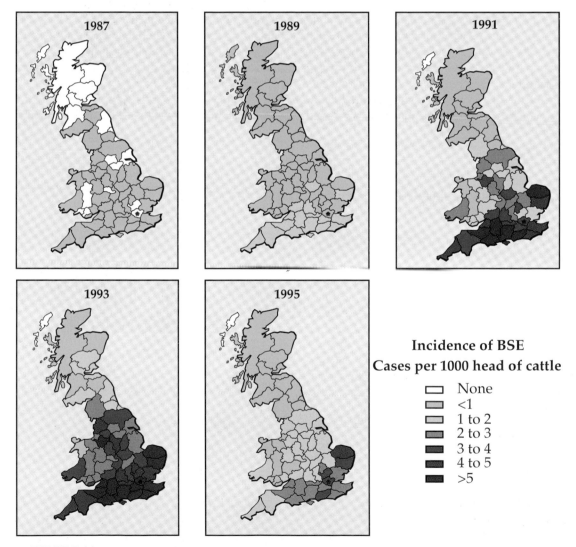

FIGURE 7.11   Spread of the BSE epidemic in the British Isles. Geographic distribution of the incidence of BSE per head of cattle by country from 1989 to 1995. [Adapted from Anderson *et al.* (1996).]

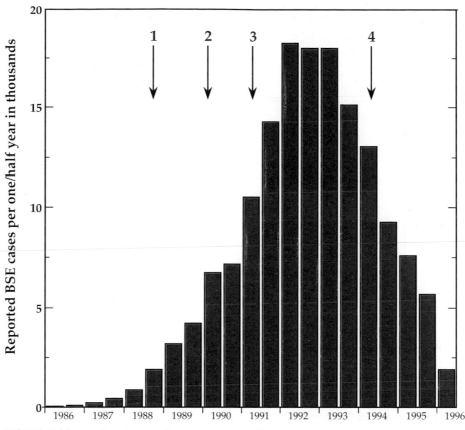

FIGURE 7.12   Confirmed cases of BSE (bovine spongiform encephalopathy) in British cattle per half year between 1986 and 1996. Arrows indicate (1) ruminant feed ban (1988); (2) specified offals ban, to prevent offal proteins from entering the human food chain (1989); (3) extended specified offals ban (prohibiting feeding of offal proteins to pigs and poultry (1991); and (4) offals ban further extended to include offals from bovines < 6 months old (1994). [Adapted from Anderson *et al.* (1996).]

possibility is that it may have arisen from infection with scrapie from infected sheep, since sheep offal was included in the rendered offal.

Once the epidemic of BSE in cattle in Britain was recognized, legislation was introduced that banned the feeding of any ruminant-derived protein to ruminants. Also introduced was legislation to make BSE a notifiable disease and to prohibit the use of brain, spinal cord, and certain other offals from any bovine animal in human food. The ruminant feed ban resulted in the waning of the epidemic in cattle in Britain (Fig. 7.12). However, new cases continued to arise, and it is not clear whether these new cases resulted from a long latent period of the infectious agent, from contaminated ruminant feed that continued to enter the system, or from alternative modes of transmission, such as passing the infection from mother to calf. With the recognition of new variant CJD in people, the issue of eradicating BSE became more pressing and, ultimately, herds containing infected cattle were destroyed. These different steps led to a marked reduction in the incidence of BSE.

New Variant CJD in Humans

At the beginning of the BSE epidemic, public health officials in Britain had little fear that the epidemic might pose a threat to human health. There is a species barrier to the transmission of the TSE from any particular animal to another animal. Even in cases where transmission does occur in experimental systems, there is a requirement for an adaptation event before the agent can be readily transmitted. Humans were thought not to be sensitive to animal TSE agents because of this species barrier. In particular, no evidence for the transmission of scrapie to man has ever been found despite the fact that people all over the world, but especially in Britain, have eaten sheep infected with scrapie for centuries.

In 1995 and 1996, however, 12 cases of a variant form of human CJD occurred in Britain. These variant CJD cases were characterized by an unusually early age of onset, with some cases in their teens, and by a different symptomology. A comparison of the ages at which people in Britain contracted sporadic CJD during the last 25 years with that of the ages of the first 21 cases of variant CJD is shown in Fig. 7.13A. Sporadic CJD is primarily a disease of people in their 50s, 60s, and 70s, with a peak of occurrence in the early 60s. Cases in people under 40 are rare. Variant CJD to date has been a disease of people in their teens, 20s, and early 30s. Symptomology also differs. Sporadic CJD is characterized by dementia as an early symptom, whereas new

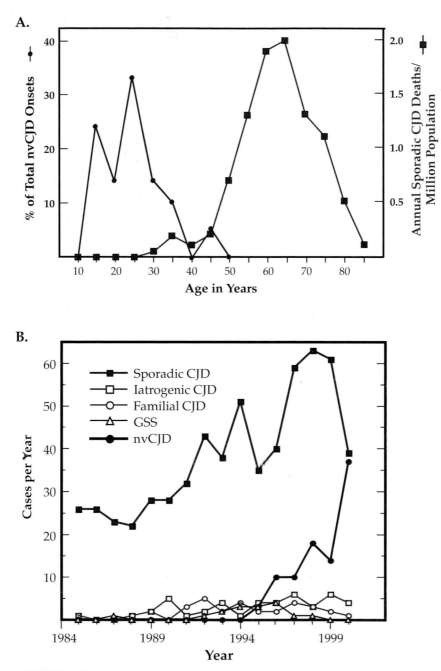

FIGURE 7.13   Creutzfeldt-Jakob disease in Britain. (A) Annual age-specific death rates for sporadic CJD (573 cases) in Britain between 1970 and 1994, compared with the age distribution of the first 21 cases of new variant CJD (nvCJD) in 1995 and 1996. The scales have been chosen to optimize comparison of the age distributions [Adapted from Nathanson *et al.* (1999, Fig. 7, p. 449).] (B) Cases of CJD of various etiolgies and Gerstmann-Straussler-Scheinker syndrome in Britain from 1985 to 2000. The data for 2000 are only available through September 4, 2000, but have been normalized to a full year. The rise in the number of cases of sporadic CJD reported is probably due to increased recognition of the disease. [Data are *Monthly CJD Statistics,* from the Department of Health of the United Kingdom.]

variant CJD is characterized by psychiatric symptoms, usually depression, and the patient is often first seen by a psychiatrist. Third, time to death averages somewhat longer in variant CJD than in sporadic CJD. Cases of variant CJD have continued to arise. As of late summer, 2000, about 70 cases have been reported in which the age of onset was from 16 to 53. The increasing rise in variant CJD cases is shown in Fig. 7.13B, which goes through August 2000. In late May, 2001, the one hundredth case was reported. Thus, the number of cases continues to increase, but the rate of increase appears to be slowing or to have leveled off.

There is now a considerable body of evidence that variant CJD is caused by infection with BSE and results from eating BSE-contaminated meat. If so, it is not clear how many cases may ultimately arise. As described above, kuru has a long latent period, with disease developing as long as 40 years after infection. Thus, there is the potential that a large number of cases may arise over time. However, unlike kuru, there is a species barrier for the transmission of BSE to humans, leaving hope that the total number of cases will remain small. A sensationalized and gripping account of kuru, CJD, and BSE is found in the book *Deadly Feasts* by Richard Rhodes.

Prion Protein

The nature of the infectious agents responsible for scrapie and other TSEs has been controversial, in part because the study of these agents has presented enormous technical difficulties. The kuru agent was shown to be transmissible to other primates many years ago, but the incubation period is very long in these animals (more than 10 years in some cases) and they are expensive to maintain, which limited early progress in the study of the molecular biology of the agents. The subsequent discovery that many TSEs could be transmitted to mice and hamsters, in which the incubation period was much shorter, as short as 60 days in some instances, speeded up progress. Transgenic mice, in particular, have been very useful because the genetic background can be controlled. However, such studies remain slow and tedious because an infectivity assay often takes more than 1 year.

Studies in mice and other animals, as well as the findings that mutations in the prion protein are associated with inherited TSEs in humans, have made clear that the prion protein, abbreviated PrP, is intimately involved in the transmission of TSE and in the disease process. The normal cellular protein is referred to as PrP^c. The structure in solution of the C-terminal half of the mouse version of this protein (residues 121–231) is illustrated in Fig. 7.14. The protein has a high content of α helix. In this half of the protein, there are three α-helical domains of 11, 15, and 18 residues, and only a short (four residues in each strand) two-stranded antiparallel β sheet. The N-terminal 98 residues of this protein form a flexible random coil in solution, as determined by nuclear magnetic resonance imaging.

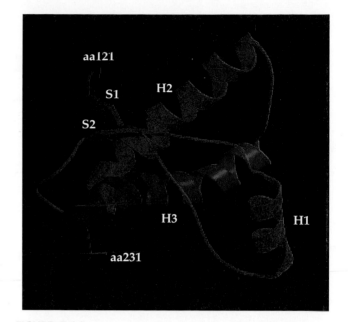

FIGURE 7.14 The structure of the prion protein. The structure of residues 121–231 of the mouse prion protein in solution as determined by NMR imaging is shown. The protein is color coded from blue at residue 121 to red at residue 231, with beta sheets shown as flat arrows and alpha helices as coils. The second and third helices are linked with a disulfide bond (not shown). (Compare with Figs. 7.10 and 7.15) [Adapted from Riek *et al.* (1996).]

The prion protein is synthesized as a larger precursor of 254 amino acids that contains both N-terminal and C-terminal extensions (Fig. 7.15). The N-terminal extension is a signal sequence that leads to the translocation of PrP into the lumen of the endoplasmic reticulum. It is removed by signal peptidase, as are most N-terminal signal sequences. The C-terminal extension is removed by another cellular protease and the protein is attached to a phosphoinositol glycolipid anchor that anchors the protein in the membrane. The protein is N-glycosylated on two sites. The processed protein is transported to the plasma membrane and transiently displayed on the surface of the cell with a half-life of about 5 hr. It is then recycled into endosomal compartments and eventually into lysosomes, where it is degraded.

The function of PrP^c is unknown. Many knockout mice that lack the gene for PrP have been constructed and most appear normal. Sleep abnormalities occur is some knockouts, however, and one strain of null mice develops abnormalities in the cerebellum. Such findings suggest that the protein might have a neurological function, but these abnormalities could arise from secondary effects of the knockout process, since other strains of knockout mice appear normal.

The brains of most humans or experimental animals exhibiting TSEs contain a conformational variant of PrP^c called PrP^{Sc} (Sc for scrapie) or PrP^{res} (res for resistant to protease). PrP^{Sc} is found in aggregates that are largely resistant to digestion with proteases. Treatment of PrP^{Sc} with

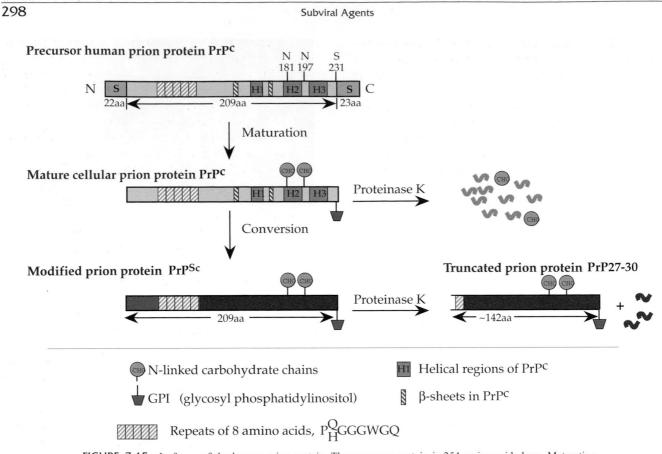

FIGURE 7.15 Isoforms of the human prion protein. The precursor protein is 254 amino acids long. Maturation involves removal of the N-terminal signal sequence and the C-terminal 23 amino acids (two boxes marked S), attachment of the new C terminus to a GPI anchor, and N-linked glycosylation at Asn181 and Asn197. After exposure to scrapie prions, the protein is converted to PrP^{Sc}, which is partially resistant to proteinase K. This conversion involves loss of some helical regions (H's) in the cellular form, and formation of new β sheets. [Adapted from Weissman (1996), Riek *et al.* (1996), and Prusiner (1998).]

proteases and subsequent disaggregation of the proteolyzed PrP^{Sc} give rise to a molecule that is truncated by about 80 amino acids at its amino terminus (Fig. 7.15). In contrast, PrP^c is completely destroyed by such protease treatment, and the normal PrP is also referred to as PrP^{sen} (for sensitive to protease). Circular dichroism and infrared spectroscopy indicate that PrP^{Sc} has a much higher content of β sheet than does PrP^c, 43 versus 3%, and a lower content of α helix, 30 versus 42%, suggesting a profound conformational rearrangement of the prion protein in the process of conversion from PrP^c to PrP^{Sc}.

Studies with Mice

Transgenic mice have been useful in the study of TSEs. Mice have been made that lack the gene for the prion protein, or that express wild-type or mutant prion proteins at levels from less than normal to several times normal. Most mice that make no PrP^c are normal, as described. However, such mice are resistant to scrapie infection. They do not become ill, and no infectious material is produced in the brain after inoculation of scrapie. In contrast, mice that overexpress PrP are more sensitive to infection with scrapie. The incubation period is shorter, and the animals die more quickly after inoculation with scrapie.

Thus, the presence of PrP is essential for the development of TSE in mice. It has also been found that individual neurons must be able to produce PrP if they are to be sensitive to scrapie-induced death. Neurografts from a donor mouse that expresses PrP have been implanted into mice that lack the PrP gene. Upon inoculation of scrapie into the brain, the neurons in the graft develop a typical scrapie-induced disease pathology. However, neurons outside the graft remain healthy.

Ex Vivo and in Vitro Studies

Cell lines have been established that are persistently infected with scrapie. These cells continuously produce PrP^{Sc}, which allows biochemical studies to be performed over a shorter time frame. The infected cells produce infectious material that causes scrapie when inoculated into mice.

Of great interest has been the development of an *in vitro* system for the conversion of PrP^c to PrP^{Sc}. In this system, radioactive PrP^c is mixed with unlabeled PrP^{Sc}, and the conversion of the labeled PrP^c to PrP^{Sc} is followed by its becoming resistant to protease. These studies make clear that PrP^c can be converted to PrP^{Sc} by exposure to PrP^{Sc} in a process that does not require the activities of intact cells. However, so much infectivity is associated with the PrP^{Sc} added to the reaction mixture that no increase in infectivity can be demonstrated. Thus, these studies do not address the question of the nature of the infectious agent. These studies have also been useful in the study of the species barrier, which can be quantitatively examined in such reactions.

Protein-Only Hypothesis

It is clear that PrP is important in the development of TSEs. There are two unresolved questions about PrP and the disease process, however. First, is the infectious agent that leads to TSE PrP^{Sc} itself or is it another entity, such as a virus? Second, does PrP^{Sc}, or some other modified form of PrP, cause the symptoms of the disease, or is it simply a side effect of the disease process?

Preparations of the infectious agent purified from scrapie-infected mouse brain consist largely of PrP^{Sc}. There is very little nucleic acid in infectious preparations of PrP^{Sc}. In particular, there is no homogeneous DNA or RNA molecule that might arise from a virus, for example. This has led to the hypothesis that PrP^{Sc} is itself the infectious agent. In this model, "infectious" PrP^{Sc} induces PrP^c to assume the PrP^{Sc} conformation, and the accumulation of PrP^{Sc} in the brain leads to the pathology associated with TSEs. Most of the experimental data are compatible with such a model. PrP^{Sc} does induce PrP^c to assume the PrP^{Sc} conformation. Mutations in PrP^c could make it easier for the protein to assume the PrP^{Sc} conformation. The species barrier could result from lowered interaction affinities between proteins of different sequence. However, it has not been possible to prove this hypothesis. PrP^{Sc} preparations have a very low specific infectivity, with at least 10^5 molecules of PrP^{Sc} required for infection. Thus it remains possible that contaminants in the preparation might be required for infectivity. It has not been possible to demonstrate an increase in infectivity associated with the conversion of PrP^c to PrP^{Sc}, as described above, which would provide solid evidence that PrP^{Sc} is infectious.

In addition to the inability to prove the protein-only hypothesis, which could be due to the technical difficulties associated with this system, there are specific conceptual difficulties with PrP^{Sc} as the infectious agent. One of the major criticisms of the protein-only hypothesis is the fact that as many as 20 different strains of scrapie exist as assayed in mice. These strains of scrapie differ in properties such as the length of the incubation period following infection before disease becomes apparent, the areas of the brain affected, and the symptoms of the disease, but these properties do not vary within a strain. Such properties are expected for an infectious entity with a nucleic acid genome, but are difficult to reconcile with the properties of an infectious protein. If the protein-only hypothesis is true, these differences in properties could only result from differences in the conformation of PrP^{Sc} in the different strains. How is it that a single, fairly small protein can take up so many different conformations and that each can induce the production of more protein having the same conformation?

Supporters of the protein-only hypothesis suggest that a limited number of conformational states of the prion protein would be sufficient to explain the multiple strains of scrapie that exist. They point to experimental data that show that at least two demonstratively different conformational states of the prion proteins of two different mammals exist that "breed true." Two different strains of transmissible mink encephalopathy that produced different disease characteristics in mink were passaged in hamsters. The PrP^{Sc} from the two strains, isolated from infected brain, are differently truncated at the amino terminus on treatment with proteases in vitro. Thus the conformations of the PrP^{Sc} in these two strains, both derived from hamster PrP, must be different. Furthermore, this difference can be reproduced in an in vitro reaction in which PrP^c is mixed with the two different types of PrP^{Sc}. Each type of PrP^{Sc} induces PrP^c to assume its own distinct conformation, as shown by the protease resistant fragment that is produced from the PrP^c on it conversion to PrP^{Sc}.

In a second example of demonstrably different prion conformations, human prions isolated from two different cases of TSE, one FFI and the second CJD, were found to be differently truncated after protease treatment. Passage of these TSEs in transgenic mice that expressed a chimeric mouse–human prion protein gave rise to prions in infected brain that reproduced the differences in truncation. Thus, these conformational differences breed true when passed in mice.

These studies support the hypothesis that PrP^{Sc} can exist in more than one conformational form, that the different conformational forms can produce different symptoms, and that the different forms are capable of propagation by inducing PrP^c to take up their own particular conformation. Thus, much of the experimental data are consistent with the protein-only hypothesis, although it has not been proven conclusively and many still doubt its validity. The hypothesis recently received a vote of confidence when its most outspoken and passionate advocate, Dr. Stanley Prusiner, was awarded the 1997 Nobel Prize for his "discovery of prions."

Transport of Infectivity to the Brain

Related to the conceptional problem of an infectious protein is how it might be transported to the brain after ingestion with food. This problem has been addressed in studies that ascertain in which tissues PrP^{Sc} is present following ingestion of PrP^{Sc}, and studies with transgenic mice that express PrP only in certain tissues. These various studies are compatible with a model in which infectivity is transported via axons following direct neuroinvasion of peripheral nerves. In the case of infection with only low doses of infectious material, amplification in follicular dendritic cells in lymphoid tissue may be required before neuroinvasion occurs. Thus, in terms of the protein-only model, PrP^{Sc} might induce the conversion of PrP^c to PrP^{Sc} in cells in Peyer's patches, which then spreads via lymphatic tissue to peripheral nerves by sequential conversion of PrP.

Formation of the PrP^{Sc} Seed

If PrP^{Sc} can transmit the disease to a new susceptible host, how is it formed in the first place? Current models propose that the conformational change resulting in PrP^{Sc} occurs rarely, but that once PrP^{Sc} is formed, it acts as a seed to induce the formation of more PrP^{Sc}. Two models to explain the conversion of PrP^c to PrP^{Sc} by PrP^{Sc} have been proposed. In one, PrP^{Sc} (which may be present in an aggregate) and PrP^c form a complex, and the PrP^{Sc} induces the conformational change in PrP^c. In the second model, PrP^c undergoes spontaneous transitions to different conformational states that are short lived and revert quickly to the native PrP^c conformational state. These conformational variants, however, can be locked into place by interaction with PrP^{Sc}. In either case, the altered PrP joins the aggregated PrP^{Sc} to form a larger aggregate. Since the aggregated PrP^{Sc} is insoluble, the reaction is essentially irreversible. Such a process could also explain the species barrier. The PrP proteins of different animals differ slightly in sequence. PrP^c that is identical in sequence to the PrP^{Sc} seed could interact with such a seed more readily than with a PrP^{Sc} seed that differs in sequence.

The protein-only hypothesis still requires a seed of PrP^{Sc} to begin the reaction. One possibility is that it can form spontaneously with a very low probability. Perhaps spontaneous changes in the conformation of PrP^c to the PrP^{Sc} conformation might be fixed if this change occurred simultaneously in a number of adjacent or interacting molecules. The effect of mutations in PrP^c might be to increase the probability of change to the PrP^{Sc} conformation, with the result that disease occurs more frequently. Such a model is compatible with data for human TSEs, where sporadic CJD occurs, albeit infrequently. However, sporadic disease has not been seen in shorter lived animals. No sporadic BSE has been described, and in countries where scrapie in sheep has been eradicated, such as New Zealand and Australia, no recurrence of disease has been observed.

Does PrP^{Sc} Cause the Disease?

If PrP^{Sc} is responsible for the pathology of TSE disease, and not simply a by-product of disease, the mechanism by which it causes disease is uncertain. An early model suggested that PrP^{Sc} itself is neurotoxic. However, it has been shown that a neuron must be able to express PrP^c before it can be killed by exposure to PrP^{Sc}. Thus simple neurotoxicity of PrP^{Sc} is not the cause of neuronal death. However, it is possible that conversion of PrP^c to PrP^{Sc} at the surface of the cell, which is known to occur, followed by accumulation of PrP^{Sc} in lysosomes as the neuron attempts to recycle it, could be toxic. In this model, it is the resistance of PrP^{Sc} to proteases in the lysosome that results in toxicity.

Recent findings have suggested another possibility. PrP can be expressed as a membrane-spanning protein as well as a protein anchored by a glycolipid anchor (Fig. 7.16). One membrane-spanning form, called ^{Ctm}PrP, has it C terminus outside the cell and the N terminus inside, with a transmembrane domain near the middle of the molecule. Preliminary data suggest that this form of PrP is neurotoxic: ^{Ctm}PrP has been found in brains of animals, including humans, suffering from TSE but not in normal brains. This has led to a model in which ^{Ctm}PrP is regularly produced at some frequency, but the normal cell has a mechanism to eliminate it. Overproduction of ^{Ctm}PrP, either by mutation or by a failure to eliminate it, leads to the symptoms of TSE. In this model, production of PrP^{Sc} might somehow result in the accumulation of ^{Ctm}PrP, perhaps by overwhelming the ability of the cell to eliminate it.

Prions of Yeast

Prions, defined as agents that possess two (or more) conformational forms, a soluble "normal" form and an aggregated form that can induce the conversion of the normal form to more of itself, have also been found in fungi. Two prions have been found in yeast (*Saccharomyces cerevisiae*) and a third in *Podospora* spp. The yeast prions have the characteristics of disease but the *Podospora* prion performs a normal cell function (controlling heterokaryon compatibility). The yeast prions are called [URE3], which affects nitrogen catabolism, and [PSI], which affects the termination of polypeptide chains during translation. A diagram of these proteins is shown in Fig. 7.17. [PSI] is a prion form of Sup35p, which is a translation release factor. In the [PSI] state, Sup35p assumes an altered conformation and aggregates, like PrP^{Sc}. The [PSI] state is dominant and can be transmitted to other yeast cells by transfer of cytoplasm containing [PSI]. Thus, the prion state induces the normal cell protein to assume the prion state, as with the model for PrP^{Sc}. The effect of the [PSI] state on the cell is to render Sup35p nonfunctional, and thus has the same effect as deletion of the gene encoding the protein. Loss of Sup35p activ-

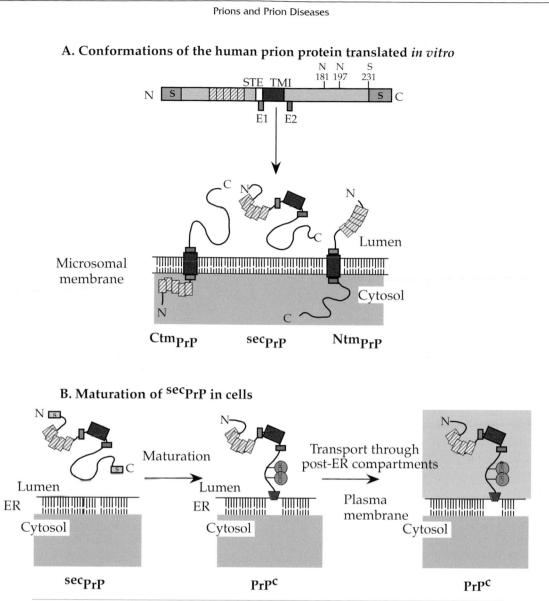

A. Conformations of the human prion protein translated *in vitro*

B. Maturation of ˢᵉᶜPrP in cells

☐ STE - stop transfer effector

■ TMI - transmembrane domain

▮ E1 (epitope for MAb 3F4)

▮ E2 (epitope for MAb 13A5)

◉ N-linked carbohydrate chains

▼ GPI- glycosyl phosphatidylinositol

▨ Repeats of 8 amino acids

FIGURE 7.16 Postulated topology of PrP proteins in membranes. (A) Topology of PrP proteins in membranes after translation in a cell-free system supplemented with microsomes. The topology was determined by a combination of protease digestion from the cytosolic compartment, and identification of the domains protected within the lumen using the two MAbs 3F4 and 13A5. Mutations have been shown to affect the ratio of the three forms shown, and greater concentrations of ᶜᵗᵐPrP are associated with neurodegenerative disease in mice. (B) Model for maturation and association with membranes of PrPᶜ in cells. [Adapted from Hegde *et al.* (1998).]

ity leads to increased read-through of stop codons during translation, and renders nonsense suppressor tRNAs much more active. [URE3] is the prion state of Ure2p, a protein involved in nitrogen catabolism. Like [PSI], [URE3] is an aggregated form of a conformational variant of Ure2p, and is dominant and transmissible. Loss of Ure2p by the cell affects the metabolism of nitrogen. Normal cells can assume the prion state with a low frequency, but once

Yeast Prion Protein   Ure2p

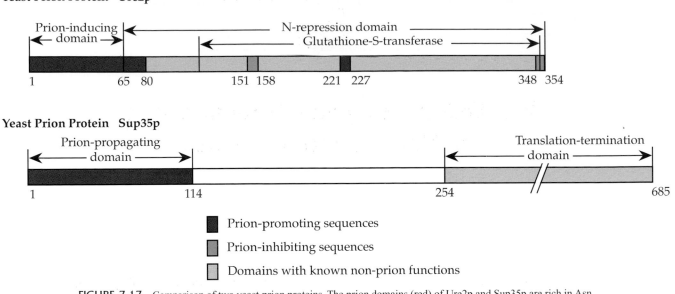

Yeast Prion Protein   Sup35p

- ■ Prion-promoting sequences
- ▦ Prion-inhibiting sequences
- ▨ Domains with known non-prion functions

FIGURE 7.17 Comparison of two yeast prion proteins. The prion domains (red) of Ure2p and Sup35p are rich in Asn and Gln residues, which are important for prion generation and propagation. [Adapted from Fig. 10 of Wickner *et al.* (1999).

assumed the prion state is retained. Cells in the prion state can be cured by certain treatments that break up the protein aggregates and causes the protein to assume a normal conformation.

FURTHER READING

Defective Interfering Particles

Roux, L. (1999). Defective interfering viruses. *In* "Encyclopedia of Virology" (A. Granoff and R. G. Webster, eds.), Vol. 1, pp. 371–375. Academic Press, San Diego, California.

Viroids

Flores, R., Di Serio, F., and Hernández, C. (1997). Viroids: The noncoding genomes. *Semin. Virol.* **8**; 65–73.

Pelchat, M., Lévesque, D., Ouellet, J., *et al.* (2000). Sequencing of peach latent mosaic viroid variants from nine North American peach cultivars shows that this RNA folds into a complex secondary structure. *Virology* **271**; 37–45.

Hepatitis Delta

Modahl, L. E., and Lai, M. M. C. (2000). Hepatitis delta virus: The molecular basis of laboratory diagnosis. *Crit. Rev. Clin. Lab. Sci.* **37**; 45–92.

Polson, A. G., Ley, H. J., III, Bass, B. L., *et al.* (1998). Hepatitis delta virus RNA editing is highly specific for amber/W site and is suppressed by hepatitis delta antigen. *Mol. Cell. Biol.* **18**; 1919–1926.

Prions and Prion Diseases

Baker, H. F., and Ridley, R. M. (1996). What went wrong in BSE? From prion disease to public disaster. *Brain Res. Bull.* **40**; 237–244.

Bons, N., Mestre-Frances, N., Belli, P., *et al.* (1999). Natural and experimental oral infection of nonhuman primates by bovine spongiform encephalopathy agents. *Proc. Natl. Acad. Sci. U.S.A.* **96**; 4046–4051.

Chesebro, B. (1999). Prion protein and the transmissible spongiform encephalopathy diseases. *Neuron* **24**; 503–506.

Hegde, R. S., Mastrianni, J. A., Scott, M. R., *et al.* (1998). A transmembrane form of the prion protein in neurodegenerative disease. *Science* **279**; 827–834.

Prusiner, S. B. (1998). Prions. *Proc. Natl. Acad. Sci U.S.A.* **95**; 13363–13383.

Prusiner, S. B., ed. (1999). "Prion Biology and Diseases." Cold Spring Harbor Laboratory Press, Cold Spring Harbor, New York.

Rhodes, R. (1997). "Deadly Feasts." Simon & Schuster, New York.

Riek, R., Hornemann, S., Wider, G., *et al.* (1996). NMR Structure of the mouse prion protein domain PrP(121–231). *Nature (London)* **382**; 180–183.

Schonberger, L. B. (1998). New-variant Creutzfeldt-Jakob disease and bovine spongiform encephalopathy: The strengthing etiologic link between two emerging diseases. *In* "Emerging Infections 2" (W. M. Scheld, W. A. Craig, and J. M. Hughes, eds.), 1–15, ASM Press; Washington, DC.

Weissmann, C. (1996). Molecular biology of transmissible spongiform encephalopathies. *FEBS Lett.* **389**; 3–11.

Yeast Prions

Wickner, R. B., Taylor, K. L., *et al.* (1999). Prions in *Saccharomyces* and *Podospora* spp.: Protein-based inheritance. *Microbiol. Mol. Biol. Rev.* **63**; 844–861.

8

Host Defenses against Viral Infection and Viral Counterdefenses

INTRODUCTION

On infection of a mammal by a microorganism, a complex response is generated that attempts to eliminate the infectious agent. These responses can be conveniently grouped into two categories, innate immune responses and adaptive immune responses. The adaptive or acquired immune response requires time to develop, is specific to the invading pathogen, and is followed by immunological memory that usually renders the host immune, or at least less susceptible, to subsequent infections by the same organism. The two most important components of this response are the production of cytotoxic T lymphocytes (CTLs) and the production of humoral antibodies. Innate responses, in contrast, generate early responses to infection, are not specific to the pathogen, and do not render the organism resistant to subsequent infection by the same pathogen. Cytokines such as interferon are among the most effective innate responses. Innate and adaptive immune responses do not function independently of one another: The proper functioning of the immune system requires the activities of both. The various activities constitute a multi-faceted, interactive, and complex series of responses to infection by a pathogen. Figure 8.1 illustrates some of the cells and other participants involved in innate and acquired immunity.

ADAPTIVE IMMUNE SYSTEM

The adaptive immune system contains two major arms. The cellular arm leads to the production of CTLs, also called killer T cells. The humoral arm leads to the production of antibodies that are secreted by B cells. T-helper cells are important players in generating these responses.

Major Histocompatibility Complex

Both cellular immunity and humoral immunity require the activation of a class of lymphocytes called T cells (T from thymus, where these cells mature). T cells recognize peptide antigens 8–20 amino acids in length that are presented to them by cell surface proteins encoded in the major histocompatibility complex (MHC) locus (in humans, the MHC is called HLA, from *h*uman *l*ymphocyte *a*ntigen). The two types of MHC molecules that present antigenic peptides are called class I and class II. Both class I and class II MHC molecules are integral membrane proteins that are composed of two polypeptide chains.

Class I and Class II MHC

The MHC class I molecule is a heterodimer composed of a heavy chain of about 350 amino acids, which is encoded within the MHC locus, and a light chain of about 100 amino acids, β_2 microglobulin, which is encoded elsewhere. The structure of an MHC class I molecule is shown schematically in Fig. 8.2A, and as determined by X-ray crystallography in Fig. 8.2B. The MHC class I heavy chain consists of three extracellular domains called α_1, α_2, and α_3, a transmembrane domain, and a cytoplasmic domain. β_2 microglobulin forms a fourth extracellular domain and is held in the complex by noncovalent interactions. The α_1 and α_2 domains, which are structurally related to one another, form a platform with helical walls. The walls form a groove in which the antigenic peptide, consisting usually of 8–10 amino acids, is anchored. The α_3 domain and β_2 microglobulin are homologous (derived from a common ancestral polypeptide by gene duplication) and are members of the immunoglobulin (Ig) superfamily. They share sequence identity and have common structural features.

Innate Immunity

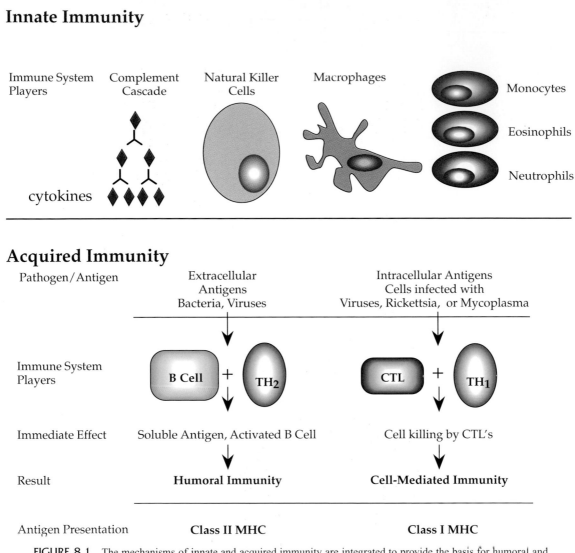

FIGURE 8.1   The mechanisms of innate and acquired immunity are integrated to provide the basis for humoral and cellular immunity. [Adapted from Mims *et al.* (1993, p. 5.8).]

The structure of MHC class II molecules is similar to that of class I molecules, as illustrated in Fig. 8.2A. Class II molecules are composed of a heterodimer of two proteins encoded within the MHC locus. These two proteins, designated α and β, each contain two extracellular domains (and thus the assembled molecule contains four extracellular domains as does class I MHC). Both proteins are anchored in the plasma membrane by membrane-spanning anchors and have cytoplasmic domains. The distal α_1 and β_1 domains form a platform with a groove that binds an antigenic peptide for presentation to T cells, but in this case the peptide is longer, usually 14–18 amino acids in length. The proximal domains, α_2 and β_2, are members of the Ig superfamily.

The number of peptides that can be presented by an individual class I or class II molecule is large. Only certain residues in the peptide, called anchor residues, interact specifically with the MHC molecule. The remainder of the peptide can vary in sequence. Furthermore, the MHC is polymorphic and there are many different alleles within the human population. The haploid number of genes in humans that encode heavy chains used in class I MHC molecules is two or three, and a diploid individual could potentially make several different class I MHC molecules with differing requirements for anchor residues. In the case of class II MHC, six α and seven β genes have been identified in humans. Thus, any individual can present very many different peptides

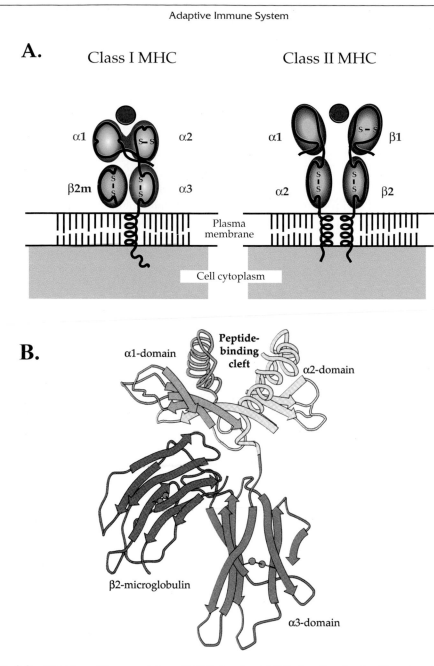

A.

Class I MHC   Class II MHC

$\alpha 1$ $\alpha 2$ $\alpha 1$ $\beta 1$

$\beta 2m$ $\alpha 3$ $\alpha 2$ $\beta 2$

Plasma membrane

Cell cytoplasm

B.

$\alpha 1$-domain   Peptide-binding cleft $\alpha 2$-domain

$\beta 2$-microglobulin $\alpha 3$-domain

FIGURE 8.2   (A) Schematic representation of MHC class I and MHC class II molecules. The orientation of these molecules at the cell surface is indicated, as are the domain structure of the extracellular portions of the proteins, the membrane-spanning domains, and the cytoplasmic domains, and how they function in the cell. The blue and green spheres represent bound peptide antigens. (B) Three-dimensional ribbon diagram of the structure of MHC-1 (HLA-A2), as determined by X-ray crystallography. [From Bjorkman *et al.* (1987) as reprinted in Kuby (1997).]

to T cells. Note that since the MHC is polymorphic, different individuals may present different peptides to T cells.

T-Cell Recognition of Peptide Antigens

T cells express a T-cell receptor on their surface, which is able to recognize a specific peptide presented in the context of class I or class II MHC molecules. The T-cell receptor is a heterodimer formed by one α and one β chain, or by one γ and one δ chain, as illustrated in Fig. 8.3. The great majority of circulating T cells possess receptors formed by $\alpha\beta$ dimers.

During maturation of T cells, three or four separate regions of the genes for α or β, respectively, are brought together by deletion of the intervening sequences. This process is illustrated in Fig. 8.4. The regions are the V (variable), D (diversity), J (joining), and C (constant) regions for β, or the V, J, and C regions for α. The V and C regions belong to the Ig superfamily, whereas D and J are

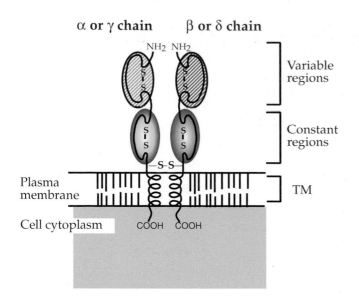

α or γ chain   β or δ chain

FIGURE 8.3 Structure of the T-cell receptor (TCR). Receptors are heterodimers of α/β or γ/δ TCR chains. Each chain has a transmembrane anchor (TM). The C-terminal domains and the TM are relatively constant in sequence, but the N-terminal domains are variable. [From Chen and Alt (1997, Figs. 7.6 and 7.7, pp. 344, 345).]

shorter, unrelated domains. V and J, or V, D, and J, are first joined at the DNA level by a process that deletes the intervening DNA. The combined VJ or VDJ is then joined to C by splicing of the pre-mRNA transcript. Multiple copies of each of the four segments exist in the germ line and combinatorial joining of these results in a very large number of possible α or β subunits. Furthermore, the

joining of VJ and VDJ is imprecise, and additional diversification results from repair of the joining regions. Combinatorial joining of α and β subunits results in the production of an even larger repertoire of T-cell receptors. An estimate of the possible diversity of T-cell receptors, sometimes referred to as the theoretical repertoire, is shown in Table 8.1. The theoretical repertoire, how many T-cell receptors could possibly be formed, is perhaps 10^{18}, which is an exceedingly large number.

Once any individual T-cell develops and expresses a T cell receptor, no further recombination occurs and the receptor does not change thereafter. However, any individual has many, many T cells and these continue to develop throughout life. The repertoire present in any human at any one time certainly exceeds 10^8 different T-cell receptors, expressed on different T cells.

Each different T-cell receptor is potentially able to recognize a different peptide epitope (the epitope is the surface of the peptide that interacts with the T-cell receptor). The T-cell receptor recognizes the peptide epitope in the context of class I or class II MHC; that is, the T-cell receptor interacts with both the peptide and the MHC molecule (Fig. 8.5). The T cell cannot recognize the peptide alone or even recognize the peptide presented by the wrong MHC molecule. Similarly, the T cell cannot recognize a class I or class II MHC molecule with the wrong peptide in its cleft. The discovery of this requirement for dual recognition, which was made using a viral system, resulted in a Nobel Prize for Doherty and Zinkernagel (Table 1.1). Such a requirement for multiple interactions is a recurring theme in the immune system, and it has evolved because this

TABLE 8.1   Comparison of Diversity in Human Immunoglobulin and T-Cell Receptor Genes

Mechanism of diversity	Immunogobulins			αβ T-cell receptors		γδ T-cell receptors	
	Heavy chains	Light chains		α chain	β chain	γ chain	δ chain
		κ chain	λ chain				
Multiple germ-line gene segments							
V	65	40	30	~70	52	12	>4
D	27	0	0	0	2	0	3
J	6	5	4	61	13	5	3
Combinatorial joining							
Combinatorial V-J-D combinations	$65 \times 27 \times 6 =$ 1.0×10^4	$40 \times 5 =$ 200	$30 \times 4 =$ 120	$70 \times 61 =$ ~430	$52 \times 2 \times 13 =$ 1.3×10^3	$12 \times 5 =$ 60	$4 \times 3 \times 3 =$ 36
D segments read in three frames	Rarely	—	—	—	Often	—	—
Joints with N and P nucleotides	2	(1)	2	1	??	??	
V gene pairs	$1.0 \times 10^4 \times 320 = \mathbf{3.2 \times 10^6}$			$430 \times 1.2 \times 10^3 = \mathbf{5.8 \times 10^6}$		$60 \times 36 = \mathbf{2.1 \times 10^3}$	
Junctional diversity	$\sim 3 \times 10^7$			$\sim 2 \times 10^{11}$		???	
Total diversity	$\mathbf{\sim 10^{14}}$			$\mathbf{\sim 10^{18}}$		???	

Source: Data from Janeway *et al.* (1999, pp. 62, 93, 158).

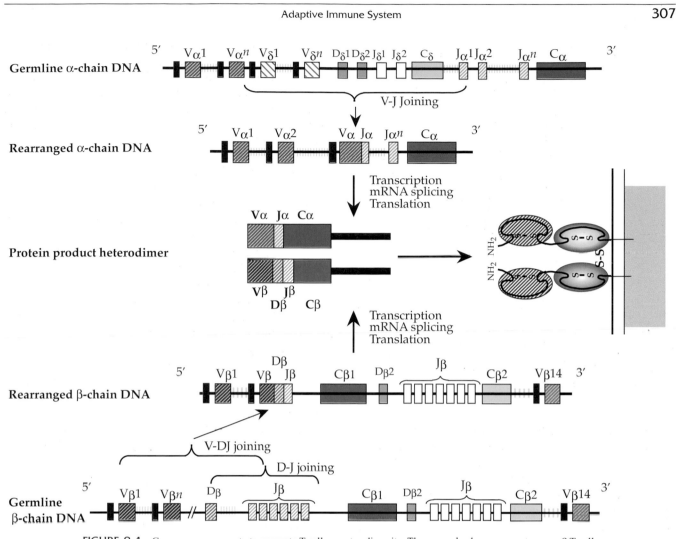

FIGURE 8.4 Gene rearrangements to generate T-cell receptor diversity. The example shown generates an $\alpha\beta$ T-cell receptor. The α chain undergoes a V_α-J_α joining, whereas the β chain undergoes two joins: D_β to J_β followed by V_β to $D_\beta J_\beta$. Primary RNA transcripts are spliced to give mRNAs in which RNA sequences encoding VJ or VDJ are joined to RNA sequences encoding constant domains, C_α or C_β. These spliced mRNAs are then translated into the α and β chains of the TCR. [Adapted from Kuby (1997, Fig. 11.6, p. 269).]

potent and potentially harmful system must be carefully regulated.

The T-cell receptor is part of a complex containing accessory molecules that are required for the function of the receptor. Two such molecules are CD4 and CD8, and mature T cells possess either CD4 or CD8 (but not both). CD8 contains one Ig domain attached to a stalk region, whereas CD4 contains four Ig domains (Fig. 8.5). $CD8^+$ T cells recognize peptides presented by class I MHC (Fig. 8.5A). $CD4^+$ T cells, in contrast, recognize peptide epitopes presented in the context of class II MHC (Fig. 8.5B). CD8 or CD4 interacts with constant regions of class I or class II MHC molecules, respectively, and increases the binding affinity of the T cell for its cognate MHC-peptide complex by about 100-fold. Class I and class II MHC molecules acquire the peptides that they present in fundamentally dif-

ferent ways and are components of two different responses to infection by microorganisms.

Cytotoxic T Cells

Most $CD8^+$ T cells are CTLs. A minority of $CD4^+$ T cells are also CTLs, but most CTLs are $CD8^+$. $CD8^+$ T cells are class I MHC restricted, as described. Because class I MHC molecules are expressed on most mammalian cells, the major exception being neurons, which express little or no class I MHC, most cells in an individual are capable of presenting peptides to T cells in a class I MHC context.

The peptides presented by class I MHC are derived from intracellular proteins and represent a sampling of all proteins being synthesized within the cell. The pathway involved is illustrated in Fig. 8.6. The peptides are generated

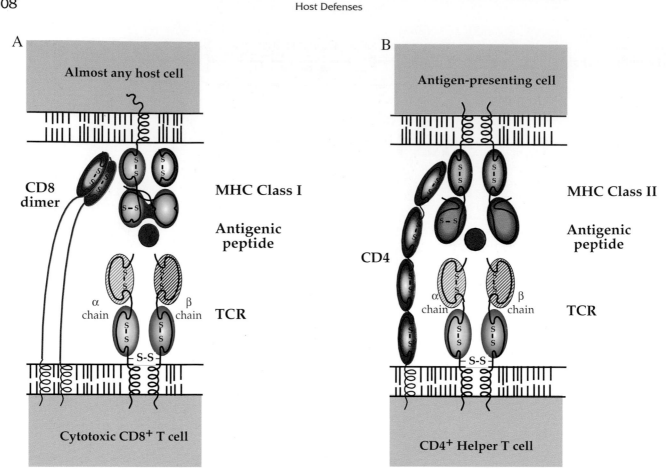

FIGURE 8.5 (A) Interaction of the TCR on a cytotoxic CD8[+] T cell with an MHC class I molecule complexed with an antigenic peptide on almost any cell. The TCR interacts with both the peptide and with the MHC molecules. The CD8 homodimer interacts with a conserved region of the MHC α_3 domain. (B) Interaction of the TCR on a CD4[+] helper T cell with an MHC class II molecule complexed with an antigenic peptide on the antigen presenting cell. The TCR interacts with both the peptide and with the MHC molecule. The membrane distal domain of CD4 recognizes a conserved region of the MHC $\beta2$ domain. [From Chen and Alt (1997, pp. 344, 345) and data from Kuby (1997, pp. 275–277).]

by proteolysis of intracellular proteins by an enzyme system referred to as the proteasome. The proteasome is a large complex, possessing many subunits, that is present in the cytoplasm. It possesses ATP-dependent proteolytic activity and is the major cellular proteolytic site other than the lysosome. In addition to its function in the immune response, the proteasome is important for turnover of many proteins within the cell and for degradation of misfolded proteins. Peptides resulting from degradation of intracellular proteins are actively secreted, in a process that requires hydrolysis of ATP, into the lumen of the endoplasmic reticulum (ER) by a transporter called TAP (TAP=*t*ransporter *a*ssociated with *a*ntigen *p*resentation). TAP is encoded in the MHC and consists of a heterodimer anchored in membranes of the ER.

Proteins secreted into the lumen of the ER during synthesis are also sampled. There is a pathway that recycles lumenal proteins back to the cytoplasm. This pathway may serve to rid the ER of misfolded proteins as well as

enabling the sampling of proteins destined for the plasma membrane or other intracellular organelles. On reentry into the cytoplasm, a cellular glyconase removes carbohydrates from glycoproteins, and the protein backbone is degraded by the proteasome. Thus viral proteins that are inserted into the lumen of the ER during synthesis, such as glycoproteins used to assemble progeny virions, are also sampled by the proteasome/TAP pathway.

A peptide delivered to the lumen of the ER by TAP can be bound by a class I MHC molecule if it has the right anchor residues. Class I MHC is associated with TAP through a protein called tapasin, which appears to allow a direct transfer from TAP to the class I MHC molecule. The class I MHC molecule also interacts with the chaperones calnexin and calreticulin. Binding of peptide stabilizes the class I molecule and facilitates its release and transport to the cell surface. Class I MHC that is transported without a peptide is unstable at 37°C and is degraded.

Infected cell   External antigen or
pathogen

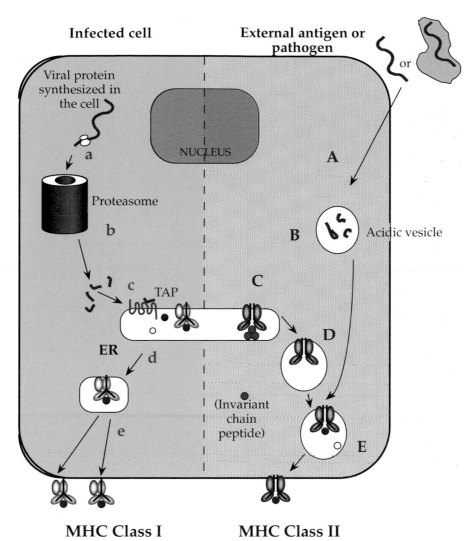

MHC Class I   MHC Class II

FIGURE 8.6   Antigen processing by the MHC. Both class I and class II pathways are shown. On the left, a viral protein is synthesized in the cell (a), degraded (b), viral peptides are bound by TAP in the ER and transported into the ER lumen, where they are bound by MHC class I molecules (c), MHC containing bound peptide is transported by cellular vesicles through the Golgi apparatus (d) to the cell surface (e) where it is expressed for inspection by T cells. On the right, an external antigen or pathogen is endocytosed into the cell (A) and is degraded in an acidic vesicle (B). Class II molecules are synthesized and imported into the ER (C) with a trimer of an invariant chain peptide in the binding site, and transported to the trans-Golgi (D) where the invariant chain is lost. The Golgi vesicle fuses with the endosome containing the antigenic peptide (E), and the peptide–MHC class II complex is transported to the cell surface. [Adapted from Fields *et al.* (1996, p. 351).]

The end result is that class I MHC presents a random sampling of peptides derived from proteins being synthesized within the cell for inspection at the cell surface by any T cell that may be in the vicinity. The peptides bound to a single isoform of class I MHC molecules present on the surface of cells in culture have been analyzed by very sensitive techniques. More than 10,000 different peptides, present at 2 to 4000 copies per cell, were identified. This great diversity of peptides consists mostly of self-peptides, but peptides derived from intracellular viruses or other intracellular pathogens will be represented if the cell is infected. If a patrolling T cell has a receptor that binds specifically to a peptide being presented by the class I MHC on another cell, the T cell may become activated and may proliferate. Once activated, CTLs kill cells that present the epitope they recognize. In different assays, the number of

MHC–peptide complexes required for recognition by a CTL has been estimated to be between one and several hundred, and may depend on the affinity of binding of the MHC–peptide target by the T-cell receptor as well as the state of activation of the T cell.

The complete activation of a T cell and its proliferation requires other signals in addition to stimulation by antigen. One such signal is the cytokine interleukin-2 (IL-2). The source of IL-2 is usually a class of T-helper (T_H) cells (most T_H cells are CD4$^+$, as described below). Activated T_H-1 cells secrete IL-2, as well as tumor necrosis factor β (TNF-β), interferon γ (IFN-γ), and other cytokines. Thus, they help CTLs to become fully active. Proliferation of T cells that recognize a specific peptide derived from a viral protein means that a vigorous CTL response against an invading pathogen ensues.

CTLs kill target cells by inducing apoptosis, a cell suicide pathway described in a later section of this chapter. One of three different mechanisms is used to induce apoptosis. These mechanisms are listed in outline form here but are described in more detail when the apoptotic pathway is considered. In one mechanism, the T cell releases the contents of granules, which contain perforins and proteases among other components, into the target cell. Perforins form pore structures in the plasma membrane of the target cell that allow ions to leach out of the cell, and the proteases participate in the activation of cell pathways leading to apoptosis. In a second mechanism, apoptosis is induced by triggering the Fas death receptor on the surface of the target cell. These two mechanisms lead to cell death within 4–6 hr. A third mechanism utilizes the TNF-α death pathway and is a slower process, leading to cell death in 18–24 hr. Killing of target cells is a drastic response and the CTL pathway is directed toward eliminating internal pathogens, usually viruses. Because early proteins encoded by the virus can be sampled by the MHC–T-cell receptor pathway, as well as late proteins, it is possible for a T cell to kill a virus-infected cell before it has time to synthesize much progeny virus. Many viruses counter this pathway by interfering with the ability of an infected cell to express class I MHC at its surface, as described below.

Although three killing mechanisms are used by different CTLs, they are not redundant. Mice that lack the perforin gene are unable to control infection by lymphocytic choriomeningitis virus and half die within a month of infection. This virus is not pathogenic in normal mice or even in immunocompromised mice. Thus death must result from immunopathology caused by an unbalanced or incomplete T-cell response.

Since the class I pathway presents peptides derived from self as well as from viruses that may have infected the cell, how then do the CTLs know not to kill cells expressing self-antigens? The answer lies in part in the selection of an appropriate repertoire of T cells. T cells bearing T-cell receptors recognizing many different possible peptide antigens are thought to arise by random combinatorial joining and diversity-inducing processes. T cells undergo their early differentiation in the thymus, where selection occurs. Only T cells that express a T-cell receptor capable of recognizing class I MHC bearing a peptide are selected (called positive selection); T cells that do not express an appropriate receptor die. However, if such a T-cell receptor has a high affinity for self-peptides present in the thymus, then that T cell also dies (called negative selection). In this process of selection only about 2% of T cells survive and most of these encode receptors that recognize non-self antigens. These T cells are released into the circulation to patrol for cells that are infected by viruses or other intracellular pathogens.

A second level of control that reduces the incidence of reaction against self lies at the level of cytokine induction. Virus infection or infection by other parasites usually leads to tissue damage that produces an inflammatory response resulting from the release of cytokines. Cytokines such as IL-2 are required as a second signal for CTL activation, and cytokines such as IFN-γ upregulate the presentation of antigens by MHC molecules as well as other aspects of the immune response. Thus, the inflammatory response makes it more likely that T cells will respond to antigens that they recognize. Furthermore, once activated, CTLs undergo apoptosis if the cytokine signals are no longer present, thus damping out autoimmune responses.

Killing of virus-infected cells by CTLs in an effort to eradicate viral infection relies on the ability of the cells of most organs to regenerate from progenitor cells. In this context it makes sense that neuronal cells express only low levels of class I MHC. These cells are terminally differentiated and nondividing, and if killed by a CTL they are unable to regenerate.

Although the ability of CTLs to kill virus-infected cells is well established, recent findings indicate that CTLs, as well as other activated cells of the immune system, may also use noncytolytic means to control and potentially clear many virus infections. This control is thought to be achieved by the secretion of cytokines such as IFN-γ and TNF-α. Dengue virus infection in the brains of mice is one example in which noncytolytic clearance appears to be important. During dengue virus infection of neurons, CTLs are actively recruited into the brain and are essential for the clearance of virus in immunized mice, at least under some conditions. Neurons are immunologically privileged, as noted above, and noncytolytic mechanisms of control are thought to be involved. Hepatitis B virus infection of hepatocytes is another system in which there is evidence that noncytolytic control is important. Other examples are also known. Noncytolytic clearance does not appear to be universal, however. It appears to be possible only in some tissues and for some viruses.

T-Helper Cells

Most CD4$^+$ T cells are helper cells. Some CD8$^+$ T cells are also helper cells, but most helper cells are CD4$^+$. CD4$^+$ T cells recognize peptides presented in the context of class II MHC molecules—they are referred to as class II-restricted cells. Class II MHC molecules, unlike class I, are present on only a restricted set of cells within the organism and are most abundant on B cells, macrophages, dendritic cells, and, in humans, activated T cells, that is, cells of the immune system itself.

The peptides presented by class II MHC molecules are derived from extracellular proteins, and thus these MHC molecules sample the extracellular environment. Proteins, whole viruses, or other microorganisms are taken up by antigen-presenting cells and degraded within intracellular organelles. Peptides derived from these sources can be bound by class II MHC molecules being transported to the cell surface. The peptides derived from this pathway are kept separate from peptides generated through the proteasome/TAP pathway, and the end result is that the class II pathway presents peptides derived from the external environment, whereas the class I pathway presents peptides derived from the intracellular environment (Fig. 8.6).

Professional antigen-presenting cells present peptides to class II-restricted T cells. They also express accessory proteins that deliver a second signal to the T cell that is required for its activation. Professional antigen-presenting cells may be macrophages, B cells, or dendritic cells. When a T_H cell is activated by the interaction of its receptor with its target peptide presented by class II MHC, and the appropriate second signals are present, the cell proliferates and secretes cytokines that are important for eliciting an immune response.

T_H cells are not homogeneous. Different T_H cells secret different panels of cytokines and have different functions in the immune response. Two main types have been recognized, which perhaps represent extremes in function. T_H-1 cells are highly effective for CTL activation and function in the cellular immune pathway. T_H-2 cells are optimal for the activation of B cells and function in the humoral immune pathway. They secrete IL-4, IL-5, IL-6, and IL-10. T_H-1 and T_H-2 cells can be mutually antagonistic. The cytokines secreted by one suppress the other, and a balanced immune response often requires a balanced activation of these two classes of helpers. T_H-2 cells usually deliver their cytokine signals directly to the B cells that they help, following cell–cell contact. T_H-1 cells, in contrast, do not deliver their signals directly to the T cells that they help.

B Cells and Secretion of Antibodies

B cells, so named because they are derived from the bone marrow in mammals, secrete antibodies, which are members of the Ig superfamily. The essential subunit of an antibody is a heterodimer of a heavy (H) chain (which has four or five Ig domains) and a light (L) chain (which has two Ig domains). This subunit is always present as a dimer in which two H-L heterodimers are linked through the H chains. The structure of a light chain, as determined by X-ray crystallography, is shown in Fig. 8.7. This structure illustrates the Ig fold that is common to all Ig domains.

The five classes of antibodies are illustrated in Fig. 8.8. An IgG molecule consists of a dimer of H-L heterodimers in which the H chain is of the γ class. IgD and IgE antibodies are also dimers of H-L heterodimers, but in this case the H chains are of the δ or ε class, respectively. IgA antibodies

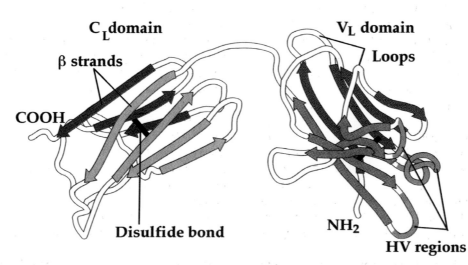

FIGURE 8.7   Ribbon diagram of an immunoglobulin light chain depicting the immunoglobulin-fold structure of its variable and constant domains. Two β pleated sheets in each domain (colored in red and pink in the constant domain and yellow and brown in the variable domain) are held together by hydrophobic interactions and a single disulfide bond (dark bar). The hypervariable regions (HV), shown in blue, form part of the antigen binding site. [Adapted from Kuby (1997, p. 113).]

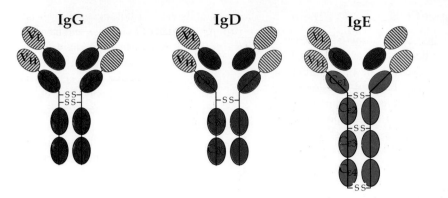

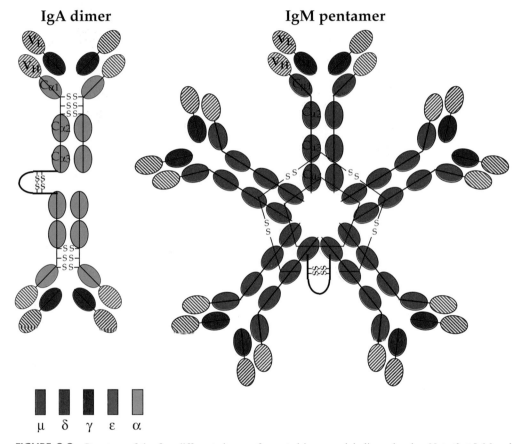

FIGURE 8.8   Structure of the five different classes of secreted immunoglobulin molecules. Note that IgM and IgA are secreted as pentamers and dimers, respectively, linked by a J chain. [Adapted from Gally (1973).]

contain four H-L heterodimers (it consists of a dimer of the dimeric unit, as shown). In this case, the H chain is of the α class. IgM contains 10 H-L heterodimers (it is a pentamer of dimeric units) formed with μ H chains.

A more detailed representation of an IgG molecule is shown in Fig. 8.9. The terminal domains of both H and L chains are variable. Within the variable domains, there are regions that are more variable than other regions, called hypervariable regions. The combining site of the antibody,

the region that specifically binds to an antigen recognized by the antibody, is formed by the variable regions of both the H and L chains.

Formation of Light and Heavy Chains

The L and H chains of antibody molecules are formed in a process that is similar to that used to form T-cell receptors. The two Ig domains of the L chain are called V (for

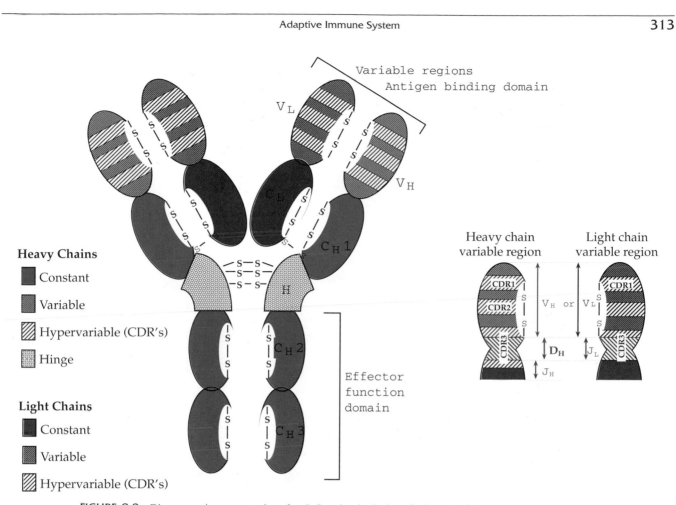

FIGURE 8.9 Diagrammatic representation of an IgG molecule. Each antibody molecule is composed of two heavy chains (which contain four domains, each consisting of an Ig fold like that shown in Fig. 8.7) and two light chains, which each have two Ig domains. The distal domains of both the light and heavy chains are the variable regions, composed of interspersed framework regions and hypervariable regions (diagonal shading) known as complementarity-determining regions or CDRs. The right part of the figure illustrates the role of the V, D, and J gene segments in encoding CDR1, 2, and 3 regions of Ig variable region genes. [From Chen and Alt (1997, pp. 340, 345).]

variable) and C (for constant), and between these two domains is a J (for joining) domain. All three domains are encoded separately in the genome. There are multiple copies of V and J in the genome (Table 8.1), and these gene segments are polymorphic in the population. During maturation of a B cell, a V-gene segment, a J-gene segment, and a C region are brought into juxtaposition to one another, as illustrated in Fig. 8.10A. In this process, a V-gene segment is fused to a J-gene segment by deleting the intervening DNA. The C region is brought into play by RNA splicing: transcription of the VJ region continues through the C region, and splicing of the pre-mRNA joins the J region to the C region. The heavy chain is formed by a similar sequence of events, but in this case there is an additional gene segment D (for diversity) that introduces additional diversity in the recombination process. As for the light chain, there are multiple copies of V, D, and J in the genome (Table 8.1), and the population is polymorphic for these

gene segments. During B-cell maturation, a D-gene segment is first joined to a J segment, and the DJ segment is then joined to a V segment (Fig. 8.10B). Recombinational rearrangements to form the antigen binding site occur only during the maturation of the B cell; once it is mature, no further rearrangements occur in this region.

The many combinatorial possibilities of V and J light-chain gene segments, and of V, D, and J heavy-chain gene segments, lead to the possible production of a very large number of light chains and heavy chains (Table 8.1). In addition, joining V with J, or joining V, D, and J, is imprecise, leading to additional diversity. Finally, joining an L chain with an H chain to form the heterodimer generates still more possible antigen recognition sites, since the antibody recognition site is formed by the V regions of both the H and the L chains. The total number of possible combinations is very large (Table 8.1). It is important to note that any individual B cell usually produces only one H chain

A. Light Chain

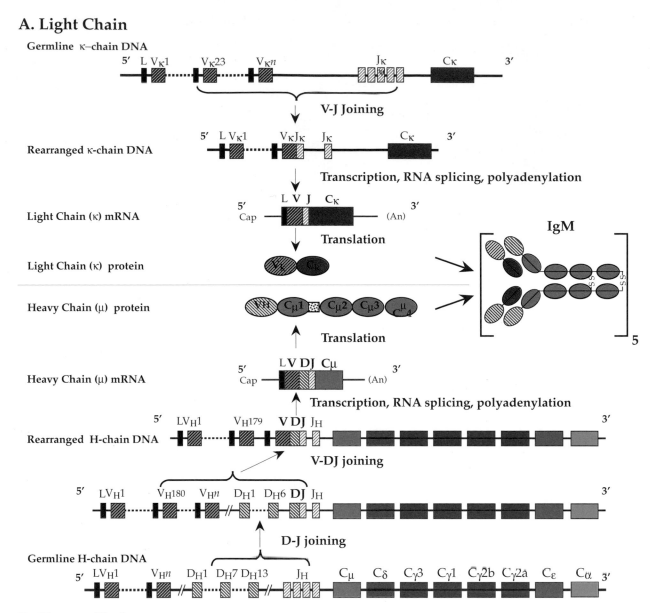

B. Heavy Chain

FIGURE 8.10   (A) Formation of the human κ immunoglobulin light chains. The top line shows the germ-line genes, and the next line illustrates the B cell genes after V-J recombination. The next line down shows the RNA transcript after splicing to join the VJ region to the C region. This mRNA is translated by cytoplasmic ribosomes into the light chain, which then combines with a heavy chain. The leader, L, is translated into a signal sequence which is removed post-translationally. (B) A comparable illustration of the heavy-chain genes in the germ line and in the B cell, after two rounds of rearrangement known as "D-J joining" and "V-DJ joining." The final IgM molecule is a pentamer held together with disulfide bonds and a "J chain", which links the Fc regions. [Adapted from Kuby (1997, Figs. 7.4 and 7.5, pp. 172, 173).]

and one L chain, and thus each B cell produces only one antibody recognition site.

Activation of B Cells

During maturation of B cells, a large population of cells results, each of which has one antibody combining site. The

theoretical repertoire, how many different types of B cells could conceivably be produced using the known mechanisms that are active during B-cell development, is thought to be much larger than the minimal estimate of 10^{14} shown in Table 8.1. In humans there may be 10^{10} B cells with differing specificities circulating at any time, and it is estimated that about 1 in 10^5 cells produces an antibody that will bind, with differing affinities, to any particular antigen that is being examined.

The antibody molecule is first produced as an integral membrane protein that is displayed on the surface of the B cell, anchored through a membrane-spanning region and containing an intracellular cytoplasmic domain. If the antibody displayed on the surface of the cell binds antigen, the B cell is activated. If the cell receives a second signal from a T_H-2 cell, it proliferates to form cells that secrete antibody. Memory cells also arise that serve to protect the organism from future infection by the same pathogen.

The T_H-2 cell signal may be delivered to the B cell either through a specific pathway or through a nonspecific pathway. Antigen, which could be in the form of a whole virus or in the form of a protein, that is bound to antibody present on the surface of the B cell can be internalized by the B cell and degraded by the MHC class II antigen-processing pathway. Peptides derived from the degraded virus or protein can then be presented on the surface of the B cell in the context of class II MHC molecules. Class II-restricted T-helper cells that recognize this peptide will secrete cytokines that stimulate the B cell to proliferate and secrete antibodies. Note that the peptide displayed by the class II molecule does not have to be related to the epitope recognized by the antibody displayed on the B-cell surface. It may, in fact, be derived from an entirely different protein. Thus, while T cells respond to peptide epitopes, the antibody molecules can recognize much more complex antigens, such as whole proteins or viruses. The epitopes recognized by antibodies are most often what are called conformational or nonlinear epitopes, which are formed by residues physically located at different places in the linear sequence of the protein but which form a contiguous surface in the protein after it folds into its three-dimensional conformation. Such discontinuous epitopes are destroyed if the protein is denatured and the different components of the epitope separated from one another. A certain fraction of antibodies, however, recognize continuous epitopes, which are formed by a linear sequence of amino acids present in the protein.

In addition to specific activation of B cells by T-helper cells, B cells can also be activated through area stimulation. If a B cell is in the vicinity of T-helper cells that are releasing cytokines to activate B cells, it may also be stimulated. The importance of area stimulation, and the frequency with which it occurs, in the context of fighting off a viral infection is not clear. In many cases of viral infection, a generalized and active inflammatory response occurs that involves the release of many cytokines, and in which many different antigens are being presented. Area stimulation could be important in developing a rapid response during such events. However, in such a process antibodies against self might also be produced, and such processes must be controlled.

Secretion of Antibodies

A B cell stimulated by exposure to its cognate antigen and by help from a T_H-2 cell proliferates and begins to secrete antibodies. The first antibodies secreted are IgM, whose structure is illustrated in Fig. 8.8. IgM antibodies, which circulate in the blood, can be detected as early as a few days after virus infection. Their production quickly wanes and over a period of weeks or months the concentration of IgM in the blood decreases to very low or undetectable levels, as illustrated schematically in Fig. 8.11. Thus the presence of IgM antibodies specific for a virus is usually a sign of acute, or at least very recent, infection.

During further maturation of the B cell, class switching occurs, as illustrated in Fig. 8.12. Recombinational events in the heavy-chain region lead to the substitution of the IgG, IgE, or IgA heavy chain for that of IgM. Homologous recombination within the intron just downstream of the J gene results in deletion of the intervening DNA such that the active VDJ gene is brought into contact with the C region of a different class of heavy chain. The order of heavy-chain genes in the mouse chromosome is $\mu\delta\gamma\varepsilon\alpha$ (Figs. 8.10 and 8.12), and class switching can happen either sequentially (IgM to IgG to IgE to IgA) or directly (e.g., IgM to IgE without passing through an IgG phase).

Production of IgG (or of IgE or IgA) thus occurs later after infection (illustrated schematically in Fig. 8.11). At least 2 weeks are required before there is production of large amounts of IgG. Once a cell begins to make IgG, it is no longer able to make IgM because the gene encoding the M heavy chain ($C\mu$ in Fig. 8.12) has been deleted. It is important to note that the antigen combining site of the IgG antibody is identical to that produced by the earlier IgM because the V region of the H chain is the same.

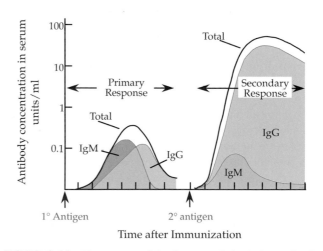

FIGURE 8.11 Time course of development of circulating antibodies after primary and secondary immunizations. No timescale has been shown, since the actual results vary depending upon the antigen, adjuvant, site of injection or infection, and animal species. [From Kuby (1997, Fig. 16.19, p. 398).]

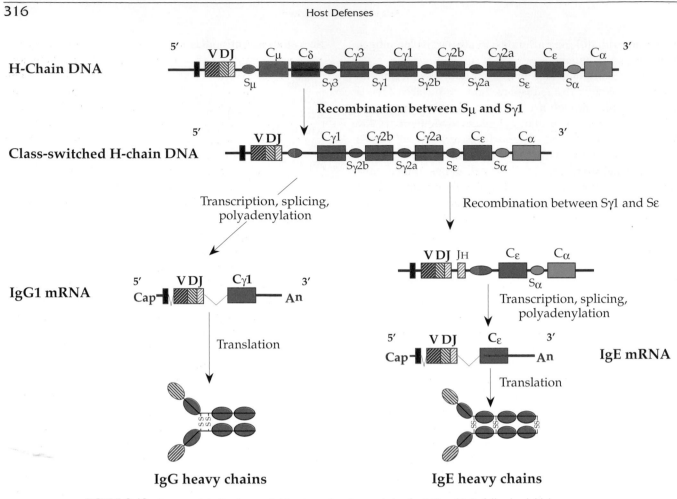

FIGURE 8.12 Immunoglobulin class switching to produce heavy chains for IgG and IgE, following initial rearrangements to produce IgM. Switch sites, designated "S" are located upstream of each CH gene except $C\delta$. There is only one allele for $C\mu$, $C\delta$, $C\varepsilon$ and $C\alpha$, but there are four alleles for $C\gamma$, designated $\gamma1$, $\gamma2a$, $\gamma2b$, and $\gamma3$, which differ somewhat in sequence. Because there is no switch site for $C\delta$, IgD is produced only in conjunction with IgM, by alternative mRNA splicing. [Adapted from Kuby (1997, Fig. 7.17, p. 185) and Chen and Alt (1997, Fig. 7.41b, p. 399).]

IgG circulates in the blood and therefore very many cells exposed to blood and blood products are exposed to IgG. Although production also wanes with time, IgG remains circulating in the blood for years or decades. Because of this, and the ease of obtaining blood samples, IgG has been extensively used to monitor past exposure to different pathogens and the immune status of individuals for any particular pathogen. As illustrated in Fig. 8.11, on a second exposure to an antigen, IgG concentrations rise dramatically and remain in the circulation at much higher levels. IgM concentrations do not show this anamnestic effect.

Because of its widespread circulation within the body, its high concentrations, and the anamnestic effect that occurs on secondary exposure, IgG is important in the control of viral diseases. It is also important in protecting the fetus and the very young. IgG crosses the placenta during pregnancy and is present in fetal blood. This transfers maternal immunity to the fetus, and this immunity lasts for the first few months of postnatal life.

IgA is also present in the blood, but its importance lies in the fact that it is secreted. It is present on mucosal surfaces where it helps prevent viral diseases such as those caused by rhinoviruses or influenza viruses in the respiratory tract or by rotaviruses or enteroviruses in the intestinal tract. It is present in secretions such as tears, saliva, and genital tract secretions, where it plays an antiparasite role. Because it is present in milk, it also serves to transfer maternal immunity to the gut of the infant.

IgE is important for control of infection by multicellular parasites. It can bind to mast cells via specific receptors and cause an inflammatory response, leading to the destruction of parasites. It is IgE that produces allergic symptoms that occur when pollen granules or mites or other comparatively large particles are attacked by IgE, producing an inflammatory response.

IgD is present in a membrane-bound form at the surface of immature B cells, along with IgM, where it helps in the activation of the cell on exposure to antigen as described above. It is also present in very small amounts in the blood. There is no class switching mechanism to express IgD and it is only produced in combination with IgM by means of differential splicing of mRNAs.

Although the B-cell repertoire first arises by combinatorial joining of the V and J light-chain gene segments and of V, D, and J heavy-chain gene segments, the antibody response is fine tuned once a B cell has been activated. There is an error-inducing mechanism during B-cell replication in which the genes encoding the V segments of both heavy and light chains undergo hypermutation. This activity takes place in germinal centers of the lymph node and there is concurrent selection for B cells that bind more tightly to the antigen. Over a period of time B cells are selected that bind to the antigen with higher and higher affinity. Hypermutation in germinal centers has been postulated to play a role in the development of Burkitt's lymphoma, as described in Chapter 6.

Antibodies may bind to a virus and neutralize its infectivity, and such neutralizing antibodies are thought to be of critical importance in controlling virus infection. However, other antibodies may bind to a virus without inactivating it and such non-neutralizing antibodies can be protective. There are several mechanisms by which non-neutralizing antibodies may protect. Aggregation of virions by antibodies leads to a reduction in the total number of infectious viruses. The maturation of enveloped viruses by budding through the cell plasma membrane can be inhibited by the binding of antibodies to viral proteins present on the surface of the cell. Antibody-dependent lysis of an infected cell can occur by a complement-mediated pathway or by a natural killer cell pathway called antibody-dependent cell-mediated cytotoxicity. These pathways are triggered by the binding of antibody to viral protein expressed on the cell surface. Complement and natural killer cells, which are components of innate immunity as well as of acquired immunity, are described below. Other mechanisms that result in protection by antibodies also exist.

Immunological Memory

In the course of the B-cell response, memory B cells are formed. Memory B cells persist after the antigen disappears from the body, and are primed to react quickly and vigorously on renewed stimulation by the cognate antigen. Renewed activity of memory B cells still requires T-helper cell stimulation, but the secondary response leads to immediate production of antibody of high affinity, because there is no need for a long maturation process, and is so vigorous that it results in the production of much larger amounts of antibody than are produced during a primary response (Fig. 8.11).

Memory T cells are also generated in the course of an immune response. After expansion of T-cell clones following stimulation by the cognate antigen, most activated T cells die by apoptosis when no longer stimulated by the presence of antigen and cytokines. Immunological memory remains, however. It has been widely thought that such memory requires a continuous supply of antigen with which T cells interact. This antigen might be present because the virus establishes a chronic infection, or antigen might be sequestered in regions of the body and made available over an extended period of time. It is even possible that T cells continue to be stimulated because of cross reactivity with self antigens. Data are now accumulating, however, that memory T cells exist in a quiescent state for long periods of time and are capable of rapid reactivation on renewed exposure to antigen. Thus, memory in T cells may be similar to memory in B cells.

It is the existence of memory cells that renders an animal immune to the virus or other pathogen that first evoked the immune response. Memory cells are primed for such a rapid and vigorous response that the invading organism is stopped early during the infection process, before disease is established. Residual antibodies circulating in the blood or present on mucosal surfaces may even prevent infection altogether (sterilizing immunity).

The immune status of an individual is often tested by examining the blood for the presence of antibodies. Such an assay is imperfect, however. The presence of antibodies in the blood is usually a good indication that a person is immune. Such antibodies fade with time, but in many cases a person remains immune despite the absence of detectable antibodies in the blood, because memory cells that are primed to react quickly are still present.

Complement System

The complement system is composed in part of more than 20 soluble proteins that circulate in the blood. These proteins are activated through a proteolytic cascade to produce effector molecules that aid in the control of viral infection or infection by other pathogens. Complement forms part of both the adaptive immune system and the innate immune system. Its activities turn antibodies into effective killers of viruses or of virus-infected cells, as well as of other pathogens, and in this role it is a component of the adaptive response. Complement can also be activated by interaction with parasites in the absence of antibody, however, and in this role it is a component of the innate responses.

The classical pathway of complement activation (an adaptive response) involves interaction of a complex of complement molecules called C1 with IgG or IgM. This could be IgG or IgM bound to antigen present at the surface of an infected cell or a bacterium, for example. The alternative pathway of complement activation (innate response) does not involve interaction with antibody, but rather requires the deposition of a molecule called C3b on the surface of a particle, such as a parasite. Once bound, C1 or C3b interacts with other components of the complement system. The result is the activation of a cascade of proteases whose cleavage activities result in the formation of effector molecules. One group of effector molecules forms a complex that inserts into the lipid bilayers of cell membranes and results in the lysis of the cell. Many enveloped viruses can also be killed by this lytic mechanism. The system must be finely regulated so that activated components of complement are produced only in response to pathogens and so that cell killing is confined to infected cells or parasites. Control of complement activation and action is therefore suitably complicated.

Other effector molecules that result from activation of complement have activities that aid in the control of viral infection or infection by other parasites by mechanisms other than cell lysis. Some products enhance the neutralization of viruses by antibody. By binding to virus that is coated with antibody, they render the virus less capable of binding to its receptor; cause aggregation of the virus, resulting in fewer infectious units; and increase the uptake of viruses by phagocytic cells (a process called opsonization). Other molecules induce an inflammatory response, in part by inducing the release of agents by mast cells and basophils, or help in the activation of B cells, or help in the clearing of immune complexes.

Numerous studies have shown that complement is required for an effective immune response. As one example, mice lacking a receptor for an early product of the complement cascade, called C3d, are unable to mount an antibody response. As a second example, depletion of complement in mice leads to a prolonged viremia and a more severe central nervous system disease on infection by Sindbis virus. As a third example, people are known who are genetically deficient for components of the complement pathway. Many suffer from immune-complex diseases as a consequence, because they are unable to effectively clear immune complexes. They also suffer from an increased incidence of bacterial infections.

Adaptive Immunity in the Control of Virus Infection

It has been conjectured that the two arms of the adaptive immune system evolved to fight off different pathogens. The CTL response seems well adapted for the control of viral infections, because these pathogens replicate intracellularly, but less well adapted for controlling extracellular pathogens such as bacteria or protozoa. Conversely, the humoral response and the associated complement system seem better adapted for the control of extracellular pathogens. Consistent with this model, children who are deficient in the production of antibodies do not in general show an increased

susceptibility to viral diseases but do show a marked increase in susceptibility to bacterial infection. Children unable to make gammaglobulin, for example, recover normally from infection by measles virus and are immune to reinfection, demonstrating the importance of T-cell immunity in this disease. Conversely, impairment of CTL function in children often leads to increased frequency and severity of virus infections. Such findings suggest that CTLs evolved primarily to deal with viral infections and remain of prime importance in dealing with viral infections, whereas the humoral system evolved to deal with infections by free-living organisms such as bacteria, protozoa, and yeast.

Although this model may well be correct, it is clear that humoral antibodies are also important in the control of viral disease. Many experiments have shown that passively transferred antibodies alone can protect against viral infection. Further, whether reinfection by a virus results in disease or asymptomatic infection is often correlated with the level of antibodies against the virus in the blood. These observations are particularly relevent to the protection of an unborn or newborn child from viral disease. Maternal antibodies are actively introduced into the fetal bloodstream during intrauterine development. Such antibodies are critical for the protection of the fetus and the newborn against viral infections early in life before its own immune system develops. It is also clear that neutralizing antibodies are of prime importance in preventing reinfection by at least some viruses, such as influenza virus. Previous infection leading to a vigorous T-cell response directed against many of the viral proteins does not protect against subsequent reinfection by variants which are altered only in their surface glycoproteins. As another example of the importance of humoral antibodies in combating viral infections, CTL-induced cytolysis is not very effective in the control of viral infection of the brain. Neurons are terminally differentiated and cannot be replaced, and express only low levels of MHC class I molecules. However, CTLs do appear to be important in control of viral infections in the brain, perhaps because they secret IFN-γ when activated. Other mechanisms involving humoral antibodies are also probably important. It has been shown in a mouse model that humoral antibodies can cure persistently infected neurons of viral infection.

Thus, a broad and varied immune response to viral infection is important for both the suppression of the original virus infection and in evoking a status of immunity to subsequent reinfection by the virus. Experiences with vaccines used in humans support this idea. Some vaccines have been found to provoke an unbalanced response that renders subsequent infection by the virulent virus more serious.

Vaccination against Viruses

For most viruses, once a person has been infected and recovered, he or she is immune to subsequent reinfection by the same virus. This is the concept behind immunization, also called vaccination, in which a person is exposed to a virus, either live or inactivated, or to components of the virus, in order to establish the immune state.

Live Virus Vaccines

Immunization has been practiced for centuries, having been introduced a millennium ago for smallpox. In a process called variolation, less virulent strains of smallpox virus were introduced into humans by intranasal inoculation. The disease induced by this procedure had a lower fatality rate than that caused by the epidemic disease (although the fatality rate was still significant), and the extent of pocking or scarring was less (which was of importance to people concerned with their appearance). This technique was greatly refined about 200 years ago by Jenner, who immunized people against smallpox with a nonhuman virus derived from cows. Cowpox virus is antigenically related to smallpox virus and induces immunity to smallpox, but does not cause a severe disease as does smallpox. This immunization procedure was very successful and smallpox has now been eradicated using modern versions of Jenner's original vaccine. The process of using a nonhuman virus to induce immunity in humans against a related human virus has been referred to as "the Jennerian approach." Jenner's use of cowpox virus to immunize against smallpox gave us the name "vaccination" and "vaccine," from the Latin word for *cow*.

Since Jenner's time, other approaches to vaccination have been developed. Rather than using a nonhuman virus as a vaccine, it is more common to use an attenuated strain of a virulent human virus. Attenuation has classically been achieved by passing the virus in animals or in cultured cells from animals. Passage selects for viruses better adapted to grow in the nonhuman host, and often results in a virus that is attenuated in humans. One of the earliest vaccines to be developed in this way was a rabies vaccine developed by Pasteur by passing the virus in rabbits. A more modern method for producing a vaccine strain, which is still often used today, was introduced by Theiler and Smith, who passaged virulent yellow fever virus in chicken tissue and chicken cells in culture. After 100 passages, a marked change in virulence of the virus occurred. Although the passaged virus retained its ability to infect humans, it no longer caused disease. Passage of virus in tissue culture cells and selection of attenuated variants has been used to produce vaccines for measles, mumps, and rubella, among others. With modern technology, it is now possible to introduce mutations into a viral genome that might be expected to attenuate the virus and to test the effects of such mutations in model systems (see Chapter 9). Although no currently licensed human vaccines have been produced in this way, it is expected that this approach will be useful for future vaccines.

In a few cases it is possible to infect humans with a virulent virus in a way that does not lead to disease. Oral vaccines have been developed for adenoviruses 4 and 7 in which the virus is encapsulated in a protective coating that does not dissolve until the virus reaches the intestine. The viruses replicate in the intestine but do not produce disease, although they do induce immunity against adenovirus respiratory disease.

Live virus vaccines in general induce a more protective and longer lasting immunity than do inactivated virus vaccines. They replicate and therefore produce large amounts of antigen over a period of days or weeks that continues to stimulate the immune system. Furthermore, the viral antigens are presented in the context of the normal viral infection and these vaccines induce the full range of immune responses, which includes production of CTLs as well as antibody. Live virus vaccines may be more effective than inactivated vaccines in eradicating the wild-type virus from a society. As described in Chapter 3, an inactivated poliovirus vaccine protects the individual from disease but it does not prevent the wild-type virus from circulating. Finally, live virus vaccines are cheaper to produce than inactivated virus vaccines because a single dose containing smaller amounts of (live) virus is usually sufficient to induce immunity.

Although they have many advantages, live virus vaccines also suffer from a number of potential problems. Attenuating the virus sufficiently so that it does not cause disease while retaining its potency for inducing immunity can be difficult to achieve. For example, many viruses quickly become overattenuated on passage, losing their ability to induce immunity on infection of humans. Another problem is the potential for reversion of the virus to virulence and the possible virulence of the attenuated virus in normal or, especially, immunocompromised people. The early poliovirus type 3 vaccine caused a number of cases of polio because the virus reverts fairly readily to virulence. Today, reversion is not a problem but about 10 vaccine-related cases of paralytic polio occur every year in the United States due to residual virulence of the virus (Fig. 3.4). Certain lots of yellow fever virus vaccine have also been found to contain partial revertants that can cause encephalitis, especially in infants. Because of this, each lot of yellow fever vaccine must be carefully monitored for virulence and vaccination of infants under 6 months of age is not recommended. Despite this care, a few deaths from vaccine-induced yellow fever have occurred very recently.

Interference caused by activation of the innate immune system can also cause problems. The effectiveness of a live virus vaccine may be diminished because of interference from a preexisting infection. Interference also makes it difficult (although not impossible) to immunize simultaneously against multiple viruses when using live virus vaccines. Problems can arise from the presence of adventitious infectious agents in the vaccine, since it has not been treated with inactivating agents. Early lots of the live Sabin poliovirus vaccine were contaminated with SV40, for example, which was present in the monkey kidney cells being used to prepare the vaccine. Millions of people were unknowingly infected with SV40, which fortunately appears to cause no disease in humans. Finally, living viruses are often unstable, making it difficult to transport vaccines over long distances and to store them so that they maintain their potency, a problem that is more acute in developing tropical countries.

Although there are potential difficulties, live virus vaccines have many advantages and have been extremely successful in the control of viral diseases. Smallpox virus has been eradicated, measles virus and poliovirus are on the brink of eradication, and many other serious diseases have been controlled by live virus vaccines. A partial list of currently licensed live virus vaccines is given in Table 8.2.

Inactivated Virus and Subunit Vaccines

The second general approach to vaccination is to use inactivated virus or subunits of the virus, such as the surface proteins of the virus. The original Salk poliovirus vaccine and current vaccines against influenza and rabies

viruses use inactivated virus. For these vaccines, virus is prepared and purified, and virus infectivity is destroyed by treatment with formalin or other inactivating agents. These inactivated viruses are injected, often intermuscularly, and induce an immune response. Subunit vaccines, on the other hand, are usually produced by expressing the surface proteins of a virus in a cell culture system. The proteins are purified and injected. Licensed subunit vaccines include the modern vaccine against hepatitis B virus.

Inactivated virus vaccines or subunit vaccines suffer from a different set of problems from live virus vaccines, but in turn have a number of advantages. Their advantages include the fact that they are usually stable; that interference by infection with other viruses does not occur; and, with proper monitoring, viral virulence is not a problem. Their stability makes transport and storage of the vaccines easier and more reliable. Their insensitivity to interference makes it possible to immunize against many viruses simultaneously. The fact that the viruses are inactivated, or that no live virus was ever present in a subunit vaccine, means that no virus infection occurs with its potential for disease. Early lots of the Salk inactivated poliovirus vaccine were contaminated with live poliovirus, which resulted in a

TABLE 8.2 Characteristics of Antiviral Vaccines

	Live attenuated virus	Inactivated virus and subunit vaccines
Currently licensed vaccines	Poliovirus (Sabin)	Poliovirus (Salk)
	Measles	Influenza
	Mumps	Rabies
	Rubella	Hepatitis B
	Yellow fever	Hepatitis A
	Vaccinia	Japanese encephalitis
	Varicella-zoster	Western equine encephalitis (experimental)
	Rotavirus[a]	
	Adenovirus (in military recruits)	
	Junin (Argentine hemorrhagic fever)	

In addition, live attenuated vaccines for the following viruses are close to release to the public: human cytomegalovirus, hepatitis A, influenza, dengue, human parainfluenza, and Japanese encephalitis.

Characteristics of the immune response[b]		
Antibody induction (B cells)	+++	+++
CD8+ cytotoxic T cells	+++	−
CD4+ helper T cells	+++	+++
Reactivity against all viral antigens	Usually	Seldom
Longevity of immunity	Years/decades	Months/years
Cross reactivity among viral strains	+++	+
Risk of viral disease	+	−

Source: Data for this table came from Granoff and Webster (1999, p. 1862) and from Fields *et al.* (1996, p. 371).
[a]The licensed rotavirus vaccine had been withdrawn from the market, pending evaluation of risk of intussusception in infants.
[b]The number of plus signs indicates the strength of the positive response; a minus sign indicates no response or no risk (last line).

number of cases of polio caused by the vaccine, but better methods of inactivation and of monitoring residual infectivity have solved this problem.

Difficulties with these vaccines include the expense of preparing the large amounts of material required, the necessity for multiple inoculations, and the failure to induce a full range of immune responses. Because no virus replication occurs, large amounts of material must be injected to induce an adquate immune response. Further, lack of virus replication means that an inflammatory response required for an efficient immune response must be obtained in some other way. In practice, inactivated virus or subunit vaccines are designed to provoke only a very limited inflammatory response, and multiple immunizations are usually required in order to achieve effective immunity. Even so, a full range of immunity is not achieved. Worse, in two cases immunization with inactivated virus resulted in more serious illness on subsequent infection with epidemic virus. Inactivated virus vaccines against measles virus and respiratory syncytial virus did not protect against infection by the respective viruses, and led to the development of atypical disease that was more severe than that caused by the viruses in nonimmume people. The measles vaccine was subsequently replaced with a live virus vaccine, which has been very successful, but no successful vaccine has yet been developed for respiratory syncytial virus.

Despite these problems, many successful vaccines have been introduced that use inactivated viruses or subunits of viruses. Some of these are listed in Table 8.2.

Modern biotechnology makes possible another approach to subunit vaccines, the use of a nonpathogenic virus as a vector to express proteins from a virulent virus. This approach potentially overcomes many of the disadvantages of inactivated virus vaccines, or of subunit vaccines based on purified components, because the antigens are presented in the context of a viral infection and in a native conformation. To date, none of the licensed vaccines use this procedure, but clinical trials are in progress or are planned to begin soon for a number of such vaccines. Two examples will be cited. In one series of trials, vaccinia virus is being used to express the HIV surface glycoproteins. Antibodies to the HIV glycoproteins are induced, but whether they are protective remains to be determined. In another series of trials that will begin shortly, the vaccine strain of yellow fever virus has been engineered to express the surface glycoproteins of Japanese encephalitis virus, of dengue virus, or of West Nile virus. Based on the successful use of yellow fever vaccine for more than 50 years, the prognosis is good that these new vaccines will be successful. The use of viruses as vectors is covered in more detail in Chapter 9.

DNA Vaccines

A new and potentially exciting approach to vaccination is to inject DNA that encodes genes whose expression leads to immunity to a virus. This approach has only been tested in model systems to date, but the results are promising. Plasmid DNA containing the gene for, say, a surface glycoprotein of influenza virus under the control of a general mammalian promoter is injected intramuscularly. Muscle cells take up DNA and import it into the nucleus, a surprising finding. Expression of the influenza gene may elicit an immune reaction against the encoded influenza protein. This method of immunization has many potential advantages. The encoded protein is made intracellularly and therefore a CTL response is generated, because peptides are presented by class I MHC molecules. In addition, the proteins are made and are present in a native conformation, which may make possible the production of antibodies against a wider range of conformational epitopes. Other advantages include the fact that multiple proteins can be encoded in plasmid DNA, which allows immunization against multiple products of one virus or against several different viruses, and that cytokine genes can also be present in the plasmid, which can lead to a stronger immune response and direct the immune response toward desired pathways. Another important feature is that the expression of the foreign protein continues for some time, potentially leading to a better immune response. Finally, plasmid DNA is easy to modify to express different genes, cheap to make in quantity, and easy to purify. Once safety trials have been conducted for a DNA vaccine, it should be possible to modify it to express antigens from other viruses with minimal safety concerns.

Vaccine Development: The Example of Measles Vaccines

Vaccines against viruses have been enormously successful in controlling viral diseases that have plagued mankind for millennia. Smallpox virus has been eradicated worldwide, poliovirus has been eradicated from the Americas and may be eradicated worldwide within the next few years, and many epidemic diseases have been controlled, at least in regions that have access to the vaccines. As an example of potential difficulties in vaccine development, however, the measles virus vaccines will be discussed. Measles virus is ubiquitous. At one time, virtually the entire population of the world was infected by the virus as children. Even today, measles is prevalent in many areas of the world (Fig. 8.13A), despite extensive vaccine coverage (Fig. 8.13B). Measles can be a serious illness, with significant mortality and debilitating sequelae (Chapter 4). Concerted efforts to produce a vaccine began many years ago. One early measles virus vaccine, used from 1963 to 1967, was made with inactivated virus. Vaccination with this vaccine potentiated a more serious illness when a person was infected with the epidemic virus, however, characterized by a more severe rash and called "atypical measles." The atypical disease is thought to be produced by an atypical or unbalanced T-cell response to the virus fol-

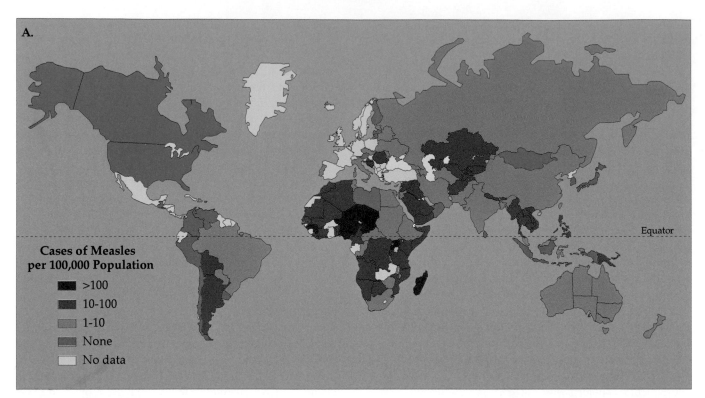

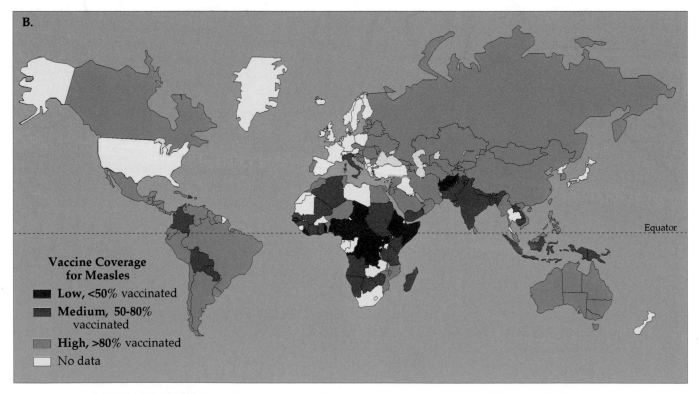

FIGURE 8.13 Worldwide incidence of measles and immunization coverage for measles. (A) Reported incidence rate of measles per country is shown as cases per 100,000 population as of August 14, 1998. (B) Coverage of infants with measles vaccine. Data shown are as of August 1998. [From maps on the web site of WHO International, http://www.who.int/vaccines-surveillance/graphics/ NY graphics/meainc.htm and http://www.who.init/vaccines-surveillance/graphics/NY graphics/meacovmap.htm.]

lowing immunization. This response controls virus infection only poorly and leads to collateral damage from T-cell attacks on virus-infected cells. An attenuated virus vaccine was more successful. The first live virus vaccine used was protective but insufficiently attenuated and thus more reactogenic than would be ideal. It was subsequently replaced with a more highly attenuated vaccine.

The attenuated virus vaccine currently in use has been highly successful in controlling measles in many developing countries as well as in the developed countries. Endemic measles has been eliminated from North America and large areas of South America and is largely controlled in many other parts of the world (Fig. 8.13A), especially in areas where vaccine coverage is high (Fig. 8.13B). This vaccine could potentially be used to eradicate measles worldwide. However, measles continues to be a major cause of infant mortality and morbidity in many less developed countries because the vaccine has one serious drawback for use in such countries. It is difficult to immunize infants with a live virus vaccine while maternal antibodies are still protective. However, in many poorer countries measles is so widely epidemic that many infants contract the disease as soon as the protection due to maternal antibodies wears off. Multiple, spaced immunizations for each infant would be required to successfully immunize the population of new susceptibles before the epidemic virus got there, which is difficult to achieve in countries with limited health care facilities. In an effort to get around this problem, very young infants were immunized with higher doses of the live virus vaccine in a trial in Senegal, on the theory that using a larger dose would overcome the effects of maternally derived immunity. This trial was a failure, however, because for unknown reasons the cumulative mortality in infants receiving the high titer vaccine was greater than that in controls (Fig. 8.14). Thus at present there is no effective way to immunize young infants in these countries against measles. The virus continues to be an important cause of infant mortality and its continued circulation in infants makes it difficult to eradicate.

INNATE IMMUNE SYSTEM

The innate immune system is composed of a large number of elements that attempt to control infection by pathogens, but this system is independent of the identity of the particular pathogen. Complement is a component of the innate system as well as the adaptive system and has been described. Other components of the innate system include the cytokines, a complicated set of small proteins that are powerful regulators of the immune system. Many cytokines are critical for the function of the adaptive immune system, as described in part above. Cytokines are also important components of the innate system. The innate system also

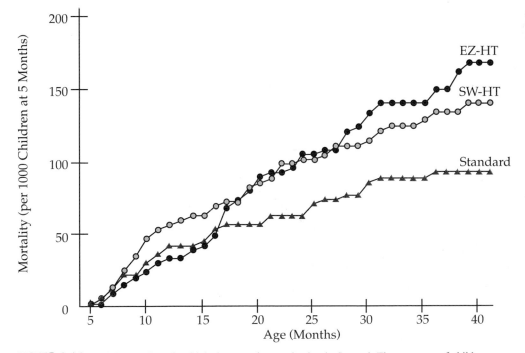

FIGURE 8.14 Child mortality after high-titer measles vaccination in Senegal. Three groups of children were given either $10^{5.4}$ PFU of Edmonston–Zagreb vaccine at 5 months of age (EZ-HT), or $10^{5.4}$ PFU Schwartz vaccine (SW-HT) at 5 months of age or the standard schedule (placebo at 5 months and $10^{3.7}$ PFU vaccine at 10 months of age). [Adapted from Garenne et al. (1991).]

includes natural killer cells, cytotoxic lymphocytes that kill cells that do not express adequate levels of class I MHC molecules. Apoptosis is another innate defense mechanism, because most cells are programmed to die if they sense that they are infected.

Natural Killer Cells

Natural killer (NK) cells are cytolytic cells that kill by an antigen-independent mechanism. They are important for the control of many virus infections, as shown by the fact that ablation of NK cells leads to more serious disease. NK cell activity increases very early after infection by a virus, within the first 2–3 days, and thereafter the number of cells declines. It is thought that one of the functions of these cells is to kill cells that do not express class I MHC or that express it in only low amounts. Such cells, of course, do not present antigen to CTLs and therefore escape normal immune surveillance. NK cells express two sets of receptors on their surface. One set of receptors interacts with MHC class I molecules on the surface of the target cell. This interaction inhibits killing by the NK cell. The second set of NK receptors interacts with activating molecules on the surface of cells. This interaction can stimulate the NK cell to kill the target cell if the NK cell is not sufficiently inhibited by its interaction of class I molecules.

Many different viruses downregulate the production of MHC in infected cells in order to escape immune surveillance. The elimination of these infected cells by NK cells could be of considerable importance to the host. It has been suggested that NK cells evolved to rid the body of cells that are not subject to normal immune surveillance, which could include tumor cells as well as virus-infected cells. In the one case known of a human with a deficiency in NK cells, as well as in experimental studies with mice in which NK cells are depleted, infection by a number of viruses that downregulate production of MHC leads to a much more severe disease.

Apoptosis

Apoptosis or programmed cell death is a defense of last resort by an infected cell. It is a cell suicide pathway in which mitochondria cease to function, chromatin in the nucleus condenses, nuclear DNA is degraded, and the cell fragments into smaller membrane-bound vesicles that are phagocytosed by neighboring cells. Apoptosis serves to eliminate cells that are no longer needed, cells that are damaged in some way, or cells that are dangerous to the host because they are infected or because their cell cycle is deregulated (and they are therefore potential tumor cells). Cells that are no longer needed include excess cells that are produced during development and must be eliminated (a normal occurrence during development) and T cells respon-

sible for the control of an infectious agent after the infection has been eliminated (and which may even be dangerous if left around). Cells that are damaged by exposure to UV light or to free radicals, or are otherwise stressed, such as by infection, are also subject to apoptosis, as are cells whose cycle is deregulated. Thus apoptosis serves many roles in an organism and the basic machinery has been conserved throughout evolution. In mammals it serves as a component of both the adaptive immune response (ridding the body of unneeded lymphocytes; killing of infected cells by CTLs) and innate immune response (suicide by infected cells), and also plays a key role in development and in ridding the body of damaged cells.

Several different mechanisms for the induction of apoptosis exist, as would be expected from the many functions served by apoptosis. Furthermore, the control of apoptosis is very complicated, which is necessary to control an event as drastic as cell death. Some of the pathways used to induce apoptosis are illustrated in Fig. 8.15. One pathway involves death receptors on the cell surface, which initiate apoptotic pathways when their ligands are present and bind to them. One such receptor is the Fas receptor (also known as CD95 or Apo-1), which can be stimulated by Fas ligand present on the surface of CTLs. Fas contains domains known as death domains in the cytoplasmic region, and trimerization of Fas induced by ligand binding results in the recruitment of other cellular proteins by these death domains. This in turn results in the activation, by cleavage, of a protease known as caspase-8. A popular model for activation is that two molecules of caspase-8 must be brought into proximity so that they cleave one another. Caspase-8 then cleaves other caspases, thereby activating them in turn. At least 10 caspases are known, and they are normally present as inactive procaspases. A cascade of caspase cleavages leads, ultimately, to the activation of effector caspases. The effector caspases cause cell death by cleaving many cellular proteins, including, for example, lamin in the nuclear membrane and a repair enzyme called poly(ADP-ribose) polymerase.

Another death receptor that may be present at the cell surface is a receptor for TNF (tumor necrosis factor). TNF can be secreted by CTLs, as well as by other lymphocytes, in response to inflammatory signals. TNF receptor oligomerization also results in cleavage of caspase-8 and the induction of the caspase cascade. The sensitivity of cells to the induction of apoptosis by these pathways thus depends on the concentration of Fas and TNF receptor in the surface, and this concentration can be regulated by other events.

Fas and TNF receptors are ubiquitous, present on most cells. Many other receptors that can lead to apoptosis are also known. Some of these are also ubiquitous, others are present on only a limited set of cells.

The caspase cascade can also be initiated by other mechanisms. Withdrawal of growth factors, such as occurs

Mechanisms of Apoptosis

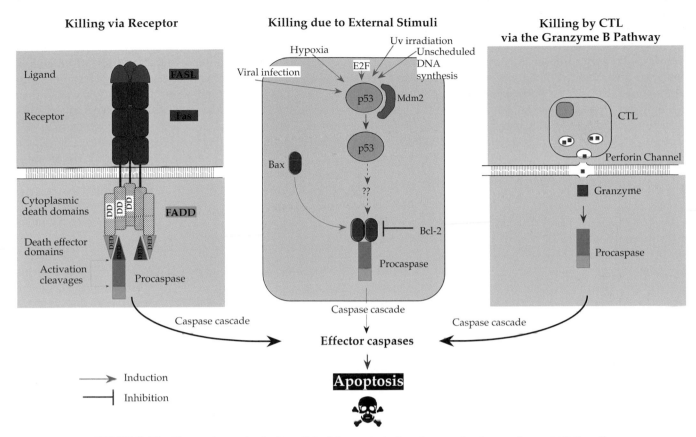

FIGURE 8.15   Three pathways that lead to cell death by apoptosis. In each case a signal comes from outside the cell. This signal may be the binding of a ligand to its receptor (left panel); stress on the cell caused by viral infection, DNA damage caused by exposure to UV light, deregulation of the cell cycle, or other stresses (center panel); or from a cytotoxic T cell (right panel). The signal leads to the activation of a procaspase by proteolytic cleavage. The activated caspase begins a cascade that leads to apoptosis. [Data from Ashkenazi and Dixit (1998) and Raff (1998).]

during development or occurs when activated T cells are no longer stimulated by interaction with their antigens, can lead to activation of caspases. Events that damage DNA or the protein synthetic machinery can lead to apoptosis. At least some of these apoptotic signals are delivered through changes in p53 concentrations. p53 is a key player in the regulation of the cell cycle (Chapter 6), and the apoptotic pathway is induced if its concentration is not tightly regulated. These mechanisms involve Bax and related molecules, as well as Bcl-2 and related molecules, as key players in the activation of the caspase cascade.

In addition to being able to initiate the caspase cascade indirectly by signaling through the Fas receptor or the TNF receptor, T cells can also initiate the caspase cascade directly. They do this by introducing granzymes into the cell by way of channels formed by perforin secreted by the T cell. Granzyme B cleaves procaspases downstream of caspase-8 to begin the cascade.

Control of apoptosis is complex. Many cells make proteins that render them less susceptible to apoptosis. One such anti-apoptotic protein is called Bcl-2, which blocks the action of Bax and interferes with the caspase cascade, among other activities. Bcl-2 is made by mature neurons, for example, in response to signals received when they form connections with the target cells that they enervate. Immature neurons do not make adequate amounts of Bcl-2 and die when nerve growth factor is removed. There is a complicated interplay between Bcl-2 and related proteins and Bax and related proteins that we are only beginning to understand.

Mitochrondria are also important players in apoptosis. A decrease in the potential across the mitochrondrial membrane makes the cell more sensitive to apoptosis. Decreased potential is accompanied by synthesis of reactive oxygen species and the release of cytochrome C into the cytoplasm, both of which are pro-apoptotic. Cytochrome C forms complexes with other cellular proteins that promote the caspase cascade. Bax, a pro-apoptotic protein, may exert its effect by lowering the mitochrondrial membrane potential, and Bcl-2 may prevent this.

The importance of apoptosis for the regulation of the organism is clear. Individuals with deficits in apoptotic pathways may suffer from many diseases, including developmental abnormalities, neurodegenerative disorders, or autoimmune diseases. In normal individuals, tumor cells can only thrive if they avoid apoptosis despite the fact that their cell cyle is deregulated, and mutations that suppress apoptosis are important for the development of tumors.

Interferons and Other Cytokines

The cytokines constitute a large family of proteins, defined by their structure, that have important regulatory roles in an animal. More than 30 cytokines have been described, some of which have multiple members. Most have a molecular mass of about 30 kDa. Many, but not all, are glycoproteins. Some function as monomers but others act as homodimers or homotrimers and at least one acts as a heterodimer. Cytokines are inducible agents and constitutive production is normally low or absent. On induction, their production is short lived. Cytokines effect their action by binding to receptors on the surface of target cells. These receptors bind cytokines with high affinity—the dissociation constants range from 10^{-9} to 10^{-12} M. Binding to the receptor induces a cascade of events in the target cell that leads to changes in gene expression within the cell. These changes in gene expression may lead to many different effects, including cell proliferation or changes in the state of differentiation. Hematopoietic cells are important targets of all cytokines, although most cytokines have a diverse range of actions. The cytokine system is redundant in that many cytokines evoke a similar spectrum of action, and it is pleiotrophic in that one cytokine may have many different target cells. This system is also complex because many different cytokines are usually induced at the same time and they may act synergistically or antagonistically to achieve a result. Cytokines are important players in both the adaptive and innate immune systems. The importance of cytokines in the maturation of T cells and B cells has been described. Figure 8.16 presents an overview of cytokine networks that illustrates their complexity, and a partial listing of cytokines is given in Table 8.3. Because of the complexity of their activities, we are only beginning to understand their many functions, but as these functions begin to be understood, attempts are being made to use different cytokines therapeutically as indicated in the table.

A second family of proteins that play an important role in the regulation of the immune response is the chemokines. Chemokines are small proteins, 70–80 amino acids in size. More than 30 are known in humans. Some serve housekeeping functions and are produced constitu-tively; others are proinflammatory and usually inducible. Chemokines serve to attract leukocytes. The housekeeping chemokines are important for the development and homeostasis of the hematopoietic system (for example, maintenance of lymphoid organs), whereas the proinflammatory cytokines recruit immune cells to sites of infection, inflammation, or tissue damage. The receptors for chemokines are distinct from those of cytokines. An example of a cytokine receptor is described below for interferon. Chemokine receptors belong to the family of seven-transmembrane-domain, G-protein-coupled receptors, and one was illustrated schematically in Fig. 1.3B. In most treatments, chemokines are considered a class of cytokines. Here, for ease of presentation, the term cytokine refers only to non-chemokine cytokines.

Importance of Interferons in the Defense against Viruses

Among the best known cytokines that function in the defense against viruses are the interferons (IFNs). There are two kinds of IFN, called type I and type II. The IFNs induce cells to become resistant to viral infection, an innate defense against viruses, but also play important roles in the adaptive immune response. The importance of IFNs in the defense of mammals against viral infection has been shown by experiments in which an IFN response is ablated. Early experiments in mice used injection of antibodies against IFN to block its activity. More stringent ablation of IFN activity has been accomplished by using transgenic mice in which the receptor for either type I or type II IFN has been abolished. In general, mice that lack a type I IFN response are extremely sensitive to infection by viruses. Viruses grow to much higher titer in such animals and, in the more dramatic examples, virus infection may be lethal, although infection with the same virus of animals able to mount a normal IFN response may be asymptomatic. Many viruses have evolved mechanisms to ablate the activity of IFNs, also demonstrating their importance in controlling viral infection. Interestingly, mice lacking a type I IFN response are able to handle bacterial infections reasonably well. Conversely, mice lacking a type II IFN response are extremely sensitive to bacterial infection but handle viral infections well. One known exception is vaccinia virus, which is lethal in mice lacking either IFN response. We should emphasize that although the relative importance of the two IFNs in the defense against viruses versus bacteria may differ, both are important in the defense against viruses.

Types of IFNs

Several characteristics of type I and type II IFNs are shown in Table 8.4. The type I IFNs are 165–172 amino acids in length and have been classified into four different

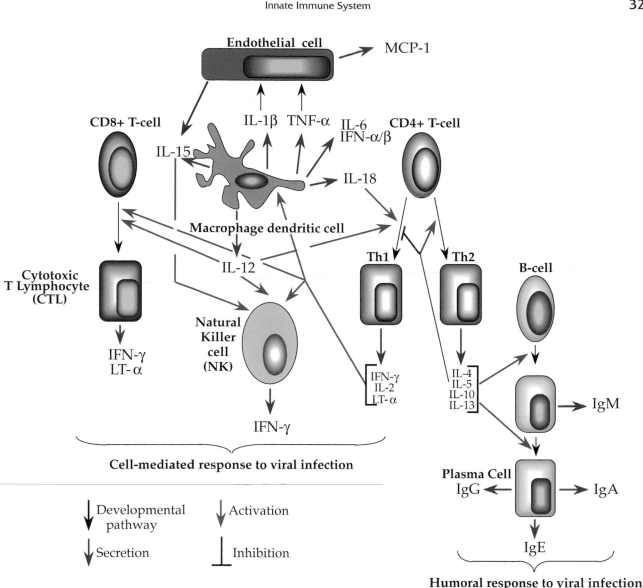

FIGURE 8.16 Overview of the cytokine networks important for innate and acquired antiviral immune responses. IFN, interferon; IL, interleukin; Ig, immunoglobulin, TNF, tumor necrosis factor, LT, lymphotoxin; MCP, monocyte chemotactic protein. [Adapted from Griffin (1999, Fig. 1, p. 340).]

subfamilies, called α, β, ω, and τ, of which α and β are the best studied. In humans there are 14 IFN-α genes and 1 IFN-β gene, none of which contain introns. IFN-β shares 25–30% amino acid sequence identity with any particular IFN-α. In humans, IFN-β is glycosylated whereas IFN-α is not, but in mice both are glycosylated. Thus glycosylation is not a fundamental property distinguishing α from β IFNs. IFN-α and IFN-β use the same receptors and therefore evoke the same responses in target cells.

Type II IFN contains only one member, IFN-γ. In man there is one IFN-γ gene, which contains three introns. The receptor used by IFN-γ is distinct from that used by type I IFNs, but the phosphorylation cascade induced by IFN-γ contains some elements that are shared with the type I cascade. Thus the responses to the two IFNs are partially overlapping.

Induction of IFN

Type I IFNs are induced by treatment with double-stranded RNA (dsRNA), by virus infection, or by the cytokine TNF-α. It is thought that the inducing agent fol-

TABLE 8.3   Some Cytokines and their Therapeutic Uses

Functional group	Name (abbreviation)	Normal biological function	Therapeutic targets	Side effects of therapy
Antiviral cytokines	Type I interferon (IFN-α,β)	Inhibits viral replication	Chronic hepatitis B, hepatitis C, herpes zoster, papilloma viruses, rhinovirus, HIV (?), warts	Fever, malaise, fatigue, muscle pain; kidney, liver, heart, bone marrow toxicity
	Type II interferon (IFN-γ)	Inhibits viral replication, upregulates expression of class I and class II MHC, enhances activity of macrophages	Lepromatous leprosy, leishmaniasis, toxoplasmosis	As above for Type I interferons
Inflammatory cytokines	Tumor necrosis factor (TNF)	Cytotoxic for tumor cells, induces cytokine secretion by inflammatory cells	Anti-TNF in septic shock	Shock with marked hypotension
	Interleukin 1 (IL-1)	Costimulates T-helper cells, promotes maturation of B cells, enhances activity of NK cells, attracts macrophages and neutrophils	Receptor antagonist in septic shock	???
	Interleukin 6 (IL-6)	Promotes differentiation of B cells, stimulates Ab secretion by plasma cells		
Regulators of lymphocyte functions	Interleukin 2 (IL-2)	Induces proliferation of T cells, B cells, and CTLs, stimulates NK cells	Leprosy, local treatment of skin lesions	Vascular leak syndrome, hypotension, edema, ascites, renal failure, hepatic failure, mental changes, and coma
	Interleukin 4 (IL-4)	Stimulates activity of B cells and proliferation of activated B cells, induces class switch to IgG and IgE		
	Interleukin 5 (IL-5)	Stimulates activity of B cells and proliferation of activated B cells, induces class switch to IgA		
	Interleukin 7 (IL-7)	Induces differentiation of stem cells, increases IL-2 in resting cells		
	Interleukin 9 (IL-9)	Mitogenic activity		
	Interleukin 10 (IL-10)	Suppresses cytokines in macrophages	Septic shock	
	Interleukin 12 (IL-12)	Induces differentiation of T cells into CTLs		
	Interleukin 13 (IL-13)	Regulates inflammatory response in macrophages		
	Transforming growth factor (TGF-β)	Chemotactically attracts macrophages, limits inflammatory response, promotes wound healing	Septic shock	Symptoms similar to those for IL-2, especially shock and hypotension

Source: Data for this table came from Mims *et al.* (1993, p. 37.3 and Fig. 7.13) and Kuby (1997, pp. 318, 319).

lowing virus infection is also dsRNA, whether infection is by a DNA- or RNA-containing virus.

Following induction, transcription of type I IFNs is regulated at both the transcriptional and post-transcriptional levels. Regulation at the transcriptional level is complex. There are both positive and negative regulatory factors, whose identity and mode of action are incompletely known. IFN mRNAs contain destabilization sequences in the 3′

TABLE 8.4   Characteristics of the Interferons

| | Type I | | Type II |
	IFN-α	IFN-β	IFN-γ
Alternative name	Leucocyte IFN	Fibroblast IFN	Immune IFN
Produced by	All cells	All cells	T lymphocytes
Inducing agent	Viral infection or dsRNA	Viral infection or dsRNA	Antigen or mitogen
Number of genes (species)	14 (human) 22 (mouse)	1	1
Chromosomal location of gene (species)	9 (human) 4 (mouse)	9 (human) 4 (mouse)	12 (human) 10 (mouse)
Number of introns	None	None	3
Size of IFN protein	165–166 aa	166 aa	146 aa, dimerizes
Receptors	Receptor for both IFN-α and INF-β consists of two polypeptides, IFN-αR1 and IFN-αR2, encoded on chromosome 21 (human) or 16 (mouse)		Receptor consists of two proteins: IFN-γR1 encoded onchromosome 6 (human) or 10 (mouse) and IFN-γR2 encoded on chromosome 21 (human) or 16 (mouse)
General functions	Antiviral activity ↑MHC class I[a]	Antiviral activity ↑MHC class I	Macrophage activation ↑MHC class I ↑ MHC class II on macrophages NK cell activation Some antiviral activity ↓ MHC class II on B cells ↓ IgE, IgG production by B cells

Source: Data for this table came form Mims *et al.* (1993, Fig. 12.9) and Fields *et al.* (1996, Table 3, p. 378).
[a]↑ indicates upregulation; ↓ indicates down regulation.

nontranslated region and have short half-lives. Thus, shutoff of IFN synthesis occurs fairly quickly once the gene is no longer transcribed.

Almost all cell types produce IFN-α or -β, and thus induction of these cytokines follows infection of almost any cell by any virus. However, some viruses are much more efficient at the induction of IFN than others, and it has been postulated that viruses may differ in the extent of production of dsRNA.

In contrast to type I IFNs, production of IFN-γ is largely restricted to T cells and NK cells. Production is induced by agents that promote T-cell activation. These include exposure to antigens to which the organism has been presensitized or exposure to IL-12, a cytokine produced by monocytes and macrophages after infection by bacteria or protozoa. Other activators of T cells are also known that induce synthesis of IFN-γ.

Many other cytokines are also induced following infection by viruses. These include TNF-α, TNF-β, IL-1, IL-2, IL-4, IL-5, IL-6, IL-8, and GM-CSF (granulocyte-macrophage colony-stimulating factor). The spectrum of cytokines induced is different for different viruses. The large DNA-containing viruses encode proteins that interfere with the action of a number of these cytokines, and with the activities of chemokines as well.

Interferon Receptors and Signal Transduction

IFNs effect their responses by binding to specific receptors on the cell surface. These receptors are composed of more than one polypeptide chain, at least one of which is an integral membrane protein that contains a cytoplasmic domain. The type I IFN receptor is present in low abundance (around 10^3 receptors per cell) on all major types of cells. It has a very high affinity for IFN (dissociation constant about 10^{-10} M).

Upon binding of type I or type II IFNs to their cognate receptors, tyrosine kinases associated with the cytoplasmic domains are activated. A phosphorylation cascade ensues that results in the production of activated transcription factors, which translocate to the nucleus and stimulate transcription of a number of genes (Fig. 8.17). Most genes that are induced by type I IFN have an upstream element referred to as ISRE, the *interferon stimulated response element*. Induction of genes controlled by this element is very rapid and happens within minutes after treatment with IFN. Activation is transient, as one or more of the proteins induced act in a feedback loop to repress the continued transcription of these genes. dsRNA can also induce the synthesis of many of the same genes through the ISRE, but the signal transduction pathway is different.

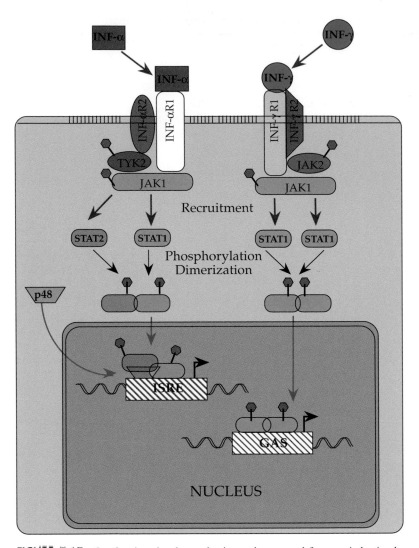

FIGURE 8.17 Overlapping signal transduction pathways used for gene induction by IFN-α and IFN-γ, which bind to different cellular receptors. Binding to their respective receptors leads to tyrosine phosphorylation of JAK1 and either TYK2 or JAK2. These in turn phosphorylate STAT1 and STAT2 proteins. The phosphorylated STAT proteins dimerize and migrate to the nucleus, where they form complexes and bind to ISRE (*inter-feron stimulated response element*) or GAS (IFN-gamma *activation site*,) which are present upstream of interferon inducible genes, resulting in transcription of these genes. [From Fields *et al.* (1996, p. 379), Kalvakolanu (1999) and Nathanson *et al.* (1996, p. 123).]

The genes that are induced by IFN-γ are under the control of several different regulatory elements, one of which is referred to as GAS, the IFN-gamma *activation site*. Induction by IFN-γ is slower and gene transcription continues for a longer period of time after induction. The complex of genes induced by type I and type II IFNs are overlapping because the tyrosine kinases associated with the type I and type II receptors are overlapping, as are the proteins activated during the tyrosine phosphorylation cascade (Fig. 8.17). In addition, many genes contain *cis*-acting regulatory elements that respond to both type I and type II IFN-induced transcriptional activators.

Biological Effects of Interferons

Type I or type II IFN induces the expression of many different genes that have important biological effects. A partial listing of such genes is given in Table 8.5. One effect is to stimulate the adaptive immune response. Both type I and type II IFNs induce increased production of class I MHC molecules, thus leading to enhanced surveillance by CTLs. Type II IFN also leads to increased production of class II MHC in macrophages, which are important players in the humoral response. The MHC response is augmented by the induction of genes in the MHC cluster that encode

TABLE 8.5   Genes Induced by Interferons

Protein	Induced by[a]			Inducible element	Functions/phenotype
	IFN-α	IFN-β	IFN-γ		
$(2'-5')$ (A_n) synthetase[b]	+++	+++	+	ISRE	$(2'-5')$ (A_n) synthesis
	(also induced by)				Induction of antiviral state especially anti-picornavirus
p68 Kinase (PKR)	+++	+++	+	ISRE	Protein kinase Induction of antiviral state
Indoleamine 2,3-dioxygenase	+	+	+++		Tryptophan degradation
γ56	+	+	+++		Trp-tRNA synthetase
GBP/g57	+	+	+++		Guanylate binding
MxA[b]	+++	+++	+		Inhibits replication of influenza and VSV
IRF1/ISGF2[b]	++	++	+++		Transcription factor
IRF2[b]	++	++	++		Transcription factor
MHC/class I	+++	+++	+++		Upregulation of antigen presentation
MHC class II			++	Not ISRE nor GAS	Upregulation of antigen presentation
RING 12					Proteasome subunit
RING 4	+++	+++			Putative TAP
β_2 Microglobulin	+++	+++	+++		MHC light chain

Source: Adapted from Nathanson et al. (1996, p. 124).
[a]The strength of the induction is indicated by the number of plus signs.
[b]Full induction also requires ds RNA

components of the proteasome. These cause the proteasome to become more active in producing peptides suitable for presentation by MHC molecules. Production of the TAP transporter system is upregulated, so that increased quantities of peptides are transferred across the ER membrane for binding by MHC molecules. Either type of IFN leads to the activation of monocytes and macrophages, the activation of natural killer cells, the activation of CTLs, and the modulation of the synthesis of Ig by B cells, all important for the immune response. IFN-γ also induces the increased expression of Fc receptors in monocytes and inhibits the growth of nonviral intracellular pathogens.

Both type I and type II IFNs are also key players in the innate immune defense against viruses. They are pyrogenic, inducing fever. High temperatures inhibit the replication of many viruses or other pathogens. It has been proposed that the major symptoms of influenza infection, which include high fever as well as muscle aches and pains, are due to the induction of IFN rather than to virus replication per se. Both IFNs also induce what has been called the antiviral state, in which cells are less susceptible to or resistant to infection by viruses.

Thus, the induction of IFNs is important for the control of viral infection at more than one level. Induction of the antiviral state results in lower yield of virus and the stimulation of immune responses leads to a more effective and faster clearance of the viral infection. Of interest is the fact that both types of IFNs inhibit cell growth, and treatment with IFN or with inducers of IFN has been effective in the treatment of at least some types of cancer.

The Antiviral State

Expression of the genes induced by interferon results in the establishment of the antiviral state in cells, in which viruses fail to replicate or replicate to much lower titers. The antiviral state is multifaceted and interference with the replication of viruses may occur at different stages of their replication cycle. The effect, therefore, depends on both the virus and the host cell. Furthermore, interference can occur at more than one stage of replication for some viruses, and such viruses tend to be more sensitive to the activities of interferon than others.

Interference with the virus replication cycle may occur very early, during penetration and uncoating of the virus. This occurs with SV40 and the retroviruses, for example. The proteins that are responsible for interference at these early stages of infection are unknown. For some viruses the transcription of the infecting viral genome is inhibited, examples being influenza, vesicular stomatitis virus, and herpes simplex virus. A host protein known as Mx, which is induced by interferon, is responsible, at least in part, for interfering with the transcription of influenza virus. In mice that are unable to produce the Mx protein, as is the case for most inbred strains of mice, infection with influenza produces a fatal outcome, demonstrating the importance of this gene. At a later stage in the infection cycle, interferon treatment results in reduced translation of many viral mRNAs, and the mechanisms by which this occurs are described below. Finally, some interferon-induced products interfere with virus assembly. This occurs with the retroviruses, vesicular stomatitis virus, and herpes simplex virus, but the proteins responsible are unknown.

Interference with Translation of Viral mRNAs

Two distinct pathways induced by IFN result in interference with the translation of viral mRNA. These are illustrated schematically in Fig. 8.18. Both of these pathways require that dsRNA be present for them to be active. Thus, not only is dsRNA a primary inducer of IFN synthesis, but these two pathways induced by IFN are also dependent on dsRNA for their activation. This dependence on dsRNA suggests that it is a common product in viral infection, whether the virus is DNA or RNA, but is not normally present in uninfected cells.

One inhibition pathway involves 2'-5'-oligo(A) synthetases. These synthetases, of which several are known, some of which have different subcellular locations, polymerize ATP into oligoadenylates that are joined in a 2'-5' linkage. 2'-5'-Oligo(A) synthetase is induced by IFN but requires dsRNA as a cofactor for activity. 2'-5'-Oligo(A), in turn, is a cofactor for a latent ribonuclease, RNase L, which is present in all animal cells, but is inactive in the absence of 2'-5' oligo(A). Once activated by its cofactor, which is bound tightly enough so that the two can be coimmunoprecipitated, RNase L can cleave single-strand mRNAs. 2'-5'-Oligo(A) is hydrolyzed by a phosphodiesterase present in cells, and activation of RNase L is transient.

Thus, RNase L results in a degradation of mRNAs only in those cells in which dsRNA is present. The picornaviruses are particularly sensitive to the action of RNase L, and it has been well established that activation of RNase L is a primary mechanism by which IFN interferes with the replication of these viruses. Other viruses seem to be less affected by RNase L, and the possible role of this enzyme in combating other virus infections remains to be determined.

The second pathway leading to the inhibition of translation of viral mRNAs involves a protein kinase known as PKR. PKR is present at low levels in most cells, but its concentration increases on IFN induction. It requires dsRNA for activity. On binding dsRNA, PKR is autophosphorylated, probably in *trans*, on serine and threonine residues. Once activated, it phosphorylates the translation initiation factor, eIF-2α. Phosphorylated eIF-2α cannot be recycled and protein synthesis is shut down. The importance of PKR

for inhibiting viral replication is shown by the fact that several viruses encode inhibitors of its activity. In particular, it is believed that reoviruses, adenoviruses, vaccinia virus, vesicular stomatitis virus, and influenza virus are sensitive to the effects of this enzyme.

It seems almost certain that PKR has important functions in the cell in addition to its activity in the inhibition of viral replication. Among other effects, if PKR is nonfunctional, unregulated cell proliferation and neoplastic transformation occur. Thus, the enzyme must be involved in the regulation of the cell cycle.

Degradation of mRNA and shutoff of protein synthesis are drastic events that represent a defense of last resort. The dependence of both pathways on dsRNA means that only cells that contain dsRNA, presumably only virus-infected cells, will shut down as a result of these pathways. To simplify somewhat, the effect of IFN is to prepare the cell for instant shutdown if it becomes infected, while leaving uninfected cells essentially alone.

Therapeutic Uses of Interferon

Since the discovery of IFN more than 30 years ago, there has been great hope and expectation that IFN therapy would be useful for the treatment of viral infections. In early studies, IFN was induced in patients by injection of dsRNA, but quantities of recombinant IFN suitable for injection are now available. In general, IFN therapy has been disappointing for treatment of viral infections in humans, although it is useful in a number of diseases. Infection with hepatitis B or hepatitis C viruses often results in chronic infection. In at least some of these chronically infected patients, the virus load can be decreased, or the disease may go into remission, upon long-term treatment with recombinant IFN. However, side effects of IFN treatment often limit the dose and duration of treatment that can be used. IFN has also been useful in treatment of infection with papilloma viruses.

During clinical trials with IFN, it was found that it is useful for the treatment of at least some cancers, including hairy cell leukemia and AIDS-related Kaposi's sarcoma. It is assumed that this control is based on the inhibition of cell growth caused by IFNs. Clinical trials using IFN, as well as other cytokines, for treatment of other viral diseases and other neoplasias continue and it is to be expected that further uses for these agents will be found (Table 8.3).

VIRAL COUNTERDEFENSES

Mammals have evolved elaborate defenses to ward off infection by viruses. Viruses, in turn, have evolved counterdefenses that allow them to persist and continue to infect mammals. These counterdefenses vary from the simple to the elaborate. The simplest counterdefense is to shut down

A) Development of the antiviral state

Induction by IFN

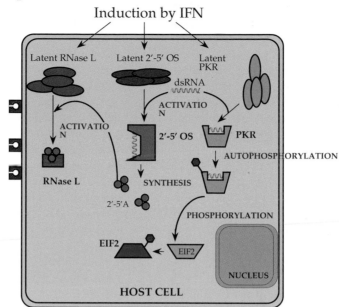

B) Virus attempts to replicate in a cell in the antiviral state

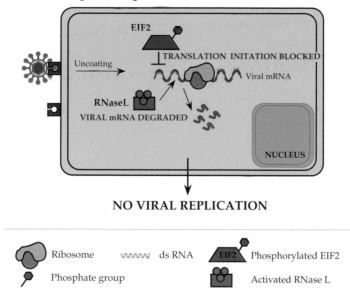

FIGURE 8.18   The antiviral state. (A) Development of the antiviral state begins with the action of interferon on an uninfected cell. The result of the signal transduction cascade shown in Fig. 8.17 is the induction of expression of up to 100 genes, of which three are shown: RNase L, the 2′-5′ oligo(A) synthetase (2′-5′ OS), and the dsRNA-dependent protein kinase (PKR). These proteins are latent until they are activated by viral infection. PKR and 2′-5′ oligo(A) synthetase are activated by dsRNA that is produced during viral infection. Once activated, PKR autophosphorylates itself, and then phosphorylates EIF2. The activated synthetase makes trimeric oligonucleotides, which in turn activate RNase L. (B) Phosphorylated EIF2 and activated RNase L are characteristic of the "antiviral state" in which a eukaryotic cell is refractory to infection by a wide variety of viruses. Phosphorylated EIF2 cannot serve to initiate translation of mRNA by ribosomes and activated RNase L degrades mRNAs, both viral and cellular, so protein synthesis stops. Without protein synthesis, no virus replication can take place, but the inhibition of protein synthesis is transient and the cell may recover. [Adapted from Nathanson *et al.* (1996, Fig. 6.8, p. 125), and Mims *et al.* (1993, Fig. 12.10, p. 12.7).]

the host cell rapidly after infection and to produce progeny virus very rapidly. Rapid shutoff of the host cell may prevent it from synthesizing IFN or other cytokines required for function of both the innate and adaptive immune systems. It will also shut off production of MHC class I molecules required for recognition of an infected cell by CTLs. Rapid growth allows the virus to go through many rounds of virus production before sufficient cytokine production has occurred and cells have responded to establish the antiviral state, and before the adaptive immune system can gear up. Many viruses take this approach, in particular RNA viruses, which have a limited coding capacity. As one example, soon after infection with poliovirus, the cap-binding protein required for cap-dependent initiation of protein synthesis is cleaved and host cap-dependent protein synthesis ceases. Translation of poliovirus RNA is not affected since it utilizes a cap-independent initiation process. This elegant mechanism for specifically interfering with host protein synthesis does not appear to be the sole pathway used by the virus. Additional mechanisms also contribute to shutdown of host protein synthesis. Shutoff of host protein synthesis not only prevents the synthesis of IFN, for example, but it also frees up the translation machinery of the cell for the virus.

As a second example, the alphaviruses also rapidly shut off host protein synthesis. As for poliovirus, more than one mechanism appears to be used. One mechanism is an interference with the cell Na/K pump, which results in an increase in the Na^+ concentration and a decrease in the K^+ concentration in the cell. This inhibits translation of cellular mRNAs, but has no effect on translation of viral mRNAs.

Some DNA viruses also inhibit host protein synthesis shortly after infection. A component of the herpes simplex virion that enters the cell during infection inhibits host-cell protein synthesis, beginning very soon after infection. As infection progresses, additional proteins contribute to a more profound inhibition of host-cell macromolecular synthesis. Viruses such as herpes simplex virus, which have large genomes, also have the luxury of encoding many proteins that interfere with host defenses by more specific mechanisms. An overview of these mechanisms, which are described below, is given in Table 8.6.

Virus Evasion of the Adaptive Immune System

More than 50 viral proteins have been identified that modulate the defense mechanisms used by mammals to control viral infection and to eliminate the viruses once infection is initiated. Most have been described only recently and more viral proteins will surely be discovered in the future. The more elaborate evasion mechanisms are utilized by viruses that contain larger genomes, and thus can afford so-called luxury functions such as specific interference with host defense mechanisms.

TABLE 8.6   Major Strategies Used by Viruses to Evade the Immune System

Rapid shutdown of host macromolecular synthesis
Evasive strategies of viral antigen production
    Restricted gene expression; virus remains latent with minimal or no expression of viral proteins
    Infection of sites not readily accessible to the immune system
    Antigenic variation; antigenic epitopes mutate rapidly
Interference with MHC class I antigen presentation
    Downregulation of transcription of class I MHC molecules
    Degradation of MHC class I molecules
    Retention of MHC class I molecules within the cell
    Downregulation of transcription of TAP
    Interference with the activity of TAP
Interference with natural killer cell function
Interference with MHC class II antigen presentation
Interference with antiviral cytokine function
    Production of viral homologs of cellular regulators of cytokines
    Neutralization of cytokine activities
    Inhibition of the function of IFN
    Production of soluble cytokine receptors
Inhibition of apoptosis

Source: Adapted from Granoff and Webster (1999, Table 3, p. 1203).

Interference with the Expression of Peptides by Class I MHC Molecules

Many viruses downregulate the expression of class I MHC molecules on the surface of infected cells, thereby rendering the cells invisible to CTLs. An overview of the mechanisms used is given in Table 8.7. Rapid shutdown of the cell can accomplish this, and this appears to occur following infection by picornaviruses, for example. More specific mechanisms are used by many viruses, however, including the lentiviruses, the adenoviruses, the herpesviruses, and, probably, the poxviruses.

The lentivirus HIV-1 uses two different mechanisms to downregulate class I MHC expression. The Tat protein interferes with the transcription of mRNA for class I MHC molecules, thereby leading to reduced synthesis of class I molecules. The Nef protein downregulates the surface expression of class I MHC molecules by relocalizing them to the *trans*-Golgi network. It does this through interactions with another cellular protein called PACS-1, which is involved in sorting proteins to the *trans*-Golgi network.

Adenoviruses prevent the presentation of peptides by class I MHC molecules by one of two different methods. These are illustrated schematically in Fig. 8.19. Some adenoviruses, such as Ad2, produce an integral membrane protein of 19 kDa, a product of the E3 gene, that binds tightly to class I MHC molecules in the lumen of the ER. This prevents their export to the cell surface. Ad12, in contrast, produces a product from the E1 gene that interferes with the transcription of genes in the MHC locus. Transcription of the genes for class I molecules, the genes for TAP, and the genes for the MHC-encoded components of the proteasome is inhibited. Thus, synthesis of new class I molecules is inhibited, and the production and transport of peptides required for MHC presentation cannot be upregulated.

The herpesviruses have learned to coexist with the immune system, enabling them to establish lifelong infections. They encode a wide variety of gene products that interfere with the presentation of peptides by class I MHC molecules. Several of these are illustrated schematically in Fig. 8.20. Human cytomegalovirus (HCMV) encodes four different proteins that interfere with class I presentation by interacting with cellular proteins that form components of the presentation pathway. One protein, called US3, binds to class I MHC molecules and retains them in the ER. Two other proteins, US2 and US11, independently cause class I MHC molecules to be immediately recycled to the cytoplasm following synthesis, where they are degraded by the proteasome. The recycling of class I molecules to the cytoplasm is thought to occur by speeding up the kinetics of the normal cellular turnover pathway that recycles products in the ER back to the cytoplasm. A fourth HCMV protein,

TABLE 8.7   Some Viruses That Alter Antigen Presentation by Class I MHC Molecules

Interference level	Virus family	Virus	Virus protein	Mechanism
Downregulation of MHC class I expression at cell surface	Adenoviridae	Ad2	E3-19K	Viral protein binds to class I molecules and keeps them in the ER
		Ad12	E1A	Inhibits transcription of MHC class I mRNA
	Herpesviridae	HCMV	US2, US11	Gene products lead to degradation of class I molecules
		HCMV	UL18	Binds to β_2 microglobulin
	Picornavirus	FMDV	??	Downregulation of class I protein at the surface of cell
	Lentivirus	HIV	Tat	Inhibits transcription of MHC class I mRNA
			Nef	Downregulates surface expression of class I proteins
Alteration of antigen processing	Adenoviridae	Ad12	E1A	Inhibits transcription of TAP1 and TAP2
	Herpesviridae	HSV	ICP47	Binds to TAP and prevents transport of antigenic peptide to MHC class I molecules
		HCMV	US6	Blocks transported by TAP
		EBV	EBNA-1	EBNA-1 protein with Gly-Ala repeat is not processed by proteasome
Alteration of spectrum of antigens presented	Hepadnavirus	HepB	Antigenic epitopes	Epitopes mutate so that they are no longer recognized by CTLs

Abbreviations: Ad2, adenovirus 2; Ad12, adenovirus 12; HCMV, human cytomegalovirus; FMDV, foot and mouth disease virus; HIV, human immunodeficiency virus; HSV, herpes simplex virus; EBV, Epstein–Barr virus; HepB, hepatitis B virus.

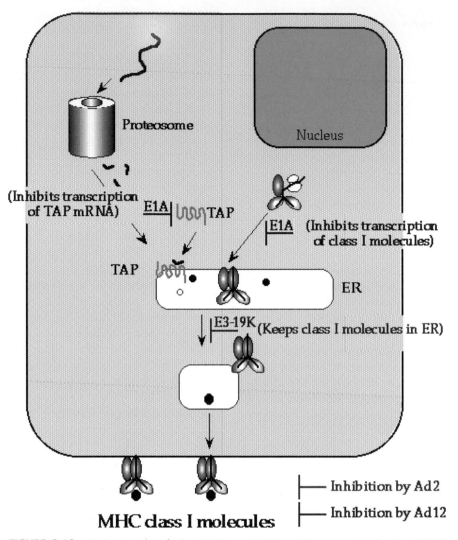

FIGURE 8.19   Mechanisms by which adenoviruses inhibit peptide presentation by class I MHC molecules on infected cells.

US6, binds to TAP and prevents transport of peptides into the lumen of the ER. Herpes simplex viruses (HSV) also encode proteins (ICP47) that block transport of peptides by TAP. Interestingly, the HSV ICP47 binds to TAP from the cytoplasmic side of this protein, which spans the ER membrane, whereas HCMV US6 binds from the lumenal side. Thus these proteins represent independent solutions to the problem of preventing transport of peptides across the ER membrane by TAP. Loss of TAP transport prevents the presentation of herpes peptides at the cell surface.

A fifth HCMV protein, UL83, is also involved in preventing the presentation of peptides by class I MHC molecules. This protein phosphorylates an early HCMV protein within an important T-cell epitope. Phosphorylation prevents the processing and presentation of this epitope.

Finally, a HCMV protein, UL18, binds β_2 microglobulin. UL18 is homologous to MHC class I molecules (the virus obviously swiped the gene from its host at some time in the past) and binds peptide. Its function is not well understood. It interacts with monocytes and presumably interferes with their function. It may have other functions as well. In any event, binding of β_2 microglobulin reduces the pool available for formation of class I MHC molecules.

Mouse CMV also downregulates expression of class I molecules but uses different mechanisms to do so. An early gene, m06, encodes a protein that binds to MHC class I molecules in the ER. The bound MHC class I molecules exit the ER and pass through the Golgi apparatus. Rather than continuing on to the cell surface, however, they are redirected to lysosomes where they are degraded.

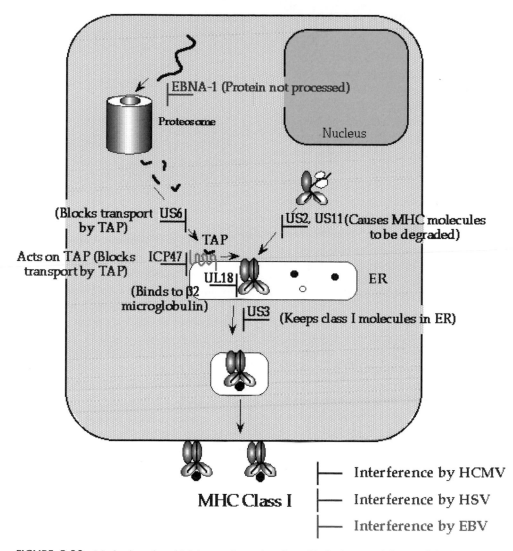

FIGURE 8.20 Mechanisms by which herpesviruses interfere with the immune defenses of the host cell, by altering peptide presentation by class I MHC molecules. Mechanisms used by human cytomegalovirus (HCMV) are shown in dark red, by herpes simplex virus (HSV) in orange, and by Epstein–Barr virus (EBV) in ochre.

Interference with the Activity of Complement

Infected cells can be killed by a complement-mediated pathway in which antibodies first bind to viral antigens present on the surface. Virus neutralization by antibodies is also enhanced by complement. Herpesviruses express proteins that interfere with complement activation. HSV expresses a protein called gC on the cell surface that binds complement component C3b, thereby preventing initiation of the complement cascade. HSV also expresses a molecule, consisting of a complex of viral proteins gE and gI, that has an Fc receptor. It binds IgG through the Fc domain, which serves to block complement activation. HCMV also expresses an Fc receptor.

Latent Infections That Avoid the Immune System

The herpesviruses have evolved means of avoiding immune recognition of latently infected cells, so that latent infection can persist for the life of the host. The alphaherpesviruses evade the class I pathway during latency by establishing a latent infection in neurons. Neurons are immunologically privileged and express only low levels of class I MHC molecules. Another mechanism is used by Epstein–Barr virus. The only viral protein expressed in latently infected cells contains a glycine-alanine tract that interferes with processing by the proteasome, rendering this protein invisible to the class I pathway.

Infection of Cells of the Immune System to Thwart an Immune Response

Another mechanism for evasion of the immune system is to infect immune effector cells. HIV lytically infects T_H cells, which leads to a profound immunosuppression because these cells are required to mount an immune response. Similarly, measles virus is known to infect many cells of the immune system, including T and B lymphocytes and also monocytes. Infection by measles virus results in immunosuppression that lasts for some weeks after infection. Other viruses are also known that infect T cells, B cells, macrophages, or other cells that are important for the immune response. In addition, if a virus infects the thymus early in the life of the animal, while the immune system is still developing, immune tolerance may result such that the virus will not be recognized as foreign. This enables the virus to establish an infection that lasts for the life of the animal.

Rapid Drift

Some viruses evade the adaptive immune system by rapid drift in the antigens that are recognized by it. Most viruses, but especially RNA viruses, undergo rapid evolution, and immune pressure can cause their sequence to drift. Viruses that are able to undergo reassortment, of which influenza virus is the classic example, are able to produce new forms very rapidly that have altered antigenic epitopes. Viruses such as the lentiviruses, of which HIV is the best known, establish a chronic infection in their host, and during this infection these viruses undergo continuing genetic drift, which may be responsible, at least in part, for their persistence.

Prevention of Killing by Natural Killer Cells

Downregulation of class I MHC expression renders a cell more sensitive to lysis by NK cells, the backup mechanism that eliminates cells that do not express class I molecules on their surface. The CMVs resists NK cell activity, however. The molecules on the infected cell surface that are required for the stimulation of the killing activity of NK cells are downregulated by mechanisms that are poorly understood. The CMV protein UL18 is also important in some way. Mouse CMV deleted for this gene replicates poorly in mice, and NK cells appear to be responsible for the poor replication.

HIV also resists NK activity. Here the mechanism seems to be a selective interference with the expression of class I MHC molecules. Human cells are protected from NK killing primarily by HLA-C and HLA-E molecules. HIV-1 selectively downregulates expression of HLA-A and HLA-B molecules, which are the human class I molecules recognized by the majority of CTLs. It does not affect expression of HLA-C or HLA-E molecules. Thus, the infected cells are resistant to killing by NK cells. They remain sensitive to killing by CTLs, but their sensitivity to CTLs is greatly reduced.

Virus Counterdefenses against Apoptosis

Infection of an animal cell by a virus often results in apoptosis of the cell, or would result in apoptosis if the virus did not block its induction. NK cells or CTLs induce apoptosis as a way of killing infected cells. However, the replication activities of the virus itself are often apoptosis inducing. For example, DNA viruses deregulate the cell and cause it to enter S phase, which is required for optimal replication of the viral DNAs. This activity would normally induce apoptosis mediated by p53. Other aspects of viral replication also have the potential to induce apoptosis.

For viruses that replicate rapidly, such as many RNA viruses, apoptosis does not seem to inhibit the production of virus. It may actually lead to increased virus production, perhaps because of less competition with cellular functions. It may also result in more rapid spread to neighboring cells because of cell fragmentation and uptake by neighboring cells, with less inflammation than would occur otherwise, since apoptosis itself does not induce inflammation. For viruses that replicate more slowly, however, such as most DNA viruses, premature apoptosis results in significant declines in virus yield. Thus, many viruses have evolved ways to inhibit or delay apoptosis in infected cells. An overview of mechanisms used by viruses in different families to interfere with apoptosis is given in Table 8.8. Most of these mechanisms interfere with the caspase activation pathway in some way or serve to regulate the activities of p53.

Production of Serpins

Many of the poxviruses produce proteins that inhibit the activity of caspases. These proteins are related to serpins (serine protease inhibitors), which are small proteins that serve as substrates for serine proteases, but which remain bound to the protease after cleavage and block their activity. Serpins are important in the regulation of inflammatory responses. The poxvirus serpins inhibit caspases, which are cysteine proteases. They appear to act similarly to host serpins in that they are cleaved by the caspase but remain bound to it and render it inactive. These viral products, of which crmA of cowpox virus is an example, appear to block apoptosis induced by any pathway requiring activation of caspases. An overview of poxvirus interference with apoptotic pathways is shown in Fig. 8.21.

TABLE 8.8   Viruses That Interfere with Apoptosis

Virus family	Virus	Viral gene	Mode of interference of gene product
Poxviridae	Cowpox	crmA	Serpin homolog, inhibits proteolytic activation of caspases
	Vaccinia	SPI-2	crmA homolog, inhibits activation of caspases
	Myxoma	M11L, T2	T2 is a homolog of TNFR, and inhibits interaction of TNA-α with TNFR; M11L has a novel function
	Molluscum contagiosum	M159, 160	Has death domains like FADD; inhibits FADD activation of caspase 8
Asfarviridae	African swine fever virus	LMW5-HL	Homolog of Bcl-2
Herpesviridae	Herpes simplex	γ34.5 gene	Prevents shutoff of protein synthesis in neuroblastoma cells
	Herpes saimiri	ORF 16 product	Homolog of Bcl-2
	HHV-8	KS Bcl-2	Homolog of Bcl-2
		K13	vFLIPS, prevents activation of caspases by death receptors
	Epstein–Barr (latent)	LMP1	Upregulates transcription of Bcl-2 and A20 mRNAs; inhibits p53-mediated apoptosis
	(lytic)	BHFR1	Inhibits p53 activity; has some sequence similarity to Bcl-2
	HCMV	IE-1, IE-2	Downregulates transcription of p53 mRNA
	Gammaherpesvirinae	??	v FLIPs, inhibits signaling from death domains to caspases
Polymaviridae	SV40	Large T antigen	Binds to and inactivates p53
Papillomaviridae	HPVs	E6	Binds to p53 and targets it for ubiquitin-mediated proteolysis
Adenoviridae	Adenovirus	E1B-55K	Binds to and inactivates p53
		E3-14.7K	Interacts with caspase-8
		E3 10.4K/14.5K	Blocks caspase-8 activation by destruction of Fas
		E4 orf 6	Binds to and inactivates p53
		E1B 19K	Functional homolog of Bcl-2; interacts with Bax, Bi, and Bak
Baculoviridae	AcMNPV	p35	Forms a complex with caspases; inhibits caspase-mediated cell death
		IAP	Like FLIPs, inhibits activation of caspases
Hepadnaviridae	HepB	pX	Binds to p53
Flaviviridae	Hep C	Core protein	Represses transcription of p53 mRNA

Source: Adapted from reviews by O'Brien (1998) and Tortorella *et al.* (2000).

Abbreviations: HHV-8, human herpesvirus 8; HCMV, human cytomegalovirus; HPV, human papillomaviruses; AcMNPV, *Autographa californica* nucleopolyhedrovirus; HepB, hepatitis B virus; HepC, hepatitis C virus; v FLIPs, *v*iral *F*LICE-like *i*nhibitors *p*roteins.

Interference with Fas or TNF Signaling

The poxvirus rabbit myxoma virus encodes a homolog of the TNF receptor (TNFR), called T2. This protein inhibits the interaction of TNF-α with TNFR and thus prevents activation of the apoptotic pathway by TNF-α (Fig. 8.21). Some other poxviruses also produce TNFRs to neutralize its activity. The adenoviruses use other mechanisms to interfere with the activities of TNF-α. They produce four different proteins, found in different parts of the infected cell, that antagonize the effects of TNF-α.

Molluscum contagiosum virus, another poxvirus, encodes proteins named MC159/160 that have death effector domains similar to those present in a cellular protein

called FADD. FADD associates with caspase-8 and recruits it to activated Fas or TNF receptors in the cell surface, resulting in the activation of caspase-8. The viral proteins disrupt this interaction and prevent the activation of caspase-8 signaled by Fas ligand or TNF (Fig. 8.21). Thus, poxviruses use a variety of mechanisms to inhibit apoptosis.

Several herpesviruses also produce proteins that interfere with the activation of caspase-8 through its interactions with FADD. These viral FLIPs thus inhibit apoptosis in a manner similar to the molluscum contagiosum virus MC159/160.

Another mechanism to interfere with Fas-induced activation of the apoptotic pathway is simply to reduce the number of Fas molecules at the cell surface. The aden-

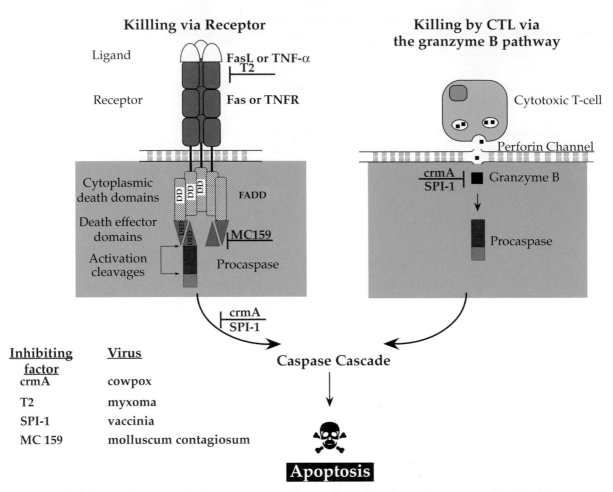

Killing via Receptor

Killing by CTL via the granzyme B pathway

Inhibiting factor	Virus
crmA	cowpox
T2	myxoma
SPI-1	vaccinia
MC 159	molluscum contagiosum

Caspase Cascade

Apoptosis

FIGURE 8.21 Poxviruses inhibit apoptosis by interfering with the ligand–receptor pathway and by interfering directly with the caspase cleavage pathway. The latter mechanism blocks induction of apoptosis by the granzyme B pathway (right) as well as by receptor signaling (left). Points of interference by products of cowpox, myxoma, vaccinia, and molluscum contagiosum viruses are shown in red. [Drawn from data in Turner and Moyer (1998).]

oviruses produce two proteins, known as RIDα and RIDβ, that cause Fas at the cell surface to be internalized and degraded. The RID proteins also inhibit apoptosis induced by TNF, but the mechanism is not known.

Production of Homologs of Bcl-2

Bcl-2 is a cellular protein that inhibits apoptosis. Several herpesviruses and adenoviruses, as well as African swine fever virus, produce homologs of Bcl-2, perhaps obtained originally from the host, that act as anti-apoptotic agents. The adenovirus homolog, called E1B-19K, is multifunctional. It appears to inhibit apoptosis not only as a Bcl-2 homolog but also by interfering with FADD-mediated activation of the caspase pathway. An overview of herpesvirus interference with the induction of apoptosis is

shown in Fig. 8.22 and of adenovirus interference in Fig. 8.23.

The herpesvirus EBV has evolved another solution to the production of Bcl-2. It does not encode a Bcl-2 homolog, but produces a protein that upregulates the production of cellular Bcl-2.

Control of p53 Concentrations

Many of the DNA viruses induce cellular DNA synthesis, which leads to an increase in p53 concentration. This antitumor protein plays a central role in the control of the cell cycle, and one of its functions is as a transcriptional activator. Its concentration in normal cells is controlled by rapid turnover and by its activation of the transcription of a gene called mdm-2, which also requires interaction with

Killing due to External Stimuli

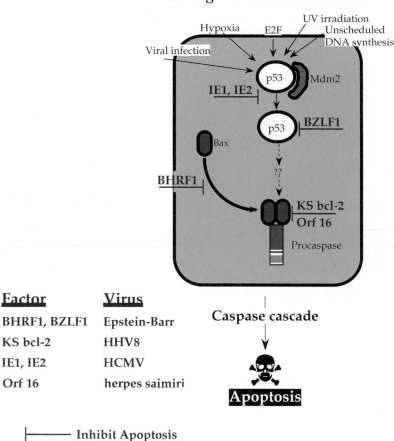

Factor	Virus
BHRF1, BZLF1	Epstein-Barr
KS bcl-2	HHV8
IE1, IE2	HCMV
Orf 16	herpes saimiri

├───── Inhibit Apoptosis

FIGURE 8.22   Various members of the herpesvirus family inhibit different steps in p53-mediated apoptosis. They interfere with Bax (an inducer of apoptosis), encode Bcl-2 homologs, and interact with p53 itself. The gammaherpesviruses also interfere with Fas-mediated signaling (not shown) by encoding products called viral FLIPs that interact with death domains and prevent signaling (similar to the action of poxvirus MC159 illustrated in Fig. 8.21). [Adapted from Hill and Masucci (1998).]

p300. The mdm-2 protein binds p53 and inhibits it ability to act as a transcription factor, thus regulating its own synthesis. This feedback loop is inactivated by DNA viruses in the process of stimulating the cell to enter S phase, and p53 accumulates. High concentrations of p53 induce apoptosis, probably as a result of its transcriptional activities. Many DNA viruses resist p53–induced apoptosis by sequestering p53 or otherwise interfering with its function as a transcriptional activator, or by causing it to be rapidly degraded. These include at least some herpesviruses, adenoviruses, papillomaviruses, and hepatitis B virus. Examples include the SV40 T antigen, the adenovirus E1B-55K protein, the hepatitis B virus pX, two proteins encoded by Epstein–Barr virus (called EBNA-5 and BZLF1) which bind p53, two proteins encoded by cytomegalovirus that interfere with p53–directed transcription (called IE1 and IE2), and the papillomavirus product E6, which binds p53 and causes it to be rapidly degraded. Many of these products were described in Chapter 6.

Other Viral Proteins That Oppose Apoptosis

There are yet other viral proteins that cause the infected cell to resist apoptosis by mechanisms that are not well understood. These include several poxvirus proteins that contain ankyrin repeats and a protein encoded by molluscum contagiosium virus that is 75% identical to human glutathione peroxidase (and may therefore prevent the accumulation of oxidizing agents that can provoke apoptosis). Herpesviruses and adenoviruses produce proteins in addition to those described above that inhibit apoptosis. It is worth considering that many individual viruses encode

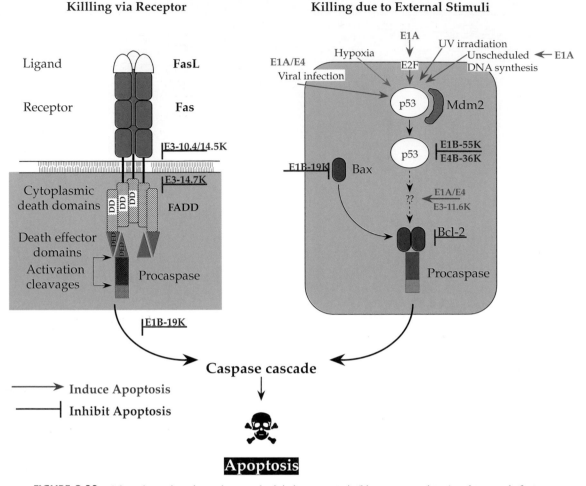

FIGURE 8.23   Adenoviruses have been shown to both induce apoptosis (blue arrows and text) and to encode factors that inhibit apoptosis (red symbols and text). These moieties interfere with "killing via receptor" and with the p53 mediated pathway of apoptosis. [Drawn from data in Chinnadurai (1998).]

multiple products that oppose apoptosis and that many of these proteins appear be multifunctional. This illustrates the importance of controlling the apoptotic pathway after infection by these viruses. It is interesting that the adenoviruses are so efficient at suppressing apoptosis that they encode another protein, the adenovirus death protein, which leads to cell death and release of virus. This protein, which is produced only late after infection, may simply allow apoptosis to proceed, or the protein may initiate some other pathway that results in the death of the cell.

Viral Counterdefenses against Cytokines and Chemokines

The cytokines and chemokines are powerful regulators of both innate and adaptive immune defenses. Because of the importance of these agents in the regulation of the immune response, and because of their potential effective-

ness against viral infection, many viruses have devised methods to disrupt their activities. These include the encoding of homologs or analogs of cytokines and chemokines or of their receptors. Some of these were acquired from the host at some time in the past and modified to meet the purposes of the virus, whereas others are viral products that have evolved to interact with the cytokine system. Because the networks of cytokine and chemokine interactions are complex, the pathways by which the viral products exert their effects are often poorly understood. It is in the interest of the virus to divert the immune responses in directions that not only favor virus growth in the present host but also the persistence of the virus in nature. These two are often antagonistic, and compromises are required. Viruses also produce products that defeat specific aspects of the innate response induced by cytokines. Of these, products that evade the antiviral state induced by IFN are among the best understood. A partial listing of strategies used by some viruses to modulate cytokine activities is given in Table 8.9.

TABLE 8.9   Virus Manipulation of Cytokine Signaling

Virus family	Virus	Cellular target or homolog	Viral factor	Mode of action
Herpesviridae	HCMV	TNF receptor	UL144	Unknown, retained intracellularly
		Chemokine receptors	US28	Competitive CC-chemokine receptor, sequesters CC-chemokines
	HHV-8	Type 1 IFNs	vIRF K9	Blocks transcription activation in response to IFN
			vIRF-2	May modulate expression of early inflammatory genes
		Chemokines and cytokines	vMIP-I, vMIP-II	T_H-2 chemoattractant, chemokine receptor antagonist
	HSV	Type 1 IFNs	$\gamma_1$34.5	Reverses IFN-induced translation block
		RNase L	2′–5′ (A) analog	RNA analog, inhibits RNase L
	EBV	Type 1 IFNs	EBNA-2	Downregulates IFN-stimulated transcription
		PKR	EBER-1	Blocks PKR activity
		Chemokine receptors	BARF-1	Secreted, sequesters CSF-1
		IL-10	BCRF-1	IL-10 homolog, antagonizes T_H-1 responses
Adenoviridae	Adenovirus	Type 1 IFNs	E1A	Blocks IFN-induced JAK/STAT pathway
		PKR	VA 1 RNA	Blocks PKR activity
		TNF-α	E3 proteins	Various mechanisms
Hepadnaviridae	HepB	Type 1 IFNs	Terminal protein	Blocks IFN signaling
Flaviviridae	HepC	PKR	E2	Inhibits PKR activation in response to type 1 IFN
Retroviridae	HIV	PKR	TAR RNA	Recruits cellular PKR inhibitor TRBP
Poxviridae	See Table 8.10			

Source: Adapted from Table 3 of Tortorella *et al.* (2000).

Abbreviations: HCMV, human cytomegalovirus; TNF, tumor necrosis factor; HHV-8, human herpesvirus eight (Kaposi's); HSV, herpes simplex virus; IFN, interferon; PKR, dsRNA-dependent protein kinase; CSF-1, colony stimulating factor; HIV, human immunodeficiency virus; TRBP, *TAR* RNA-binding protein.

A listing of defense molecules encoded by poxviruses is given in Table 8.10.

Evasion of the Antiviral State

Many viruses encode products that specifically interfere with the activation of the PKR pathway that leads to shutdown of protein synthesis. This suggests that this pathway must be important for the control of viruses. Several different mechanisms are used: synthesis of competitive RNA that binds PKR but does not activate it; synthesis of products that sequester dsRNA; production of a protein that binds PKR and prevents it from phosphorylating eIF-2α; and activation of cellular inhibitors or degradation of PKR.

Several viruses, including adenoviruses, Epstein–Barr virus, and HIV, encode small RNA products that bind to PKR in lieu of dsRNA, but which do not activate PKR. The adenoviral RNA is called VA1 RNA and was the first of these products to be described. It has been hypothesized that activation of PKR requires two PKR molecules to bind the same dsRNA, so that they are in proximity and can phosphorylate one another in *trans*. In this model, the viral RNAs can be bound by only one PKR molecule. Regardless of mecha-

nism, the net result is that the viral RNAs act as inhibitors of the cofactor dsRNA and prevent activation of PKR.

Other viruses, including vaccinia virus and reoviruses, encode protein products that bind dsRNA, thus making it unavailable as a cofactor for activation of PKR. Furthermore, vaccinia virus appears to limit the production of dsRNA by using an arrangement of genes and stop transcription signals that reduces transcription from both strands, at least early after infection. Sequestering of dsRNA also inhibits the induction of IFN, since dsRNA is a primary inducer of interferon. Vaccinia virus interferes with the PKR pathway in yet another way. The virus encodes a protein that binds activated PKR and prevents phosphorylation of eIF-2α.

Influenza virus, poliovirus, and HIV also interfere with the PKR pathway. They activate cellular inhibitors of the enzyme or make products that leads to degradation of PKR.

The RNase L pathway is also inhibited by viral products that bind dsRNA or by decoy RNAs. However, herpes simplex virus targets RNase L activity directly. It makes an analog of 2′-5′-oligo(A) that binds to RNase L but which does not activate the enzyme. By preventing the binding of authentic 2′-5′-oligo(A), the analog prevents the activation of RNase L.

TABLE 8.10   Pox Defense Molecules

System	Target	Virus	Gene	Homolog	Properties
Complement	C4B and C3B	Vaccinia	C3L	C4B binding protein	4SCRs, secreted, binds and inhibits C4B and C3B, virulence factor
	?	Vaccinia	B5R	Complement control proteins	4 SCRs, EEV class I membrane glycoprotein, for virus egress
	?	Variola	B6R		
Interferon	Type 1 IFN	Vaccinia	B18R	IFN receptor	Binds to and inhibits IFN-α
	PKR	Vaccinia	K3L	eIF-2α	Binds PKR, inhibits phosphorylation of eIF-2α, IFN resistance
		Variola	C3L		
		Swinepox	K3L		
	dsRNA	Vaccinia	E3L	PKR	Binds dsRNA, nuclear localization inhibits activation of PKR, IFN resistance
		Variola	E3L		
	IFN-γ	Myxoma	T7	IFN-γ receptor	Secreted, binds and inhibits IFN-γ
		Vaccinia	B8R		
		Variola	B8R		
		Swinepox	C61		
IL-1	ICE	Cowpox	crmA	SERPIN	Prevents proteolytic activation of IL-1β, inhibits inflammatory response, inhibits apoptosis
		Vaccinia	B14R		
		Variola	B12R		
	IL-1β	Vaccinia	B15R	IL-1 receptor	Secreted glycoprotein, binds and inhibits IL-1β
		Cowpox			
IL-8	IL-8	Swinepox	ecrf3	IL-8 receptor	Binds IL-8
		Swinepox	K2R		
TNF	TNF-α, TNF-β	Myxoma	T2	TNF receptor	Secreted, binds and inhibits TNF-α, TNF-β
		Vaccinia	G2R		
		Variola	Truncated		
		Cowpox	crmB		
CC-chemokines		Myxoma	p35	Chemokine receptor	Secreted, binds to CC-chemokines
		Vaccinia			
		Variola			
		Cowpox			

Abbreviations: SCR, 60 amino acid sequence called short consensus sequence; EEV, extracellular enveloped virions; PKR, dsRNA-dependent protein kinase; IFN, interferon; SERPIN, serine protease inhibitor superfamily.
Source: Adapted from Fields *et al.* (1996, p. 2657), Evans (1996), and a review by Tortorella *et al.* (2000).

Interference with Signal Transduction Pathways

Adenoviruses, Epstein–Barr virus, and hepatitis B virus make products that interfere with signal transduction by the IFN receptor on binding of the ligand. The mechanisms by which signal transduction is blocked are not understood.

Production of Cytokine-Binding Proteins

A number of viruses make proteins that bind to cytokines. Many of these cytokine-binding proteins are homologs of cellular cytokine receptors and have certainly been acquired from the host. Most of these viral proteins are secreted from the cell as soluble proteins that neutralize the activity of cytokines by binding to them in a nonproductive fashion. Others function at the surface of the infected cells.

Various poxviruses encode receptors for IFN-γ, IFN-α/β, IL-1, IL-6, IL-8, and TNF (Table 8.10). Most poxviruses secrete a receptor for IFN-γ that is distantly related to the human receptor. This receptor neutralizes the activity of IFN-γ and presumably functions to prevent IFN-γ-induced events. The potential efficacy of interference with IFN-γ is shown by the receptor secreted by rabbit myxoma virus. This virus causes an infection of European rabbits that has a 99% fatality rate (Chapter 6). Mutants that lack the IFN-γ receptor cause nonfatal illness in these rabbits.

Many poxviruses also produce a receptor for IFN-α/β and for TNF-α. TNF-α is a cytokine that has multiple roles in the control of virus-infected cells. It plays a role in apoptosis, but is also important in noncytolytic clearing of virus infection. The TNF-α receptor encoded by rabbit myxoma virus is multifunctional. It is secreted in part and binds TNF-α to neutralize it. However, it is partially retained within the cell where it interferes with signal transduction that induces apoptosis. The importance of TNF-α in control of viral infections is shown by the fact that some poxviruses produce two different TNF receptors to neutralize its activity, and by the fact that adenoviruses produce four different proteins, found in different parts of the infected cell, that antagonize the effects of TNF.

IL-1, IL-6, and IL-8 are also neutralized gene products encoded by various poxviruses. Deletion of any of these poxvirus genes usually results in an attenuation of virus growth in experimental animals.

Poxviruses make a number of proteins that bind various chemokines. Neutralization of the activity of chemokines results in damping the inflammatory response to viral infection. Some of these chemokine-binding proteins are soluble proteins and some are expressed on the surface of the infected cell. Rabbit myxoma virus, for example, produces two proteins that bind chemokines. One protein binds with high affinity to a subset of chemokines called CC-chemokines. The second protein is the IFN-γ receptor described above. The rabbit myxoma virus IFN-γ receptor, but not that of other poxviruses, binds a number of chemokines through their heparin-binding domains. This binding is of low affinity. Thus, this protein is also multifunctional, and binding of IFN-γ and chemokines is independent of one another.

Many beta- and gammaherpesviruses also produce cytokine or chemokine receptors. Human CMV, for example, produces four chemokine receptors, whereas HHV-8 produces one. Epstein–Barr virus encodes a receptor for the cytokine macrophage colony-stimulating factor (CSF-1).

Secretion of Virokines

Many beta- and gammaherpesviruses secrete cytokine or chemokine analogs, called virokines. As examples, HHV-8 produces three chemokines and one cytokine (IL-6), and Epstein–Barr virus encodes a homolog of IL-10. The chemokines may serve to attract target cells, since these viruses infect B cells. IL-6 and IL-10 are necessary for the growth of B cells and presumably serve to expand the target cell population. Furthermore, as described earlier, these cytokines skew the immune response toward a B-cell

response, helped by T_H-2 cells, and away from a CTL response, helped by T_H-1 cells. Thus, while expanding the number of host cells available for infection by the virus, these cytokines also depress the number of CTLs that control the infected cell population.

In most cases, deletion of viral genes that interfere with cytokine or chemokine activity attenuates the virus in experimental animals. However, in some cases deletion of such genes leads to an increase in the virulence of the virus. The increased virulence appears to result from a more severe inflammatory response to virus infection.

INTERACTIONS OF VIRUSES WITH THEIR HOSTS

The interaction of viruses with their hosts is intimate and the product of a long period of evolution during which viruses coevolved with their hosts. Humans cannot survive without a functioning immune system to protect them from viruses. However, this is the result of the long evolutionary history during which hosts and viruses adapted to one another, because viruses in turn cannot survive without their hosts. The example of rabbit myxoma virus demonstrates that the virulence of a virus diminishes if it kills too large a proportion of its hosts too rapidly. We can even speculate that the exceptional virulence of the influenza virus responsible for the 1918 pandemic might have been made possible because of active warfare ongoing at the time. Very ill and dying soldiers continue to be moved about and the virus could continue to spread, perhaps could even spread more readily, if it incapacitated its hosts. On the other hand, the many examples of ways in which viruses modify the immune and cytokine defenses of the host in order to replicate demonstrate that viruses are capable of evolving more virulent forms if it is to their advantage. The end result is an interplay in which viruses and their hosts exist in an uneasy equilibrium punctuated by the emergence of new viruses or the spread of new epidemics accompanied by changes in the immune system that protect against these viruses.

The virulence of a virus for its host depends in part on the epidemiology of the virus, how it gets from one host to another. The herpesviruses set up a lifelong infection in which they are effectively transferred once per generation. It is in the virus's interest not to incapacitate the host so that the host can pass it on perhaps 20 years later, and herpesviruses cause minor illness or no illness in most humans. On the other hand, arboviruses must cause a viremia (virus circulating in the blood) high enough to infect an insect taking a blood meal. Because many of these viruses are RNA viruses that do not encode functions to ablate the immune response, rapid and vigorous replication is required to establish the viremia before immunity is established, and this is often harmful to the host because many cells are killed in the process. To take another example, respiratory viruses that are transmitted as aerosols or in respiratory secretions must produce enough virus in the respiratory tract so that respiratory droplets expelled by coughing or sneezing will contain sufficient virus to infect a person nearby. These viruses are transmitted in epidemics that can spread rapidly and that require close contact between individuals, and one infected individual can infect dozens or even hundreds of others in a very short time. Thus these viruses need be transmissible only over a short period. Sexually transmitted viruses have different hurdles to overcome. Because the potential for sexual transmission is usually infrequent and one person interacts with a limited number of others, these viruses need to establish infections that last for long periods of time and that do not incapacitate the infected individual, at least not early in the infection.

The close interplay between viruses and their hosts means that the study of viruses continues to tell us much about the hosts. We now know much more about the adaptive immune system, the cytokine system, and about apoptosis because of recent studies that started with viruses. Continuing studies on viruses have told us much about the function of regulatory genes and cancers. We are confident that the study of viruses will continue to teach us much about human biology.

FURTHER READING

General Treatment of the Vertebrate Immune Response

Janeway, C. A., P. Travers, Walport, M., *et al.* (1999). "Immunobiology: The Immune System in Health and Disease." Elsevier Science, London.

Kuby, J. (1997). "Immunology." W. H. Freeman, New York.

Interferons and Other Cytokines

Guidotti, L. G., and Chisari, F. V. (1999). Cytokine-induced viral purging—role in viral pathogenesis. *Curr. Opin. Microbiol.* **2**; 388–391.

Landolfo, S., Gribaudo, G., Angeretti, A., *et al.* (1995). Mechanisms of viral inhibition by interferons. *Pharmacol. Ther.* **65**; 415–442.

Tortorella, D., Gewurz, B. E., Furman, M. H., Schust, D. J., and Ploegh, H. L. (2000). Viral subversion of the immune system. *Ann. Rev. Immun.* **18**; 861–926.

How Viruses Interfere with the Immune Response

Blair, G. E., and Hall, K. T., (1998). Human adenoviruses: Evading detection by cytotoxic T lymphocytes. *Semin. Virol.* **8**; 387–398.

Davis-Poynter, N. J., and Farrell, H. E., (1998). Human and murine cytomegalovirus evasion of cytotoxic T lymphocyte and natural killer cell-mediated immune responses. *Semin. Virol.* **8**; 369–376.

Gale, M. J., and Katze, M. G., (1998). Molecular mechanisms of interferon resistance mediated by viral-directed inhibition of PKR, the interferon-induced protein kinase. *Pharmacol. Ther.* **78**; 29–46.

Kalvakolanu, D. V. (1999). Virus interception of cytokine-regulated pathways. *Trends Microbiol.* **7**; 166–171.

Smith, G. L., Symons, J. A., and Alcami, A., (1998). Poxviruses: Interfering with Interferon. *Semin. Virol.* **8**; 409–418.

Tortorella, D., Gewurz, B. E., Furman, M. H., *et al.* (2000). Viral subversion of the immune system. *Annu. Rev. Immunol.* **18**; 861–926.

Apoptosis

Ashkenazi, A., and Dixit, V. M., (1998). Death receptors: Signaling and modulation. *Science* **281**; 1305–1308.

O'Brien, V. (1998). Viruses and apoptosis. *J. Gen. Virol.* **79**; 1833–1845.

Raff, M. (1998). Cell suicide for beginners. *Nature, (London)* **396**; 119–122.

How Viruses Interfere with Apoptosis

Chinnadurai, G. (1998). Control of apoptosis by human adenovirus genes. *Semin. Virol.* **8**; 399–408.

Hill, A. B., and Masucci, M. G. (1998). Avoiding immunity and appoptosis: Manipulation of the host environment by herpes simplex virus and Epstein-Barr virus. *Semin. Virol.* **8**; 361–368.

McFadden, G., and Barry, M. (1998). How poxviruses oppose apoptosis. *Semin. Virol.* **8**; 429–442.

Miller, L. K., and White, E., eds. (1998). Apoptosis in virus infection *Semin. Virol.* 8(6)

O'Brien, V. (1998). Viruses and apoptosis. *J. Gen. Virol.* **79**; 1833–1845.

CHAPTER
9

Gene Therapy

INTRODUCTION

Molecular genetic studies during the last decades have led to an enormous increase in our understanding of the molecular biology of the replication of viruses. The complete nucleotide sequences of many virus genomes have been determined. Information on the origins required for the replication of these genomes, the promoters used to express the information within them, and the packaging signals required for packaging progeny genomes into virions have been established for many. The mechanisms by which viral mRNAs are preferentially translated have been explored. Together with methods for cloning and manipulating viral genomes, this information has made possible the use of viruses as vectors to express foreign genes. In principle, any virus can be used as a vector, and systems that use a very wide spectrum of virus vectors have been described. DNA viruses were first developed as vectors, since it is possible to manipulate the entire genome in the case of smaller viruses, or to use homologous recombination to insert a gene of interest in the case of larger viruses. When complete cDNA clones of RNA viruses were obtained, it became straightforward to rescue plus-strand viruses from clones because the viral RNA itself is infectious. The use of minus-strand RNA viruses as vectors has only recently become possible, because the virion RNA itself is not infectious and rescue of virus from cloned DNA requires coexpression of the appropriate proteins.

A sampling of expression systems and their uses is given here to illustrate the approaches that are being followed. Every virus system has advantages and disadvantages as a vector, depending on its intended use. One of the more exciting uses has been the development of viruses as vectors for gene therapy, that is, to correct genetic defects in humans. Although results have been disappointingly slow

in coming, such systems offer great promise. This use represents an example of taking these infectious agents that have been the source of much human misery and developing them for the betterment of mankind.

VIRUS VECTOR SYSTEMS

A representative sampling of viruses that are being developed as vectors is described below in order to illustrate some of the strengths and weaknesses of the different systems. The viruses used in most clinical trials to date have been the poxviruses, the adenoviruses, and the retroviruses, and these are described here. Several other virus systems that may be used in the future for treatment of humans are also described.

Vaccinia Virus

Vaccinia virus is a poxvirus with a large dsDNA genome of 200 kb (Chapter 6). This genome is too big to handle in one piece in a convenient fashion, and homologous recombination has been used to insert foreign genes into it. The large size of the viral genome, however, does mean that very large pieces of foreign DNA can be inserted, while leaving the virus competent for independent replication and assembly. Another advantage of the virus is that it has been used to vaccinate hundreds of millions of humans against smallpox. Thus, there is much experience with the effects of the virus in humans. Although the vaccine virus did cause serious side effects in a small fraction of vaccinees, highly attenuated strains of vaccinia have been developed for use in gene therapy by deleting specific genes associated with virulence. A new approach to the use of poxvirus vectors has been the development of nonhuman poxviruses, such as canarypox virus, as vectors. Canarypox virus infection of mammals is abortive and essentially asymptomatic, but foreign genes incorporated into the canarypox virus genome are expressed in amounts that are sufficient to obtain an immunological response.

A variety of approaches have been used to obtain recombinant vaccinia viruses that express a gene of interest, but only the first such method to be used, and one that remains in wide use, is described here. This method is illustrated in Fig. 9.1. The thymidine kinase (TK) gene of vaccinia virus is nonessential for growth of the virus in tissue culture. Furthermore, deletion of the TK gene results in attenuation

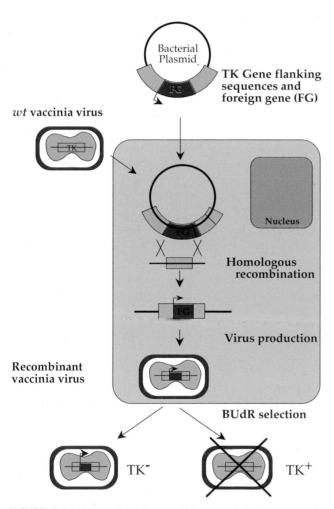

FIGURE 9.1 Construction of a recombinant vaccinia virus expression vector. The foreign gene (red) is inserted into a bacterial plasmid adjacent to a vaccinia promoter (black arrow), flanked with sequences from the vaccinia thymidine kinase gene (turquoise). Plasmid DNA is transferred into cells infected with wild-type (TK+) vaccinia virus. Recombinant progeny from homologous recombination are all TK- due to interruption of the TK gene, and can be selected by growth in bromodeoxyuridine (BUdR), since incorporation of BUdR into TK+ vaccinia is lethal. These TK- vaccinia will infect normally and express the foreign gene under control of the vaccinia promoter. [Adapted from Strauss and Strauss (1997, Fig. 2.25, p. 115).]

of the virus in humans, which is a desirable trait. Finally, the TK gene can be either positively or negatively selected by using different media for propagation of the virus. The starting point is a plasmid clone that contains a copy of the TK gene that has a large internal deletion. In the region of the deletion a vaccinia virus promoter is inserted upstream of a polylinker. The gene of interest is inserted into the polylinker using standard cloning technology. Thus, we have the foreign gene downstream of a vaccinia promoter, and the entire insert is flanked by sequences from the vaccinia TK gene. The plasmid containing the cloned TK gene with its foreign gene insert is transfected into cells that have been infected by wild-type vaccinia virus. Homologous recombination between the TK gene in the virus and the TK flanking sequences in the plasmid occurs with a sufficiently high frequency that a reasonable fraction of the progeny have the gene of interest incorporated. These viruses have an inactive TK gene (they are TK-), because the TK gene has been replaced by the deleted version containing the inserted foreign gene. The next step, then, is to select for viruses that are TK- by growing virus in the presence of bromodeoxyuridine (BUdR). An active TK enzyme will phosphorylate BUdR to the monophosphate form, which can be further phosphorylated by cellular enzymes to the triphosphate and incorporated into the viral nucleic acid during replication. Incorporation of BUDR is lethal under the appropriate conditions, and thus viruses that survive this treatment are those in which the TK gene has been inactivated.

It is usually necessary to select among the surviving progeny for those that possess the gene of interest, because inactivation of the TK gene can occur spontaneously through deletion or mutation. Selection can be accomplished by a plaque lift hybridization assay in which virus in plaques is transferred to filter paper. Virus plaques on the filter paper are probed with radiolabeled hybridization probes specific for the inserted gene. Virus in plaques that hybridize to the probe is recovered and further passaged. In this way can be isolated a pure virus stock that will express the gene of interest.

Adenoviruses

Adenovirus infections of humans are common and normally cause only mild symptoms. Deletion of virulence genes from adenovirus vectors further attenuates these viruses. In addition, adenovirus vaccines have been used by the military for some years and, therefore, some experience has been gained in the experimental infection of humans by adenoviruses, although gene therapy trials use a different mode of delivery of adenovirus vectors. Because of their apparent safety, adenoviruses have been developed for use as vectors in gene therapy trials or for vaccine purposes.

Two approaches have been used. In one, infectious aden-oviruses have been produced that express a gene of interest. In the second approach, suicide vectors are produced that can infect a cell and express the gene of interest, but which are defective and cannot produce progeny virus. Suicide vectors cannot spread to neighboring cells, and the infection is therefore limited in scope and in duration.

The genome of adenoviruses is dsDNA of 36 kbp (Chapter 6). Thus, the genome is smaller than that of poxviruses and can accommodate correspondingly smaller inserts. However, inserts large enough for most applications can be accommodated. The genome is small enough that the virus can be reconstituted from DNA clones. Such an approach is inconvenient, however, and homologous recombination is often used to insert the gene of interest into the virus genome.

The foreign gene is inserted into the region occupied by either the adenovirus E1 or E3 genes, one or both of which are deleted in the vector construct. Virus lacking E1 cannot replicate, and such viruses form suicide vectors. For gene therapy, suicide vectors are normally used so as to prevent the spread of the infection. To prepare the stock of virus lacking E1, the virus must be grown in a cell line that expresses E1. An overview of this process is shown in Fig. 9.2. The complementing cell line, which produces E1 constitutively, supplies the E1 needed for replication of the defective adenovirus. The cells are transfected with the defective adenovirus DNA and a full yield of progeny virions results. The progeny virus is defective and cannot replicate in normal cells, but it can be amplified by infection of the complementing cell line. On introduction of the virus into a human, the virus will infect cells and express the foreign gene, but the infection is abortive and no progeny virus is formed. The stock of defective virus must be tested to ensure that no replication-competent virus is present, since such virus can arise by recombination between the vector and the E1 gene in the complementing cell line.

Adenoviruses with only E3 deleted are often used to express proteins for vaccine purposes. These E3-deleted viruses possess intact E1 and will replicate in cultured cells and in humans, but are attenuated. Because the virus replicates, expression of the immunizing antigen persists for a long time and a good immune response usually results.

The procedure for insertion of the gene of interest by homologous recombination resembles that used for the poxviruses. The gene is inserted into a plasmid containing flanking sequences from the E1 or E3 region, and transfected into cells infected with adenovirus. Recombinant viruses containing the gene of interest are selected and stocks prepared. It is also possible to transfect cells with the E1 or E3 expression cassette together with DNA clones encoding the rest of the adenovirus genome, in which case homologous recombination results

in the production of virus. In the case of insertions into E1, cells that express E1 must be used to produce the recombinant virus.

Retroviruses

Retrovirus-based expression systems offer great promise because the retroviral genome integrates into the host-cell chromosome during infection and, in the case of the simple retroviruses at least, remains there as a Mendelian gene that is passed on to progeny cells on cell division. Thus, there is the potential for permanent expression of the inserted gene of interest. The essential components of a retrovirus vector are the long terminal repeats (LTRs), the packaging sequences known as ψ, the primer binding site, and the sequences required for jumping by the reverse transcriptase during reverse transcription to form the dsDNA copy of the genome (Chapter 5).

The process of creating and packaging a retrovirus-based expression cassette is illustrated in Fig. 9.3. A packaging cell line is created that expresses the retroviral *gag*, *pol*, and *env* genes, but whose mRNAs do not contain the packaging signal and so cannot be packaged. The vector DNA/RNA is created by modifying a DNA clone of a retrovirus to contain the gene of interest in place of the *gag–pol–env* genes. In the process, all of the essential *cis*-acting signals required for packaging, reverse transcription, and integration are retained. The foreign gene can be under the control of the LTRs, or it can be under the control of another promoter positioned in the insert upstream of it. The resulting DNA clone is transfected into the packaging cell line, and a producer cell line isolated that expresses the vector DNA as well as the helper DNA. Vector RNA transcribed from the vector DNA is packaged into retroviral particles, using the proteins expressed from the helper DNA. These particles are infectious and can be used to infect other cells or to transfer genes into a human. On infection of cells by the packaged vector, the vector RNA is reverse transcribed into DNA that integrates into the host-cell chromosome, where it can be expressed under the control of the promoters that it contains. The limitation on the size of the insert is about 10 kb, the upper limit of RNA size that can be packaged.

Although murine leukemia viruses are not known to cause disease in man, it has been found that these viruses will cause tumors in immunosuppressed subhuman primates. Thus, it is thought to be essential that there be no replication-competent virus in stocks used to treat humans. Replication-competent virus can arise during packaging of the vector by recombination between the vector and the retroviral sequences used to produce Gag–Pol–Env. At the current time, preparations of packaged vectors are screened to ensure that replication-competent viruses are not present.

A. Wild type Adenovirus Genome

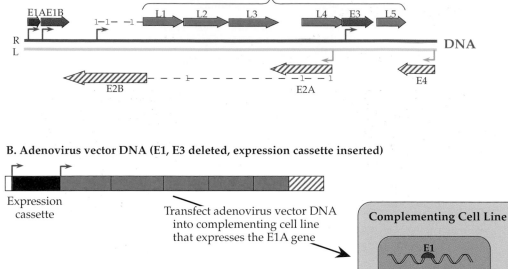

B. Adenovirus vector DNA (E1, E3 deleted, expression cassette inserted)

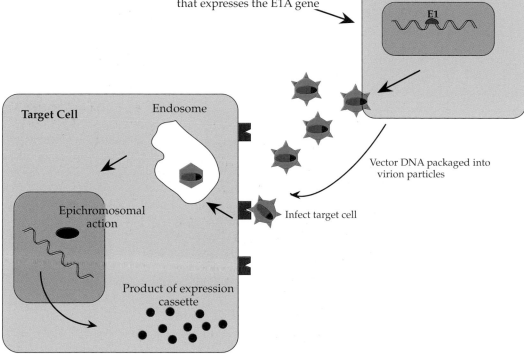

FIGURE 9.2 Generation of a nonreplicating adenovirus expression vector. From the wild-type adenovirus genome, the E1 and E3 genes are removed. The E1 genes are replaced with an expression cassette. This adenovirus DNA is transfected into a complementing cell line that produces E1 protein. The transfection produces particles that are able to infect cells, but which are E1⁻ and non-replicating. The DNA genome is delivered to the nucleus where it functions as an epichromosome and directs expression of the inserted foreign gene. [From Crystal (1995).]

Efforts are being made to reduce the incidence of recombination during packaging in order to simplify the procedure. One approach is to develop vectors that have very little sequence in common with the helper sequences, in order to reduce the incidence of homologous recombination. A second approach is to separate the Gag–Pol sequences from the Env sequences in the helper cell. In this case, recombination between three separate DNA fragments in the pro- ducer cell (that encoding Gag–Pol, that encoding Env, and sequences in the vector) are required in order to give rise to replication-competent retrovirus.

In gene therapy trials that use retroviruses, it has been found that the expression of the foreign gene in humans is often downregulated after a period of months. Attempts are being made to identify promoters that will not be downregulated. Different promoters might be required for different

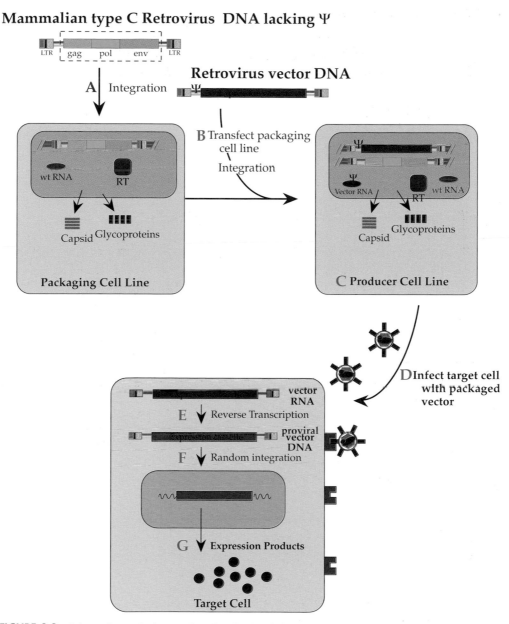

FIGURE 9.3 Scheme for producing a packaged replication-defective retrovirus expression vector. (A) A "packaging" cell line is generated by introduction of DNA encoding *gag, pol,* and *env* genes into the chromatin of a fibroblast cell line. This DNA lacks the packaging signal ψ and RNA transcribed from it is not packaged. (B) The packaging cell line is transfected with a second retroviral DNA, in which the foreign gene (expression cassette) replaces *gag, pol,* and *env,* but which has an intact ψ packaging signal and intact LTRs, to form a "producer" cell line. This producer cell line (C) releases vector particles containing the expression cassette genome packaged with the proteins from the helper genome. (D) These particles enter target cells via specific cell surface receptors, (E) are reverse transcribed, (F) randomly integrate, and (G) produce expression products. [Adapted from Dunbar (1996).]

uses, and promoters that target transcription to particular cell types would be useful.

A major problem with retroviral vectors is that simple retroviruses will only infect dividing cells. Although they enter cells and are reverse transcribed to DNA, the DNA copy of the genome can enter the nucleus only during cell division. In many gene therapy treatments, it is desirable to infect stem cells in order to maintain expression of the therapeutic gene indefinitely. Because stem cells divide relatively infrequently, it is difficult to infect a high proportion of them by vectors used to date. Attempts are being made to identify methods to stimulate stem cells to divide during *ex vivo* treatment, so that a larger fraction of them can be infected. A second approach is to develop lentivirus vectors. Lentiviruses, which include HIV, can infect non-replicating cells and could potentially infect nondividing stem cells during *ex vivo* treatment. Lentivirus vectors could also be useful for therapy involving other nondividing cells, such as neurons.

It would be of considerable utility to be able to target retroviruses to specific cells. One possible approach to this is to replace all or part of the external domains of the retroviral surface glycoprotein with a monoclonal antibody that is directed against an antigen expressed only on the target cells. In principle, this approach is feasible, but whether it can be developed into something practical is as yet an open question. If specific cells could be infected, it would allow protocols in which the therapeutic gene would be expressed only in cells where it would be most useful. It would also allow the specific killing of cells such as tumor cells or HIV-infected cells. For example, the retrovirus could express a gene that rendered the cell sensitive to toxic drugs such as BUdR. A retrovirus vector that expressed such a gene could also be useful for conventional gene therapy, because it would allow the infected cells to be killed if the infection process threatened to get out of hand.

Alphaviruses

The genomes of plus-strand RNA viruses are self-replicating molecules that replicate in the cytoplasm, and they can express very high levels of protein. These properties make them potentially valuable as expression vectors.

The alphaviruses possess a genome of single-strand RNA of about 12 kb (Chapter 3). Their genomes can be easily manipulated as cDNA clones, and infectious RNA can be transcribed from these clones by RNA polymerases, either *in vivo* or *in vitro*. RNA transcribed *in vitro* can be transfected into cells and give rise to a full yield of virus, whereas RNA transcribed *in vivo* will begin to replicate and produce virus. The structural proteins are made from a subgenomic mRNA, making it easy to insert a foreign gene under the control of the subgenomic promoter. Two approaches that have been used are illustrated in Fig. 9.4. In one approach, a second subgenomic promoter is inserted into the genome downstream of the structural proteins, or between the structural proteins and the nonstructural proteins (Fig. 9.4C). Two subgenomic mRNAs are transcribed, one for the structural proteins and the second for the gene of interest. The size of the insert must be relatively small, on the order of 2000 nucleotides or less, because longer RNAs are not packaged efficiently. However, this system has the advantage that the resulting double subgenomic virus is an infectious virus that can be propagated and maintained without helpers.

A second approach is to delete the viral structural proteins and replace them with the gene of interest. In this case, there is room for an insert of about 5 kb that will still allow the resulting replicon to be packaged. The replicon is capable of independent replication, and transcription of a subgenomic messenger results in expression of the gene of interest. The replicon constitutes a suicide vector. It cannot be packaged unless the cells are coinfected with a helper to supply the structural proteins, or unless a packaging cell line that expresses the viral structural proteins is used.

Alphavirus replicons can be extremely efficient in expressing a foreign gene. In some cases as much as 25% of the protein of a cell can be converted to the foreign protein expressed by the replicon over a period of about 72 hr. Wild-type replicons are cytolytic in vertebrate cells, inducing apoptosis, and the infection dies out. However, replicons have been produced with mutations in the replicase proteins that are not cytolytic and will produce the protein of interest indefinitely. Thus, a wide sprectrum of choices is available, and the system chosen can be adapted to the needs of a particular experiment or treatment.

Viral expression systems would be more useful if they could be directed to specific cell types. An approach that uses monoclonal antibodies to direct Sindbis virus to specific cells has been described. Protein A, produced by *Staphylococcus aureas*, binds with high affinity to IgG. It is an important component of the virulence of the bacterium because it interferes with the host immune system. The IgG binding domain of protein A has been inserted into one of the viral glycoproteins. Virions containing this domain are unable to infect cells using the normal receptor. However, the virus will bind IgG monoclonal antibodies. If an antibody directed against a cell surface component is bound, the virus will infect cells expressing this protein at the cell surface. Thus, this system has the potential to direct the virus to a specific cell type. One of the advantages of this approach is that the virus, once made, can be used with many different antibodies and thus directed against a variety of cell types. This approach is potentially applicable to any enveloped virus, and perhaps to nonenveloped viruses as well.

A modification of the alphavirus system is to use a DNA construct containing the replicon downstream of a promoter for a cellular RNA polymerase, rather than using packaged RNA replicons. On transfection of a cell with the DNA, the replicon RNA is launched when it is transcribed from the DNA by cellular enzymes. Once produced, the RNA replicates independently and produces the subgenomic mRNA that is translated into the gene of interest. As described in Chapter 8, naked DNA can be used to transfect muscle cells and perhaps other cells.

Polioviruses

Plus-strand viruses that do not produce subgenomic mRNAs, such as the picornaviruses and flaviviruses, present different problems for development as vectors. The

A. Alphavirus genome organization

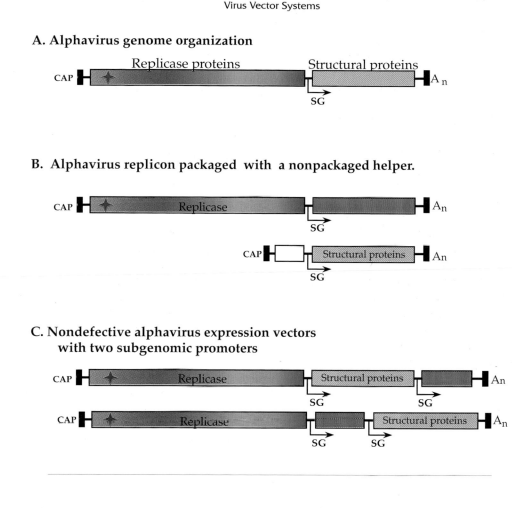

B. Alphavirus replicon packaged with a nonpackaged helper.

C. Nondefective alphavirus expression vectors with two subgenomic promoters

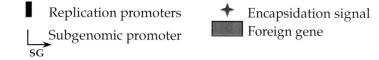

FIGURE 9.4 Alphavirus expression vectors. (A) The genome organization of a typical alphavirus with the location of the promoters for replication and production of subgenomic RNA as well as the RNA packaging signal indicated. (B) A simple alphavirus replicon. The structural proteins of the virus have been replaced with the foreign gene to be expressed. If packaging of the replicon is required, the structural proteins of the virus are supplied on a DI RNA lacking a packaging signal. (C) Packaged expression vectors with two subgenomic promoters. These constructs are unstable if the foreign gene is much larger than 2 kb. [Adapted from Strauss and Strauss (1994, Fig. 23).]

translated product from the gene of interest must either be incorporated into the polyprotein produced by the virus and provisions made for its excision, or tricks must be used to express the gene of interest independently. Two approaches with poliovirus will be described as examples of how such viruses might be used as vectors.

Poliovirus replicons have been constructed by deleting the region encoding the structural proteins and replacing this sequence with that for a foreign gene. The foreign gene must be in phase with the remainder of the poliovirus polyprotein, and the cleavage site recognized by the viral 2A protease is used to excise the foreign protein from the polyprotein. Because the poliovirus replicon lacks a full complement of the structural genes (it is a suicide vector), packaging to produce particles requires infection of a cell that expresses the polioviral structural proteins by some mechanism. A construct that uses this approach to express the cytokine tumor necrosis factor alpha (TNF-α) is illustrated in Fig. 9.5. A poliovirus "infectious clone" in which a DNA copy of the viral genome is positioned downstream of a promoter for T7 RNA polymerase is modified by replacing the genes for VP3 and VP1 with the gene for TNF-α. Recognition sites for the poliovirus 2A protease are positioned on both sides of the TNF-α gene. The TNF-α

A) Poliovirus infectious clone

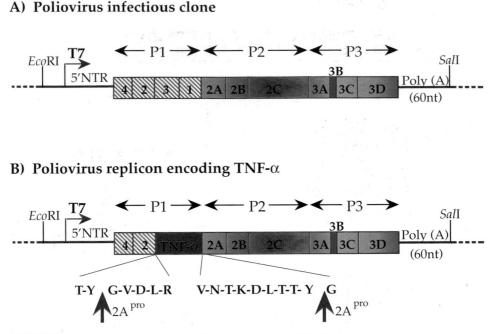

B) Poliovirus replicon encoding TNF-α

FIGURE 9.5 Generation of poliovirus replicons for expression of foreign genes in motor neurons. Based on an earlier construct to express interleukin-2 via a poliovirus replicon, the gene for wild-type murine (TNF-α) was positioned between the VP0 and 2A proteins of poliovirus, replacing VP3 and VP1. It was flanked on either side by sites for cleavage by the poliovirus 2A protease. These constructs were injected into transgenic mice expressing the poliovirus receptor, and expression of murine TNF-α was monitored. [Adapted from Bledsoe *et al.* (2000).]

protein is produced as part of the poliovirus polyprotein, and cleaved from the polyprotein by the 2A protease. Packaged replicons were used to infect transgenic mice that expressed the polio receptor (Chapter 1). One of the interests of this system is that poliovirus exhibits an extraordinary tropism for motor neurons in the central nervous system (CNS) (Chapter 3). The packaged replicons, on introduction into the CNS, infected only motor neurons, and therefore the foreign gene was expressed only in motor neurons. Such replicons may be useful to treat CNS diseases in which motor neurons are affected.

A second approach to the use of poliovirus replicons is to use a second internal ribosome entry site (IRES) (Chapter 1) to initiate the synthesis of the nonstructural proteins. If the foreign gene replaces the structural genes, it will be translated from the 5′ end of the genome. If the poliovirus nonstructural genes are placed downstream of a second IRES, internal initiation at this IRES results in production of a polyprotein for the nonstructural proteins. This approach is similar to the approach shown in Fig. 3.3, where the structural proteins are replaced by a gene of interest.

Rhabdoviruses

In minus-strand RNA viruses, the genomic RNA is not itself infectious. Ribonucleoprotein containing the N, P, and L genes is required for replication of the viral RNA, and thus for infectivity, and only recently have methods been devised to recover virus from cDNA clones. A schematic diagram of how virus can be recovered from DNA clones of the rhabdovirus vesicular stomatitis virus (VSV) (Chapter 4) is shown in Fig. 9.6. A cell is transfected with a set of cDNA clones that together express N, P, and L as well as the genomic or antigenomic RNA. The antigenomic RNA usually works better, probably because it does not hybridize to the mRNAs being produced from the plasmids. Encapsidation of the antigenomic RNA by N, P and L to form nucleocapsids allows it to replicate and produce genomic RNA that is also encapsidated. Synthesis of mRNAs from the genomic RNA, together with continued replication, results in a complete virus replication cycle and production of infectious progeny virus that have as their genome the RNA supplied as a cDNA clone. The yield of infectious virus is small, but sufficient to isolate individual plaques and thus obtain viruses from the cDNA clones.

The ability to rescue virus from a cDNA clone makes it possible to manipulate the viral genome. Since the rhabdovirus genome is transcribed into multiple mRNAs, one for each gene, and the transcription signals recognized by the enzyme are well understood, it is relatively simple to add or delete genes. A modified VSV that was produced by using DNA clones is illustrated in Fig. 9.7. In this VSV, the surface

A. Rhabdovirus genome organization

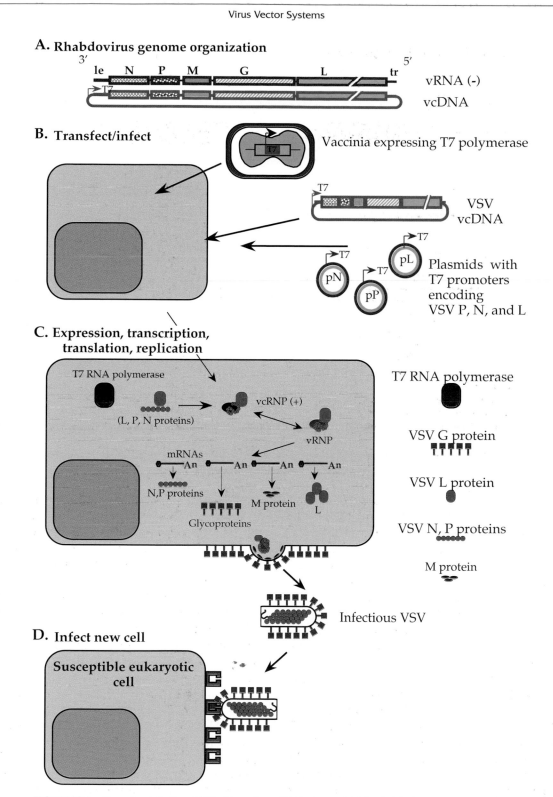

B. Transfect/infect

C. Expression, transcription, translation, replication

D. Infect new cell

FIGURE 9.6   Rescuing infectious VSV virions from cDNA clones. (A) The rhabdovirus genome organization, and a schematic of a cDNA clone containing the genome sequence (cDNA copy of vRNA). (B) A susceptible cell is infected with vaccinia virus expressing the T7 RNA polymerase, and four separate plasmids: the genome plasmid from which plus-strand antigenome RNA is transcribed, and three individual plasmids expressing VSV N protein, P protein, and L protein, all under the control of T7 promoters. (C) Plus-strand vcRNA is transcribed and encapsidated with N, P, and L. The RNP then replicates and viral proteins are expressed from individual mRNAs transcribed from genome sense RNPs. Virions bud from the cell and (D) can infect a new susceptible cell. [Adapted from Conzelmann and Meyers (1996).]

A. Genome organization of plasmid with CD4 and CXCR4 in place of VSV G protein

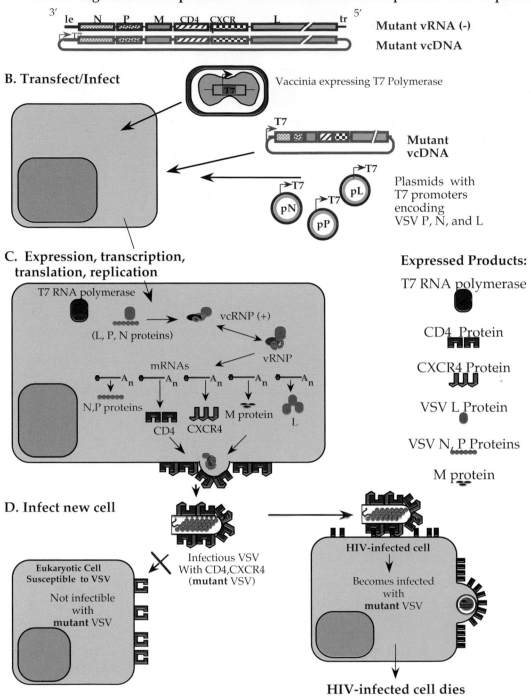

FIGURE 9.7 Producing a mutual VSV targeted to kill HIV-infected cells. (A) Genome of a rhabdovirus in which the glycoprotein G gene has been replaced with sequences encoding CD4 and CXCR4, the HIV primary receptor and coreceptor, and a schematic of a cDNA clone containing the genome sequence (cDNA copy of vRNA). (B) A susceptible cell is infected with vaccinia virus expressing the T7 RNA polymerase, and four separate plasmids: the mutant genome plasmid from which full-length vc (plus-strand) RNA is transcribed, and three individual plasmids expressing VSV N protein, P protein, and L protein, all under the control of T7 promoters. (C) Plus-strand mutant vcRNA is transcribed and encapsidated with N, P, and L. The RNP then replicates and both viral proteins and CD4 and CXCR4 are expressed from individual mRNAs transcribed from genome sense RNPs. Virions bud from the cell and (D) cannot infect a new susceptible cell as before in Fig. 9.6, but can infect an HIV-infected cell expressing HIV *env* proteins on its surface. Infection with VSV is cytolytic and the HIV-infected cell dies. [Adapted from Conzelmann and Meyers (1996).]

glycoprotein present on the VSV particle, called G, has been deleted and replaced with CD4, the cell surface protein that is used as a receptor by HIV. In addition a new gene has been inserted, the gene encoding the HIV coreceptor CXCR4, so that the virus now contains six genes. The virions produced are unable to infect the cells normally infectable by VSV because they lack the G protein. However, because they contain the HIV receptor and coreceptor on their surface, they do infect cells that express the HIV glycoprotein on their surface, such as cells infected by HIV. Since VSV is a lytic virus, the HIV-infected cells are killed.

USE OF VIRUSES AS EXPRESSION VECTORS

Viruses have been widely used as vectors to express a variety of genes in cultured cells. This use is of long standing and has led to important results. Of perhaps more interest are efforts to develop viruses as vectors for medical purposes. The manipulation of virus genomes to develop new vaccines is very promising. Although no licensed human vaccines have been introduced using this technology, clinical trials are ongoing. There is also expectation that viruses will be useful as vectors for gene therapy, and numerous clinical trials are taking place. The results to date have been disappointing, but the promise remains.

Expression of Proteins in Cultured Cells

The use of viruses to express foreign genes in cultured cells is well established and only a few examples are cited to illustrate the range and purpose of such use. In addition to the expression systems described above, which are based upon vertebrate viruses, expression systems based upon baculoviruses have also been widely used to exprss proteins in cultured cells. Baculoviruses are large DNA-containing viruses of insects (Chapter 6). The gene of interest is inserted by homologous recombination in a procedure that resembles that used for vaccinia virus. The virus is grown in continuous lines of insect cells, and large amounts of protein derived from the inserted gene can be obtained. The protein is often expressed in a way that leads to its secretion from the cell, which makes purification of the desired protein easier.

Hepatitis C virus (HCV) does not grow in cultured cells to titers sufficient to allow studies on the expression of viral proteins. The only experimental model for the virus is the chimpanzee, which severely restricts the number and nature of experiments that can be done. Thus, most of what we know about the expression of the HCV genome has been obtained through expression of parts of the genome by virus vectors, often by recombinant vaccinia virus. These studies have resulted in an understanding of the two viral proteases within the HCV genome, the processing pathway through which the polyprotein translated from the genome is processed, the function of the viral IRES, and the function of the viral replicase, among other results. The use of virus vectors means that such studies on HCV can be conveniently conducted in mammalian cells under conditions that are related to the natural growth cycle of the virus.

Norwalk virus is another virus for which there is no cell culture system. The virus can be grown only in human volunteers, again limiting the range of studies that can be done. Virus particles isolated from the stools of infected volunteers are often degraded and difficult to purify to homogeneity. Thus, structural studies of infectious virus have been limited. Expression of cDNA copies of the structural proteins of the virus in baculovirus vectors has allowed the production of large amounts of viral structural proteins that spontaneously assemble into virus-like particles. These virus-like particles have been studied by cryo-electron microscopy, and detailed information on the structure of the virus has been obtained in this way.

Baculoviruses are also widely used to prepare large amounts of protein for crystallographic studies. Such studies require 20 mg or more of protein, and the baculovirus system can be used to prepare such quantities. An advantage of the system is that the protein is made in a eukaryotic cell, which can be important for obtaining the protein folded into its correct three-dimensional conformation.

Even for viruses for which cell culture systems exist, the use of virus vectors that express to higher levels can be advantageous. There are cell culture systems in which rubella virus will grow and plaque, and there is a full-length cDNA clone of rubella virus from which infectious RNA can be recovered. However, the cell culture systems produce only low amounts of virus proteins, especially of the nonstructural proteins, and it has been difficult to study the expression and processing of the nonstructural polyprotein. Expression of the nonstructural region of rubella virus in vaccinia virus vectors or in Sindbis virus vectors has allowed the production of much larger quantities of the polyprotein precursor. This has been used to determine the processing pathways the identification of the virus nonstructural protease, and the identification of the cleavage sites that are cleaved by this protease.

As a final example, vaccinia virus vectors and Sindbis virus vectors have been used to map T-cell epitopes for a number of viruses (Chapter 8). For this, defined regions of a viral protein are expressed in order to determine whether a particular T-cell epitope lies within that region.

Viruses as Vectors to Elicit an Immune Response

Much effort is being put into the development of viruses as agents to immunize against other infectious agents, including other viruses. Such an approach has a number of advantages. There is a large body of experience in the use of attenuated or avirulent viruses as vaccines. Many of these, such as vaccinia virus or the yellow fever 17D virus, both of which have been used to immunize many millions of people, can be potentially developed as vectors to express other antigens, such as those in HCV or HIV. Use of a live virus as a vector to express antigens of other pathogens has many of the advantages of live virus vaccines. This includes the fact that only low initial doses are required, and therefore the expense of vaccine production may be less; that subsequent virus replication leads to the expression of large amounts of the antigen over an extended period of time, and the antigen folds in a more or less native conformation; and that a full range of immunity, including production of CTLs as well as of humoral immunity, usually develops.

No human vaccines have been licensed that use such recombinant viruses, but there are ongoing clinical trials of several potential vaccines. Several trials of candidate vaccines against HIV have been conducted that use vaccinia virus or retrovirus vectors to express the HIV surface glycoprotein. These trials have been moderately successful in the sense that immune responses to HIV glycoprotein were obtained, but it is not known if the immune response is protective. Studies in monkeys with related vaccines against simian immunodeficiency virus have given mixed results. In most such trials, immune responses were generated, but these were not fully protective. One recent trial did generate a protective response, however, giving hope that continued efforts in this direction will ultimately work out. A very recent study with anti-HIV drugs given very soon after infection found that limiting the replication of the virus early appears to allow the generation of a protective immune response in some patients. Of eight patients treated with anti-HIV drugs very early and then taken off the drugs, five have no detectable virus 8–11 months after stopping therapy. Although these studies are preliminary and involve only a few patients, they do suggest that a nonsterilizing immune response that restricts virus replication early might prove to be protective.

Other clinical trials have also tested poxviruses as vectors. Vaccinia virus has been used in an attempt to immunize against Epstein–Barr virus, and canarypox virus has been used as a vector for potential immunization against rabies virus.

Although no licensed human vaccines use poxvirus vectors, veterinary vaccines that are based on poxvirus vectors are in use. One such vaccine consists of vaccinia virus that expresses the rabies surface glycoprotein. This vaccine has been used to immunize wildlife. The recombinant vaccinia viruses are spread in baits that are eaten by wild animals that serve as reservoirs of the virus, such as skunks, raccoons, foxes, and coyotes. This approach has been useful in limiting the spread of rabies in wildlife populations. Other poxvirus-based vaccines include vaccinia virus vectors to protect cattle against vesicular stomatitis virus and rinderpest virus, and to immunize chickens against influenza virus; pigeonpox virus vectors to immunize chickens against Newcastle disease virus; fowlpox

virus vectors to immunize chickens against influenza, Newcastle disease , and infectious bursal disease viruses; a capripox virus vector to immunize pigs against pseudorabies virus; and a canarypox virus vector used to immunize dogs against canine distemper virus. Thus, it should be possible to develop human vaccines based on poxvirus vectors.

In a quite different approach, clinical trials of a novel vaccine against Japanese encephalitis (JE) virus have begun recently. JE is a scourge in parts of Asia, causing a large number of deaths and neurological sequelae in people that survive the encephalitis (Chapter 3). Vaccines in widespread use are inactivated virus vaccines, and the difficulties in preparing the large amounts of material required and delivering it to large segments of the population are significant. An attenuated virus vaccine, SA14-14-2, has been prepared in China by passing the virus in cultured cells and in rodent tissues. This vaccine is safe but overattenuated, so that the effectiveness is only 80% after a single dose. In contrast, the yellow fever virus (YF) 17D vaccine has an effectiveness of virtually 100% after a single dose, and immunity is long lasting, probably lifelong. A candidate JE vaccine has been developed that consists of the 17D strain of YF virus in which the prM and E genes have been replaced with those of JE, as illustrated in Fig. 9.8. Four chimeric viruses were tested. The JE structural proteins were taken from either the virulent Nakayama strain or from the attenuated SA-14-14-2 strain. In both cases, chimeras containing all three structural proteins from JE were tested as well as chimeras that contained only prM and E from JE. Chimeras containing C, prM, and E from JE were not viable, whereas chimeras containing only prM and E from JE were viable and grew well in culture (Fig. 9.8).

The viable chimeras were first tested in mice. The chimera containing the Nakayama strain proteins caused lethal encephalitis in mice, as does the YF 17D virus (even though it is safe for use in humans). However, the chimera containing prM and E from the attenuated JE strain was fully attenuated in mice and did not cause illness. The fully attenuated chimera was chosen for testing in monkeys, and was found to be safe and to protect monkeys against challenge with JE virus.

Clinical trials of this candidate vaccine have begun in humans. There is every reason to believe that this vaccine will be safe and more effective than the JE vaccines now in use. Furthermore, this approach should be applicable to other flaviviruses, such as the dengue viruses, for which no licensed vaccines exist, or West Nile virus, which recently spread to the Americas and caused a number of fatal cases of human encephalitis in the New York area.

Gene Therapy

A number of genetic diseases result from the failure to produce a specific protein. One of the more exciting poss-

A. Original constructs
Infectious 17D yellow fever cDNA clone

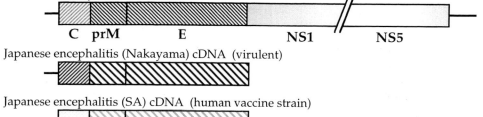

C   prM   E   NS1   NS5

Japanese encephalitis (Nakayama) cDNA (virulent)

Japanese encephalitis (SA) cDNA (human vaccine strain)

B. Chimeric constructs

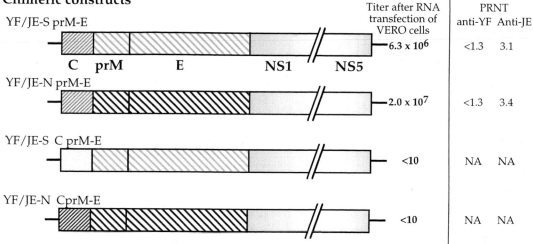

	Titer after RNA transfection of VERO cells	PRNT anti-YF	Anti-JE
YF/JE-S prM-E	6.3 x 10⁶	<1.3	3.1
YF/JE-N prM-E	2.0 x 10⁷	<1.3	3.4
YF/JE-S C prM-E	<10	NA	NA
YF/JE-N CprM-E	<10	NA	NA

FIGURE 9.8 Construction of yellow fever/Japanese encephalitis chimeric viruses. Starting with the full-length cDNA clone for 17D yellow fever virus, a number of chimeric viruses were constructed in which the M and E proteins were replaced with those of different strains of Japanese encephalitis virus. However, when C, M, and E of JE were put into the yellow fever clone, no viable virus was obtained. Both prM-E chimeras grew well in tissue culture, and were neutralized by anti-JE antiserum. YF/JE-S prM-E was attenuated, and did not kill adult mice by intracerebral inoculation, but YF/JE-N prM-E was neurovirulent. PRNT is the log reciprocal of the dilution yielding 50% plaque reduction neutralization, based on 100 PFU on LLV-MK cells, using either YF or JE hyperimmune ascitic fluid. [Adapted from Chambers *et al.* (1999).]

ible uses for virus vectors is for the expression of a missing protein as a cure for the genetic defect associated with its absence. Some genetic diseases that might be curable through the use of gene therapy are listed in Table 9.1. For successful treatment, expression of the missing protein must be long-term and preferably lifelong, the levels of protein produced must be sufficient to alleviate the symptoms of the disease, the protein must be expressed in or translocated to those cells that require the normal protein for function, and infection with the virus vector must be free of disease symptoms. Because of the requirement for long term expression, viruses whose DNA integrates into the host chromosome, which include retroviruses and adeno-associated viruses, offer the most promising system for many diseases. To date, several hundred patients have been treated with vectors based on Moloney murine leukemia virus in clinical trials. Clinical trials have also been conducted that use adenovirus vectors and adeno-associated virus vectors. A comparison of various virus systems that are being considered for gene therapy is shown in Table 9.2. Naked DNA has also been used in a recent trial for coronary artery disease, and the properties of this system are included in the table.

In addition to the possible treatment of genetic defects, virus vectors may also be useful for the treatment of a number of acquired diseases. These include cancer, HIV infection, Parkinson's disease, injuries to the spinal cord, and vascular diseases such as restenosis and arteriosclerosis. A partial listing of candidate diseases for gene therapy is given in Table 9.1.

A partial listing of clinical trials that attempt to treat several different genetic defects, the diseases rheumatoid

TABLE 9.1   Candidate Diseases for Gene Therapy

Disease	Defect	Incidence	Target cells
Genetic			
Severe combined immunodeficiency (SCID)	Adenosine deaminase (ADA) in 25% of SCID patients	Rare	Bone-marrow cells or T lymphocytes
Hemophilia A	Factor VII deficiency	1:10,000 males	Liver, muscle, fibroblasts or bone marrow cells
Hemophilia B	Factor IX deficiency	1:30,000 males	Same as for hemophilia A
Familial hypercholesterolemia	Deficiency of low-density lipoprotein (LDL) receptor	1:1 million	Liver
Cystic fibrosis	Faulty transport of salt in lung epithelium	1:3000 Caucasians	Airways in the lungs
Hemoglobinopathies, thalassemias	(Structural) defects in the α or β globin gene	1:600 in certain ethnic groups	Bone marrow cells that are precursors to red blood cells
Gaucher's disease	Defect in the enzyme glucocerebrosidase	1:450 in Ashkenazi Jews	Bone marrow cells, macrophages
α_1 antitrypsin deficiency, inherited emphysema	Lack of α_1 antitrypsin	1:3500	Lung or liver cells
Duchenne muscular dystrophy	Lack of dystrophin	1:3000 males	Muscle cells
Acquired			
Cancer	Many causes, including genetic and environmental	1 million/year in USA	Variety of cancer cell types, in liver, brain, pancreas, breast, kidney
Neurological diseases	Parkinson's, Alzheimer's spinal-cord injury	1 million Parkinson's and 4 million Alzheimer's patients in USA	Neurons, glial cells, Schwann cells
Cardiovascular	Restenosis, arteriosclerosis	13 million in USA	Arteries, vascular endothelia walls
Infectious diseases	AIDS	> 43 million worldwide	T cells, liver, macrophages
	Hepatitis B	350 million worldwide	Liver cells
Rheumatoid arthritis	Autoimmune inflammation of joints	Increasing numbers with aging population	Intra-auricular delivery and expression of IL-1 and TNF-α inhibitors

arthritis and restenosis or AIDS by using virus vectors, cationic lipids, or naked DNA to deliver specific genes, is given in Table 9.3. There have been few successes to date and the table is more of a compendium of the variety of genes and diseases, as well as the variety of delivery schemes, that are being examined. Table 9.4 contains a list of clinical trials that attempt to treat various forms of cancer by using viruses, or in a few cases other means, to deliver genes to patients. These trials are all in early stages.

Retrovirus Vectors to Genetically Mark Cells

Retroviruses have been used in a number of clinical trials to genetically tag cells. Although this use does not fall within the narrow definition of gene therapy, it does provide background experience in the use of retrovirus vectors in humans. One such use has been in bone marrow transplantation for leukemia. Severe forms of leukemia can sometimes be treated by ablation of the hematopoietic system with chemotherapy and/or X rays in order to kill all tumor cells, followed by reconstitution of the system by transplan-

tation of bone marrow from a compatible donor. Although often successful, the leukemia sometimes recurs and it is desirable to know whether it recurs because of incomplete destruction of the patient's leukemic cells or whether the donor cells are the source of the leukemia. Experiments in which the donor cells have been tagged using retroviruses that express a marker gene have been used to answer this question, which is important for the design of transplantation protocols.

Gene Therapy for ADA Deficiency

Patients who lack the enzyme adenosine deaminase (ADA) will die early in life unless treated. Lack of ADA results in the failure to clear adenosine from the body and, consequently, the accumulation of adenosine in cells throughout the body. Adenosine is toxic at high concentrations, producing a variety of symptoms. The most serious symptom results from the extreme sensitivity of T cells to elevated adenosine concentrations. Loss of T cells results

TABLE 9.2   Comparison of Properties of Various Vector Systems

Features	Vectors based on RNA viruses			Vectors based on DNA viruses				Naked DNA
	Retroviral	Lentiviral	Alphaviral	Adenoviral	AAV	Herpesviral	Vaccinia	
Maximum insert size	7–7.5 kb	7–7.5 kb	5 kb	7.5 kb	4.5 kb	~30 kb	>25 kb	Unlimited size
Concentrations, viral particles/ml	$>10^8$	$>10^8$	$>10^9$	$>10^{10}$	$>10^{12}$	$>10^8$	10^7–10^9	No limitation
Route of gene delivery	*Ex vivo*	*Ex/in vivo*	*In vivo*	*Ex/in vivo*	*Ex/in vivo*	*Ex vivo*	*Ex/in vivo*	*Ex/in vivo*
Integration	Yes	Yes	No	No	Yes/no	No	No	Very poor
Duration of expression *in vivo*	Shorter than theorized	Long	Short	Short	Long	Short/ long in CNS	Short	Short
Stability	Good	Not tested	Good	Good	Good	Unknown	Good	Very good
Ease of preparation scale-up	Pilot scale-up up to 20–50 liters	Unknown	Unknown	Easy to scale up	Difficult to purify, difficult to scale up	Not yet tried	Vaccine production facilities exist	Easy to scale up
Immunological problems	Few	Few	Unknown	Extensive	Unknown	Unknown	Extensive	None
Preexisting host immunity	Unlikely	Unlikely, except in AIDS patients	No	Yes	Yes	Yes	Diminishing as unvaccinated population grows	No
Safety problems	Insertional mutagenesis?	Insertional mutagenesis?	Few	Inflammatory response, toxicity	Inflammatory response, toxicity	Neurovirulence? Insertional mutagenesis	Disseminated vaccinia in immunocompromised hosts	None?

Source: Verma and Somia (1997) and Jolly (1994).

TABLE 9.3   Gene Therapy in Clinical Trials in the United States as of 2000

Disease[a]	Gene	Vector	Number of trials	Number of patients	Results
Monogenic Diseases					
Gaucher's disease	GC	Retrovirus	3	9	One trial shows long-term elevation of GC expression, other trials primarily phase I
OTC deficiency	OTC	Adenovirus	1	14	Trial suspended after one fatality (see text)
ADA-SCID	ADA + NeoR	Retrovirus	1	6	Ongoing since 1990
Cystic fibrosis	CFTR	Adenovirus	9	83	Some correction of defect in 30% of patients, but inflammation at clinical doses, and reduction in therapeutics with repeated injection
	CFTR	AAV	4	36	Some correction of defect; phase II study started
	CFTR	Cationic lipids	4	25	30–50% of patients showed improvement
Chronic granulomatosus	p47 phox/ gp91 phox	Retrovirus	3	9	Phase I/II; study closed in 1998
Familial hypercholesterolemia	LDLR	Retrovirus	1	5	Phase I; closed in 1994
Other Diseases					
Rheumatoid arthritis	IRAP	Retrovirus	1	7	?
Artery disease and restenosis	VEGF	Naked DNA	2	29	?
Infectious Diseases					
AIDS	HIV-IT(V)	Retrovirus	3	298	Most gene trials for HIV are in phase I, with a few in phase II; few results reported
	CD4-Zeta TcR	Retrovirus	3	54	
	Anti-HIV ribozyme	Retrovirus	2	12	
	TK + HyR	Retrovirus	2	14	
	Antisense to TAR	Retrovirus	3	17	

Source: Data from the Wiley Journal of Gene Medicine/Clinical Trial Database at the URL: http://www.wiley.co.uk/wileychi/genmed/clinical.
Abbreviations: GC, glucocerebrosidase; OTC, ornithine transcarbamylase; ADA, adenosine deaminase; SCID, severe combined immunodeficiency; CTFR, cystic fibrosis transmembrane conductance regulator; AAV, adeno-associated virus; p47, phox gene for antimicrobial oxidant. LDLR, low-density lipoprotein receptor; IRAP, interleukin 1 receptor antagonist; VEGF, vascular epithelial growth factor.
[a]Gene therapy trials for cancer are given in Table 9.4.

in SCID, *severe combined immunodeficiency*. Both CTL responses (which are T cell based) and humoral responses (which require T-helper cells) are impaired. People with SCID syndrome are unable to mount an immunological response to infectious agents, and SCID is invariably fatal early in life unless treated in some way. ADA deficiency accounts for about 25% of SCID syndromes in humans.

SCID can be treated by bone marrow transplantation if a suitable donor can be found. In the case of SCID due to ADA deficiency, weekly or twice weekly injections of ADA mixed with polyethylene glycol (PEG) have been used to successfully treat about 60 patients in whom bone marrow transplantation cannot be used because of the lack of com-

patible donors. Of these, about 10 patients have also been treated with retroviral vectors that express ADA. In these experiments, T cells were taken from the patient (or in the case of three newborns, umbilical cord cells were used), infected *ex vivo* with the retrovirus vector using a number of different cell culture and infection protocols, and the cells reinfused into the patient. Many of the patients continue to produce ADA from the vector several years after treatment. However, all of the patients continue to receive ADA–PEG injections, which is known to be an effective treatment. Although some patients who have received retroviral therapy have been partially weaned from the supplementary ADA–PEG, it appears that some of these, and

TABLE 9.4 Clinical Trials of Gene Transfers for Cancer Therapy in the United States as of 2000

Location	Gene	Vector	Number of trials	Number of patients	Phase[a]
Brain cancers					
Neuroblastoma	IFNγ	Retrovirus	1	4	II
	IL-2	Retrovirus	1	12	I
	IL-2	Adenovirus	1	6	I
Central nervous system	TK	Adenovirus	2	22	I
Pediatric tumor	TK	Retroviral producing cells	1	2	I
Adult brain tumor	TK	Retroviral producing cells	1	15	I
Ovarian cancer	HSV-TK	Adenovirus	1	10	I
	TK	Retroviral producing cells	3	42	I
	BRCA-1	Retrovirus	1	40	I/II
	p53	Adenovirus	1	16	I
Small cell lung cancer	IL-2 + NeoR	Lipofection	1	8	I
	Anti-sense to k-ras	Retrovirus	1	9	I
	p53	Adenovirus	2	59	I/II
Prostate cancer	GM-CSF	Retrovirus	1	8	I/II
	PSA	Poxvirus	1	3	I
	HSV-TK	Adenovirus	1	18	I
Breast cancer	BRCA-1	Retrovirus	1	21	I
	E1A	Lipofection	1	16	I
	MDR-1+NeoR	Retrovirus	4	39	I
	CD80	Lipofection	1	15	I
	CEA	Poxvirus	4	53	I
	CEA	RNA transfer	1	30	I
Melanoma	GM-CSF	Gene gun	1	17	I
	GM-CSF	Retrovirus	2	29	I
	HLA-B7/b2m	Lipofection	8	165	I/II
	IL-2 + NeoR	Retrovirus	5	115	I
	IFNγ	Retrovirus	3	91	I
	TNF+NeoR	Retrovirus	1	12	I/II
	MART-1	Adenovirus	1	33	I
	MART-1	Poxvirus	2	16	I
	gp100	Poxvirus	1	19	I
	gp100	Adenovirus	1	7	I
	CD80	Lipofection	1	17	I
Miscellaneous carcinomas	p53	Adenovirus	1	26	I
	HLA-B7/b2m	Lipofection	4	76	II
	IL-2	Lipofection	1	11	I
	CEA	Poxvirus	1	8	I
Lymphomas and solid tumors	IL-2	Retrovirus	2	29	I
	TK	Retrovirus	1	11	I
	IL-12 + NeoR	Retrovirus	1	31	I
Bladder cancer	p53	Adenovirus	1	5	I
Colo/rectal, renal, and liver cancers	GM-CSF	Retrovirus	1	18	I
	HLA-B7/b2m	Lipofection	4	53	I/II
	CD	Adenovirus	1	6	I
	IL-4	Retrovirus	1	18	I
	TNF+NeoR	Retrovirus	1	12	I

Source: Same as for Table 9.3.

Abbreviations: IFNγ, interferon gamma; IL-2, interleukin 2; TK, thymidine kinase (sometimes used with bromodeoxyuridine); HSV-TK, herpes simplex thymidine kinase, often coupled with gancyclovir treatment; BRCA-1, breast cancer 1, early onset; PSA, puromycin-sensitive aminopeptidase; CEA, carcinoembryonic antigen; GM-CSF, granulocyte-marcrophage colony-stimulating factor; MDR-1, multidrug resistance protein 1 (used to insert chemotherapy-resistance genes into the hematopoetic lineage); CD80, protein involved in T-cell activation; CD, cytosine deaminase; TNF, tumor necrosis factor.

[a]Definitions of phases in a clinical trial: Phase I usually has fewer than 100 healthy volunteers, primarily to gauge adverse reactions, and to determine optimal dose and best route of administration. Phase II generally consists of pilot efficacy studies involving 200–500 volunteers randomly assigned to control and study groups. Phase II will test for immunogenicity in the case of vaccines, and duration of expression and amelioration of symptoms for gene therapy. Note that none of these trials has proceeded beyond phase II, and most are in phase I.

perhaps all, do not produce enough ADA to be cured. Thus, although no cures have been effected, the results to date have been encouraging and suggest that future protocols may be more successful. Two areas of retroviral therapy that need improvement are to increase the efficiency with which stem cells are infected, and the need to prevent the retroviral promoter from being downregulated.

Cystic Fibrosis

Cystic fibrosis results from loss of the cystic fibrosis transmembrane conductance regulator (CFTR), which regulates epithelial transport of ions and water. Although lack of this protein results in damage to the epithelium in many parts of the body, the most serious manifestation is lung disease accompanied by chronic bacterial infection of the airways. Clinical trials using adenoviral vectors, which infect respiratory epithelium, to express CFTR in the lungs have been conducted. The first such studies were encouraging, but a recent trial that was carefully controlled found no relief of symptoms. Inflammation produced by the high doses of adenovirus used in trials is also a problem. More recent trials have begun that use adeno-associated virus. The results of these trials are preliminary but encouraging. Cationic lipids have also been used to deliver the gene.

Rheumatoid Arthritis

Rheumatoid arthritis is a chronic, progressive inflammatory disease of the joints. An estimated 5 million people in the United States suffer from it. There is no cure. Drugs therapies are used that ameliorate the symptoms, but most of these drugs have side effects and cannot be taken indefinitely. If the disease progresses far enough, joint replacement may be required. The disease is associated with the release of inflammatory cytokines in the affected joints. Clinical trials have started that use retroviruses to deliver the gene for an anti-arthritic cytokine gene to the joints. The gene encodes the interleukin (IL)-1 receptor antagonist, which inhibits the biological actions of both IL-1a and IL-1b. It is hoped that such treatment might damp out the disease or at least keep it from progressing.

A Gene Therapy Failure

Patients who have deficiencies in enzymes that participate in the urea cycle have increased concentrations of ammonia in the blood. High concentrations of ammonia result in various symptoms, which can include behavioral disturbances or coma. Severe deficiencies in these enzymes result in early death, but moderate deficiencies can result in delayed appearance of symptoms and may be partially controlled by diet. One such enzyme is ornithine transcarbamylase (OTC), which is found on the X chromosome.

Deficiencies in OTC are therefore more common in males than in females.

Gene therapy trials that use virus vectors recently received a major setback when a relatively fit 18-year-old male with an inherited deficiency for OTC died 4 days after an adenovirus vector was injected into his liver. A high dose of adenovirus (4×10^{10}) that expressed OTC was injected in an effort to achieve adequate levels of enzyme production. The virus unexpectedly spread widely and a systemic inflammatory response developed, inducing a fever of $40.3°C$. He went into a coma, his lungs filled with fluid, and he died of asphyxiation. This unfortunate result makes clear the possible drawbacks to experimental treatments and the difficulties in designing protocols that allow an adequate margin of safety while trying to achieve a clinically relevant result.

A Gene Therapy Success: Treatment of Restenosis

A recent gene therapy trial in patients with heart disease has given very encouraging results. Although this study did not involve virus vectors, a brief description will be given since it serves as an incentive for continuation of gene therapy trials. Coronary artery disease is common in older people. Angioplasty or bypass surgery is used to open clogged arteries, but in many patients the arteries close up again (a process called restenosis). Thirteen patients with chronic chest pain who had failed angioplasty or bypass surgery or both were injected in the heart muscle with DNA encoding vascular endothelial growth factor. This factor promotes the growth of blood vessels, a process called angiogenesis. Two months after treatment, all patients exhibited an improvement in vascularization of damaged areas of the heart, as shown by imaging and mapping studies. All patients reported a decrease in disease symptoms, and all had an improved performance in treadmill tests. Although the number of patients is small, the uniformly positive results are encouraging.

Viruses as Anticancer Agents

There is hope that viruses can be developed as anticancer agents. Table 9.4 lists a number of malignacies that are being considered as candidates for treatment using gene therapy approaches, and that are now in at least phase I trials. Although more than 1000 patients are participating in the trials listed in the table, this field is still in its infancy, and only a brief summary of approaches is presented.

In most of the trials in Table 9.4, viruses are used to express proteins that control the growth of tumors or that are toxic to tumor cells. A number of different cytokines are being tried, such as IFN-γ, IL-2, TNF and GM-CSF. Another approach is to try to repair the defective regulatory gene in the tumor cell, which is often p53. Many other gene products are also being tested. All of these trials represent preliminary attempts, and it will be some time before we know if any of them represent successful approaches.

Further afield, thought is being given to the possibility of using viruses to express proteins overexpressed in tumor cells in an attempt to stimulate the immune system to respond by killing tumor cells. This is in essence an attempt to vaccinate a person against a tumor. For this approach to succeed, an antigen overproduced by a tumor cell, such as a melanoma cell, must be identified, inserted into a suitable vector, and the person with the tumor infected with the virus vector in an attempt to stimulate the immune system. In principle, this approach may be feasible, but only time will tell whether it is in fact practical.

Another approach is to try to direct the virus, more or less specifically, to infect the tumor cells, so that upon infection the cells are killed. Cell death might result either because the virus itself is cytolytic or because the virus expresses a protein that renders the cell sensitive to a toxic agent such as BUdR. A number of the trials listed in Table 9.4 use the TK gene for this, since cells that express TK are sensitive to BUdR.

A number of possible vectors have been suggested as a way to specifically kill tumor cells. One concept is to try to cure or control brain tumors, especially gliomal tumors, by using either herpesviruses or retroviruses. Simple retroviruses can only replicate in dividing cells. Thus, they should be able to infect only tumor cells in the brain, since most neuronal cells are terminally differentiated and do not divide. If the retroviruses express a protein that renders the cells sensitive to a toxin, it might be possible to kill replicating cells and therefore only the tumor cells. A different approach has been to try to use herpesviruses as antitumor viruses. These viruses set up latent infections in neurons and might in principle be used to control brain tumors.

FURTHER READING

Infectious Clones

Bridgen, A., and Elliott, R. M. (1996). Rescue of a segmented negative-strand RNA virus entirely from cloned complementary DNAs. *Proc. Natl. Acad. Sci. U.S.A.* **93**; 15400–15404.

Conzelmann, K.-K., and Meyers, G. (1996). Genetic engineering of animal RNA viruses. *Trends Microbiol.* **4**; 386–393.

Rice, C. M., Levis, R., Strauss, J. H., *et al.* (1987). Production of infectious RNA transcripts from Sindbis virus cDNA clones: Mapping of lethal mutations, rescue of a temperature sensitive marker, and *in vitro* mutagenesis to generate defined mutants. *J. Virol.* **61**; 3809–3819.

Schnell, M. J., Mebatsiion, T., and Conzelmann, K.-K. (1994). Infectious rabies viruses from cloned cDNA. *EMBO J.* **13**; 4195–4203.

Expression Systems

Mackett, M., Smith, G. L., and Moss, B. (1982). Vaccinia virus: A selectable eukaryotic cloning and expression vector. *Proc. Natl. Acad. Sci. U.S.A.* **79**; 7415–7419.

Pinto, V. B., Prasad, S., Yewdell, J., *et al.* (2000). Restricting expression prolongs expression of foreign genes introduced into animals by retroviruses. *J. Virol.* **74**; 10202–10206.

van Craenenbroeck, Vanhoenacker, K., P., and Haegeman, G. (2000). Review: Episomal vectors for gene expression in mammalian cells. *Eur. J. Biochem.* **267**; 5665–5678.

DNA Vaccines

Cohen, E. P., de Zoeten, E. F., and Schatzman, M. (1999). DNA vaccines as cancer treatment. *Am. Sci.* **87**; 328–335.

Viruses as Vectors for Heterologous Antigens

Basak, S., McPherson, S., Kang, S., *et al.* (1998). Construction and characterization of encapsidated poliovirus replicons that express biologically active murine interleukin-2. *J. Interferon Cytokine Res.* **18**; 305–313.

Bledsoe, A. W., Jackson, C. A., McPherson, S., *et al.* (2000). Cytokine production in motor neurons by poliovirus replicon vector gene delivery. *Nat. Biotechnol.* **18**; 964–969.

Guirakhoo, F., Zhang, Z. X., Chambers, T. J., *et al.* (1999). Immunogenicity, genetic stability, and protective efficacy of a recombinant, chimeric yellow fever-Japanese encephalitis virus (ChimeriVax-JE) as a live, attenuated vaccine candidate against Japanese encephalitis. *Virology* **257**; 363–372.

Monath, T. P., Levenbook, I., Soike,K., *et al.* (2000). Chimeric yellow fever virus 17D-Japanese encephalitis virus vaccine: Dose-response effectiveness and extended safety testing in rhesus monkeys. *J. Virol.* **74**; 1742–1751.

Schneider, U., Bullough, F., Vongpunsawad, S., *et al.* (2000). Recombinant measles viruses efficiently entering cells through targeted receptors. *J. Virol.* **74**; 9928–9936.

Gene Therapy

Athanasopoulos, T., Fabb, S., and Dickson, G. (2000). Gene therapy vectors based on adeno-associated virus: Characteristics and applications to acquired and inherited diseases. *Int. J. Mol. Med.* **6**; 363–375.

Kohn, D. B., Hershfield, M. S., Carbonaro, D., *et al.* (1998). T lymphocytes with a normal ADA gene accumulate after transplantation of transduced autologous umbilical cord blood CD34[+] cells in ADA-deficient SCID neonates. *Nat. Med.* **4**; 775–780.

Robbins, P. D., and Ghivizzani, S. C. (1997). Viral vectors for gene therapy. *Pharmacol. Ther.* **80**; 35–47.

Russell, W. C. (2000). Update on adenovirus and its vectors. *J. Gen. Virol.* **81**; 2573–2604.

Verma, I. M., and Somia, N., (1997). Gene therapy—promise, problems, and prospects. *Nature (London)* **389**; 239–242.

Walther, W., and Stein, U. (2000). Viral vectors for gene transfer: A review of their use in the treatment of human diseases. *Drugs* **60**; 249–271.

A

References for Figures

CHAPTER 1

Coffin, J. M., Hughes, S. H., and Varmus, H. E., eds. (1997). "Retroviruses". Cold Spring Harbor Laboratory Press, Cold Spring Harbor, New York.

Ding, J., McGrath, W. J., Sweet, R. M., and Mangei, W. F. (1996). Crystal structure of the human adenovirus proteinase with its 11 amino acid cofactor. *EMBO J.* **15**; 1778–1783.

Fields, B. N., Knipe, D. M., and Howley, P. M., eds. (1996). "Fields Virology," 3rd ed. Lippincott-Raven, New York.

Flint, S. J., Enquist, L. W., Krug, R. M., Racaneillo, V. R., and Skalka, A. M. (2000). "Principles of Virology, Molecular Biology, Pathogenesis, and Control." ASM Press, Washington DC.

Goff, S. P. (1997). Retroviruses: Infectious genetic elements. *In* "Exploring Genetic Mechanisms" (M. Singer and P. Berg, eds.), pp. 133–202. University Science Books, Sausalito, California.

Matthews, D. A., Smith, W. W., Ferre, R. A. *et al.* (1994). Structure of the human rhinovirus 3C protease reveals a trypsin-like polypeptide fold, RNA-binding site, and means for cleaving precursor polyprotein. *Cell (Cambridge, Mass.)* **77**; 761–771.

Mims, C. A., Playfair, J. H. L., Roitt, I. M., Wakelin, D., and Williams, R. (1993). "Medical Microbiology." Mosby Europe Limited, London.

Rutenber, E., Fauman, E. B., Keenan, R. J., Fong, S., Furth, P. S., Ortiz de Montellano, P. R., Meng, E., Kuntz, I. D., DeCamp, D. L., and Salto, R. (1993). Structure of a non-peptide inhibitor complexed with HIV-1 protease. Developing a cycle of structure-based drug design. *J. Biol. Chem.* **268**; 15343–15346.

Strauss, E. G., and Strauss, J. H. (1997). Eukaryotic RNA viruses, a variant genetic system. *In* "Exploring Genetic Mechanisms," (M. Singer and P. Berg, eds.), pp. 71–132. University Science Books, Sausalito, California.

Wimmer, E., Hellen, C. U. T., and Cao, X. (1993). Genetics of poliovirus. *Annu. Rev. Genet.* **27**; 353–436.

CHAPTER 2

Athappilly, F. K., Murali, R., Rux, J. J., Cai, Z., and Burnett, R. M. (1994). The refined crystal-structure of hexon, the major coat protein of adenovirus type 2, at 2.9 angstrom resolution. *J. Mol. Biol.* **242**; 430–455.

Baker, T. S., Olson, N. H., and Fuller, S. D. (1999). Adding the third dimension to virus life cycles: Three-dimensional reconstruction of icosahedral viruses from cryo-electron micrographs. *Microbiol. Mol. Biol. Rev.* **63**; 862–922.

Birdwell, C. R., and Strauss, J. H. (1974). Maturation of vesicular stomatitis virus: Electron microscopy of surface replicas of infected cells. *Virology* **69**; 587–590.

Coffin, J. M., Hughes, S. H., and Varmus, H. E., eds. (1997). "Retrovirus". Cold Spring Harbor Laboratory Press, Cold Spring Harbor, New York.

Compans, R. W., and Dimmock, N. J. (1969). An electron microscopic study of single cycle infection of chick embryo fibroblasts by influenza virus. *Virology* **39**; 499–515, as shown in Dalton and Haguenau (1973, p. 219).

Compans, R. W., Holmes, J. V., Dales, S., and Choppin, P. W. (1966). An electron microscopic study of moderate and virulent virus-cell interactions of the parainfluenza virus SV5. *Virology* **30**; 411–426, as shown in Dalton and Haguenau (1973, p. 227).

Dalton, A. J., and Haguenau, F. eds. (1973). "Ultrastructure of Animal Viruses and Bacteriophages; An Atlas." Academic Press, New York.

Fenner, F., and Nakano, J. H. (1988). *Poxviridae*, the poxviruses. *In*. "The Laboratory Diagnosis of Infectious Diseases: Principles and Practice" (E. H. Lennette, P. Halonen, and F. A. Murphy, eds.), Vol. II, pp. 177–210. Springer-Verlag, New York.

Fields, B. N., Knipe, D. M., and Howley, P. M., eds. (1996). "Fields Virology," 3rd ed. Lippincott-Raven, New York.

Granoff, A., and Webster, R. G., eds. (1999). "Encyclopedia of Virology," 2nd ed. Academic Press, San Diego, California.

Higashi, N., Arimura, H., Matsumoto, A., and Nagatomo, Y. (1976). "Septieme Congrés International de Microscopie Electronique, Grenoble," Vol. III, p. 295 (as seen in Dalton and Haguenau, 1973, p. 194).

Johnson, J. E. (1996). Review: Functional implications of protein-protein interactions in icosahedral viruses. *Proc. Natl. Acad. Sci. U.S.A.* **93**; 27–33.

Kolatkar, P. R., Bella, J., Olson, N. H., Bator, C. M., Baker, T. S., and Rossmann, M. G. (1999). Structural studies of two rhinovirus serotypes complexed with fragments of their cellular receptor. *EMBO J.* **18**; 6249–6259.

Murphy, F. A., Webb, P. A., Johnson, K. M., and Whitfield, S. G. (1969). Morphological comparison of Machupo with lymphocytic choriomeningitis virus: Basis for a new taxonomic group. *J. Virol.* **4**; 535–541, as shown in Dalton and Haguenau (1973, p. 329).

Murphy, F. A., Fauquet, C. M., Bishop, D. H. L., Ghabrial, S. A., Jarvis, A. W., Martelli, G. P., Mayo, M. A., and Summers, M. D., eds. (1995).

"Virus Taxonomy" 6th Report of the International Committee on Taxonomy of Viruses. Springer-Verlag, Vienna.

Spear, P. and Roizman, B. Unpublished studies, as shown in Dalton and Haguenau (1973, p. 93).

Stehle, T., Gamblin, S. J., Yan, Y., and Harrison, S. C. (1996). The structure of simian virus 40 refined at 3.1Å resolution. *Structure* **4**; 165–182.

Stewart, P. L., Burnett, R. M., Cyrklaff, M., and Fuller, S. D. (1991). Image reconstruction reveals the complex molecular organization of adenovirus. *Cell (Cambridge, Mass.)* **4**; 145–154.

Strauss, J. H., Strauss, E. G., and Kuhn, R. J. (1995). Budding of Alphaviruses. *Trends Microbiol.* **3**; 346–350.

CHAPTER 3

Barnes, G. L., Uren, E., Stevens, K. B., and Bishop, R. F. (1998). Etiology of acute gastroenteritis in hospitalized children in Melbourne, Australia, from April 1980 to March 1993. *J. Clin. Microbiol.* **35**; 133–138.

Bartenschlager, R., and Lohmann, V. (2000). Replication of hepatitis C virus. *J. Gen. Virol.* **81**; 1631–1648.

Bowen, G. S. (1988). Colorado Tick Fever. *In* "The Arboviruses" (T. P. Monath) ed., Vol. 2, pp. 159–176. CRC Press, Boca Raton, Florida.

Brandt, C. D., Kim, H. W., Rodriguez, W. J., Arrobio, J. O., Jeffries, B. C., Stallings, E. P., Lewis, C., Miles, A. J., Chanock, R. M., Kapikian, A. Z., and Parrott, R. H. (1983). Pediatric viral gastroenteritis during eight years of study. *J. Clin. Microbiol.* **18**; 71–78.

Chambers, T. J., Hahn, C. S., Galler, R., and Rice, C. M. (1990). Flavivirus genome organization, expression, and replication. *Annu. Rev. Microbiol.* **44**; 649–688.

den Boon, J. A., Snijder, E. J., Chirnside, E. D., de Vries, A. A. F., Horzinek, M. C., and Spaan, W. J. M. (1991). Equine arteritis virus is not a togavirus but belongs to the coronavirus-like superfamily. *J. Virol.* **65**; 2910–2920.

de Vries, A. A. F., Horzinek, M. C., Rottier, P. M., and de Groot, R. J. (1997). The genome organization of the *Nidovirales*. Similarities and differences between arteri-, toro-, and coronaviruses. *Semin. Virol.* **8**; 33–47.

Duncan, R. (1999). Extensive sequence divergence and phylogenetic relationships between the fusogenic and nonfusogenic orthoreoviruses: A species proposal. *Virology* **260**; 316–328.

Fields, B. N., Knipe, D. M., and Howley, P. M., eds. (1996). "Fields Virology," 3rd ed. Lippincott-Raven, New York.

Glass, R. I., Bresee, J. S., Parashar, U., Miller, M., and Gentsch, J. R. (1997). Rotavirus vaccines at the threshold. *Nat. Med.* **3**; 1324–1325.

Halstead, L. S. (1998). Post-polio syndrome. *Sci. Am.* **278**; April, 42–47.

Heinz, F. X., Auer, G., Stiasny, K., Holzmann, H., Mandl, C., Guirakhoo, F., and Kunz, C. (1994). Interactions of the flavivirus envelope proteins – implications for virus entry and release. *Arch. Virol., Suppl.* **9**; 339–348.

Kapikian, A. Z. (1993). Viral gastroenteritis. *J. Am. Med. Assoc.*, **269**; 627–630.

Koonin, E. V., and Dolja, V. V. (1993). Evolution and taxonomy of positive-strand RNA viruses – Implication of comparative analysis of amino-acid seuqences. *Crit. Rev. Biochem. Mol. Biol.* **28**; 375–430.

Lu, H.-H., Li, X., Cuconati, A., and Wimmer, E. (1995). Analysis of picornavirus 2A^pro proteins: separation of proteinase from translation and replication functions. *J. Virol.* **69**; 7445–7452.

Marin, M. S., Zanotto, P. M. D., Gritsun, T. S., and Gould, E. A. (1995). Phylogeny of TYU, SRE, and CFA virus – different evolutionary rates in the genus Flavivirus. *Virology* **206**; 1133–1139. Reprinted in Strauss and Strauss (1996, p. 115).

Meyers, G., and Thiel, H.-J. (1996). Molecular characterization of pestiviruses. *Adv. Virus Res.* **47**; 53–118.

Molla, A., Paul, A. V., Schmid, M., Jang, S. K., and Wimmer, E. (1993). Studies on dicistronic polioviruses implicate viral proteinase 2A^pro in RNA replication. *Virology* **196**; 739–747.

T.P. Monath, ed. (1988). "The Arboviruses," Vol. I. CRC Press, Boca Raton, Florida.

Murphy, F. A., Fauquet, C. M., Bishop, D. H. L., Ghabrial, S. A., Jarvis, A. W., Martelli, G. P., Mayo, M. A., and Summers, M. D., eds. (1995). "Virus Taxonomy," 6th Report of the International Committee on Taxonomy of Viruses." Springer-Verlag, Vienna.

Nathanson, N., Ahmed, R., Gonzalez-Scarano, F., Griffin, D. E., Holmes, K. V., Murphy, F. A., and Robinson, H. L., eds. (1996). "Viral Pathogenesis." Lippincott-Raven, Philadelphia.

Niebert, M. L., and Fields, B. N. (1995). Early steps in reovirus infection of cells. *In* "Cellular Receptors for Animal Viruses," (E. Wimmer, ed.) pp. 341–364. Cold Spring Harbor Laboratory Press, Cold Spring Harbor, New York.

Numata, K., Hardy, M. E., Nakata, S., Chiba, S., and Estes, M. K. (1997). Molecular characterization of morphologically typical human calicivirus Sapporo. *Arch. Virol.* **142**; 1537–1552.

Poidinger, M., Hall, R. A., and MacKenzie, J. S. (1996). Molecular characterization of the Japanese encephalitis serocomplex of the Flavivirus genus. *Virology* **218**; 417–421.

Porterfield, J. S., ed. (1995). "Exotic Viral Infections." Chapman & Hall Medical, London.

Rauscher, S., Flamm, C., Mandl, C. W., Heinz, F. X., and Stadler, P. F. (1997). Secondary structure of the 3' noncoding region of flavivirus genomes: Comparative analysis of base pairing probabilities. *RNA* **3**; 779–791.

Rey, F., Heinz, F. X., Mandl, C., Kunz, C., and Harrison, S. C. (1995). The envelope glycoprotein from tick-borne encephalitis virus at 2 Å resolution. *Nature (London)* **375**; 291–298.

Saitou, N., and Nei, M. (1986) Polymorphism and evolution of influenza A viruses. *Mol. Biol. Evol.* **3**; 57–74.

Schlauder, G. G., Dawson, G. J., Erker, J. C., Kwo, P. Y., Knigge, M. F., Smalley, D. L., Rosenblatt, J. E., Desai, S. M., and Mushahwar, I. K. (1998). The sequence and phylogenetic analysis of a novel hepatitis E virus isolated from a patient with acute hepatitis reported in the United States. *J. Gen. Virol.* **79**; 447–456.

Shi, P.-Y., Brinton, M. A., Veal, J. M., Zhong, Y. Y., and Wilson, W. D. (1996). Evidence for the existence of a pseudoknot structure at the 3 terminus of the flavivirus genomic RNA. *Biochemistry* **35**; 4222–4230.

Strauss, E. G., and Strauss, J. H. (1988). Evolution of RNA Viruses. *Annu. Rev. Microbiol.* **42**; 657–683.

Strauss, E. G., and Strauss, J. H. (1997). Eukaryotic RNA viruses: A variant genetic system. *In* "Exploring Genetic Mechanisms" (M. Singer and P. Berg, eds.), pp. 71–132. University Science Books, Sausalito, California.

Strauss, J. H., and Strauss, E. G. (1994). The alphaviruses: Gene expression, replication and evolution. *Microbiol. Rev.* **58**; 491–562.

Strauss, J. H., and Strauss, E. G. (1996). Current advances in yellow fever research. *In* "Microbe Hunters – Then and Now" (H. Koprowski and M. B.A. Oldstone, eds.), pp. 113–137. Medi-Ed Press, Bloomington, Illinois.

Takeda, N., Tanimura, M., and Miyamura, K. (1994). Molecular evolution of the major capsid protein VPI of enterovirus 70. *J. Virol.* **68**; 854–862.

Török, T. J., Kilgore, P. E., Clarke, M. I., Holman, R. C., Bresee, J. S., and Glass, R. I. (1997). Visualizing geographic and temporal trends in rotavirus activity in the United States, 1991 to 1996. *Pediatr. Infect. Dis. J.* **16**; 941–946.

Tsai, T. F. (1991). Arboviral infections in the United States. *Infect. Dis. Clin. North Am.* **5**; 73–102; reprinted in Fields, *et al.* (1996, p. 1753).

Tsarev, S. A., Binn, L. N., Gomatos, P. J., Arthur, R. R., Monier, M. K., van Cuyck-Gandre, H., Longer, C. F., and Innis, B. L. (1999). Phylogenetic analysis of hepatitis E virus isolates from Egypt. *J. Med. Virol.* **57**; 68–74.

Weaver, S. C., Kang, W. L., Shirako, Y., Rümenapf, T., Strauss, E. G., and Strauss, J. H. (1997). Recombinational history and molecular evolution of Western equine encephaloymyelitis complex alphaviruses. *J. Virol.* **71**; 613–623.

World Health Organization (WHO) (2000). "Weekly Epidemiological Record for 21 January 2000," Vol. 75, pp. 17–28. WHO, Geneva.

Yamashita, T., Sakae, K., Tsuzuki, H., Suzuki, Y., Ishikawa, N., Takeda, N., Miyamura, T., and Yamazaki, S. (1998). Complete nucleotide sequence and genetic organization of Aichi virus, a distinct member of the *Picornaviridae* associated with acute gastroenteritis in humans. *J. Virol.* **72**; 8408–8412.

CHAPTER 4

Black, F. L. (1996). Measles endemicity in insular populations: Critical community size and its evolutionary implication. *J. Theor. Biol.* **11**; 207–211.

Chua, K. B., Bellini, W. J., Rota, P. A., Harcourt, B. H., Tamin, A., Lam, S. K., Ksiazek, T. G., Rollin, P. E., Zaki, S. R., Shieh, W. J., Goldsmith, C. S., Gubler, D. J., Roehrig, J. T., Eaton, B., Gould, A. R., Olson, J., Field, H., Daniels, P., Ling, A. E., Peters, C. J., Anderson, L. J., and Mahy, B. W. J. (2000). Nipah virus: A recently emergent deadly paramyxovirus. *Science* **288**; 1432-1435.

Clegg, J. C. S., Wilson, S. M., and Oram, J. D. (1990). Nucleotide sequence of the S RNA of Lassa virus (Nigerian Strain) and comparative analysis of arenavirus gene products. *Virus Res.* **18**; 151-164.

Crosby, A. W. (1989). "America's Forgotten Pandemic: The Influenza of 1918," pp. 22–24. Cambridge University Press, Cambridge England.

Georges-Courbot, M.-C., Sanchez, A., Lu, C.-Y., Baize, S., Leroy, E., Lansout-Soukate, J., Tévi-Bénissan, C., Georges, A. J., Trappeir, S. G., Zaki, S. R., Swanepoel, R., Leman, P. A., Rollin, P. E., Peters, C. J., Nichol, S. T., and Ksiazek, T. G. (1997). Isolation and phylogenetics characterization of Ebola viruses causing different outbreaks in Gabon. *Emerg. Infect. Dis.* **3**; 59–62.

Gorman, O. T., Bean, W. J., Kawaoka, Y., and Webster, R. G. (1990). Evolution of the nucleoprotein gene of influenza A virus. *J. Virol.* **64**; 1487–1497.

Lukashevich, I. S., Djavani, M., Shapiro, K., Sanchez, A., Ravkov, E., Nichol, S. T., and Salvato, M. S. (1997). The Lassa fever virus L gene: Nucleotide sequence, comparison, and precipitation of a predicted 250 kDa protein with monospecific antiserum. *J. Gen. Virol.* **78**; 547–551.

Peters, C. J. (1998a). Hemorrhagic fevers: How they wax and wane. *In* "Emerging Infections 1" (W. M. Scheld, D. Armstrong, and J. M. Hughes, eds.), pp. 15–25. ASM Press, Washington, DC.

Peters, C. J. (1998b). Hantavirus pulmonary syndrome in the Americas. *In* "Emerging Infections 2" (W. M. Scheld, W. A. Craig, and J. M. Hughes, eds.), pp. 17–64. ASM Press, Washington, DC.

Peters, C. J., and Khan, A. S. (1999). Filovirus diseases. *Curr. Top. Microbiol.* **235**; 85–95.

Porterfield, J. S., ed. (1995). "Exotic Viral Infections." Chapman & Hall Medical, London.

Rose J. K., and Schubert, M. (1987). Rhabdovirus genomes and their products. *In* "The Rhabdoviruses" (R. R. Wagner, ed.), pp. 129–166. Plenum Press, New York.

Rota, P. A., Rota, J. S., and Bellini, W. J. (1995). Molecular epidemiology of measles virus. *Semin. Virol.* **6**; 379–386.

Smith, J. S., Orciari, L. A., and Yager, P. A. (1995). Molecular epidemiology of rabies in the United States. *Semin. Virol.* **6**, 387–400.

Strauss, E. G., and Strauss, J. H. (1991). RNA viruses: Genome structure and evolution. *Curr. Opin. Genet. Dev.* **1**; 485–493.

Strauss, E. G., and Strauss, J. H. (1997). Eukaryotic RNA viruses: A variant genetic system. *In* "Exploring Genetic Mechanisms" (M. Singer and P. Berg eds.), pp. 171–132. University Science Books, Sausalito, California.

Strauss, E. G., Strauss, J. H., and Levine, A. J. (1996). Virus evolution. *In* "Fields Virology" (B. N. Fields, D. M. Knipe, and P. M. Howley, eds.), 3rd ed., pp. 153–171. Lippincott-Raven, Philadelphia.

CHAPTER 5

Buckwold, V. E., and Ou, J.-H. (1999). Hepatitis B virus C-gene expression and function: The lessons learned from viral mutants. *Curr. Top. Virol.* **1**; 71–81.

Coffin, J. M., Hughes, S. H., and Varmus, H. E. eds. (1997). "Retroviruses." Cold Spring Harbor Laboratory Press, Cold Spring Harbor, New York.

Eickbush, T. H. (1994). The origin and evolutionary relationships of retroelements. *In* "The Evolutionary Biology of Viruses" (S. S. Morse, ed.), pp. 121–157. Raven Press, New York.

Fields. B. N., Knipe, D. M., and Howley, P. M., eds. (1996). "Fields Virology," 3rd ed. Lippincott-Raven, New York.

Goff, S. P. (1997). Retroviruses: Infectious genetic elements. *In* "Exploring Genetic Mechanisms" (M. Singer, and P. Berg, eds.), pp. 133–202. University Science Books, Sausalito, California.

Hindmarsh, P., and Leis, J. (1999). Retroviral DNA integration. *Microbiol. Mol. Biol. Rev.* **63**; 836–843.

Hu, J., and Seeger, C. (1997). RNA signals that control DNA replication in hepadnaviruses. *Semin. Virol.* **8**; 205–211.

Locarnini, S. A., Civitico, G. M., and Newbold, J. E. (1996). Hepatitis B: new approaches for antiviral chemotherapy. *Antiviral Chem. Chemother.* **7**; 53–64.

Mellors, J. W., Rinaldo, C. R., Jr., Gupta, P., White, R. M., Todd, J. A., and Kingsley, L. A. (1996). Prognosis in HIV-1 infection predicted by the quantity of virus in plasma. *Science* **272**; 1167–1170.

Parkin, D. M., Pisani, P., and Ferlay, J. (1999). Estimates of worldwide incidence of 25 major cancers in 1990. *Int. J. Cancer* **80**; 827–841.

Whetter, L. E., Ojukwu, I. C., Novembre, F. J., and Dewhurst, S. (1999). Pathogenesis of simian immunodeficiency virus infection *J. Gen. Virol.* **80**; 1557–1568.

Yen, T. S. B. (1998). Posttranscriptional regulation of gene expression in hepadnaviruses. *Semin. Virol.* **8**; 319–326.

CHAPTER 6

Berg, P., and Singer, M. (1997). Mammalian DNA viruses: Papoviruses as models of cellular genetic function and oncogenesis. *In* "Exploring Genetic Mechanisms" (M. Singer and P. Berg, eds.), pp. 1–70. University Science Books, Sausalito, California.

Brady, J. N., and Salzman, N. P. (1986). The papoviruses: General properties of polyoma and SV40. *In* "The Papovaviridae" (N. P. Salzman, ed.) Vol. 1, pp. 1–26. Plenum Press, New York.

Brister, J. R., and Muzyczka, N. (2000). Mechanism of rep-mediated adeno-associated virus origin nicking. *J. Virol.* **74**; 7762–7771.

Carter, B. J., Trempe, J. P., and Mendelson, E. (1990). Adeno-associated virus gene expression and regulation. *In* "Handbook of Parvoviruses," (P. Tijssen, ed.), Vol. 1, pp. 227–254. CRC Press, Boca Raton, Florida.

Chee, M. S., Bankier, A. T., Beck, S., Bohni, R., Brown, C. M., Cerny, R., Horsnell, T., Hutchinson, C. A., III, Kouzarides, T., Martignetti, J. A., Preddie, E., Satchwell, S. C., Tomlinson, P., Weston, K. M., and Barrell, B. G. (1990). Analysis of the protein-coding content of the sequence of human cytomegalovirus strain AD169. *Curr. Top. Microbiol. Immunol.* **154**; 125–169.

Cotmore, S. F. (1990). Gene expression in the Autonomous parvoviruses. *In* "Handbook of Parvoviruses," (P. Tijssen, ed.) Vol. 1, pp. 141–154. CRC Press Boca Raton, Florida.

DiMaio, D., Lai, C.-C., and Klein, O. (1998). Virocrine transformation: The intersection between viral transforming proteins and cellular signal transduction pathways. *Annu. Rev. Microbiol.* **52**; 397–421.

Dingle, J., and Langmuir, A. D. (1968). Epidemiology of acute respiratory diseases in military recruits. *Am. Rev. Respir. Des.* **97**; 1–65.

Fenner, F. (1983). The Florey Lecture, 1983. Biological control, as exemplified by smallpox eradication and myxomatosis. *Proc. R. Soc. London, Ser. B* **218**; 259–285.

Fenner, F., Wittek, R., and Dumbell, K. R., eds. (1988a). "The Orthopoxviruses." Academic Press, Orlando, Florida.

Fenner, F., Henderson, D. A., Arita, I., Jezek, A., and Ladnyi, I. D. (1988b). "Smallpox and its Eradication," World Health Organization, Geneva.

Fields, B. N., Knipe, D. M., and Howley, P. M., eds. (1996). "Fields Virology," 3rd ed. Lippincott-Raven, New York.

Fox, J. P., Brandt, C. D., Wasserman, F. E., Hall, C. E., Spigland, I., Kogon, A., and Elveback, L. R. (1969). The Virus Watch Program: A continuing surveillance of viral infections in metropolitan New York families. VI. Observations of adenovirus infections: Virus excretion patterns, antibody response, efficiency of surveillance, patterns of infection and relation to illness. *Am. J. Epidemiol.* **89**; 25–50.

Ginzberg, H. S. (1984) An overview. *In* "The Adenoviruses" (H. S. Ginzberg, ed.), pp. 2–4. Plenum Press, New York.

Johnson, R. E., Nahmias, A. J., Magder, L. S., Lee, F. K., Brooks, C., and Snowden, C. (1989). A seroepidemiological survey of the prevalence of herpes simplex virus type 2 infection in the United States. *N. Engl. J. Med.* **321**; 7–12.

McBride, A. A., Byrne, J. C., and Howley, P. M. (1989). E2 polypeptides encoded by bovine papilloma virus type-1 form dimers through the common carboxyl-terminal domain—transactivation is mediated by the conserved amino-terminal domain. *Proc. Natl. Acad. Sci. U.S.A.* **86**; 510–514.

McGeoch, D. J., and Cook, S. (1994). Molecular phylogeny of the Alphaherpesvirinae subfamily and a proposed evolutionary timescale. *J. Mol. Biol.* **238**; 9–38.

Moyer, R. M., and Graves, R. L. (1981). Mechanism of cytoplasmic orthopox DNA replication. *Cell (Cambridge, Mass.)* **27**; 391–401.

Murphy, F. A., Fauquet, C. M., Bishop, D. H. L., Ghabrial, S. A., Jarvis, A. W., Martelli, G. P., Mayo, M. A., and Summers, M. D., eds. (1995). "Virus Taxonomy," 6th Report of the International Committee on Taxonomy of Viruses. Springer-Verlag, Vienna.

Nahmias, A. J., Lee, F. K., and Bechman-Nahmias, S. (1990). Sero-epidemiological and sociological patterns of herpes simplex virus infection in the world. *Scand. J. Infect Dis., Suppl.* **69**; 19–30.

Nathanson, N., Ahmed, R., Gonzalez-Scarano, F., Griffin, D. E., Holmes, K. V., Murphy, F. A., and Robinson, H. L., eds. (1996). "Viral Pathogenesis." Lippincott-Raven, Philadelphia.

Parrish, C. R. (1997). How canine parvovirus suddenly shifted host range. *ASM News* **63**; 307–311.

Rhode, S. L., and Iversen, P. (1990). Parvovirus genomes: DNA sequences. *In* "Handbook of Parvoviruses," (P. Tijssen, ed.) Vol. 1, pp. 31–57. CRC Press, Boca Raton, Florida.

Roizman, B., and Sears A. E. (1990). The replication of herpes simplex viruses. *In* "The Human Herpesviruses" (B. Roizman, C. Lopez, and R. J. Whitley, eds.), pp. 11–68. Raven Press, New York.

Strauss, S. E. (1984). Adenovirus infections in humans. *In* "The Adenovirus", (H. Z. Ginzberg, ed.), pp. 451–496. Plenum Press, New York.

Traktman, P. (1990). The enzymology of poxvirus DNA replication. *Curr. Top. Microbiol. Immun.* **163**; 93–123.

Wold, W. S. M., and Golding, L. R. (1991). Region E3 of adenovirus: A cassette of genes involved in host immunosurveillance and virus-cell interactions. *Virology* **184**; 1–8.

CHAPTER 7

Anderson, R. M., Donnelly, C. A., Ferguson, N. M., Woolhouse, M. E. J., Watt, C. J., Udy, H. J., MaWhinney, S., Dunstan, S. P., Southwood, T. R. E., Wilesmith, J. W., Ryan, J. B. M., Hoinville, L. J., Hillerton, J. E., Austin, A. R., and Wells, G. A. H. (1996). Transmission dynamics and epidemiology of BSE in British cattle. *Nature (London)* **382**; 779–788.

Brian, D. A., and Spaan, W. J. M. (1997). Recombination and coronavirus defective interfering RNAs. *Semin. Virol.* **8**; 101–111.

Fields, B. N., Knipe, D. M., and Howley, P. M., eds. (1996). "Fields Virology," 3rd ed. Lippincott-Raven, New York.

Flores, R., Di Serio, F., and Hernández, C. (1997). Viroids: The noncoding genomes. *Semin. Virol.* **8**; 65–73.

Granoff, A., and Webster, R. G., eds. (1999). "Encyclopedia of Virology," 2nd ed. Academic Press, San Diego, California.

Hegde, R. S., Mastrianni, J. A., Scott, M. R., DeFea, K. A., Tremblay, P., Torchia, M., DeArmond, S. J., Prusiner, S. B., and Lingappa, V. R. (1998). A transmembrane form of the prion protein in neurodegenerative disease. *Science* **279**; 827–834.

Modahl, L. E., and Lai, M. M. C. (2000). Hepatitis delta virus: The molecular basis of laboratory diagnosis. *Crit. Rev. Clin. Lab. Sci.* **37**; 45–92.

Monthly CJD Statistics, from the Department of Health of the United Kingdom. From URL: http://www.doh.gov.uk/cjd/stats/julo1.htm.

Nathanson, N., Wilesmith, J., Wells, G. A., and Griot, C. (1999). Bovine spongiform encephalopathy and related diseases. *In* "Prion Biology and Diseases" (S. Prusiner, ed.). Cold Spring Harbor Laboratory Press, Cold Spring Harbor, New York.

Pelchat, M., Lévesque, D., Ouellet, J., Laurendeau, S., Lévesque, S., Lehoux, J., Thompson, D. A., Eastwell, K. C., Skrzeczkowski, L. J., and Perreault, J. P. (2000). Sequencing of peach latent mosaic viroid variants from nine North American peach cultivars shows that this RNA folds into a complex secondary structure. *Virology* **271**; 37–45.

Prusiner, S. (1998). Prions. *Proc. Natl. Acad. Sci. U.S.A.* **95**; 13363–13383.

Riek, R., Hornemann, S., Wider, G., Billeter, M., Glockshuber, R., and Wüthrich, K. (1996). NMR structure of the mouse prion protein domain PrP (121–231). *Nature (London)* **382**; 180–183.

Strauss, J. H., and Strauss, E. G. (1997). Recombination in alphaviruses. *Semin. Virol.* **8**; 85–94.

Whelan, S. P. J., and Wertz, G. W. (1997). Defective interfering particles of vesicular stomatitis virus: Functions of the genomic termini. *Semin. Virol.* **8**; 131–139.

Wickner, R. B., Taylor, K. L., Edskes, H. K., Maddelein, M.-L., Moriyama, H., and Roberts, B. T. (1999). Prions of *Saccharomyces* and *Podospora* ssp.: Protein-based inheritance. *Microbiol. Mol. Biol. Rev.* **63**; 844–861.

CHAPTER 8

Ashkenazi, A., and Dixit, V. M. (1998). Death receptors: Signaling and modulation. *Science* **281**; 1305–1308.

Bjorkman, P. J., Saper, M. A., Samraoui, B., Bennett, W. S., Strominger, W. L., and Wiley, D. C. (1987). Structure of the human class I histocompatibility antigen, HLA-A2. *Nature (London)* **329**; 506–512, as reprinted in Kuby, (1997).

Chen, J., and Alt, F. W. (1997). Generation of antigen receptor diversity. *In* "Exploring Genetic Mechanisms" (M. Singer and P. Berg, eds.), pp. 337–404. University Science Books, Sausalito, California.

Chinnadurai, G. (1998). Control of apoptosis by human adenovirus genes. *Semin. Virol.* **8**; 399–408.

Fields, B. N., Knipe, D. M., and Howley, P. M., eds. (1996). "Fields Virology," 3rd ed. Lippincott-Raven, New York.

Gally, J. (1993). Structure of Immunoglobulins. *In* "The Antigen" (M. Sela, ed.), p. 209. Academic Press, San Diego, California.

Garenne, M., Leroy, O., Beau, J. P., and Sene, I. (1991). Child-mortality after high-titer measles vaccines: prospective study in Senegal. *Lancet* **338**; 903–907.

Griffin, D. E. (1999). Cytokines and chemokines. *In* "Encyclopedia of Virology," 2nd ed. (A. Granoff, and R. G. Webster, eds.), pp. 339–344. Academic Press, San Diego, California.

Hill, A. B., and Mascucci, M. G. (1998). Avoiding immunity and apoptosis: Manipulation of the host environment by herpes simplex virus and Epstein-Barr virus. *Semin. Virol.* **8**; 361–368.

Kalvakolanu, D. V. (1999). Virus interception of cytokine regulated pathways. *Trends Microbiol.* **7**; 166–171.

Kuby, J. (1997). "Immunology," 3rd ed. Freeman, New York.

Mims, C. A., Playfair, J. H. L., Roitt, I. M., Wakelin, D., and Williams, R. (1993). "Medical Microbiology." Mosby Europe Limited, London.

Nathanson, N., Ahmed, R., Gonzalez-Scarano, F., Griffin, D. E., Holmes, K. V., Murphy, F. A., and Robinson, H. L., eds. (1996). "Viral Pathogenesis." Lippincott-Raven, Philadelphia.

Raff, M. (1998). Cell suicide for beginners. *Nature (London)* **396**; 119–122.

Turner, P. C., and Moyer, R. W. (1998). Control of apoptosis by poxviruses *Semin. Virol.* **8**; 453–469.

CHAPTER 9

Bledsoe, A. W., Jackson, C. A., McPherson, S., and Morrow, C. D. (2000). Cytokine production in motor neurons by poliovirus replicon vector gene delivery. *Nat. Biotechnol.* **18**; 964–969.

Chambers, T. J., Nestorowicz, A., Mason, P. W., and Rice, C. M. (1999). Yellow-fever/Japanese encephalitis chimeric viruses: Construction and biological properties. *J. Virol.* **73**; 3095–3101.

Conzelmann, K.-K., and Meyers, G. (1996). Genetic engineering of animal RNA viruses. *Trends Microbiol.* **4**; 386–393.

Crystal, R. G. (1995). Transfer of genes to humans—early lessons and obstacles to success. *Science* **270**; 404–410.

Dunbar, C. E. (1996). Gene transfer to hemapoietic stem cells: Implications for gene therapy of human disease. *Annu. Rev. Med.* **47**; 11–20.

Strauss, E. G., and Strauss, J. H. (1997). Eukaryotic RNA viruses, a variant genetic system. *In* "Exploring Genetic Mechanisms" (M. Singer and P. Berg, eds.), pp. 71–132. University Science Books, Sausalito, California.

Strauss, J. H. and Strauss, E. G. (1994). The Alphaviruses: Gene expression, replication, and evolution. *Microbiol. Rev.* **58**; 491–562.

References for Tables

CHAPTER 1

Granoff, A., and Webster, R. G., eds. (1999). "Encyclopedia of Virology," 2nd ed., pp. 1745–1748. Academic Press, San Diego, California.

CHAPTER 3

Granoff, A., and Webster, R. G., eds. (1999). "Encyclopedia of Virology," 2nd ed., p. 442. Academic Press, San Diego, California.

Joklik, W. K., and Roner, M. R. (1996). Molecular recognition in the assembly of the segmented reovirus genome. *Prog. Nucleic Acid Res.* **53**; 249–281.

CHAPTER 4

Fields, B. N., Knipe, D. M., and Howley, P. M., eds. (1996). "Fields Virology", pp. 1355 and 1522. Lippincott-Raven, Philadelphia.

Granoff, A., and Webster, R. G., eds. (1999). Tables of respiratory viruses. "Encyclopedia of Virology," 2nd ed., pp. 1493 and 1494. Academic Press, San Diego, California.

Nathanson, N., Ahmed, R., Gonzalez-Scarano, F., Griffin, D. E., Holmes, K. V., Murphy, F. A., and Robinson, H. L., eds. (1996). "Viral Pathogenesis", p. 780. Lippincott-Raven, Philadelphia.

Porterfield, J. S., eds. (1995). "Exotic Viral Infections," p. 228. Chapman & Hall Medical, London.

CHAPTER 5

Coffin, J. M., Hughes, S. H., and Varmus, H. E. eds. (1997). "Retroviruses," pp. 38, 36, and 597. Cold Spring Harbor Laboratory Press, Cold Spring Harbor, New York.

Eickbush, T. (1994). The evolution of retroelements. *In* "The Evolutionary Biology of Viruses" (S. S. Morse, ed.), pp. 121–157. Raven Press, New York.

Fields, B. N., Knipe, D. M., and Howley, P. M., eds. (1996). "Fields Virology," 3rd ed., pp. 309 and 2708. Lippincott-Raven, New York.

Levy, J. A. (1994). "HIV and the Pathogenesis of AIDS," p. 8. ASM Press, Washington, DC.

CHAPTER 6

Alani, R. M., and Münger, K. (1998). Human papillomaviruses and associated malignancies. *J. Clin. Oncol.* **16**; 330–337.

Fields. B. N., Knipe, D. M., and Howley, P. M., eds. (1996). "Fields Virology," 3rd ed., pp. 2030, 2048, 2049, 2085, 2122, 2433, 2643, and 2645. Lippincott-Raven, New York.

Goebel, S. J., Johnson, G. P., Perkus, M. E., Davis, S. W., Winslow, J. P., and Paoletti, E. (1990). The complete sequence of vaccinia virus. *Virology* **179**; 247–266.

Walker, D. L., and Frisque, R. J. (1986). The biology and molecular biology of JC virus. *In* "The Papovaviridae" (N. P. Salzman, ed.), Vol. 1, pp. 327–377. Plenum Press, New York.

CHAPTER 7

Granoff, A., and Webster, R. G., eds. (1999). "Encyclopedia of Virology," 2nd ed., p. 1389. Academic Press, San Diego, California.

CHAPTER 8

Evans, C. H. (1996). Cytokines and viral anti-immune genes. *Stem Cells* **14**; 177–184.

Fields. B. N., Knipe, D. M., and Howley, P. M., eds. (1996). "Fields Virology," 3rd ed., pp. 371, 378, and 2657. Lippincott-Raven, New York.

Granoff, A., and Webster, R. G., eds. (1999). "Encyclopedia of Virology," 2nd ed., pp. 1203 and 1862. Academic Press, San Diego, California.

Janeway, C. A., Travers, P., Walport, M., and Capra, J. D. (1999). "Immunobiology: the Immune System in Health and Disease," pp. 62, 93, and 158. Elsevier Science, Amsterdam.

Kuby, J. (1997). "Immunology," 3rd ed. Freeman, New York.

Mims, C. A., Playfair, J. H. L., Roitt, I. M., Wakelin, D., and Williams, R. (1993). "Medical Microbiology." Mosby Europe Limited, London.

Nathanson, N., Ahmed, R., Gonzalez-Scarano, F., Griffin, D. E., Holmes, K. V., Murphy, F. A., and Robinson, H. L., eds. (1996). "Viral Pathogenesis," p. 124. Lippincott-Raven, Philadelphia.

O'Brien, V. (1998). Viruses and apoptosis. *J. Gen. Virol.* **79**; 1833–1845.

Tortorella, D., Gewurz, B. E., Furman, M. H., Schust, D. J., and Ploegh, H. L. (2000). Viral subversion of the immune system. *Annu. Rev. Immunol.* **18**; 861–926.

CHAPTER 9

Jolly, D. (1994). Viral vector systems for gene-therapy. *Cancer Gene Ther.* **1**; 51–64.

Verma, I, M., and Somia, N. (1997). Gene therapy—promises, problems and prospects. *Nature (London)* **389**; 239–242.

Index